Excel

Get the Results You Want!

YEAR 11 & 12 MATHEMATICS EXTENSION 1

Allyn Jones

PASCAL PRESS

CONTENTS

YEAR 11

TOPIC: FUNCTIONS

CHAPTER 1: FURTHER WORK WITH FUNCTIONS 4

1 Graphical relationships 4
2 Quadratic inequalities 6
3 Inequalities with unknown in denominator 7
4 Inequalities involving absolute values 9
5 Inverse functions 10
6 Parametric form of a function or relation 12
Year 11 exam-type questions 14
Answers 16

CHAPTER 2: POLYNOMIALS 28

1 Terminology 28
2 Division of polynomials 28
3 Remainder and factor theorems 29
4 Sums and products of roots of polynomials 30
5 Multiplicity of roots 31
6 Graphs of polynomial functions 32
Year 11 exam-type questions 35
Answers 36

TOPIC: TRIGONOMETRIC FUNCTIONS

CHAPTER 3: FURTHER TRIGONOMETRIC IDENTITIES 42

1 Sums and differences of angles 42
2 Double-angle formulae 43
3 The t-formulae $\left(t = \tan\dfrac{A}{2}\right)$ 44
4 Trigonometric products as sums and differences ... 45
Year 11 exam-type questions 47
Answers 48

CHAPTER 4: INVERSE TRIGONOMETRIC FUNCTIONS 53

1 Domain, range and graphs 53
2 Properties of inverse trigonometric functions 54
Year 11 exam-type questions 57
Answers 59

TOPIC: CALCULUS

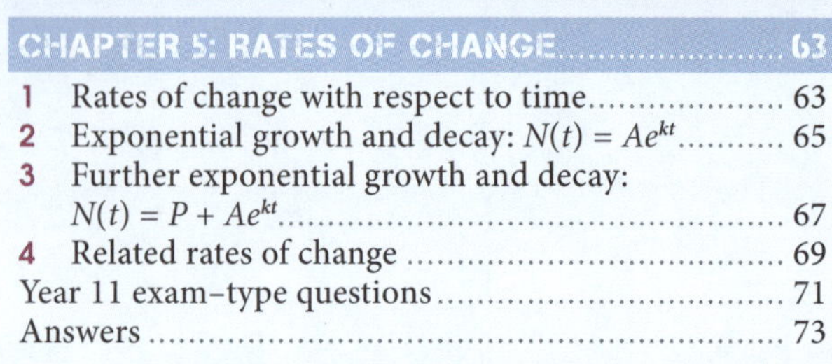

CHAPTER 5: RATES OF CHANGE 63

1 Rates of change with respect to time 63
2 Exponential growth and decay: $N(t) = Ae^{kt}$ 65
3 Further exponential growth and decay: $N(t) = P + Ae^{kt}$ 67
4 Related rates of change 69
Year 11 exam-type questions 71
Answers 73

TOPIC: COMBINATORICS

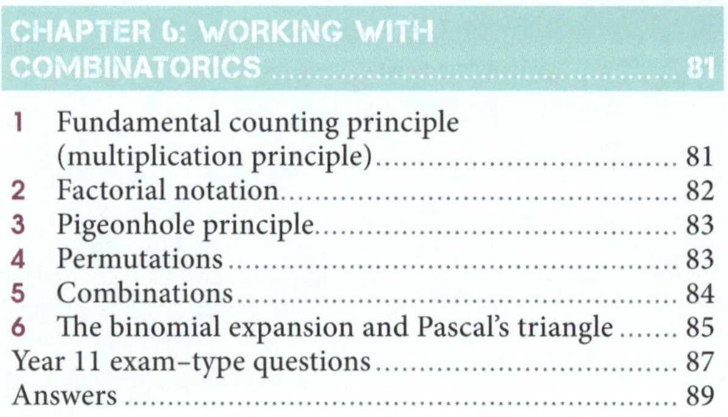

CHAPTER 6: WORKING WITH COMBINATORICS 81

1 Fundamental counting principle (multiplication principle) 81
2 Factorial notation 82
3 Pigeonhole principle 83
4 Permutations 83
5 Combinations 84
6 The binomial expansion and Pascal's triangle 85
Year 11 exam-type questions 87
Answers 89

SAMPLE YEAR 11 EXAMINATION PAPER 93
Answers 96

YEAR 12

TOPIC: PROOF

CHAPTER 8: PROOF BY MATHEMATICAL INDUCTION 100

1 Mathematical induction relating to series 100
2 Mathematical induction relating to divisibility 101
3 False 'proofs by induction' 101
HSC Exam-type questions 103
Answers 104

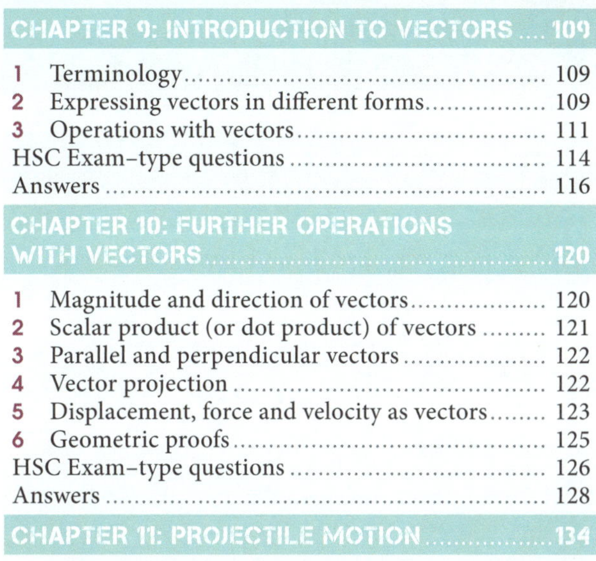

TOPIC: VECTORS

CHAPTER 9: INTRODUCTION TO VECTORS 109

1 Terminology 109
2 Expressing vectors in different forms 109
3 Operations with vectors 111
HSC Exam-type questions 114
Answers 116

CHAPTER 10: FURTHER OPERATIONS WITH VECTORS 120

1 Magnitude and direction of vectors 120
2 Scalar product (or dot product) of vectors 121
3 Parallel and perpendicular vectors 122
4 Vector projection 122
5 Displacement, force and velocity as vectors 123
6 Geometric proofs 125
HSC Exam-type questions 126
Answers 128

CHAPTER 11: PROJECTILE MOTION 134

1 Position vector as a function of time 134
2 Velocity and acceleration vectors as functions of time 135
3 Projectile motion 136
HSC Exam-type questions 141
Answers 143

TOPIC: TRIGONOMETRIC FUNCTIONS

CHAPTER 12: TRIGONOMETRIC EQUATIONS ...150

1 Expressing in the form $r\sin(\theta \pm \alpha)$, $r\cos(\theta \pm \alpha)$ 150
2 Equations of the form $a\cos x + b\sin x = c$ 151
3 Equations involving sums, differences and double angles 151
4 Using t-formulae to solve equations 152
5 Expressing $\sin 3\theta$ and $\cos 3\theta$ in terms of θ 153
HSC Exam-type questions 154
Answers 155

TOPIC: CALCULUS

CHAPTER 13: FURTHER CALCULUS SKILLS ... 160

1 Integration involving products of trigonometric functions 160
2 Integration using substitution 160
3 Integration of $\sin^2 nx$ and $\cos^2 nx$ 161
4 Using the result $\frac{dy}{dx} = \frac{1}{\frac{dx}{dy}}$ 162
5 Differentiation of inverse trigonometric functions . 163
6 Integration resulting in inverse trigonometric functions 165
HSC Exam-type questions 167
Answers 168

CHAPTER 14 FURTHER AREA AND VOLUMES OF SOLIDS OF REVOLUTION 176

1 Area between two curves 176
2 Volumes of solids formed by rotating about the x-axis 177
3 Volumes of solids formed by rotating about the y-axis 180
4 Volumes of solids formed by rotating regions between two curves 182
HSC Exam-type questions 186
Answers 188

CHAPTER 15: DIFFERENTIAL EQUATIONS 196

1 Verifying solutions of differential equations 196
2 Direction (slope) field of a differential equation.. 196
3 Solving differential equations of the form $\frac{dy}{dx} = f(x)$ .. 200
4 Solving differential equations of the form $\frac{dy}{dx} = g(y)$ .. 201
5 Solving differential equations of the form $\frac{dy}{dx} = f(x)g(y)$ 202
6 Exponential growth and decay models 202
7 Mixing problems 204
8 The logistic growth model 205
HSC Exam-type questions 206
Answers 208

TOPIC: STATISTICAL ANALYSIS

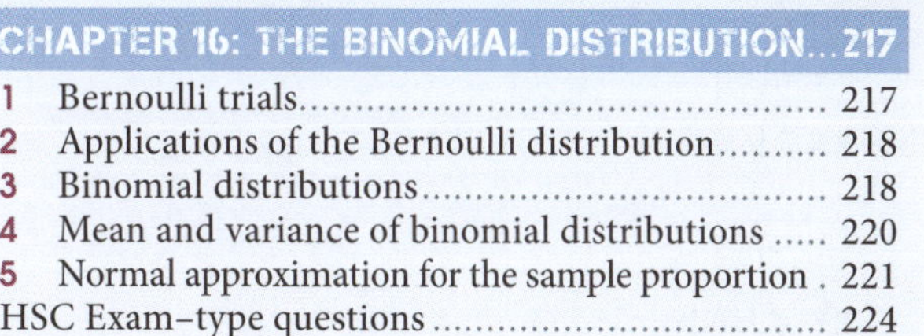

CHAPTER 16: THE BINOMIAL DISTRIBUTION ... 217

1 Bernoulli trials 217
2 Applications of the Bernoulli distribution 218
3 Binomial distributions 218
4 Mean and variance of binomial distributions 220
5 Normal approximation for the sample proportion . 221
HSC Exam-type questions 224
Answers 226

SAMPLE HSC EXAMINATION PAPERS 231

Sample HSC Examination 1 231
Sample HSC Examination 2 234
Answers 237

Index 246
Reference Sheet 248

CHAPTER 1 FURTHER WORK WITH FUNCTIONS

TOPIC: FUNCTIONS

YEAR 11

1 Graphical relationships

- **If $y = f(x)$ then related graphs** have equations such as $y = f(x + c)$, $y = |f(x)|$, $y = \frac{1}{f(x)}$, $y = \sqrt{f(x)}$, and so on.
- **The graph of a related graph to $y = f(x)$** can be developed by selecting critical points.

EXAMPLE 1

A sketch of $y = f(x)$ is drawn on the number plane.

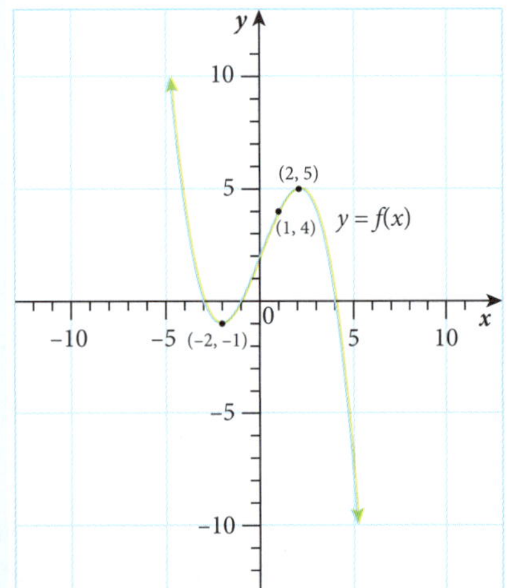

Use the sketch to graph:

a $y = f(x - 2)$

b $y = |f(x)|$

c $y = f(|x|)$

d $y = \frac{1}{f(x)}$

e $y = \sqrt{f(x)}$

f $y^2 = f(x)$

a

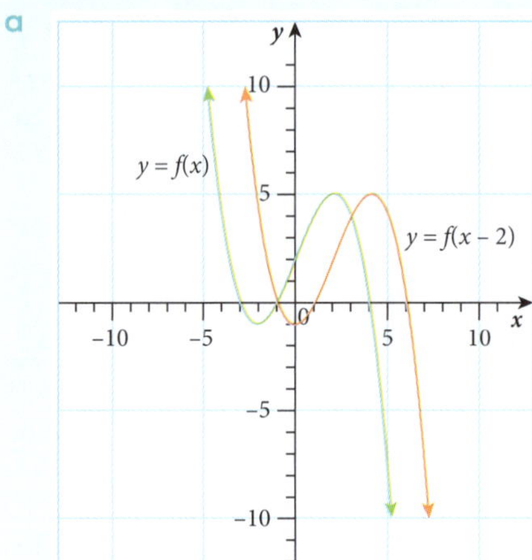

The graph of $y = f(x - c)$ is a translation of the graph of $y = f(x)$ to the right c units.

b

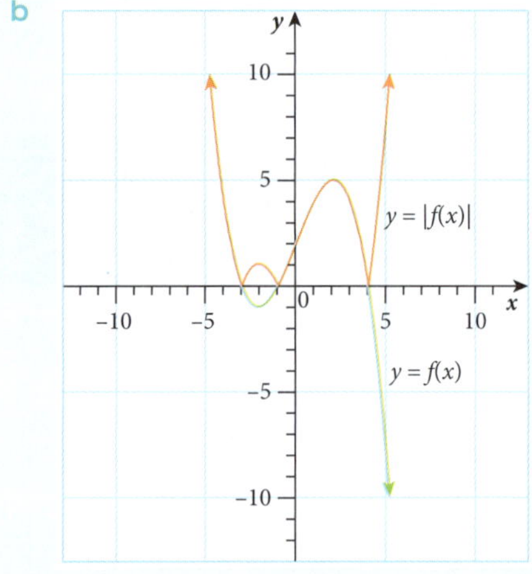

In the graph of $y = |f(x)|$, any part of $y = f(x)$ below the x-axis is reflected above the x-axis.

c

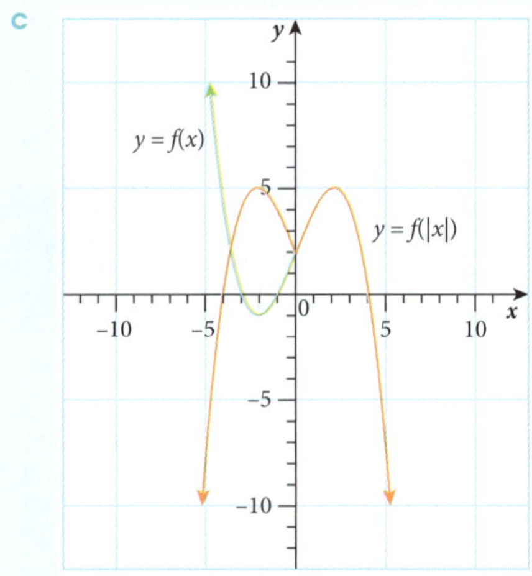

In the graph of $y = f(|x|)$, any part of $y = f(x)$ to the right of the y-axis is reflected about the y-axis.

d

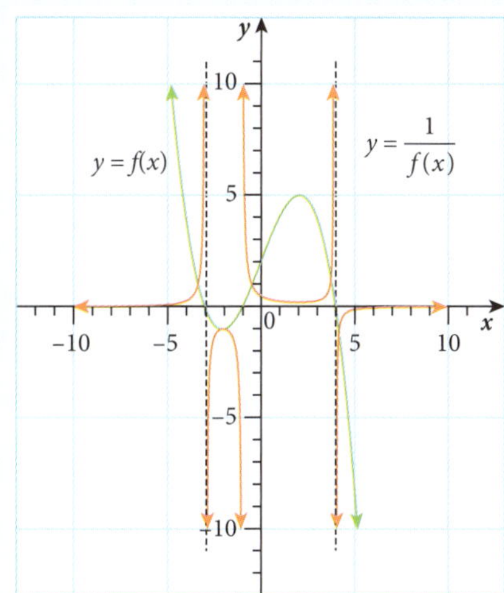

As $\frac{1}{0}$ is undefined, in the graph of $y = \frac{1}{f(x)}$ there are vertical asymptotes when $f(x) = 0$.

If $y = f(x)$ and $y = \frac{1}{f(x)}$ intersect, will be at $y \pm 1$.

e

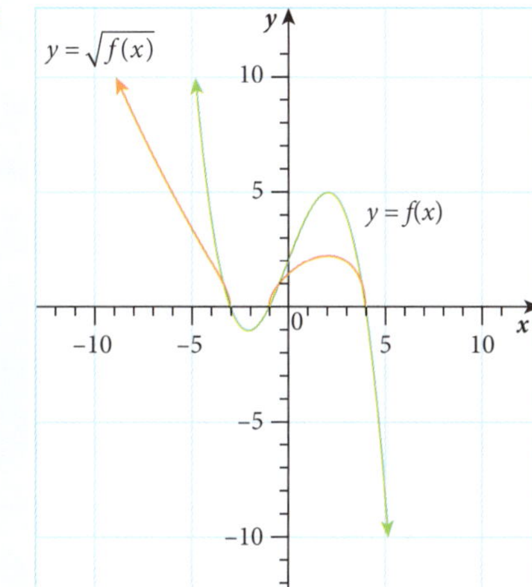

$y = \sqrt{f(x)}$ is undefined when $f(x) < 0$.

f

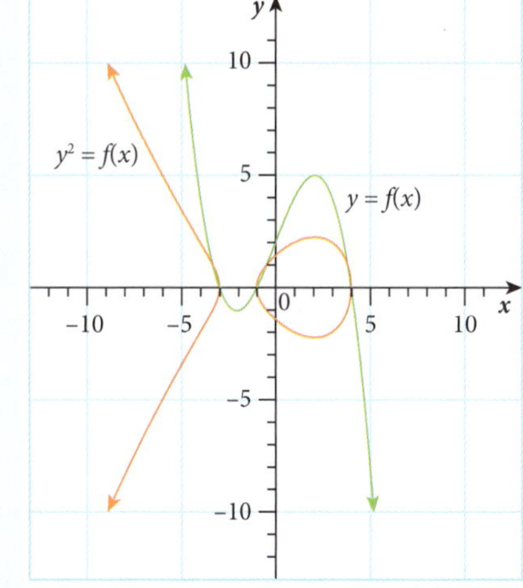

If $y^2 = f(x)$, then $y = \pm\sqrt{f(x)}$.

EXAMPLE 2

Given $f(x) = 3x - 3$ and $g(x) = x^2 - 4$, write down the:

a **y-intercept of the graph of $y = f(x) + g(x)$**

b **y-intercept of the graph of $y = f(x).g(x)$**

a Sum of the two intercepts $= -3 + (-4)$

$= -7$ — Add the constants.

$\therefore$ y-intercept of $y = f(x) + g(x)$ is -7.

b Product of the two intercepts $= (-3) \times (-4)$

$= 12$ — Multiply the constants.

$\therefore$ y-intercept of $y = f(x).g(x)$ is 12.

EXAMPLE 3

Given $f(x) = x^2$ and $g(x) = x - 2$, on the same number plane sketch $y = f(x)$, $y = g(x)$, $y = f(x) + g(x)$ and $y = f(x).g(x)$.

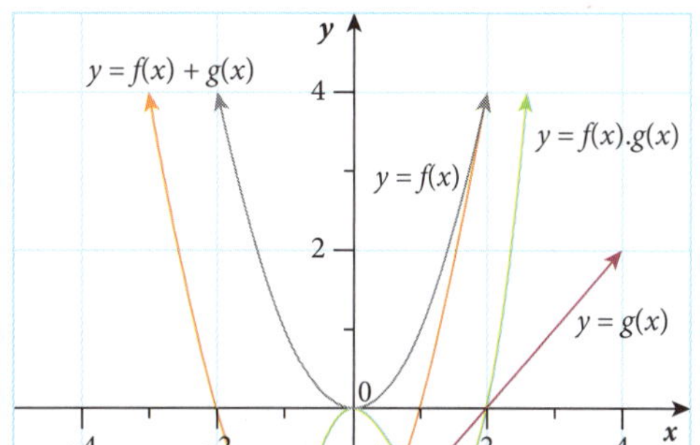

Find expressions for $f(x) + g(x)$ and $f(x).g(x)$.

→ PRACTICE

1 i **A sketch of $y = f(x)$ is drawn on the number plane.**

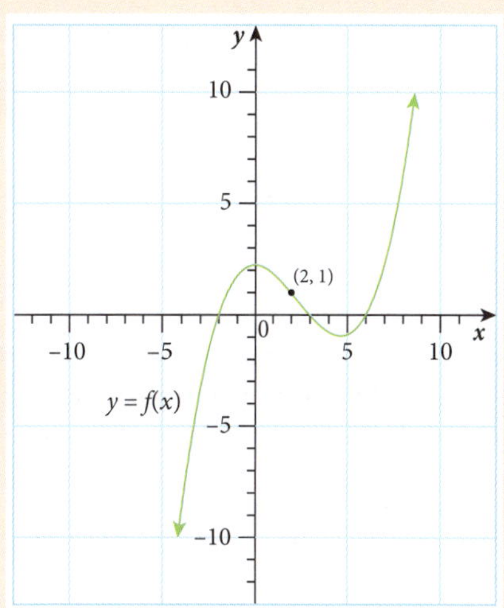

Use the sketch to graph:

a $y = f(x + 3)$

b $y = |f(x)|$

c $y = f(|x|)$

d $y = \dfrac{1}{f(x)}$

e $y = \sqrt{f(x)}$

f $y^2 = f(x)$

ii A sketch of $y = f(x)$ is drawn on the number plane.

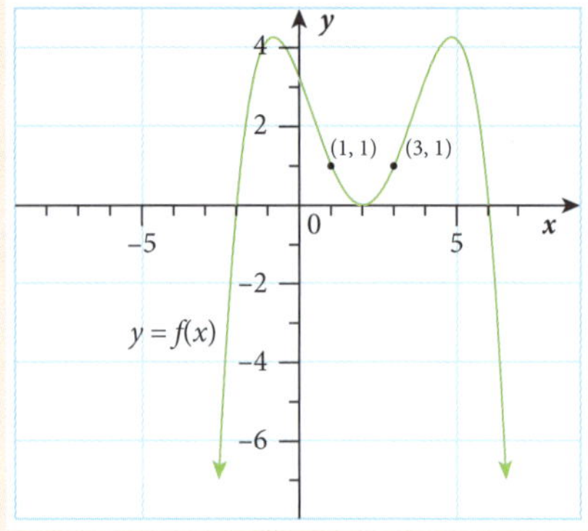

Use the sketch to graph:

a $y = f(x) - 3$

b $y = |f(x)|$

c $y = |f(|x|)|$

d $y = \dfrac{1}{f(x)}$

e $y = \sqrt{f(x)}$

f $y = -f(x)$

2 i Given $f(x) = 2x + 6$ and $g(x) = x^2 - 3x - 4$, write down the:

a y-intercept of the graph of $y = f(x) + g(x)$

b y-intercept of the graph of $y = f(x).g(x)$

ii Given $p(x) = 2 - 2x$ and $q(x) = 7x - 12 - x^2$, write down the:

a y-intercept of the graph of $y = p(x) + q(x)$

b y-intercept of the graph of $y = p(x).q(x)$

3 i Given $f(x) = x^2 - 1$ and $g(x) = x - 1$, on the same number plane sketch $y = f(x)$, $y = g(x)$, $y = f(x) + g(x)$ and $y = f(x).g(x)$.

ii Given $f(x) = x^2 - 3x$ and $g(x) = 4 - 2x$, on the same number plane sketch $y = f(x)$, $y = g(x)$, $y = f(x) + g(x)$ and $y = f(x).g(x)$.

Answers ➲ p. 16

2 Quadratic inequalities

➔ **When an inequality is solved** there is a range of values.

➔ **A quadratic inequality** is an inequality involving a quadratic expression or function.

EXAMPLE 1

Solve $2x^2 < 8x$.

$2x^2 - 8x < 0$
$2x(x - 4) < 0$
Let $2x(x - 4) = 0$
$x = 0, 4$
Test $x = 1$: $2(1)^2 < 8(1)$? … Yes!

When solving, make sure all terms are on LHS of inequality symbol.

$\therefore$ the solution is $0 < x < 4$.

EXAMPLE 2

Solve $2x^2 + x - 6 \leq 0$.

$(2x - 3)(x + 2) \leq 0$
Let $(2x - 3)(x + 2) = 0$
$x = \dfrac{3}{2}, -2$
Test $x = 0$; $(-3)(2) \leq 0$? … Yes!

Factorise first to solve the inequality.

$\therefore$ the solution is $-2 \leq x \leq \dfrac{3}{2}$.

EXAMPLE 3

Sketch $y = 8 + 2x - x^2$ and hence, or otherwise, solve $8 + 2x - x^2 \leq 0$.

Consider $y = 8 + 2x - x^2$
$\therefore y = (4 - x)(2 + x)$

Look for values of x where the parabola is on or below the x-axis.

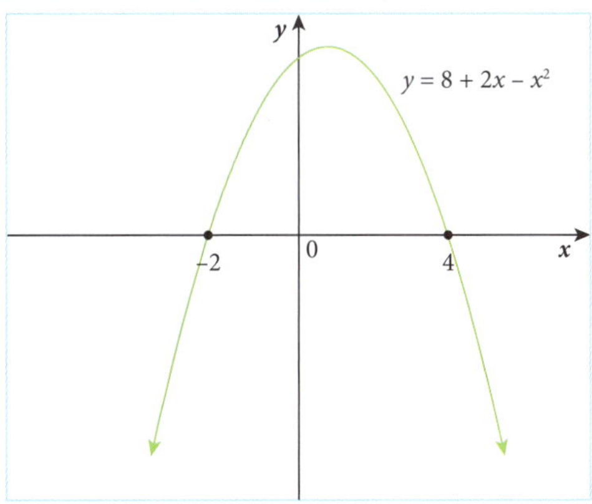

$\therefore$ the solution is $x \le -2, x \ge 4$.

EXAMPLE 4

Sketch $y = x^2 - 2$ and $y = 2x + 1$ on the same number plane and hence, or otherwise, solve $x^2 - 2 > 2x + 1$.

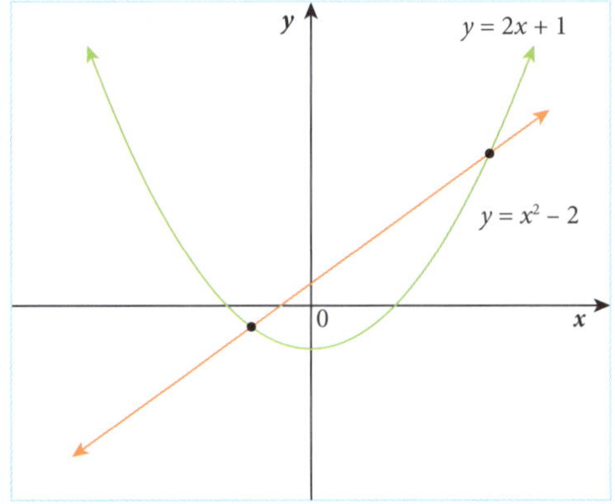

Consider $x^2 - 2 = 2x + 1$

$x^2 - 2x - 3 = 0$

$(x - 3)(x + 1) = 0$

$x = 3, -1$

Use algebra for accuracy.

We find where the curve is above the straight line:

$\therefore$ the solution is $x < -1, x > 3$.

EXAMPLE 5

For what values of x is $2x^2 - 8x - 24$ negative?

Consider $2x^2 - 8x - 24 < 0$

Let $2x^2 - 8x - 24 = 0$

$2(x^2 - 4x - 12) = 0$

$2(x - 6)(x + 2) = 0$

$x = 6, -2$

You can use a number line or sketch a parabola.

Test $x = 0$: $2(0 - 6)(0 + 2) < 0$ … yes!

$\therefore -2 < x < 6$

➔ PRACTICE

1 Solve:

i $3x^2 > 6x$

ii $6a \ge 4a^2$

2 Solve:

i $2x^2 + 7x - 4 \ge 0$

ii $6x^2 + x - 2 < 0$

3 i Sketch $y = 3 + 5x - 2x^2$ and hence or otherwise solve $3 + 5x - 2x^2 \le 0$.

ii Sketch a parabola on a number plane to solve $2x^2 + 3x - 20 < 0$.

4 i Sketch $y = x^2$ and $y = 4 - 3x$ on the same number plane and hence, or otherwise, solve $x^2 \le 4 - 3x$.

ii Using the graphs of $y = x^2$ and $y = 8 - x^2$, solve $x^2 \ge 8 - x^2$.

5 For what values of x is:

i $4x - 2x^2$ negative?

ii $6x^2 + 5x - 6$ positive?

Answers ➲ pp. 18–19

3 Inequalities with unknown in denominator

➔ **We cannot simply multiply by the denominator** because some values of x will make the denominator positive and other values negative.

➔ **First square the denominator** and then use it to multiply both sides of the inequality.

EXAMPLE 1

Solve $\frac{4}{x+1} \ge 3$.

Multiply both sides by $(x + 1)^2$:

$$(x + 1)^2 \times \frac{4}{x+1} \ge 3(x + 1)^2$$

$$4x + 4 \ge 3x^2 + 6x + 3$$

$$3x^2 + 2x - 1 \le 0$$

$$(3x - 1)(x + 1) \le 0$$

Let $(3x - 1)(x + 1) = 0$

$$x = \frac{1}{3}, -1$$

Test $x = 0$: $\frac{4}{0+1} \geq 3$? … Yes!

But $x \neq -1$.

$\therefore$ the solution is $-1 < x \leq \frac{1}{3}$.

You will need to exclude the x-value when the denominator = 0.

EXAMPLE 2

Solve $\frac{x+3}{2x} > 1$.

Mulitply both sides by $4x^2$:

$$4x^2 \times \frac{x+3}{2x} > 1 \times 4x^2$$
$$2x^2 + 6x > 4x^2$$
$$2x^2 - 6x < 0$$
$$x^2 - 3x < 0$$
$$x(x - 3) < 0$$

Let $x(x - 3) = 0$

$$x = 0, 3$$

Test $x = 1$; $\frac{1+3}{2(1)} > 1$? … Yes!

You could use x^2, $2x^2$, etc.

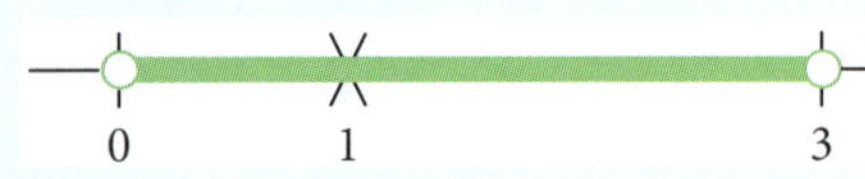

$\therefore$ the solution is $0 < x < 3$.

EXAMPLE 3

Solve $\frac{3}{2x-1} < x$.

Multiply both sides by $(2x - 1)^2$:

$$(2x - 1)^2 \times \frac{3}{2x-1} < x(2x - 1)^2$$
$$3(2x - 1) < x(2x - 1)^2$$
$$x(2x - 1)^2 - 3(2x - 1) > 0$$
$$(2x - 1)[x(2x - 1) - 3] > 0$$
$$(2x - 1)(2x^2 - x - 3) > 0$$
$$(2x - 1)(2x - 3)(x + 1) > 0$$

Let $(2x - 1)(2x - 3)(x + 1) = 0$

$$\therefore x = \frac{1}{2}, \frac{3}{2}, -1$$

Sketch $y = (2x - 1)(2x - 3)(x + 1)$

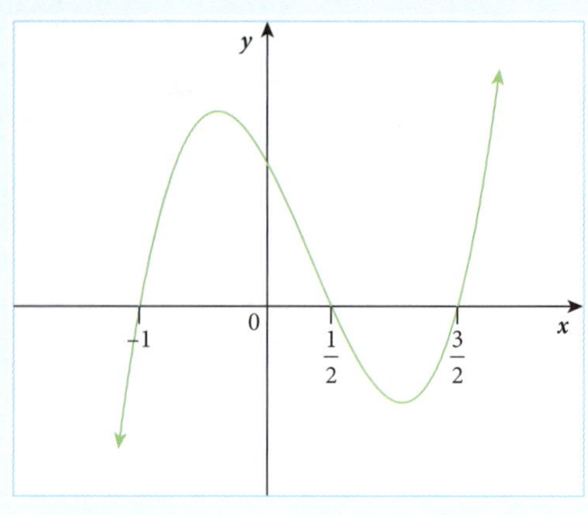

Sketch the cubic.

Consider $y > 0$: $-1 < x < \frac{1}{2}$ or $x > \frac{3}{2}$.

$\therefore$ the solution is $-1 < x < \frac{1}{2}$ or $x > \frac{3}{2}$.

EXAMPLE 4

Solve $\frac{1}{a} \geq \frac{1}{2a-1}$.

Multiply both sides by $a^2(2a - 1)^2$:

$$a^2(2a - 1)^2 \times \frac{1}{a} \geq \frac{1}{2a-1} \times a^2(2a - 1)^2$$
$$a(2a - 1)^2 \geq a^2(2a - 1)$$
$$a(2a - 1)^2 - a^2(2a - 1) \geq 0$$
$$a(2a - 1)[2a - 1 - a] \geq 0$$
$$a(2a - 1)(a - 1) \geq 0$$

Factorising

Let $a(2a - 1)(a - 1) = 0$

$$\therefore a = 0, \frac{1}{2}, 1$$

Sketch $f(a) = a(2a - 1)(a - 1)$

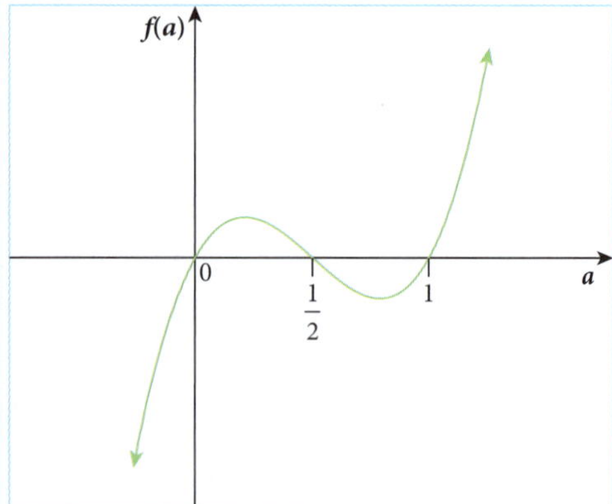

Consider $f(a) \geq 0$: $0 \leq a \leq \frac{1}{2}$ or $a \geq 1$.

But $a \neq 0$ $\therefore$ the solution is $0 < a \leq \frac{1}{2}$ or $a \geq 1$.

→ PRACTICE

1 Solve:

i $\dfrac{5}{x-1} \geq 2$

ii $\dfrac{2}{3x+1} \leq 1$

2 Solve:

i $\dfrac{2x+1}{x} > 1$

ii $\dfrac{1-4x}{2x} \leq 3$

3 Solve:

i $\dfrac{2}{3x+1} > x$

ii $x + \dfrac{1}{1-2x} \leq 0$

4 Solve:

i $\dfrac{1}{2p} \leq \dfrac{1}{p+1}$

ii $\dfrac{1}{y-2} \geq \dfrac{1}{y+3}$

Answers ➲ pp. 19–20

4 Inequalities involving absolute values

→ **We consider the two cases when solving an equation or an inequality involving absolute values.**

EXAMPLE 1

Solve $|2x - 1| < 5$.

$-5 < 2x - 1 < 5$
$-4 < 2x < 6$
$-2 < x < 3$

If possible, use the double inequality approach.

EXAMPLE 2

Solve $|3x + 4| \geq 1$.

$3x + 4 \leq -1$ or $3x + 4 \geq 1$
$3x \leq -5$ or $3x \geq -3$
$x \leq -\dfrac{5}{3}$ or $x \geq -1$
$\therefore x \leq -\dfrac{5}{3}$ or $x \geq -1$.

If an inequality has the $\geq$ symbol then it may be best to rewrite it as two inequalities.

EXAMPLE 3

Solve $\dfrac{|2-3x|}{4} \leq 1$.

$-1 \leq \dfrac{2-3x}{4} \leq 1$
$-4 \leq 2 - 3x \leq 4$
$-6 \leq -3x \leq 2$
$2 \geq x \geq -\dfrac{2}{3}$
$\therefore -\dfrac{2}{3} \leq x \leq 2$

Change symbols when dividing by a negative.

EXAMPLE 4

Solve $\dfrac{|x-1|}{x} > 3$.

$\dfrac{x-1}{x} < -3$ or $\dfrac{x-1}{x} > 3$

Use two cases.

$x^2 \times \dfrac{x-1}{x} < -3 \times x^2$ or $x^2 \times \dfrac{x-1}{x} > 3 \times x^2$
$x(x - 1) < -3x^2$ or $x(x - 1) > 3x^2$
$x^2 - x < -3x^2$ or $x^2 - x > 3x^2$
$4x^2 - x < 0$ or $2x^2 + x < 0$
$x(4x - 1) < 0$ or $x(2x + 1) < 0$
Let $x(4x - 1) = 0$ Let $x(2x + 1) = 0$
$x = 0, \dfrac{1}{4}$ $x = 0, -\dfrac{1}{2}$

By testing points: $x = -1$: no; $x = -\dfrac{1}{3}$: no; $x = \dfrac{1}{8}$: yes; $x = 1$: no.

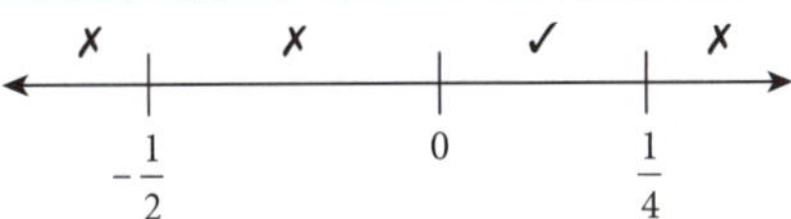

$\therefore 0 < x < \dfrac{1}{4}$

→ PRACTICE

1 Solve:

i $|3x - 2| < 1$

ii $|2x + 5| < 3$

2 Solve:

i $|3 + 2a| \geq 1$

ii $|4m - 3| > 2$

3 Solve:

i $\frac{|4x+2|}{5} \leq 1$

ii $\frac{|3-2x|}{2} > 3$

4 Solve:

i $\frac{|4x+2|}{x} < 2$

ii $\frac{|2x-1|}{x+1} \geq 2$

Answers ➲ pp. 21–22

5 Inverse functions

- **An inverse function** is a function obtained by expressing the dependent variable of one function as the independent variable of another so if $y = f(x)$, swap x and y to find $y = f^{-1}(x)$.
- **If the graph of $y = f(x)$** is such that any horizontal line intersects it at most only one point, then $f(x)$ is said to be a one-to-one function and it has an inverse function.
- **The graphs of inverse functions** reflect about the line $y = x$.

EXAMPLE 1

Determine whether the following function has an inverse function:

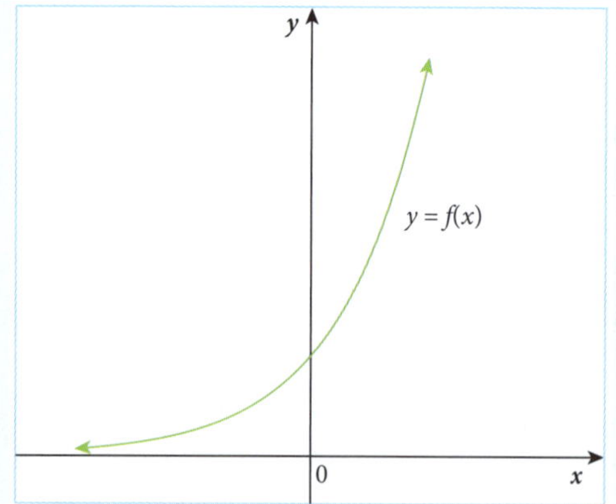

Yes, it would have an inverse function, as any horizontal line only cuts once.

Draw a horizontal line.

EXAMPLE 2

Find the largest possible domain for which $f(x) = 4x - x^2$ is a one-to-one increasing function.

Consider $y = 4x - x^2$.

$= x(4 - x) = 0$

$x = 0, 4$

As concave down, maximum value is at $x = 2$.

$\therefore$ one-to-one increasing function when $x \leq 2$.

Find the turning point of the function.

EXAMPLE 3

Find the inverse function, $f^1(x)$, of $f(x) = x^2 - 6x + 9$, for $x \geq 3$. Hence, or otherwise, find the domain and range of $f^1(x)$.

Let $y = x^2 - 6x + 9$.

Swap $y \leftrightarrow x$:

$x = y^2 - 6y + 9$

$x = (y - 3)^2$

$\sqrt{x} = y - 3$ (as domain for $f(x)$ is $x \geq 3$)

$y = \sqrt{x} + 3$

$\therefore f^1(x) = \sqrt{x} + 3$

Domain is $x \geq 0$, range is $y \geq 3$.

The domain and range of $f(x)$ can be used to find the range and domain of $f^1(x)$ respectively.

EXAMPLE 4

If $f(x) = (x - 2)^2$, where $x \geq 2$, find the points of intersection of $y = f(x)$ and $y = f^{-1}(x)$ and hence sketch $y = f(x)$ and $y = f^{-1}(x)$ on the same number plane.

Solve $(x - 2)^2 = x$

$x^2 - 4x + 4 - x = 0$

$x^2 - 5x + 4 = 0$

$(x - 4)(x - 1) = 0$

$x = 4, 1$

$\therefore$ the points of intersection are (1, 1) and (4, 4).

But $x \geq 2$, hence, the point of intersection is (4, 4).

Solve $f(x) = x$, as $y = f(x)$ and $y = f^{-1}(x)$ intersect on $y = x$.

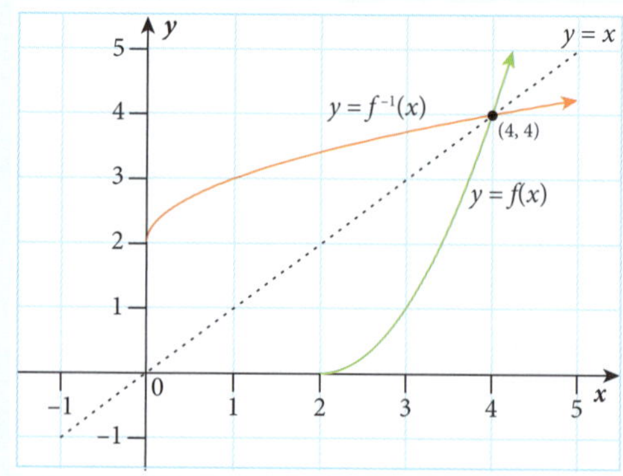

EXAMPLE 5

Consider the function $f(x) = (x + 2)^2 - 1$.

a Find the largest possible domain for which $y = f(x)$ is a one-to-one increasing function.

b Using the restricted domain found in part a, find the inverse function $f^{-1}(x)$, stating its domain and range.

c On the same number plane, draw a sketch of the curves $y = f(x)$ (with restricted domain), $y = f^{-1}(x)$ and the line $y = x$.

a Consider $y = (x + 2)^2 - 1$.

As concave up and vertex at $(-2, -1)$, the minimum value at $x = -2$.

Use the minimum value.

$\therefore$ one-to-one increasing function when $x \geq -2$.

b Let $y = (x + 2)^2 - 1$ (Using domain: $x \geq -2$, range: $y \geq -1$)

Swap $y \leftrightarrow x$:

$$x = (y + 2)^2 - 1$$

$$(y + 2)^2 = x + 1$$

$$y + 2 = \pm\sqrt{x+1}$$

$$y = -2 \pm \sqrt{x+1}$$

Use the range to eliminate positive case.

But range is $y \geq -2$:

$\therefore f^{-1}(x) = -2 + \sqrt{x+1}$

Domain: $x \geq -1$, range: $y \geq -2$.

c

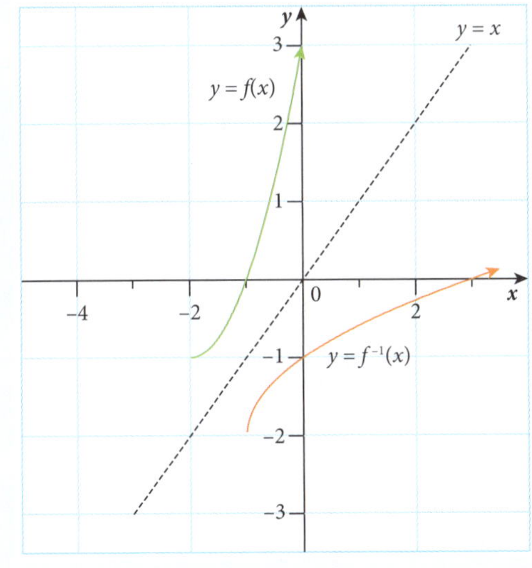

Curves reflect about $y = x$.

→ PRACTICE

1 Determine whether the following functions have an inverse function:

i

y
y = f(x)
0
x

ii

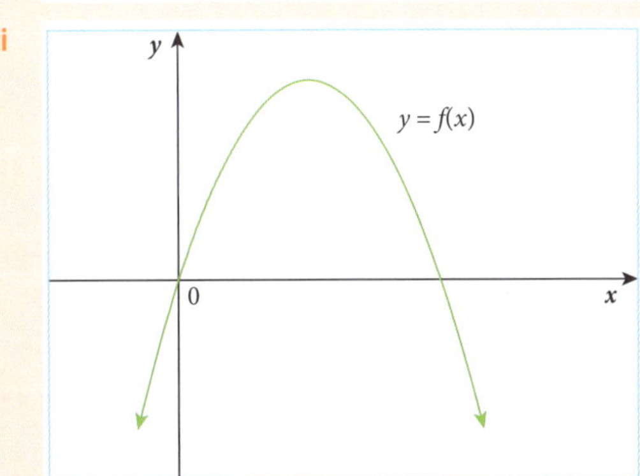

2 Find the largest possible domain for which $y = f(x)$ is a one-to-one increasing function if:

i $f(x) = 2x^2 - 12x$

ii $f(x) = 15 - 2x - x^2$

3 For the following functions find the inverse function, $f^{-1}(x)$ and hence, or otherwise, find the domain and range of $f^{-1}(x)$:

i $f(x) = x^2 - 4x + 4$, for $x \geq 2$

ii $f(x) = \sqrt{2x-3}$

4 Find the point(s) of intersection of $y = f(x)$ and $y = f^{-1}(x)$, and hence sketch $y = f(x)$ and $y = f^{-1}(x)$ on the same number plane:

i $f(x) = 3 + 2\sqrt{x}$

ii $f^{-1}(x) = x^2 - 2$, for $x \geq 0$

5 i Consider the function $f(x) = (x - 1)^2 + 2$.

a Find the largest possible domain for which $y = f(x)$ is a one-to-one increasing function.

b Using the restricted domain found in part a, find the inverse function $f^{-1}(x)$, stating its domain and range.

c On the same number plane, draw a sketch of the curves $y = f(x)$ (with restricted domain), $y = f^{-1}(x)$ and the line $y = x$.

ii Consider the function $f(x) = 1 - (x + 2)^2$.

a Find the largest possible domain for which $y = f(x)$ is a one-to-one increasing function.

b Using the restricted domain found in part a, find the inverse function $f^{-1}(x)$, stating its domain and range.

c On the same number plane, draw a sketch of the curves $y = f(x)$ (with restricted domain), $y = f^{-1}(x)$ and the line $y = x$.

Answers ➲ pp. 22–23

6 Parametric form of a function or relation

- **An equation involving x and y such as $y = 3x^2 - 2x - 1$** is called a cartesian equation.
- **Parametric form** is when both x and y are written in terms of a parameter.
- **The pronumerals p or t** are often used as parameters.
- **Any cartesian equation** can be written in parametric form.

EXAMPLE 1

Find the cartesian equation of the line represented by the parametric equations $x = t - 3$ and $y = 2t + 5$.

$x = t - 3 \ldots$ ①

$y = 2t + 5 \ldots$ ②

From ①: $t = x + 3$

Subs. in ②: $y = 2(x + 3) + 5$

$= 2x + 6 + 5$

$= 2x + 11$

Use simultaneous equations.

EXAMPLE 2

Find the cartesian equation of the parabola represented by the parametric equations $x = 12p$ and $y = 6p^2$.

$x = 12p \ldots$ ①

$y = 6p^2 \ldots$ ②

From ①: $p = \dfrac{x}{12}$

Subs. in ②: $y = 6\left(\dfrac{x}{12}\right)^2$

$= \dfrac{x^2}{24}$

Use the 'simpler' equation to make the parameter the subject.

EXAMPLE 3

Find the cartesian equation of the circle represented by the parametric equations $x = 3 \cos t$ and $y = 3 \sin t$, and hence graph the circle showing the coordinates of the centre of the circle and its radius.

$\cos t = \dfrac{x}{3}$, and hence, $\cos^2 t = \dfrac{x^2}{9} \ldots$ ①

$\sin t = \dfrac{y}{3}$, and hence, $\sin^2 t = \dfrac{y^2}{9} \ldots$ ②

Hence, $\dfrac{x^2}{9} + \dfrac{y^2}{9} = 1$, as $\sin^2 t + \cos^2 t = 1$.

(using ① and ②)

Look for a trigonometric result to use.

$\therefore x^2 + y^2 = 9$, which is a circle, centre (0, 0) and radius 3 units.

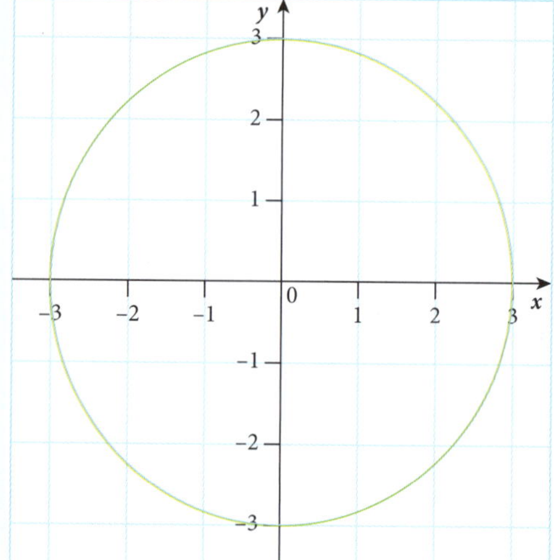

EXAMPLE 4

Determine the cartesian equation of the circle with the parametric equations $x = 5 + 2 \cos \theta$ and $y = -4 + 2 \sin \theta$ and hence find the centre and radius.

As $x = 5 + 2 \cos \theta$

$2 \cos \theta = x - 5$

$\cos \theta = \dfrac{x-5}{2}$

$\cos^2\theta = \dfrac{(x-5)^2}{4}$

Also, $y = -4 + 2 \sin \theta$

$2 \sin \theta = y + 4$

$\sin \theta = \dfrac{y+4}{2}$

$\sin^2\theta = \dfrac{(y+4)^2}{4}$

Circle will be in the form $(x - h)^2 + (y - k)^2 = r^2$.

As $\cos^2\theta + \sin^2\theta = 1$, then

$$\frac{(x-5)^2}{4} + \frac{(y+4)^2}{4} = 1$$

$(x - 5)^2 + (y + 4)^2 = 4$

Circle, centre (5, −4) and radius 2 units.

→ PRACTICE

1 Find the cartesian equation of the line represented by the parametric equations:

i $x = t + 1$ and $y = 3t + 2$

ii $x = 5 - p$ and $y = \frac{4p-2}{3}$

2 Find the cartesian equation of the parabola represented by the parametric equations:

i $x = 4t$ and $y = 2t^2$

ii $x = \frac{p-3}{2}$ and $y = \frac{p^2}{5}$

3 Find the cartesian equation of the circle represented by the parametric equations:

i $x = 5 \cos t$ and $y = 5 \sin t$, and hence graph the circle

ii $x = 4 \cos t - 2$ and $y = 1 + 4 \sin t$, and hence graph the circle

4 i Determine the cartesian equation of the circle with the parametric equations $x = 2 - 3 \cos \theta$ and $y = 4 + 3 \sin \theta$ and hence find the centre and radius.

ii Determine the cartesian equation of the circle with the parametric equations $x = 2 - 5 \cos \theta$ and $y = 5(1 - \sin \theta)$ and hence sketch the curve.

Answers ➲ pp. 23–24

A QUICK NOTE! To help you study effectively we have made the **Examples** and **Practice** questions with the same numbers very similar to each other or you will require very similar skills to solve them.

- Please see the sample below. It shows that the **Example 1** question is similar to the **Practice 1** i and ii questions, the **Example 2** question is similar to the **Practice 2** i and ii questions and the **Example 3** question is similar to the **Practice 3** i and ii questions.
- This is to enable you to **practise** what you have just read in an **Example**.
- You can also refer to the **Examples** when you are answering the **Practice** questions as their worked solutions and tips will help you answer the questions.

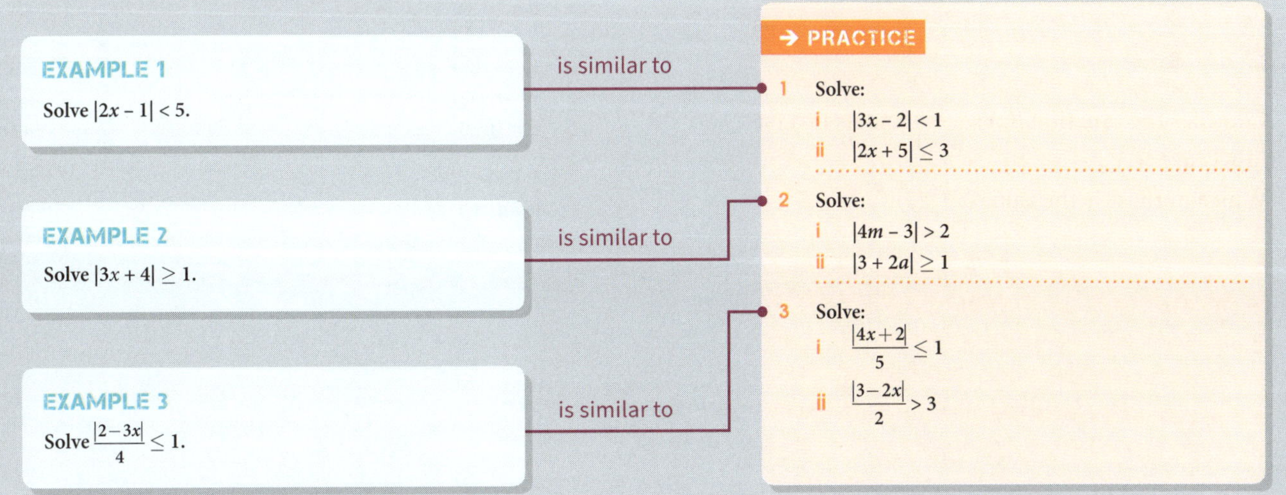

YEAR 11 EXAM-TYPE QUESTIONS

Now for the real thing! The following questions are modelled on the types of questions you will face in the yearly examination in Year 11. Think about it: if you get extensive practice at answering these sorts of questions, you will be more confident in answering them in the yearly examination. It makes sense, doesn't it?

Remember the Answer section is located straight after this section so that you can check the answers readily. There are worked solutions and also ticks that indicate the mark allocation for all questions. Both these features will help you to strive for full marks.

Make sure also that when you check your work you highlight any questions you found difficult and earmark these areas for extra study.

Total marks: 35

PART A

(Suggested time: 10 minutes)

(1 mark each)

1 Which of these is the inverse function of $y = 3x - 2$?

A $y = 2x - 3$ B $y = 2 - 3x$

C $y = \frac{x-2}{3}$ D $y = \frac{x+2}{3}$

2 Which of these is the graph of the solution of $|4 - 2x| \leq 6$?

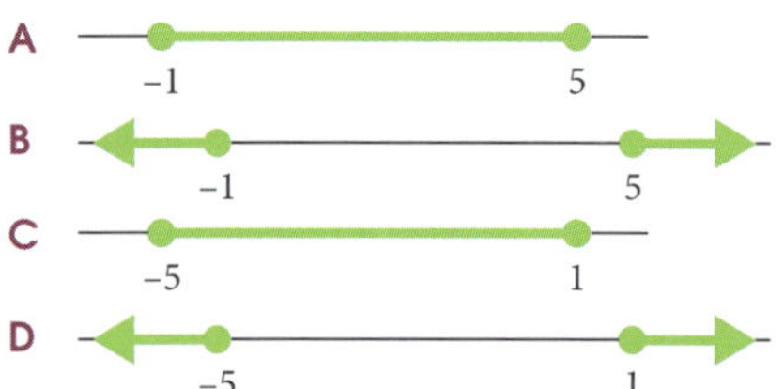

3 A curve has parametric equations $x = \frac{t}{2}$ and $y = 3t^2$.

Which of these is the Cartesian equation for this curve?

A $y = \frac{x^2}{3}$

A $y = 6x^2$

C $y = 12x^2$

D $y = 18x^2$

4 Consider the function $f(x) = \frac{2x}{x+1}$ and its inverse function $f^{-1}(x)$.
Which of these is the value of $f^{-1}(3)$?

A -3 B $-\frac{1}{3}$

C $\frac{2}{3}$ D $1\frac{1}{2}$

5 Which of these is the solution of $\frac{1}{1-x} \leq 1$?

A $0 \leq x \leq 1$

B $0 \leq x < 1$

C $x \leq 0$ or $x > 1$

D $x > 1$

PART B

(Suggested time: 45 minutes)

Show all working.

6 The graph of the function $y = f(x)$ is shown on the number plane below.

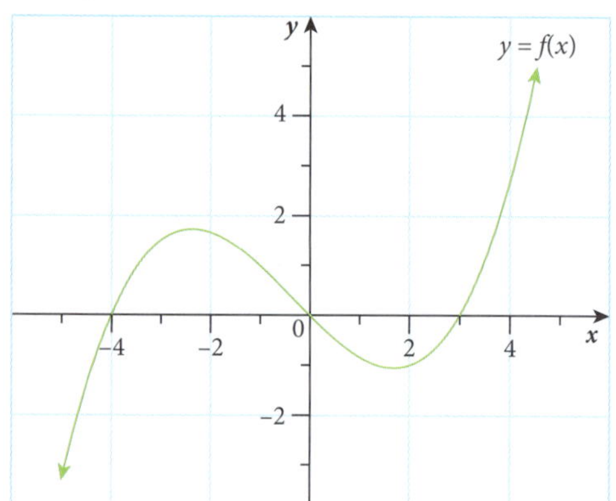

On separate number planes, graph:

a $y = \frac{1}{f(x)}$ (2 marks)

b $y = -f(x)$ (2 marks)

c $y = f(|x|)$ (2 marks)

d $y = |f(x)|$ (2 marks)

e $y = \sqrt{f(x)}$ (2 marks)

f $y = f(x + 2)$ (2 marks)

7 Find the point of intersection of the curve with parametric equations $x = \frac{t}{2}$, $y = \frac{t^2}{2}$ and the line with parametric equations at $x = p + 1$, $y = 4p + 2$. (3 marks)

8 Sketch the graph of $y = |8 - 2x - x^2|$ and hence find the number of solutions of $|8 - 2x - x^2| = 4$. (2 marks)

9 Solve:

a $2x^2 - 5x > 3$ (2 marks)

b $\frac{4x-5}{2x+1} \leq 1$ (3 marks)

c $|3x - 2| > 1$ (2 marks)

10 Consider the function $f(x) = (x - 3)^2$.

a Sketch $y = f(x)$. (1 mark)

b Explain why $f(x)$ does not have an inverse function for all x in its domain. (1 mark)

c State a largest possible domain for which $f(x)$ has an inverse function $f^{-1}(x)$. (1 mark)

d Determine the equation of the inverse function $f^{-1}(x)$ for your domain in part (c). (2 marks)

e Sketch $y = f^{-1}(x)$ onto your graph on part (a). (1 mark)

PRACTICE

1 Graphical relationships ➲ pp. 5–6

1 i a

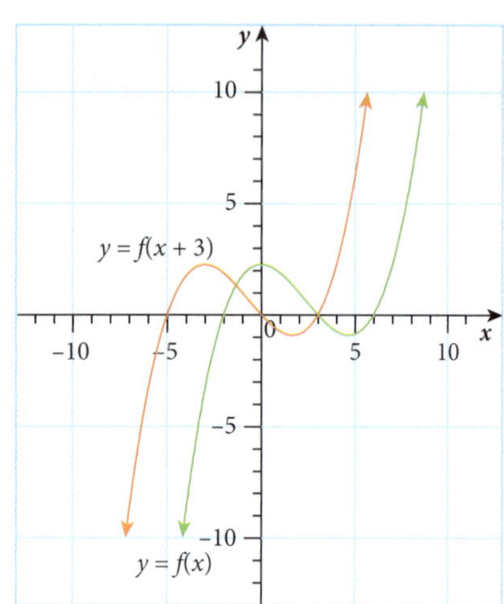

b

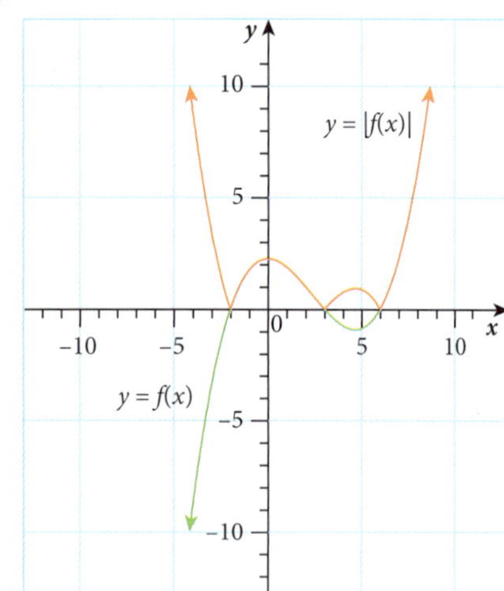

c

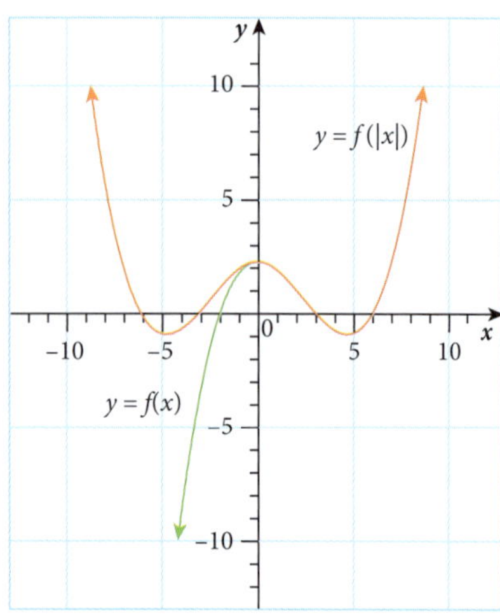

d

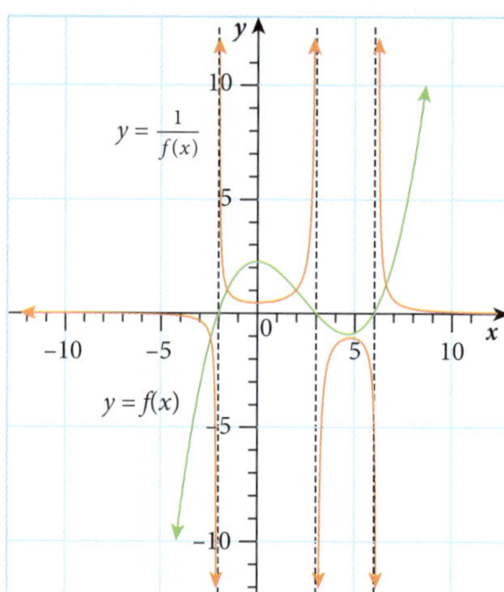

e

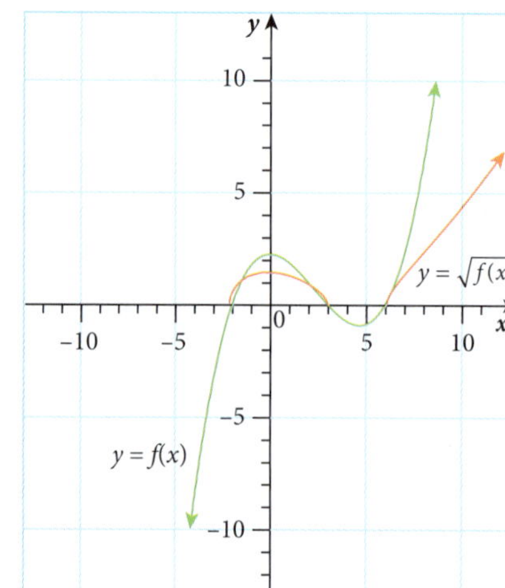

f

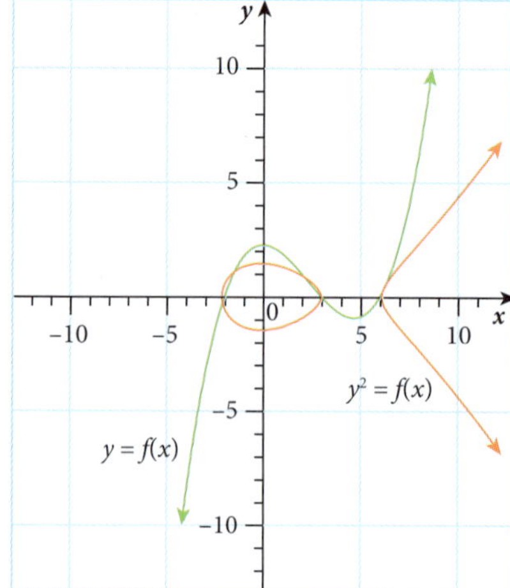

ii a

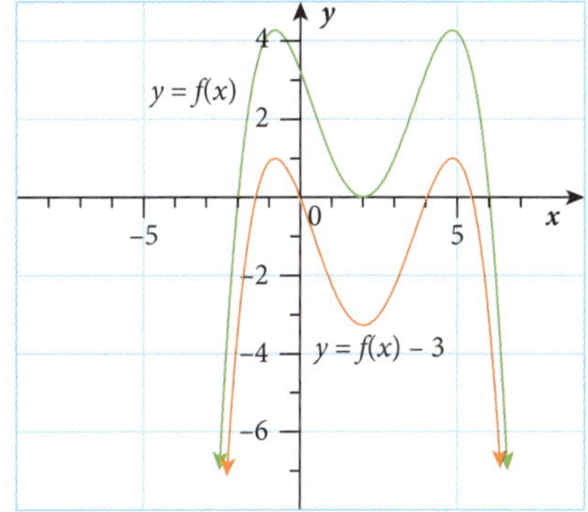

b

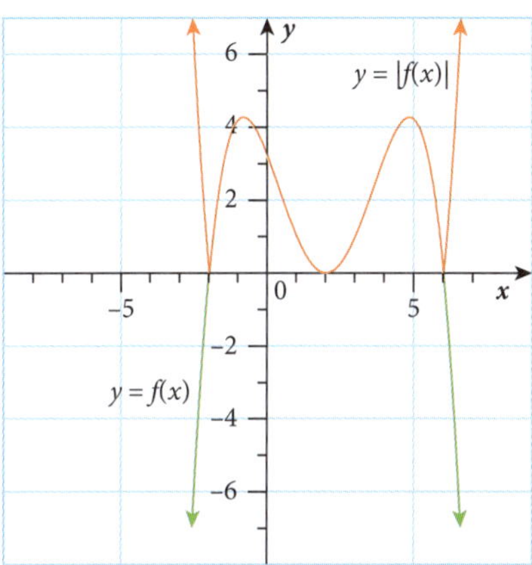

c

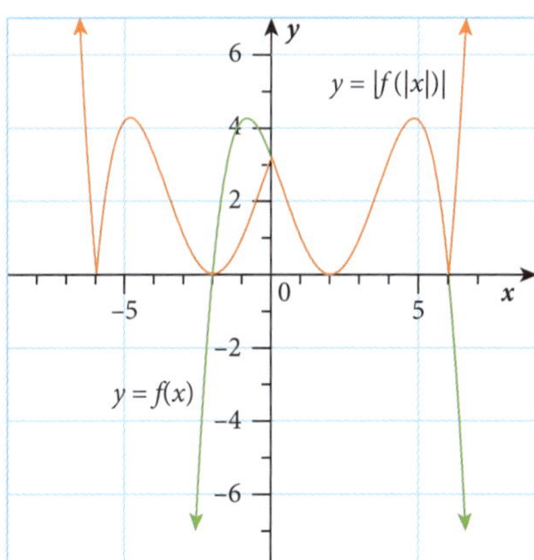

d

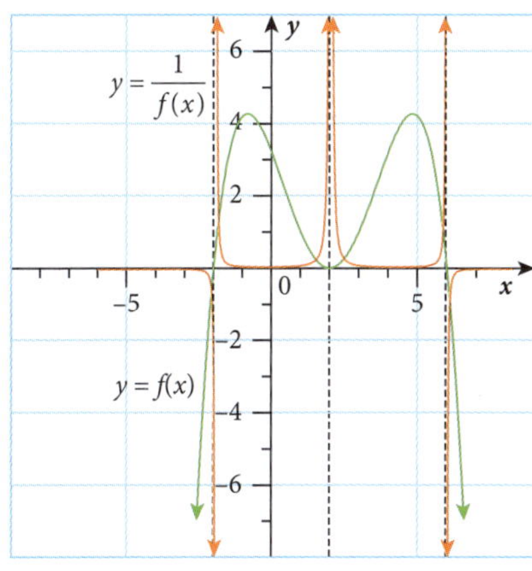

e

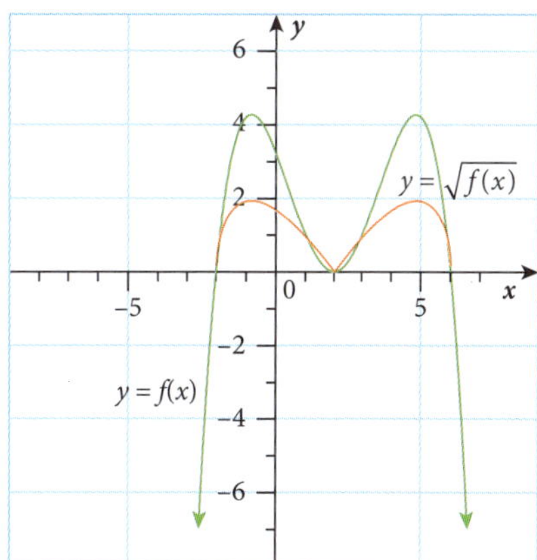

f

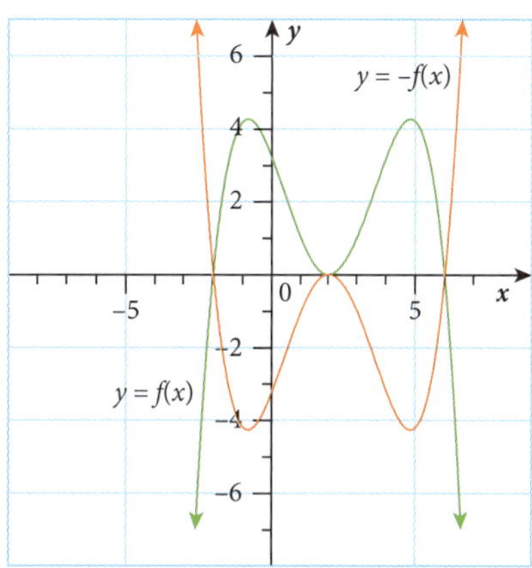

2 i a Sum of the two intercepts $= 6 + (-4)$
$= 2$

$\therefore$ **the y-intercept of $y = f(x) + g(x)$ is 2.**

b Product of the two intercepts $= 6 \times (-4)$
$= -24$

$\therefore$ **the y-intercept of $y = f(x).g(x)$ is –24.**

ii a Sum of the two intercepts $= 2 + (-12)$
$= -10$

$\therefore$ **the y-intercept of $y = p(x) + q(x)$ is –10.**

b Product of the two intercepts $= 2 \times (-12)$
$= -24$

$\therefore$ **the y-intercept of $y = p(x).q(x)$ is –24.**

3 i

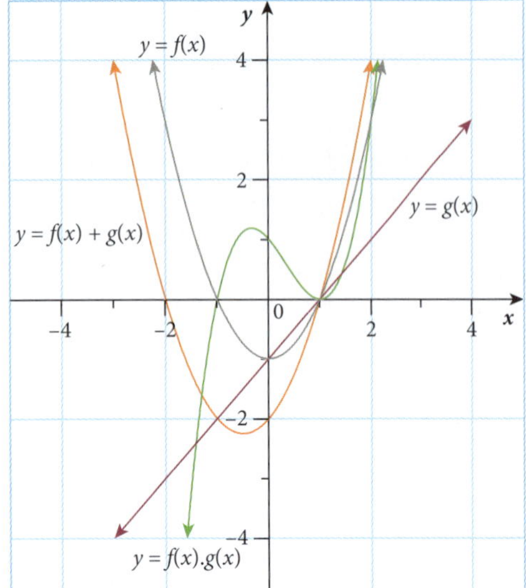

ii

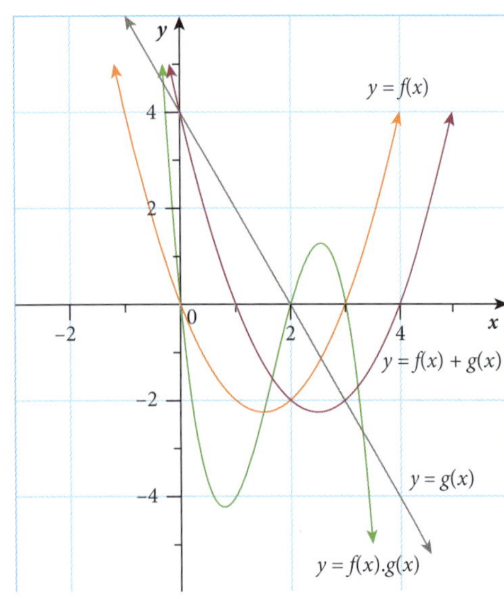

2 Quadratic inequalities ⊃ p. 7

1 i

$$3x^2 > 6x$$
$$3x^2 - 6x > 0$$
$$3x(x - 2) > 0$$
Let $3x(x - 2) = 0$
$$x = 0, 2$$
Test $x = 1$; $3(1)(1 - 2) > 0$ … No!

0 1 2

$\therefore \boldsymbol{x < 0, x > 2}$

ii

$$6a \geq 4a^2$$
$$4a^2 - 6a \leq 0$$
$$2a(2a - 3) \leq 0$$
Let $2a(2a - 3) = 0$
$$a = 0, \frac{3}{2}$$
Test $a = 1$; $2(1)(2 - 3) \leq 0$ … Yes!

0 1 $\frac{3}{2}$

$\therefore \boldsymbol{0 \leq a \leq \frac{3}{2}}$

2 i

$$2x^2 + 7x - 4 \geq 0$$
$$(2x - 1)(x + 4) \geq 0$$
Let $(2x - 1)(x + 4) = 0$
$$x = \frac{1}{2}, -4$$
Test $x = 0$; $(0 - 1)(0 + 4) \geq 0$ … No!

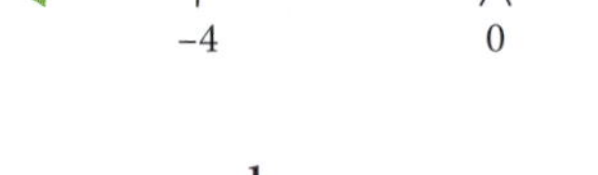

$\therefore \boldsymbol{x \leq -4, x \geq \frac{1}{2}}$

ii

$$6x^2 + x - 2 < 0$$
$$(3x + 2)(2x - 1) < 0$$
Let $(3x + 2)(2x - 1) = 0$
$$\boldsymbol{x = -\frac{2}{3}, \frac{1}{2}}$$
Test $x = 0$; $(0 + 2)(0 - 1) < 0$ … Yes!

$-\frac{2}{3}$ 0 $\frac{1}{2}$

$\therefore \boldsymbol{-\frac{2}{3} < x < \frac{1}{2}}$

3 i $y = 3 + 5x - 2x^2$
$= (3 - x)(1 + 2x)$
Consider $(3 - x)(1 + 2x) = 0$:
$$\therefore x = 3, -\frac{1}{2}$$

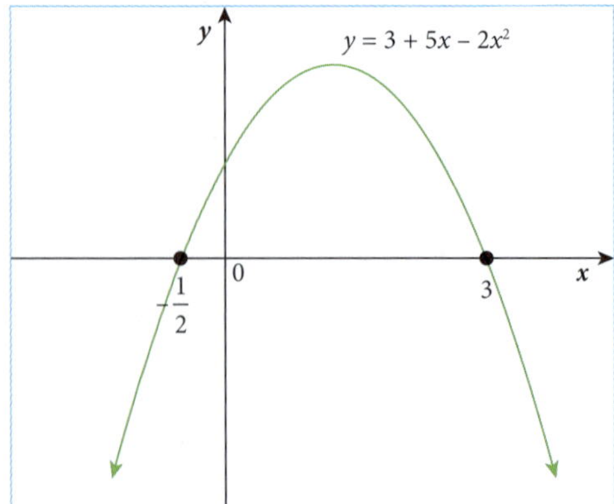

As $3 + 5x - 2x^2 \leq 0$, then
$\therefore \boldsymbol{x \leq -\frac{1}{2}, x \geq 3}$.

ii Let $y = 2x^2 + 3x - 20$
$= (2x - 5)(x + 4)$

Consider $(2x - 5)(x + 4) = 0$:

$\therefore x = \frac{5}{2}, -4$

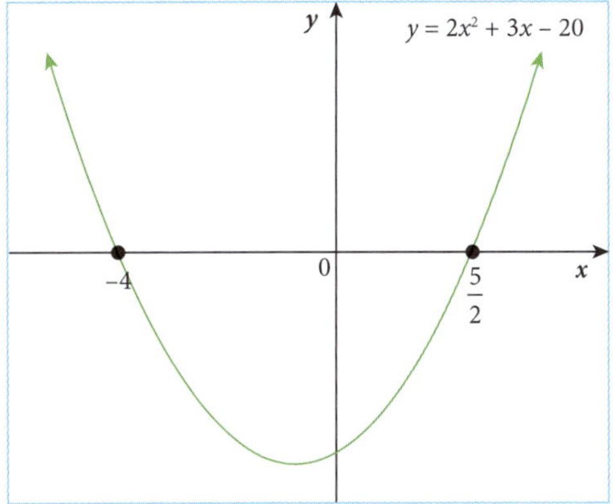

As $2x^2 + 3x - 20 < 0$, then

$\therefore \mathbf{-4 < x < \frac{5}{2}}$

4 i

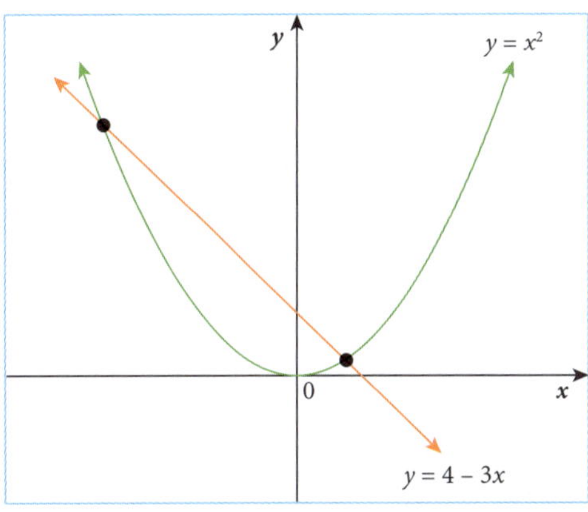

Let $x^2 = 4 - 3x$

$x^2 + 3x - 4 = 0$

$(x + 4)(x - 1) = 0$

$x = -4, 1$

For $x^2 \leq 4 - 3x$:

$\therefore \mathbf{-4 \leq x \leq 1}$

ii

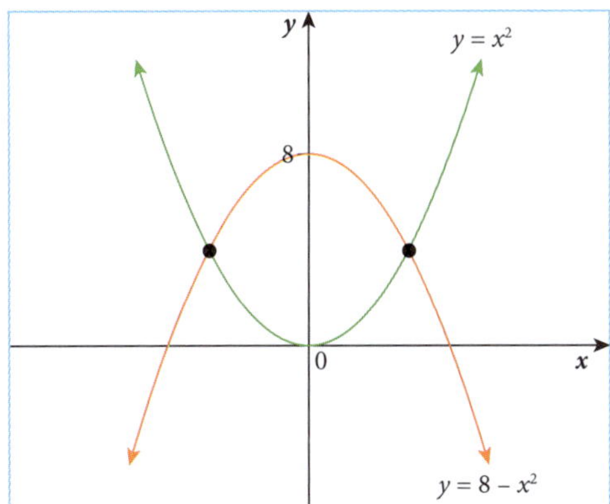

Let $x^2 = 8 - x^2$

$2x^2 = 8$

$x^2 = 4$

$x = \pm 2$

For $x^2 \geq 8 - x^2$:

$\therefore \mathbf{x \leq -2, x \geq 2}$.

5 i $4x - 2x^2 < 0$

Let $4x - 2x^2 = 0$

$2x(2 - x) = 0$

$x = 0, 2$

Test $x = 1$; $2(1)(2 - 1) < 0$ … No!

$\therefore \mathbf{x < 0, x > 2}$

ii $6x^2 + 5x - 6 > 0$

Let $6x^2 + 5x - 6 = 0$

$(3x - 2)(2x + 3) = 0$

$x = \frac{2}{3}, -\frac{3}{2}$

Test $x = 0$: $(0 - 2)(0 + 3) > 0$ … No!

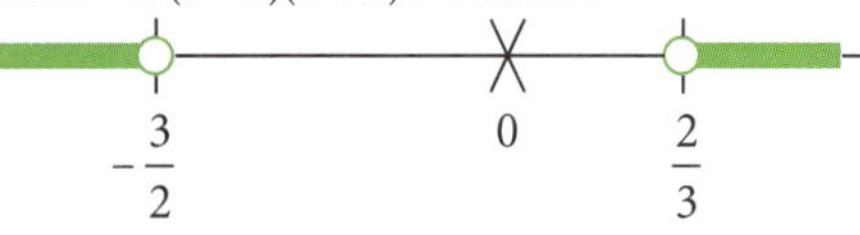

$\therefore \mathbf{x < -\frac{3}{2}, x > \frac{2}{3}}$

3 Inequalities with unknown in denominator ⊃ p. 9

1 i $\frac{5}{x-1} \geq 2$

Multiply both sides by $(x - 1)^2$:

$(x - 1)^2 \times \frac{5}{x-1} \geq 2(x - 1)^2$

$5x - 5 \geq 2x^2 - 4x + 2$

$2x^2 - 9x + 7 \leq 0$

$(2x - 7)(x - 1) \leq 0$

Let $(2x - 7)(x - 1) = 0$

$x = \frac{7}{2}, 1$

Test $x = 2$: $\frac{5}{2-1} \geq 2$ … Yes!

But $x \neq 1$

$\therefore \mathbf{1 < x \leq \frac{7}{2}}$

ii $\frac{2}{3x+1} \leq 1$

Multiply both sides by $(3x + 1)^2$:

$(3x + 1)^2 \times \frac{2}{3x+1} \leq (3x + 1)^2$

$6x + 2 \leq 9x^2 + 6x + 1$

$9x^2 - 1 \geq 0$

$(3x + 1)(3x - 1) \geq 0$

Let $(3x + 1)(3x - 1) = 0$

$$x = -\frac{1}{3}, \frac{1}{3}$$

Test $x = 0$: $\frac{2}{3(0)+1} \leq 1$ ... No!

But $x \neq -\frac{1}{3}$

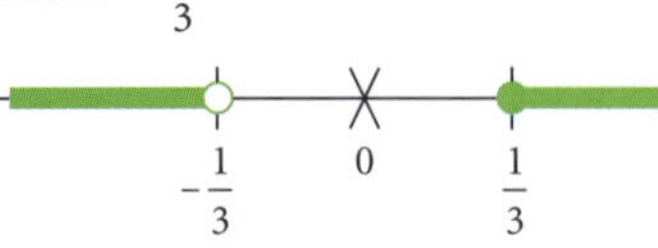

$\therefore \boldsymbol{x < -\frac{1}{3}, x \geq \frac{1}{3}}$

2 i $\frac{2x+1}{x} > 1$

Multiply both sides by x^2:

$$x^2 \times \frac{2x+1}{x} > x^2$$
$$x(2x + 1) > x^2$$
$$2x^2 + x > x^2$$
$$x^2 + x > 0$$
$$x(x + 1) > 0$$

Let $x(x + 1) = 0$

$$x = 0, -1$$

Test $x = 1$: $\frac{2+1}{1} > 1$ ... Yes!

$\therefore \boldsymbol{x < -1, x > 0}$

ii $\frac{1-4x}{2x} \leq 3$

Multiply both sides by $2x^2$:

$$2x^2 \times \frac{1-4x}{2x} \leq 2x^2 \times 3$$
$$x(1 - 4x) \leq 6x^2$$
$$x - 4x^2 \leq 6x^2$$
$$10x^2 - x \geq 0$$
$$x(10x - 1) \geq 0$$

Let $x(10x - 1) = 0$

$$x = 0, \frac{1}{10}$$

Test $x = 1$: $\frac{1-4}{2} \leq 3$ ... Yes!

But $x \neq 0$.

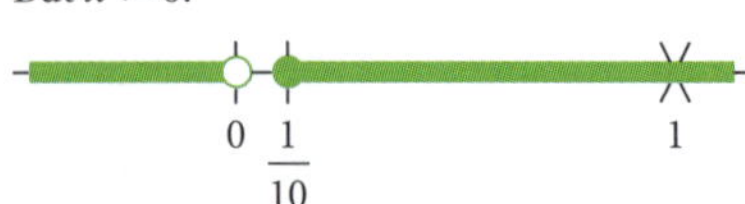

$\therefore \boldsymbol{x < 0, x \geq \frac{1}{10}}$

3 i $\frac{2}{3x+1} > x$

Multiply both sides by $(3x + 1)^2$:

$$(3x + 1)^2 \times \frac{2}{3x+1} > x(3x + 1)^2$$
$$2(3x + 1) > x(3x + 1)^2$$
$$x(3x + 1)^2 - 2(3x + 1) < 0$$
$$(3x + 1)[x(3x + 1) - 2] < 0$$
$$(3x + 1)[3x^2 + x - 2] < 0$$
$$(3x + 1)(3x - 2)(x + 1) < 0$$

Let $(3x + 1)(3x - 2)(x + 1) = 0$

$$\boldsymbol{x = -\frac{1}{3}, \frac{2}{3}, -1}$$

$x = -\frac{1}{3}, \frac{2}{3}, -1$

Now sketch $y = (3x + 1)(3x - 2)(x + 1)$.

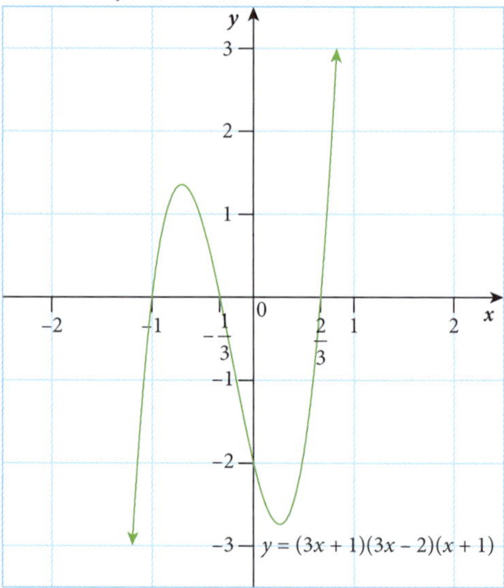

Consider $y < 0$: $x < -1$, or $-\frac{1}{3} < x < \frac{2}{3}$.

$\therefore$ the solution is $x < -1$, or $-\frac{1}{3} < x < \frac{2}{3}$.

$\boldsymbol{x < -1}$ **or** $\boldsymbol{-\frac{1}{3} < x < \frac{2}{3}}$

ii $x + \frac{1}{1-2x} \leq 0$

Multiply both sides by $(1 - 2x)^2$:

$$x(1 - 2x)^2 + (1 - 2x)^2 \times \frac{1}{1-2x} \leq 0$$
$$x(1 - 2x)^2 + (1 - 2x) \leq 0$$
$$(1 - 2x)[x(1 - 2x) + 1] \leq 0$$
$$(1 - 2x)[x - 2x^2 + 1] \leq 0$$
$$-(1 - 2x)[2x^2 - x - 1] \leq 0$$
$$(2x - 1)(2x + 1)(x - 1) \leq 0$$

Let $(2x - 1)(2x + 1)(x - 1) = 0$

$$x = \frac{1}{2}, -\frac{1}{2}, 1$$

$x = \frac{1}{2}, -\frac{1}{2}, 1$

Now sketch $y = (2x - 1)(2x + 1)(x - 1)$.

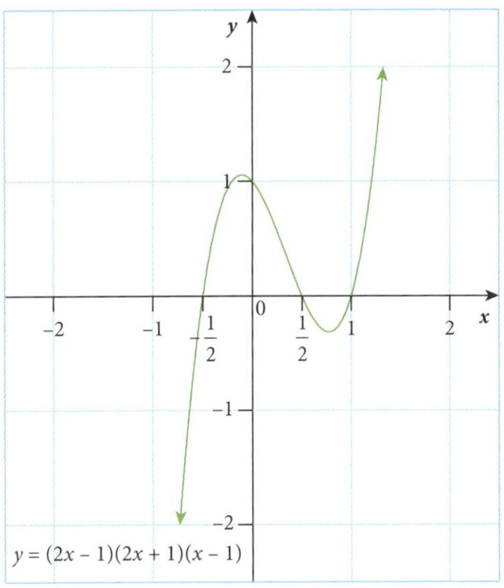

Consider $y \le 0$: $x \le -\frac{1}{2}$, or $\frac{1}{2} \le x \le 1$.

But $x \ne \frac{1}{2}$.

$\therefore$ **the solution is** $\boldsymbol{x \le -\frac{1}{2}}$**, or** $\boldsymbol{\frac{1}{2} < x \le 1}$.

4 i
$$\frac{1}{2p} \le \frac{1}{p+1}$$
$$2p^2(p+1)^2 \times \frac{1}{2p} \le 2p^2(p+1)^2 \times \frac{1}{p+1}$$
$$p(p+1)^2 \le 2p^2(p+1)$$
$$p(p+1)^2 - 2p^2(p+1) \le 0$$
$$p(p+1)(p+1-2p) \le 0$$
$$p(p+1)(1-p) \le 0$$
Let $p(p+1)(1-p) = 0$
$$p = 0, -1, 1$$

$p = 0, -1, 1$

Now sketch $f(p) = p(p+1)(1-p)$.

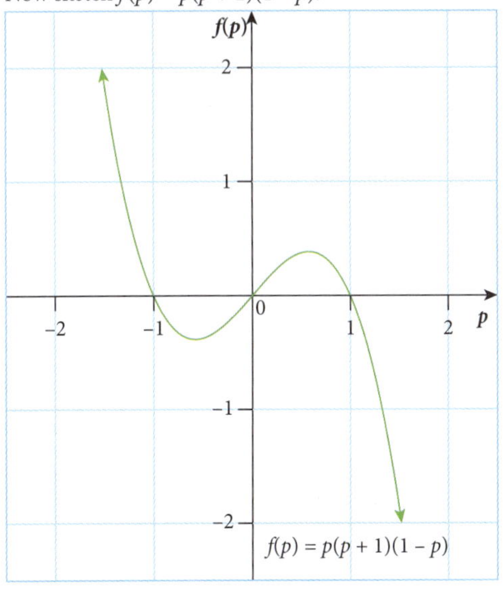

Consider $f(p) \le 0$: $-1 \le p \le 0$, $p \le 1$

But $p \ne 0, 1$.

$\therefore$ **the solution is** $\boldsymbol{-1 \le p < 0, p > 1}$.

ii
$$\frac{1}{y-2} \ge \frac{1}{y+3}$$
$$(y-2)^2(y+3)^2 \times \frac{1}{y-2} \ge (y-2)^2(y+3)^2 \times \frac{1}{y+3}$$
$$(y-2)(y+3)^2 \ge (y-2)^2(y+3)$$
$$(y-2)(y+3)^2 - (y-2)^2(y+3) \ge 0$$
$$(y-2)(y+3)[y+3-(y-2)] \ge 0$$
$$5(y-2)(y+3) \ge 0$$
Let $(y-2)(y+3) = 0$
$$y = 2, -3$$
Test $y = 0$; $\frac{1}{0-2} \ge \frac{1}{0+3}$ … Yes!

But $y \ne 2, -3$

−3 0 2

$\boldsymbol{y < -3, y > 2}$

4 Inequalities involving absolute values ➲ pp. 9–10

1 i $|3x-2| < 1$

$-1 < 3x - 2 < 1$

$1 < 3x < 3$

$\boldsymbol{\frac{1}{3} < x < 1}$

ii $|2x+5| < 3$

$-3 < 2x + 5 < 3$

$-8 < 2x < -2$

$\boldsymbol{-4 < x < -1}$

2 i $|3+2a| \ge 1$

$3 + 2a \le -1$ or $3 + 2a \ge 1$

$2a \le -4$ or $2a \ge -2$

$a \le -2$ or $a \ge -1$

$\therefore \boldsymbol{a \le -2}$ **or** $\boldsymbol{a \ge -1}$

ii $|4m-3| > 2$

$4m - 3 < -2$ or $4m - 3 > 2$

$4m < 1$ or $4m > 5$

$m < \frac{1}{4}$ or $m > \frac{5}{4}$

$\therefore \boldsymbol{m < \frac{1}{4}}$ **or** $\boldsymbol{m > \frac{5}{4}}$

3 i $\frac{|4x+2|}{5} \le 1$

$-1 \le \frac{4x+2}{5} \le 1$

$-5 \le 4x + 2 \le 5$

$-7 \le 4x \le 3$

$\boldsymbol{-\frac{7}{4} \le x \le \frac{3}{4}}$

ii $\frac{|3-2x|}{2} > 3$

$\frac{3-2x}{2} < -3$ or $\frac{3-2x}{2} > 3$

$3 - 2x < -6$ or $3 - 2x > 6$

$-2x < -9$ or $-2x > 3$

$x > \frac{9}{2}$ or $x < -\frac{3}{2}$

$\therefore \boldsymbol{x < -\frac{3}{2}}$ **or** $\boldsymbol{x > 4\frac{1}{2}}$

4 i $\frac{4x+2}{x} > -2$ or $\frac{4x+2}{x} < 2$

$x^2 \times \frac{4x+2}{x} > -2 \times x^2$ or $x^2 \times \frac{4x+2}{x} < 2 \times x^2$

$x(4x + 2) > -2x^2$ or $x(4x + 2) < 2x^2$

$4x^2 + 2x > -2x^2$ or $4x^2 + 2x < 2x^2$

$6x^2 + 2x > 0$ or $2x^2 + 2x < 0$

$2x(3x + 1) > 0$ or $2x(x + 1) < 0$

Let $2x(3x + 1) = 0$ Let $2x(x + 1) = 0$

$x = 0, -\frac{1}{3}$ $x = 0, -1$

By testing points: $x = -2$: yes; $x = -\frac{2}{3}$: yes; $x = -\frac{1}{6}$: yes; $x = 2$: no.

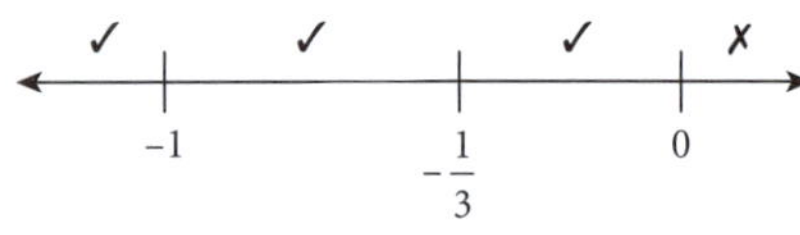

$\therefore$ **$x < 0$**

ii $\frac{|2x-1|}{x+1} \geq 2$

$\frac{2x-1}{x+1} \leq -2$ or $\frac{2x-1}{x+1} \geq 2$

$(x + 1)^2 \times \frac{2x-1}{x+1} \leq -2 \times (x + 1)^2$ or

$(x + 1)^2 \times \frac{2x-1}{x+1} \geq 2 \times (x + 1)^2$

$(x + 1)(2x - 1) \leq -2(x + 1)^2$ or

$(x + 1)(2x - 1) \geq 2(x + 1)^2$

$2(x + 1)^2 + (x + 1)(2x - 1) \leq 0$ or

$(x + 1)(2x - 1) - 2(x + 1)^2 \geq 0$

$(x + 1)[2(x + 1) + 2x - 1] \leq 0$ or

$(x + 1)[(2x - 1) - 2(x + 1)] \geq 0$

$(x + 1)(4x + 1) \leq 0$ or $(x + 1)(-3) \geq 0$

Let $(x + 1)(4x + 1) = 0$ Let $x + 1 = 0$

$x = -1, -\frac{1}{4}$ $x = -1$

By testing points $x = -1\frac{1}{2}$: no; $x = -\frac{1}{2}$: yes; $x = 0$: no.
But $x \neq -1$

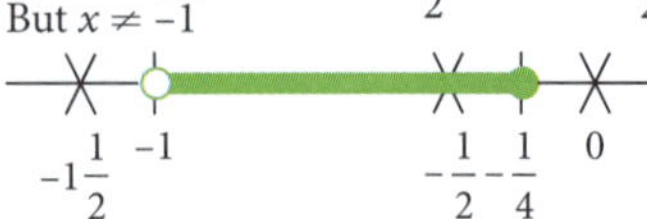

$\therefore$ **$-1 < x \leq -\frac{1}{4}$**

5 Inverse functions ⊃ pp. 11–12

1 i

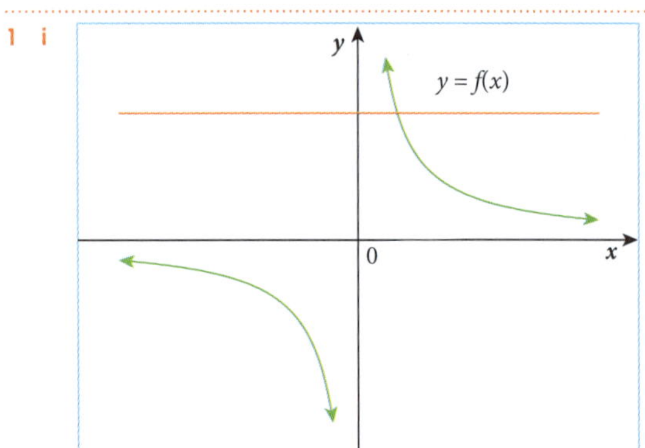

Horizontal line cuts in one place only.

Yes, has an inverse function.

ii

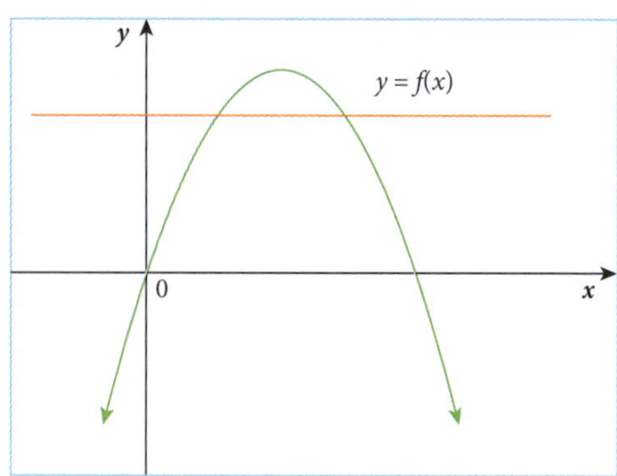

Horizontal line cuts in two places.

No inverse function as not one-to-one relationship.

2 i $f(x) = 2x^2 - 12x$

Consider $y = 2x(x - 6) = 0$

$x = 0, 6$

As concave up, minimum value at $x = 3$.

$\therefore$ **one-to-one increasing function when $x \geq 3$.**

ii $f(x) = 15 - 2x - x^2$

Consider $y = (5 + x)(3 - x) = 0$

$x = -5, 3$

As concave down, maximum value at $x = -1$.

$\therefore$ **one-to-one increasing function when $x \leq -1$.**

3 i $f(x) = x^2 - 4x + 4$, for $x \geq 2$

Let $y = x^2 - 4x + 4$

Swap $y \leftrightarrow x$:

$x = y^2 - 4y + 4$

$x = (y - 2)^2$

$\sqrt{x} = y - 2$ (as domain for $f(x)$ is $x \geq 2$)

$y = \sqrt{x} + 2$

$\therefore f^{-1}(x) = \sqrt{x} + 2$

Domain is $x \geq 0$, range is $y \geq 2$.

ii $f(x) = \sqrt{2x-3}$

Let $y = \sqrt{2x-3}$ (domain: $x \geq \frac{3}{2}$, range: $y \geq 0$)

Swap $y \leftrightarrow x$:

$x = \sqrt{2y-3}$

$x^2 = 2y - 3$

$2y = x^2 + 3$

$y = \frac{x^2+3}{2}$

$\therefore f^{1}(x) = \frac{x^2+3}{2}$, for $x \geq 0$

Domain is $x \geq 0$, range is $y \geq \frac{3}{2}$.

4 i $f(x) = 3 + 2\sqrt{x}$

Solve $3 + 2\sqrt{x} = x$.

$2\sqrt{x} = x - 3$

$4x = (x - 3)^2$

$4x = x^2 - 6x + 9$

$x^2 - 10x + 9 = 0$

$(x - 1)(x - 9) = 0$

$x = 1, 9$

$\therefore$ the points of intersection are (1, 1) and (9, 9).

But as $x \geq 3$ hence the point of intersection is (9, 9).

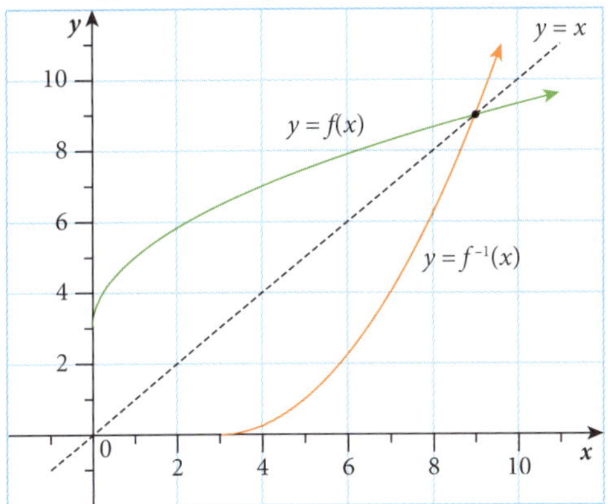

ii $f^{-1}(x) = x^2 - 2$

Solve $x^2 - 2 = x$

$x^2 - x - 2 = 0$

$(x - 2)(x + 1) = 0$

$x = 2, -1$

$\therefore$ the point of intersection is (2, 2) and (−1, −1).

But as $x \geq 0$ hence the point of intersection is (2, 2).

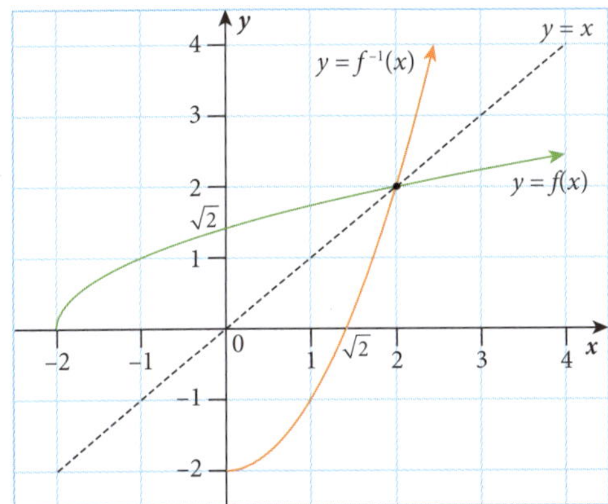

5 i a $y = (x - 1)^2 + 2$

As concave up and vertex at (1, 2), the minimum value at $x = 1$.

$\therefore$ **there is a one-to-one increasing function when $x \geq 1$.**

b Let $y = (x - 1)^2 + 2$ (using domain: $x \geq 1$, range: $y \geq 2$)

Swap $y \leftrightarrow x$:

$x = (y - 1)^2 + 2$

$(y - 1)^2 = x - 2$

$y - 1 = \pm\sqrt{x-2}$

$y = 1 \pm \sqrt{x-2}$

But range is $y \geq 1$:

$\therefore f^{1}(x) = 1 + \sqrt{x-2}$

Domain: $x \geq 2$, range: $y \geq 1$.

c

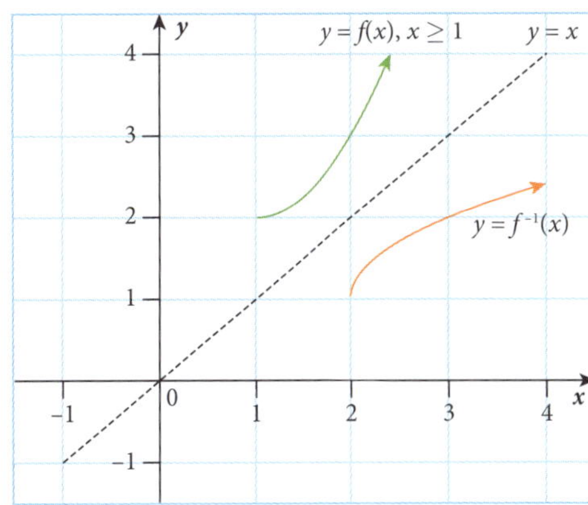

ii a $f(x) = 1 - (x + 2)^2$

As concave down and vertex at (−2, 1), the maximum value at $x = -2$.

$\therefore$ **there is a one-to-one increasing function when $x \leq -2$.**

b Let $y = 1 - (x + 2)^2$ (using domain: $x \leq -2$, range: $y \leq 1$)

Swap $y \leftrightarrow x$:

$x = 1 - (y + 2)^2$

$(y + 2)^2 = 1 - x$

$y + 2 = \pm\sqrt{1-x}$

$y = -2 \pm \sqrt{1-x}$

But range is $y \leq -2$:

$\therefore f^{1}(x) = -2 - \sqrt{1-x}$

Domain: $x \leq 1$, range: $y \leq -2$.

c

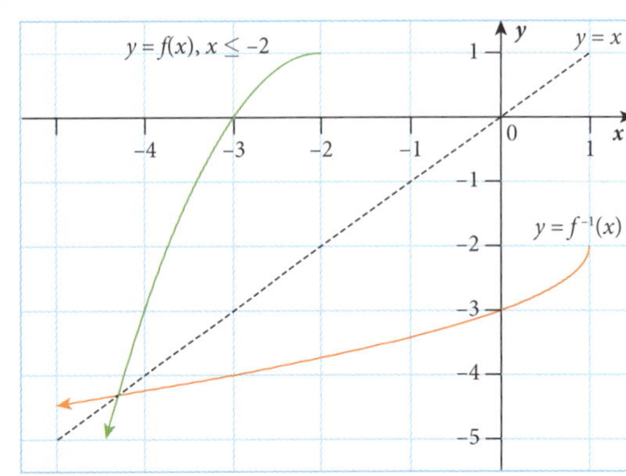

6 Parametric form of a function or relation ➲ p. 13

1 i $x = t + 1$ … ①

$y = 3t + 2$ … ②

From ①: $t = x - 1$

Subs. in ②: $y = 3(x - 1) + 2$

$= 3x - 3 + 2$

$= 3x - 1$

$\therefore$ **$y = 3x - 1$**

ii $x = 5 - p \ldots$ ①

$y = \dfrac{4p-2}{3} \ldots$ ②

From ①: $p = 5 - x$

Subs. in ②: $y = \dfrac{4(5-x)-2}{3}$

$= \dfrac{20-4x-2}{3}$

$= \dfrac{18-4x}{3}$

$\therefore \boldsymbol{y = \dfrac{18-4x}{3}}$

2 i $x = 4t \ldots$ ①

$y = 2t^2 \ldots$ ②

From ①: $t = \dfrac{x}{4}$

Subs. in ②: $y = 2\left(\dfrac{x}{4}\right)^2$

$\boldsymbol{y = \dfrac{x^2}{8}}$

ii $x = \dfrac{p-3}{2} \ldots$ ①

$y = \dfrac{p^2}{5} \ldots$ ②

From ①: $2x = p - 3$

$p = 2x + 3$

Subs. in ②: $\boldsymbol{y = \dfrac{(2x+3)^2}{5}}$

3 i $\cos t = \dfrac{x}{5}$, and hence, $\cos^2 t = \dfrac{x^2}{25} \ldots$ ①

$\sin t = \dfrac{y}{5}$, and hence, $\sin^2 t = \dfrac{y^2}{25} \ldots$ ②

Hence, $\dfrac{x^2}{25} + \dfrac{y^2}{25} = 1$, as $\sin^2 t + \cos^2 t = 1$ (using ① and ②).

$\therefore$ $\boldsymbol{x^2 + y^2 = 25}$, which is a circle, centre (0, 0) and radius 5 units.

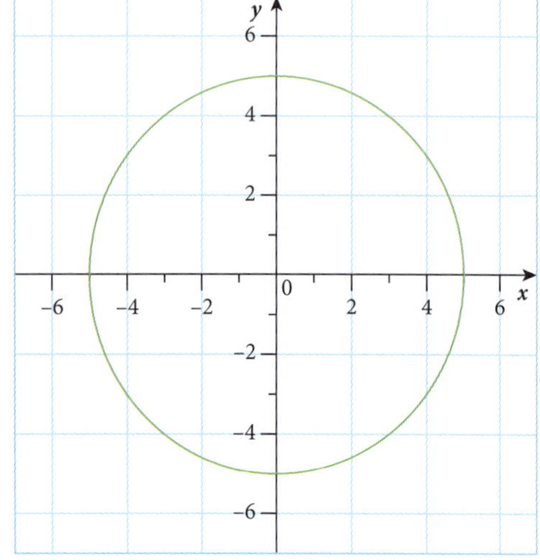

ii $4 \cos t = x + 2$, hence, $\cos t = \dfrac{x+2}{4}$ $\therefore \cos^2 t = \dfrac{(x+2)^2}{16} \ldots$ ①

$4 \sin t = y - 1$, hence, $\sin t = \dfrac{y-1}{4}$ $\therefore \sin^2 t = \dfrac{(y-1)^2}{16} \ldots$ ②

Hence, $\dfrac{(x+2)^2}{16} + \dfrac{(y-1)^2}{16} = 1$, as $\sin^2 t + \cos^2 t = 1$

(using ① and ②).

$\therefore$ $\boldsymbol{(x+2)^2 + (y-1)^2 = 16}$, which is a circle, centre (−2, 1) and radius 4 units.

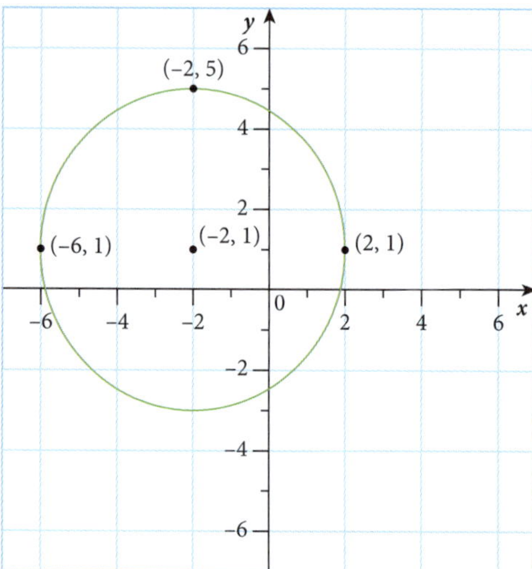

4 i As $x = 2 - 3\cos\theta$

$3\cos\theta = 2 - x$

$\cos\theta = \dfrac{2-x}{3}$

$\cos^2\theta = \dfrac{(2-x)^2}{9}$

Also, $y = 4 + 3\sin\theta$

$3\sin\theta = y - 4$

$\sin\theta = \dfrac{y-4}{3}$

$\sin^2\theta = \dfrac{(y-4)^2}{9}$

As $\cos^2\theta + \sin^2\theta = 1$, then

$\dfrac{(2-x)^2}{9} + \dfrac{(y-4)^2}{9} = 1$

$(2-x)^2 + (y-4)^2 = 9$

$(x-2)^2 + (y-4)^2 = 9$

Circle, centre (2, 4) and radius 3 units.

ii As $x = 2 - 5\cos\theta$

$5\cos\theta = 2 - x$

$\cos\theta = \dfrac{2-x}{5}$

$\cos^2\theta = \dfrac{(2-x)^2}{25}$

Also, $y = 5(1 - \sin\theta)$

$y = 5 - 5\sin\theta$

$5\sin\theta = 5 - y$

$\sin\theta = \dfrac{5-y}{5}$

$\sin^2\theta = \dfrac{(5-y)^2}{25}$

As $\cos^2\theta + \sin^2\theta = 1$, then

$$\frac{(2-x)^2}{25} + \frac{(5-y)^2}{25} = 1$$

$(2 - x)^2 + (5 - y)^2 = 25$

$(x - 2)^2 + (y - 5)^2 = 25$

Circle, centre (2, 5) and radius 5 units.

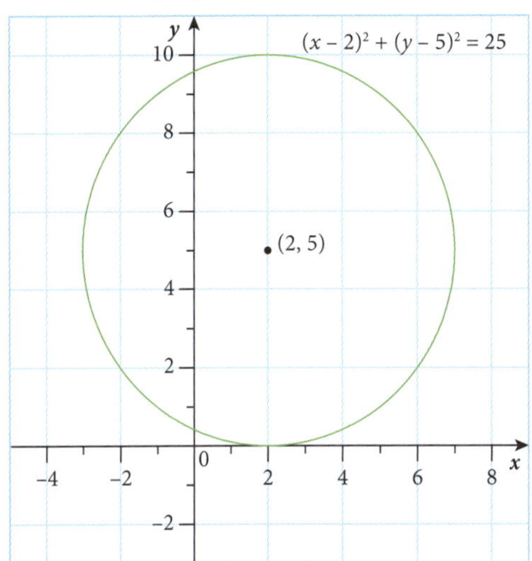

YEAR 11 EXAM-TYPE QUESTIONS

Part A

1 D. Swap $y \leftrightarrow x$:

$x = 3y - 2$

$3y = x + 2$

$y = \frac{x+2}{3}$

$\therefore$ inverse function is $y = \frac{x+2}{3}$.

2 A. $|4 - 2x| \le 6$

$-6 \le 4 - 2x \le 6$

$-10 \le -2x \le 2$

$5 \ge x \ge -1$

$\therefore -1 \le x \le 5$

3 C. $x = \frac{t}{2}$ … ①

$y = 3t^2$ … ②

From ①: $t = 2x$

Subs. in ②: $y = 3(2x)^2$

$= 12x^2$

4 A. To find $f^{-1}(3)$, need to solve $f(x) = 3$.

$\frac{2x}{x+1} = 3$

$3(x + 1) = 2x$

$3x + 3 = 2x$

$3x - 2x = -3$

$x = -3$

$\therefore f^{-1}(3) = -3$

5 C. $\frac{1}{1-x} \le 1$

$(1 - x)^2 \times \frac{1}{1-x} \le 1(1 - x)^2$

$1 - x \le (1 - x)^2$

$(1 - x)^2 - (1 - x) \ge 0$

$(1 - x)[1 - x - 1] \ge 0$

$-x(1 - x) \ge 0$

Let $-x(1 - x) = 0$

$x = 0, 1$

Test $x = 2$; $\frac{1}{1-2} \le 1$? … Yes!

Also test at $x = -1$ and $x = \frac{1}{2}$

By testing points $x = -1$: yes; $x = \frac{1}{2}$: no; $x = 2$: yes.

But $x \ne 1$

$\therefore x \le 0$ or $x > 1$

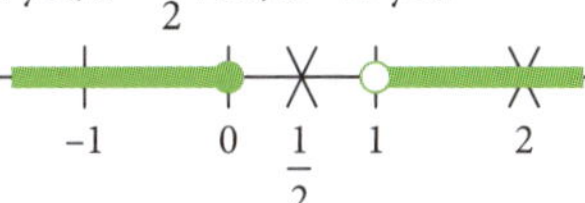

Part B

6 a

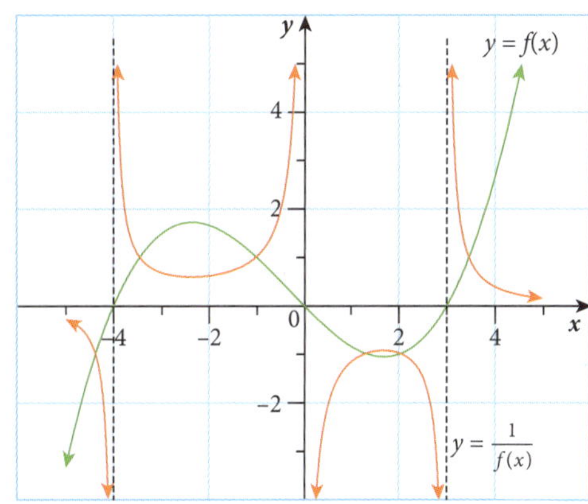

✓✓

b

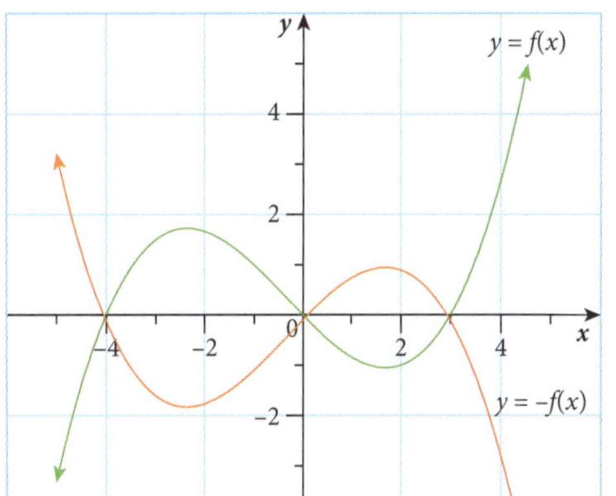

✓✓

c $y = f(|x|)$

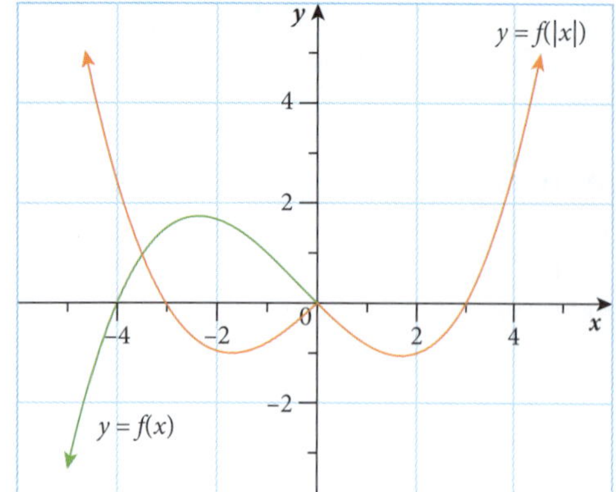

✓✓

d $y = |f(x)|$

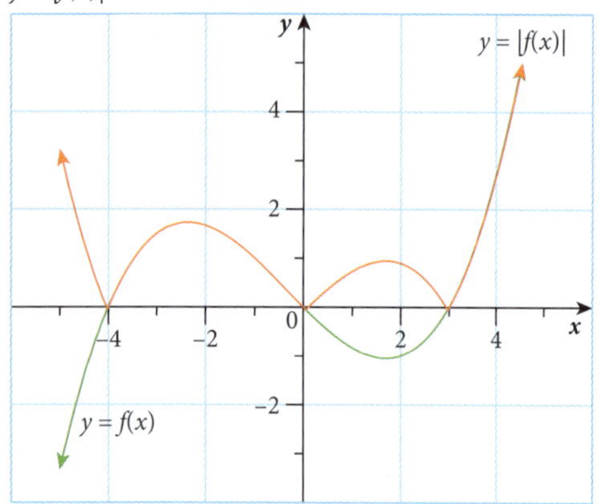

✓✓

e $y = \sqrt{f(x)}$

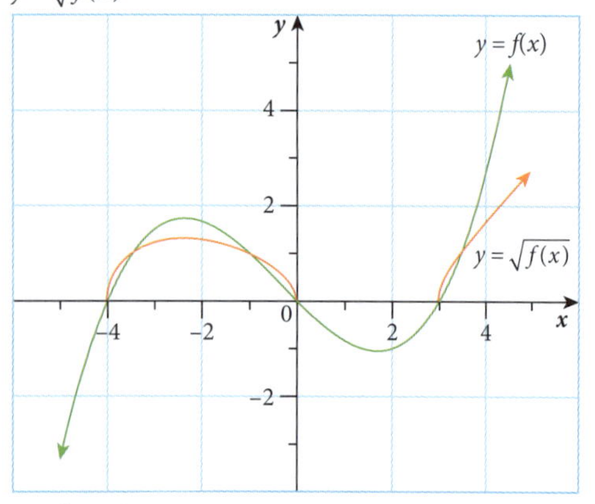

✓✓

f $y = f(x + 2)$

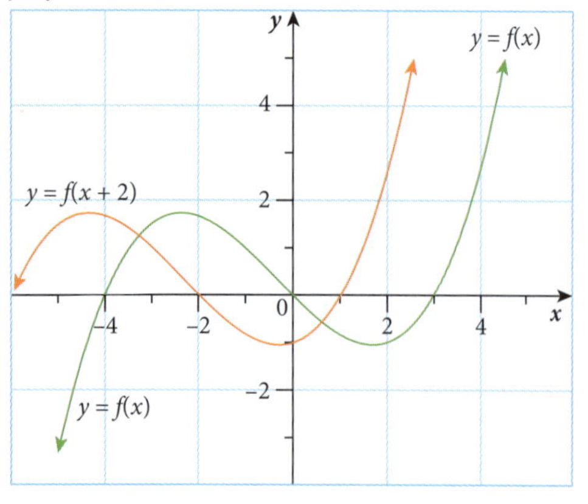

✓✓

7 $x = \frac{t}{2} \ldots$ ①

$y = \frac{t^2}{2} \ldots$ ②

From ①: $t = 2x$

Subs. in ②: $y = \frac{(2x)^2}{2}$

$= 2x^2 \quad \therefore y = 2x^2 \ldots$ ③ ✓

Also, $x = p + 1 \ldots$ ④

$y = 4p + 2 \ldots$ ⑤

From ④: $p = x - 1$

Subs. in ⑤: $y = 4(x - 1) + 2$

$= 4x - 4 + 2$

$= 4x - 2 \quad \therefore y = 4x - 2 \ldots$ ⑥ ✓

Let ⑤ = ⑥: $2x^2 = 4x - 2$

$2x^2 - 4x + 2 = 0$

$x^2 - 2x + 1 = 0$

$(x - 1)^2 = 0$

$x = 1$

Subs. in ③: $y = 2(1)^2$

$y = 2$

$\therefore$ **(1, 2)** ✓

8 Consider $8 - 2x - x^2 = 0$

$(4 + x)(2 - x) = 0$

$x = -4, 2$

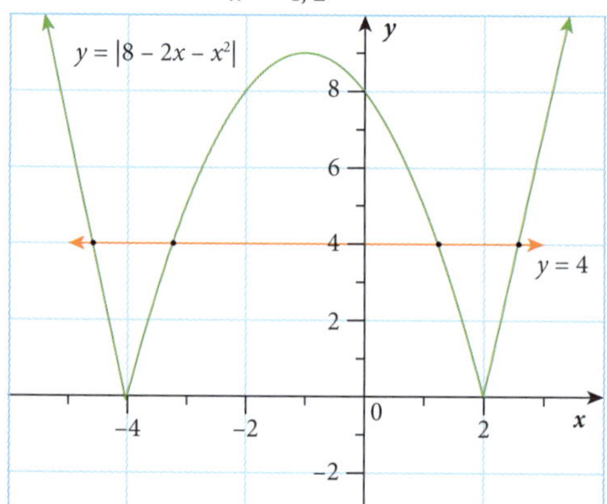

✓

$\therefore$ **4 solutions** ✓

9 a $2x^2 - 5x > 3$

$2x^2 - 5x - 3 > 0$

$(2x + 1)(x - 3) > 0$

Let $(2x + 1)(x - 3) = 0$

$x = -\frac{1}{2}, 3$ ✓

Test $x = 0$; $(1)(-3) > 0$? … No!

$\therefore \boldsymbol{x < -\frac{1}{2}, x > 3}$ ✓

b $\frac{4x-5}{2x+1} \leq 1$

Multiply both sides by $(2x + 1)^2$:

$(2x + 1)^2 \times \frac{4x-5}{2x+1} \leq (2x + 1)^2 \times 1$ ✓

$(2x + 1)(4x - 5) \leq (2x + 1)^2$

$(2x + 1)^2 - (2x + 1)(4x - 5) \geq 0$

$(2x + 1)[(2x + 1) - (4x - 5)] \geq 0$

$(2x + 1)(-2x + 6) \geq 0$

$2(2x + 1)(3 - x) \geq 0$

Let $2(2x + 1)(3 - x) = 0$

$x = -\frac{1}{2}, 3$ ✓

Test $x = 0$; $\frac{-5}{1} \leq 1$ … Yes!

But $x \neq -\frac{1}{2}$

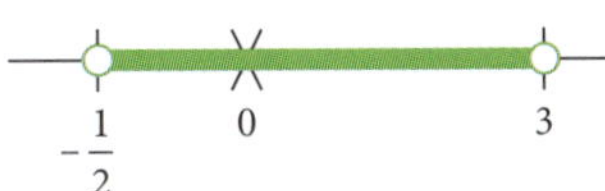

$\therefore \boldsymbol{-\frac{1}{2} < x \leq 3}$ ✓

c $|3x - 2| > 1$

$3x - 2 < -1$ or $3x - 2 > 1$ ✓

$3x < 1$ or $3x > 3$

$x < \frac{1}{3}$ or $x > 1$

$\therefore \boldsymbol{x < \frac{1}{3}}$ **or** $\boldsymbol{x > 1}$ ✓

10 Consider the function $f(x) = (x - 3)^2$.

a, e

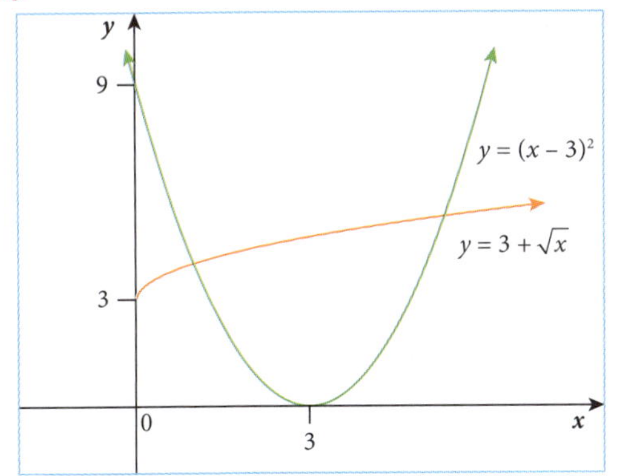

✓✓

b **A horizontal line would cut in more than one point—there is no one-to-one relationship between *x* and *y*.** ✓

c $\boldsymbol{x \geq 3}$ ✓

d Consider $y = (x - 3)^2$

Let $y \leftrightarrow x$:

$x = (y - 3)^2$ ✓

$y - 3 = \pm\sqrt{x}$

$y = 3 \pm \sqrt{x}$

But domain of $y = f(x)$ was restricted as $x \geq 3$, hence

$y = 3 + \sqrt{x}$.

$\therefore \boldsymbol{f^{-1}(x) = 3 + \sqrt{x}}$ ✓

CHAPTER 2 POLYNOMIALS

TOPIC: FUNCTIONS

YEAR 11

1 Terminology

➔ **The general form of a polynomial** is $P(x) = a_n x^n + a_{n-1}x^{n-1} + \ldots + a_1 x + a_0$, where n is a positive integer or zero and $a_n \neq 0$.

➔ **For the polynomial $P(x)$ above**, the degree is n, leading coefficient is a_n and the constant term is a_0.

EXAMPLE 1

For the polynomial $P(x) = 5x^3 - 3x^4 + 2x - 1$, state the:

a degree

b leading coefficient

c constant

a The degree is 4.

> The degree is the highest power in the polynomial.

b The leading coefficient is −3.

> The leading coefficient is the number in front of the term with the highest power.

c The constant is −1.

> The constant term is the term independent of x.

EXAMPLE 2

Find the number of real roots of the equation $P(x) = 0$ if $P(x) = (x - 1)(x + 4)(x^2 - 9)$.

$$(x - 1)(x + 4)(x^2 - 9) = 0$$
$$(x - 1)(x + 4)(x - 3)(x + 3) = 0$$
$$\therefore x = 1, -4, 3, -3$$

$\therefore$ there are four real roots.

> The maximum number of roots in a polynomial equation is its degree.

➔ PRACTICE

1 i **For the polynomial $P(x) = 6 - 2x^4 - x^5 + x$, state the:**

a degree

b leading coefficient

c constant

ii **For the polynomial $P(x) = (3x - 4)(1 - 2x^3)$, state the:**

a degree

b leading coefficient

c constant

2 **Find the number of real roots of the equation $P(x) = 0$, if:**

i $P(x) = x(x + 7)(x^2 - 25)(2x + 1)$

ii $P(x) = (x^2 - 2)(x^2 + 9)(x^2 - 5x + 7)$

Answers ➲ p. 36

2 Division of polynomials

➔ **$P(x) = A(x).Q(x) + R(x)$** where $A(x)$ = divisior, $Q(x)$ = quotient and $R(x)$ = remainder.

EXAMPLE 1

Divide $P(x) = 2x^3 - 3x^2 - 4x + 2$ by $A(x) = x - 3$, and hence write $P(x)$ in the form $P(x) = A(x).Q(x) + R(x)$.

- How many x's in $2x^3$?
- The answer is $2x^2$, which is the first term written in the quotient.
- Then, $2x^2$ times $(x - 3)$ is $2x^3 - 6x^2$.
- Now subtracting, gives $3x^2$.
- Bring down the next term which is $-4x$.
- Start again … How many x's in $3x^2$?, etc.

$$\begin{array}{r} 2x^2+3x+5 \\ x-3\overline{)2x^3-3x^2-4x+2} \\ \underline{2x^3-6x^2} \\ 3x^2-4x \\ \underline{3x^2-9x} \\ 5x+2 \\ \underline{5x-15} \\ 17 \end{array}$$

$\therefore 2x^3 - 3x^2 + 4x + 2 = (x - 3)(2x^2 + 3x + 5) + 17.$

> When the remainder is 0, then the division must be a factor of $P(x)$.

EXAMPLE 2

Express $P(x)$ in the form $P(x) = A(x).Q(x) + R(x)$ if $P(x) = 3x^4 + 2x^2 + 4$ and $Q(x) = x + 2$.

$$\begin{array}{r|l} & 3x^3-6x^2+14x-28 \\ x+2 & 3x^4+0x^3+2x^2+0x+4 \\ & 3x^4+6x^3 \\ & \quad -6x^3+2x^2 \\ & \quad -6x^3-12x^2 \\ & \qquad 14x^2+0x \\ & \qquad 14x^2+28x \\ & \qquad\quad -28x+4 \\ & \qquad\quad -28x-56 \\ & \qquad\qquad 60 \end{array}$$

It is helpful with the division to include, say, $0x^3$ when x^3 is absent in the polynomial $P(x)$.

$\therefore 3x^4 + 2x^2 + 4 = (x + 2)(3x^3 - 6x^2 + 14x - 28) + 60$.

→ PRACTICE

1 **Divide $P(x)$ by $A(x)$ and write $P(x)$ in the form $P(x) = A(x).Q(x) + R(x)$ where:**
   - i $P(x) = x^3 - 2x^2 + 3x + 1$ and $A(x) = x - 1$
   - ii $P(x) = 3x^3 + 4x^2 - 2x - 5$ and $A(x) = x + 3$

2 **Express $P(x)$ in the form $P(x) = A(x).Q(x) + R(x)$ if:**
   - i $P(x) = 2x^4 - 5x + 3$ and $Q(x) = x - 2$
   - ii $P(x) = 3x^3 - 4x^2 + 2$ and $Q(x) = x^2 - 3$

Answers ➲ p. 36

3 Remainder and factor theorems

→ **You may be required to prove** the remainder theorem and the factor theorem.

→ **Remainder theorem**: If a polynomial $P(x)$ is divided by $x - a$ then the remainder is $P(a)$.

- **Proof**: For $P(x) = A(x).Q(x) + R(x)$, let $A(x) = (x - a)$.

  $P(x) = (x - a).Q(x) + R(x)$

  As $(x - a)$ has degree 1, then $R(x)$ has degree 0 (i.e. $R(x) = k$, a constant).

  $\therefore P(x) = (x - a).Q(x) + k$

  Subs. $x = a$: $P(a) = (a - a).Q(x) + k$

  $= k$

$\therefore$ the remainder is $P(a)$.

→ **Factor theorem:** For polynomial $P(x)$, if $P(a) = 0$, then $x - a$ is a factor of the polynomial.

- **Proof**: For $P(x) = A(x).Q(x) + R(x)$, let $A(x) = (x - a)$.

  $P(x) = (x - a).Q(x) + R(x)$

The remainder when $P(x)$ is divided by $x - a$ is $P(a)$ (from remainder theorem).

$\therefore P(x) = (x - a).Q(x) + P(a)$

But $P(a) = 0$,

$\therefore P(x) = (x - a).Q(x)$

$\therefore x - a$ is a factor of the polynomial.

EXAMPLE 1

Find the remainder if $2x^3 - 3x^2 + 2x + 15$ is divided by $(x + 2)$.

Let $P(x) = 2x^3 - 3x^2 + 2x + 15$

$P(-2) = 2(-2)^3 - 3(-2)^2 + 2(-2) + 15$

$= -16 - 12 - 4 + 15$

$= -17$

Show the step of substitution into the polynomial.

$\therefore$ the remainder is -17.

EXAMPLE 2

Show that $x - 2$ is a factor of $x^3 + x^2 - 5x - 2$.

Let $P(x) = x^3 + x^2 - 5x - 2$

$P(2) = (2)^3 + (2)^2 - 5(2) - 2$

$= 8 + 4 - 10 - 2$

$= 0$

If $P(a) = 0$, then $x = a$ is a factor.

$\therefore x - 2$ is a factor of $x^3 + x^2 - 5x - 2$.

EXAMPLE 3

It is known that $x + 1$ is a factor of $P(x) = x^3 + ax^2 + bx - 1$ and there is a remainder of -4 when $P(x)$ is divided by $(x - 3)$. Find the values of a and b.

$P(-1) = (-1)^3 + a(-1)^2 + b(-1) - 1 = 0$

$-1 + a - b - 1 = 0$

$a - b - 2 = 0$

$a - b = 2$ ... ①

$P(3) = 3^3 + a(3)^2 + b(3) - 1 = -4$

$27 + 9a + 3b - 1 = -4$

$9a + 3b + 26 = -4$

$9a + 3b = -30$

$3a + b = -10$ ... ②

Use simultaneous equations.

① + ②: $4a = -8$

$a = -2$

Subs. in ①: $-2 - b = 2$

$b = -4$

$\therefore a = -2, b = -4$.

EXAMPLE 4

When a polynomial $P(x)$ is divided by $(x - 1)(x + 2)$ the remainder is $-4x + 5$. What is the remainder when $P(x)$ is divided by $(x + 2)$?

Consider $P(x) = (x - 1)(x + 2).Q(x) + (-4x + 5)$

$$P(-2) = (-2 - 1)(-2 + 2).Q(-2) + (-4(-2) + 5) = 13$$

$\therefore$ the remainder is 13.

Use $P(x) = A(x).Q(x) + R(x)$.

EXAMPLE 5

Solve $x^3 - 2x^2 - 11x + 12 = 0$.

Let $P(x) = x^3 - 2x^2 - 11x + 12$

Try $P(1) = 1^3 - 2(1)^2 - 11(1) + 12$
$= 1 - 2 - 11 + 12$
$= 0$

Substitute factors of the constant term to find a root.

$\therefore x = 1$ is a root and hence $x - 1$ is a factor.

$$\begin{array}{r|l} & x^2 - x - 12 \\ \hline x - 1 & x^3 - 2x^2 - 11x + 12 \\ & x^3 - x^2 \\ \hline & -x^2 - 11x \\ & -x^2 + x \\ \hline & -12x + 12 \\ & -12x + 12 \\ \hline & 0 \end{array}$$

$\therefore (x - 1)(x^2 - x - 12) = 0$

$(x - 1)(x - 4)(x + 3) = 0$

$\therefore x = 1, 4, -3$

→ PRACTICE

1 Find the remainder if:
- i $x^3 + 2x^2 + 3x + 6$ is divided by $(x + 4)$
- ii $x^4 - x^2 + 2$ is divided by $(2x + 1)$

2 Show that:
- i $x - 4$ is a factor of $x^3 - 5x^2 + 2x + 8$
- ii $x + 3$ is a factor of $2x^3 + 3x^2 - 4x + 15$

3 i It is known that $x + 3$ is a factor of $P(x) = x^3 + ax^2 + bx - 24$ and there is a remainder of -8 when $P(x)$ is divided by $(x + 2)$. Find the values of a and b.
- ii When $P(x) = x^4 - 3x^3 + 2x^2 + px + q$ is divided by $(x + 1)$ the remainder is 9 and when divided by x the remainder is -2. What is the remainder when $P(x)$ is divided by $(1 - x)$?

4 i When a polynomial $P(x)$ is divided by $(x + 3)(x - 1)$ the remainder is $-2x + 3$. What is the remainder when $P(x)$ is divided by $(x - 1)$?
- ii The remainder when the polynomial $P(x) = x^4 + kx^2 - 10x - 5$ is divided by $(x^2 - 2x)$ is $kx - 5$. What is the value of k?

5 i Solve $x^3 + 3x^2 - 10x - 24 = 0$.
- ii Find the zeros for the polynomial $6x^3 + 7x^2 - x - 2$.

Answers ➲ pp. 36–37

4 Sums and products of roots of polynomials

➔ **Quadratics:** If α and β are roots of $ax^2 + bx + c = 0, a \neq 0$ then $\alpha + \beta = -\frac{b}{a}, \alpha\beta = \frac{c}{a}$.

- **Proof:** If the roots of a quadratic are α and β, then the equation can be wriiten as:

$$x^2 + \frac{b}{a}x + \frac{c}{a} = (x - \alpha)(x - \beta)$$
$$= x^2 - \beta x - \alpha x + \alpha\beta$$
$$= x^2 - (\alpha + \beta)x + \alpha\beta$$
$$\therefore \alpha + \beta = -\frac{b}{a}, \alpha\beta = \frac{c}{a}.$$

➔ **Cubics:** If α, β and γ are roots of $ax^3 + bx^2 + cx + d = 0$, $a \neq 0$ then $\alpha + \beta + \gamma = -\frac{b}{a}, \alpha\beta + \alpha\gamma + \beta\gamma = \frac{c}{a}, \alpha\beta\gamma = -\frac{d}{a}$.

- **Proof:** If the roots of a cubic are α, β and γ, then the equation can be written as:

$$x^3 + \frac{b}{a}x^2 + \frac{c}{a}x + \frac{d}{a}$$
$$= (x - \alpha)(x - \beta)(x - \gamma)$$
$$= (x^2 - \beta x - \alpha x + \alpha\beta)(x - \gamma)$$
$$= x^3 - \gamma x^2 - \beta x^2 + \beta\gamma x - \alpha x^2 + \alpha\gamma x + \alpha\beta x - \alpha\beta\gamma$$
$$= x^3 - (\alpha + \beta + \gamma)x^2 + (\alpha\beta + \alpha\gamma + \beta\gamma)x - \alpha\beta\gamma$$
$$\therefore \alpha + \beta + \gamma = -\frac{b}{a}, \alpha\beta + \alpha\gamma + \beta\gamma = \frac{c}{a}, \alpha\beta\gamma = -\frac{d}{a}.$$

➔ **Quartics:** If α, β, γ and δ are roots of $ax^4 + bx^3 + cx^2 + dx + e = 0$ then $\alpha + \beta + \gamma + \delta = -\frac{b}{a}$,

$\alpha\beta + \alpha\gamma + \alpha\delta + \beta\gamma + \beta\delta + \gamma\delta = \frac{c}{a}$,

$\alpha\beta\gamma + \alpha\beta\delta + \alpha\gamma\delta + \beta\gamma\delta = -\frac{d}{a}$, $\alpha\beta\gamma\delta = \frac{e}{a}$.

EXAMPLE 1

The roots of $2x^3 + 4x^2 - 6x + 3 = 0$ are α, β and γ. Find the value of:

a $\alpha + \beta + \gamma, \alpha\beta + \alpha\gamma + \beta\gamma$ and $\alpha\beta\gamma$

b $\frac{1}{\alpha} + \frac{1}{\beta} + \frac{1}{\gamma}$

c $\alpha^2 + \beta^2 + \gamma^2$

d $(\alpha + 1)(\beta + 1)(\gamma + 1)$

a $\alpha + \beta + \gamma = -\frac{b}{a} = -2$, $\alpha\beta + \alpha\gamma + \beta\gamma = \frac{c}{a} = -3$ and

$\alpha\beta\gamma = -\frac{d}{a} = -\frac{3}{2}$

b Now, $\frac{1}{\alpha} + \frac{1}{\beta} + \frac{1}{\gamma} = \frac{\beta\gamma + \alpha\gamma + \alpha\beta}{\alpha\beta\gamma}$

$= -3 \div -\frac{3}{2}$

$= 2$

c Now, $\alpha^2 + \beta^2 + \gamma^2 = (\alpha + \beta + \gamma)^2 - 2(\alpha\beta + \alpha\gamma + \beta\gamma)$

$= (-2)^2 - 2(-3)$

$= 10$

Remember $\alpha^2 + \beta^2 = (\alpha + \beta)^2 - 2\alpha\beta$.

d Now $(\alpha + 1)(\beta + 1)(\gamma + 1)$

$= (\alpha\beta + \alpha + \beta + 1)(\gamma + 1)$

$= \alpha\beta\gamma + \alpha\beta + \alpha\gamma + \alpha + \beta\gamma + \beta + \gamma + 1$

$= \alpha\beta\gamma + (\alpha\beta + \alpha\gamma + \beta\gamma) + (\alpha + \beta + \gamma) + 1$

$= -\frac{3}{2} + (-3) + (-2) + 1$

$= -5\frac{1}{2}$

EXAMPLE 2

In the equation $x^3 + 6x^2 - x - 30 = 0$, one of the roots is equal to the sum of the other two roots. Find the values of the three roots.

Let the roots be α, β and $\alpha + \beta$.

$\alpha + \beta + \alpha + \beta = -6$

$2(\alpha + \beta) = -6$

$\alpha + \beta = -3$

Express the third root in terms of the first two.

$\therefore$ the third root is -3.

$\therefore x = -3$ is a root and hence $(x + 3)$ is a factor.

$$\begin{array}{r} x^2 + 3x - 10 \\ x+3 \overline{) x^3 + 6x^2 - x - 30} \\ \underline{x^3 + 3x^2} \\ 3x^2 - x \\ \underline{3x^2 + 9x} \\ -10x - 30 \\ \underline{-10x - 30} \\ 0 \end{array}$$

$\therefore (x + 3)(x^2 + 3x - 10) = 0$

$(x + 3)(x + 5)(x - 2) = 0$

$x = -3, -5, 2$

→ PRACTICE

1 i **The roots of $2x^3 - 2x^2 - 4x + 6 = 0$ are α, β and γ. Find the value of:**

a $\alpha + \beta + \gamma$, $\alpha\beta + \alpha\gamma + \beta\gamma$ **and** $\alpha\beta\gamma$

b $\frac{1}{\alpha} + \frac{1}{\beta} + \frac{1}{\gamma}$

c $\alpha^2 + \beta^2 + \gamma^2$

d $(\alpha + 1)(\beta + 1)(\gamma + 1)$

ii **The roots of $2x^2 - 3x + 4 = 0$ are α and β. Find the value of:**

a $\frac{\alpha}{\beta} + \frac{\beta}{\alpha}$

b $\frac{1}{\alpha^2} + \frac{1}{\beta^2}$

c $(3\alpha + 4\beta)(3\beta + 4\alpha)$

d $\left(\frac{1}{\alpha} + \frac{1}{\beta}\right)^2$

2 i **In the equation $x^3 - 10x^2 + 29x - 20 = 0$, one of the roots is equal to the sum of the other two roots. Find the values of the three roots.**

ii **One of the roots of $x^3 + bx^2 + 36x - 36 = 0$ is the product of the other two roots. If the roots are positive integers, find the value of b and hence all the roots.**

Answers ➲ pp. 37–38

5 Multiplicity of roots

→ **If $P(x) = (x - a)^k.Q(x)$, and $Q(a) \neq 0$, $k > 0$,** the polynomial has k roots at $x = a$. It is said to have a root of multiplicity of k.

→ **You may be required to prove** the relationship between the multiplicity of roots of $P(x)$ and $P'(x)$:

- **If $P(x) = (x - a)^k.Q(x)$ has a root of multiplicity k at $x = a$, then $P'(x)$ has a root of multiplicity $k-1$ at $x = a$.**

  Let $P(x) = (x - a)^k.Q(x)$, which has a root of multiplicity k.

Now, $P'(x) = k(x - a)^{k-1}.Q(x) + (x - a)^k.Q'(x)$

$P'(a) = k(a - a)^{k-1}.Q(a) + (a - a)^k.Q'(a) = 0$

$\therefore P'(x)$ has a root of multiplicity $k-1$.

EXAMPLE 1

A polynomial $P(x)$ has a triple root at $x = 2$. Show that $P'(2) = 0$.

Use the product rule.

Let $P(x) = (x - 2)^3.Q(x)$

$P'(x) = 3(x - 2)^2.Q(x) + (x - 2)^3.Q'(x)$

$P'(2) = 3(2 - 2)^2.Q(2) + (2 - 2)^3.Q'(2)$

$= 0$

EXAMPLE 2

The polynomial $P(x) = x^3 + ax^2 + bx + 18$ has a double root at $x = 3$. Find the values of a and b.

$P(3) = 3^3 + a(3)^2 + b(3) + 18$
$\quad = 27 + 9a + 3b + 18 = 0$
$9a + 3b = -45$
$3a + b = -15 \ldots$ ①
$P'(x) = 3x^2 + 2ax + b$
$P'(3) = 3(3)^2 + 2a(3) + b = 0$
$27 + 6a + b = 0$
$6a + b = -27 \ldots$ ②
② − ①: $3a = -12$
$a = -4$
Subs. in ①: $3(-4) + b = -15$
$-12 + b = -15$
$b = -3$
$\therefore a = -4, b = -3$.

Solve simultaneous equations.

➔ PRACTICE

1 i A polynomial $P(x)$ has a double root at $x = -1$. Show that $P'(-1) = 0$.
ii A polynomial $P(x)$ has a root of multiplicity 3 at $x = 4$. Show that $P''(4) = 0$.

2 i The polynomial $P(x) = x^3 + ax^2 + bx - 16$ has a double root at $x = -2$. Find the values of a and b.
ii The polynomial $P(x) = x^4 + x^3 + px^2 + qx + r$ has a root of multiplicity 3 at $x = -1$. Find the values of p, q and r.

Answers ⊃ p. 38

6 Graphs of polynomial functions

➔ $y = (x - a)(x + b)$

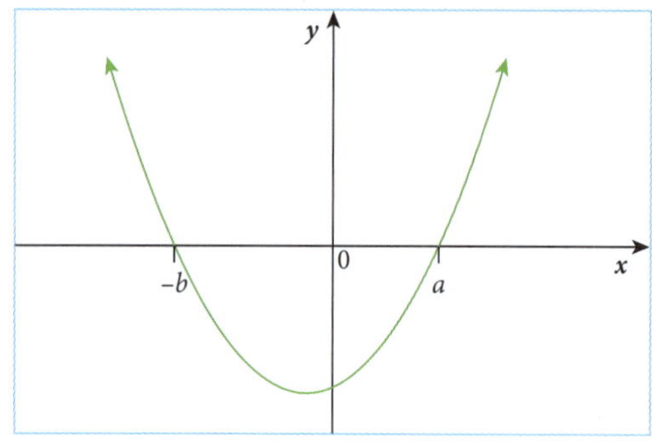

➔ $y = (x - a)^2$

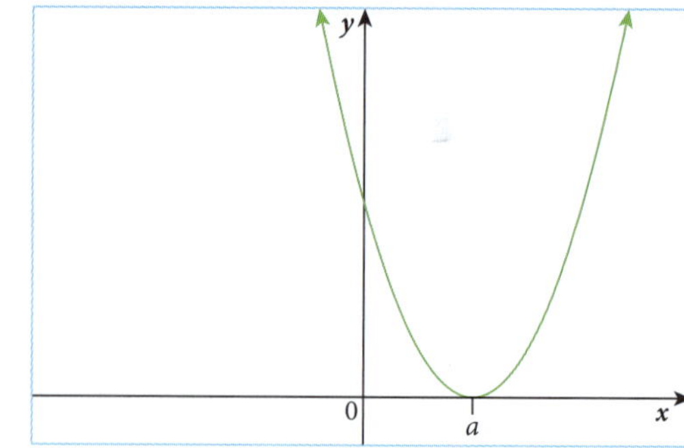

➔ $y = (x - b)^3$

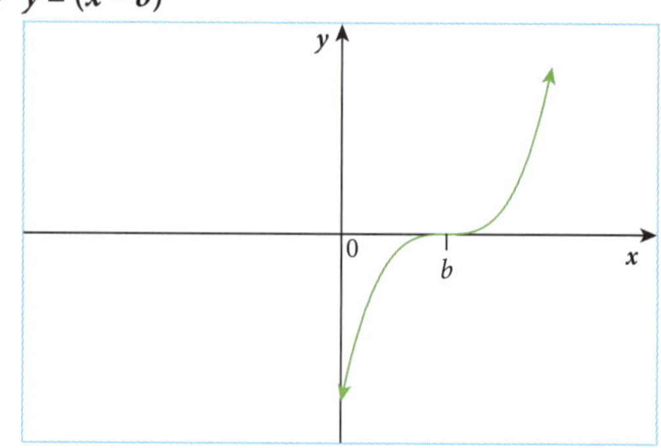

EXAMPLE 1

Sketch $y = P(x)$ if $P(x) = (x + 2)(x - 3)^2$, showing the intercepts.

Roots at $x = -2$ and $x = 3$ (twice)
y-intercept of $2(-3)^2 = 18$
Also consider the value of the polynomial for $x > 3$:
for example, $P(4) = (4 + 2)(4 - 3)^2 > 0$

Solve $P(x) = 0$ to find the x-intercepts.

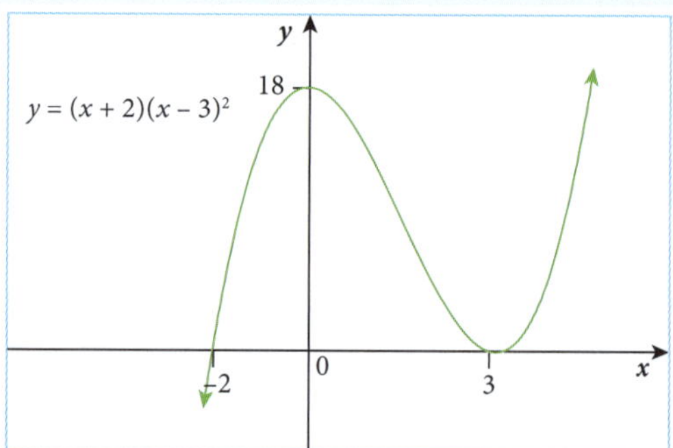

EXAMPLE 2

Draw an example of a polynomial $P(x)$ with a root of multiplicity 3 at $x = -1$ with leading term $-x^4$.

e.g. $P(x) = (x + 1)^3(2 - x)$

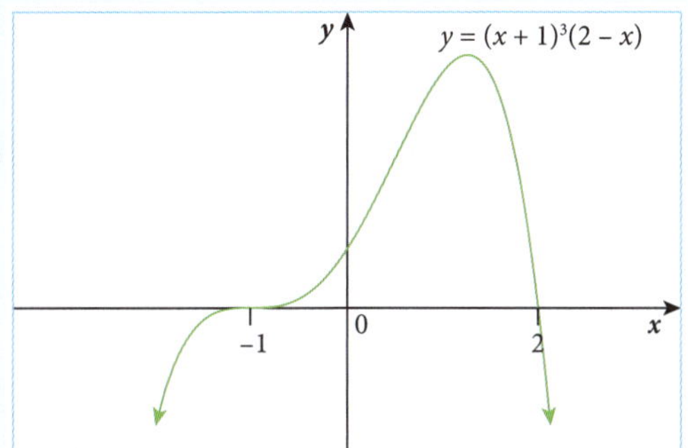

If the degree is 4 there needs to be another root.

EXAMPLE 3

Write down an example of a polynomial $P(x)$ with the following graph.

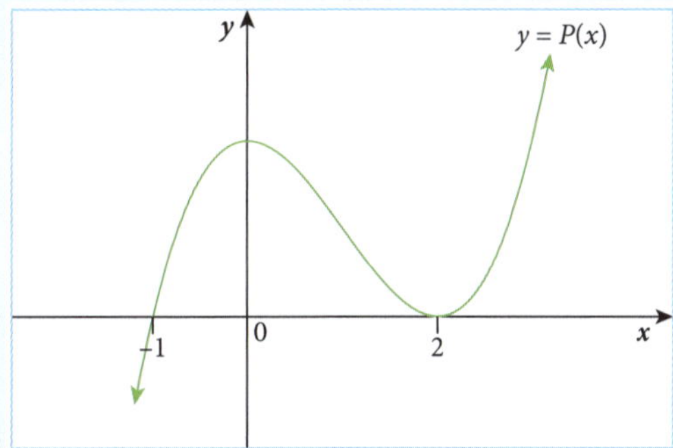

Double root at $x = 2$ and root at $x = -1$. Also $P(3) > 0$.
Possibly $P(x) = (x - 2)^2(x + 1)$.

Consider the multiplicity of each of the roots.

EXAMPLE 4

Sketch $y = P(x)$ if $P(x) = x^3 - 5x^2 + 8x - 4$, showing all x- and y-intercepts.

Consider $P(2) = 2^3 - 5(2)^2 + 8(2) - 4$
$= 8 - 20 + 16 - 4$
$= 0$

$\therefore (x - 2)$ is a factor.

$$\begin{array}{r|l} & x^2 - 3x + 2 \\ x - 2 & x^3 - 5x^2 + 8x - 4 \\ & x^3 - 2x^2 \\ & \quad -3x^2 + 8x \\ & \quad -3x^2 + 6x \\ & \qquad 2x - 4 \\ & \qquad 2x - 4 \\ & \qquad\quad 0 \end{array}$$

$\therefore P(x) = (x - 2)(x^2 - 3x + 2)$
$= (x - 2)(x - 2)(x - 1)$
$= (x - 2)^2(x - 1)$

The y-intercept is the constant term.

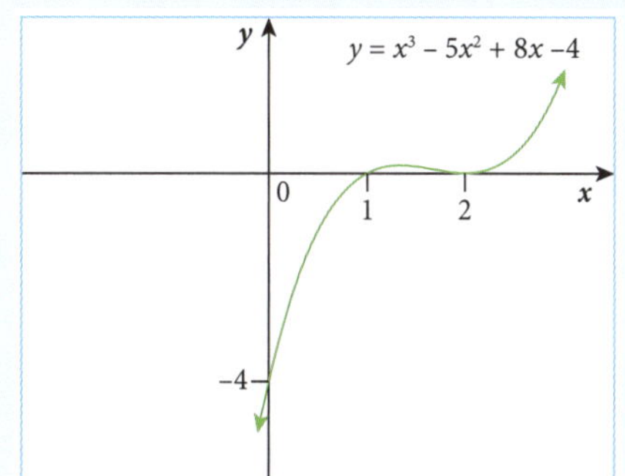

→ PRACTICE

1 Sketch $y = P(x)$, showing intercepts, if:
 i $P(x) = (x - 4)^2(x + 2)$
 ii $P(x) = x(x + 1)(1 - x)^3$

2 Draw an example of a polynomial $P(x)$ with:
 i a root of multiplicity 3 at $x = -2$ with leading term $-x^4$
 ii double root at $x = 4$, triple root at $x = 0$, with leading term $2x^6$

3 Write down an example of a polynomial $P(x)$ with the following graph:
 i

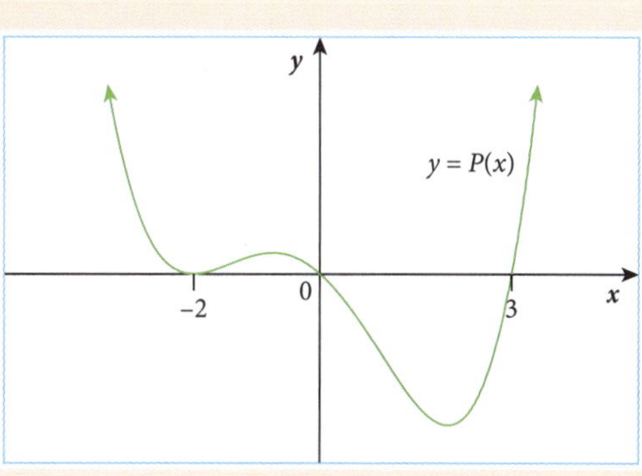

ii

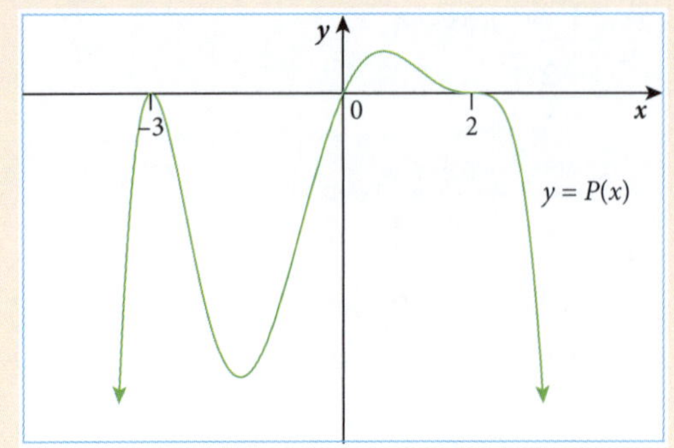

4 **Sketch $y = P(x)$ showing all x- and y-intercepts if:**

i $P(x) = x^3 + x^2 - 5x + 3$

ii $P(x) = -x^4 - 4x^3 + 3x^2 + 18x$

Answers ➲ pp. 38–39

YEAR 11 EXAM-TYPE QUESTIONS

Total marks: 35

PART A

(Suggested time: 10 minutes)

(1 mark each)

1 What is the remainder when $2x^3 - 5x^2 + 3x - 4$ is divided by $(x + 1)$?

A -14 B -4
C $2x^2 - 5x + 3$ D $2x^2 - 7x + 10$

2 Which group of three numbers could be the zeros of the polynomial $x^3 + ax^2 - 18x + 72$?

A 2, 2, 18 B 6, 6, −2
C −3, −8, 8 D 3, −4, 6

3 When the polynomial $P(x)$ is divided by $(x + 2)(x - 1)$, the remainder is $x + 9$. What is the remainder when $P(x)$ is divided by $(x - 1)$?

A −2 B 2 C 7 D 10

4 Which polynomial has a root of $x = 1$ of multiplicity 2?

A $x^5 - x^4 - x^3 + 1$ B $x^5 - x^3 - x^2 + 1$
C $x^5 - x^4 + x + 1$ D $x^5 - x^4 + x - 1$

5 Which diagram best describes the graph of $y = -x(2 - x)^2(x + 3)^3$?

A
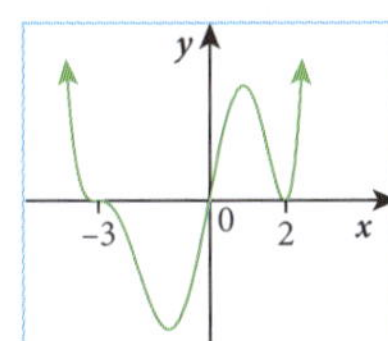

B
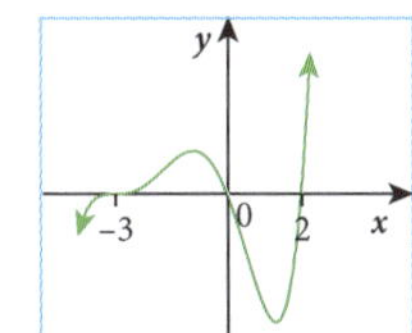

C
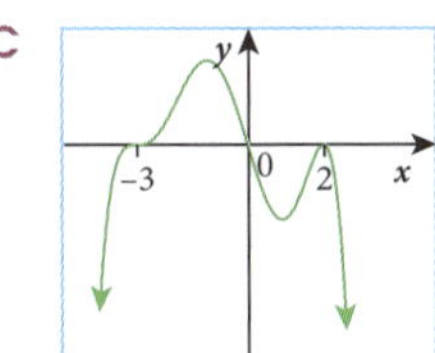

D
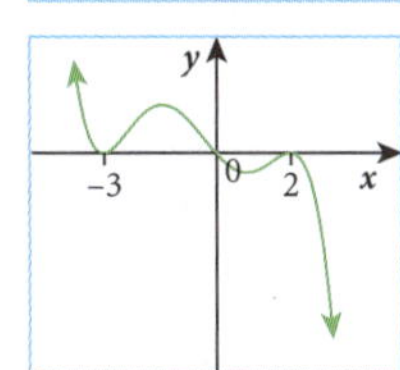

PART B

(Suggested time: 45 minutes)

Show all working.

6 The polynomial $P(x) = x^3 - 2x^2 - kx + 32$ has roots α, β and γ.

a Find the value of $\alpha + \beta + \gamma$. (1 mark)
b Find the value of $\alpha\beta\gamma$. (1 mark)
c It is known that two of the roots are equal in magnitude but opposite in sign. Find the third root and hence find the value of k. (2 marks)

7 Let $P(x) = (x + 2)(x - 1)Q(x) + ax + b$ where $Q(x)$ is a polynomial and a and b are real numbers. The polynomial $P(x)$ has a factor of $x + 2$. When $P(x)$ is divided by $(x - 1)$ the remainder is 6.

a Find the values of a and b. (2 marks)
b Find the remainder when $P(x)$ is divided by $(x + 2)(x - 1)$. (1 mark)

8 The polynomial $P(x) = ax^3 - 4x^2 + cx + 60$, where a and c are constants. The three zeros of $P(x)$ are α, 3 and −4. Find the value of α. (3 marks)

9 Consider the polynomials $P(x) = x^3 - x^2 - kx + 24$ and $A(x) = x + 4$.

a Given that $P(x)$ is divisible by $A(x)$, verify that $k = 14$. (1 mark)
b Find all the zeros of $P(x)$ when $k = 14$. (2 marks)

10 The equation $P(x) = 4x^3 - 3x + k$ has a double root. Find the possible values of k. (2 marks)

11 Suppose $x^3 - 3x^2 + 2x + d = (x - 3)Q(x) + 2$, where $Q(x)$ is a polynomial.
Find the value of d. (2 marks)

12 Suppose α, β, γ and δ are the roots of $x^4 + px^3 + qx^2 + rx + s = 0$.

a Find the values of $\alpha + \beta + \gamma + \delta$ and $\alpha\beta\gamma + \alpha\beta\delta + \alpha\gamma\delta + \beta\gamma\delta$ in terms of p, q, r or s. (2 marks)
b Show that $\alpha^2 + \beta^2 + \gamma^2 + \delta^2 = p^2 - 2q$. (2 marks)

13 The polynomial $P(x) = x^3 + bx^2 + cx + d$ has roots 0, 2 and −2.

a Find the values of b, c and d. (2 marks)
b Sketch the graph of $y = P(x)$. (1 mark)
c Hence, or otherwise, solve the inequality $\dfrac{x^2 - 4}{x} > 0$. (1 mark)

14 If two of the roots of the cubic $x^3 + bx^2 + cx + d = 0$ are equal but opposite in sign, prove that $d = bc$. (2 marks)

15 Let $P(x) = x^3 + bx^2 + cx - 12$, where b and c are real numbers. Find the values of b and c given that $(x - 2)^2$ is a factor of $P(x)$. (3 marks)

PRACTICE

1 Terminology ➲ p. 28

1 i a **5**
b **−1**
c **6**
ii a **4**
b **−6**
c **−4**

2 i $x(x + 7)(x^2 - 25)(2x + 1) = 0$
$x(x + 7)(x - 5)(x + 5)(2x + 1) = 0$
$\therefore x = 0, -7, 5, -5, -\frac{1}{2}$
∴ there are 5 real roots.

ii $(x^2 - 2)(x^2 + 9)(x^2 - 5x + 7) = 0$
Now, for $x^2 - 5x + 7$, $\Delta = b^2 - 4ac$
$= -3 < 0 \therefore$ no real roots
$\therefore (x - \sqrt{2})(x + \sqrt{2})(x^2 + 9)(x^2 - 5x + 7) = 0$
$x = \pm\sqrt{2}$
∴ there are 2 real roots.

2 Division of polynomials ➲ p. 29

1 i

$$\begin{array}{r}x^2-x+2\\ x-1\overline{)x^3-2x^2+3x+1}\\ \underline{x^3-x^2}\\ -x^2+3x\\ \underline{-x^2+x}\\ 2x+1\\ \underline{2x-2}\\ 3\end{array}$$

$\therefore \mathbf{x^3 - 2x^2 + 3x + 1 = (x - 1)(x^2 - x + 2) + 3}$

ii

$$\begin{array}{r}3x^2-5x+13\\ x+3\overline{)3x^3+4x^2-2x-5}\\ \underline{3x^3+9x^2}\\ -5x^2-2x\\ \underline{-5x^2-15x}\\ 13x-5\\ \underline{13x+39}\\ -44\end{array}$$

$\therefore \mathbf{3x^3 + 4x^2 - 2x - 5 = (x + 3)(3x^2 - 5x + 13) - 44}$

2 i

$$\begin{array}{r}2x^3+4x^2+8x+11\\ x-2\overline{)2x^4+0x^3+0x^2-5x+3}\\ \underline{2x^4-4x^3}\\ 4x^3+0x^2\\ \underline{4x^3-8x^2}\\ 8x^2-5x\\ \underline{8x^2-16x}\\ 11x+3\\ \underline{11x-22}\\ 25\end{array}$$

$\mathbf{2x^4 - 5x + 3 = (x - 2)(2x^3 + 4x^2 + 8x + 11) + 25}$

ii

$$\begin{array}{r}3x-4\\ x^2-3\overline{)3x^3-4x^2+0x+2}\\ \underline{3x^3\qquad-9x}\\ -4x^2+9x+2\\ \underline{-4x^2\qquad+12}\\ 9x-10\end{array}$$

$\therefore \mathbf{3x^3 - 4x^2 + 2 = (x^2 - 3)(3x - 4) + (9x - 10)}$

3 Remainder and factor theorems ➲ p. 30

1 i Let $P(x) = x^3 + 2x^2 + 3x + 6$
$P(-4) = (-4)^3 + 2(-4)^2 + 3(-4) + 6$
$= -64 + 32 - 12 + 6$
$= -38$
∴ the remainder is −38.

ii Let $P(x) = x^4 - x^2 + 2$
$P\left(-\frac{1}{2}\right) = \left(-\frac{1}{2}\right)^4 - \left(-\frac{1}{2}\right)^2 + 2$
$= \frac{1}{16} - \frac{1}{4} + 2$
$= \frac{29}{16}$
∴ the remainder is $\frac{29}{16}$.

2 i Let $P(x) = x^3 - 5x^2 + 2x + 8$
$P(4) = 4^3 - 5(4)^2 + 2(4) + 8$
$= 64 - 80 + 8 + 8$
$= 0$
$\therefore$ **$x - 4$ is a factor of $x^3 - 5x^2 + 2x + 8$.**

ii Let $P(x) = 2x^3 + 3x^2 - 4x + 15$
$P(-3) = 2(-3)^3 + 3(-3)^2 - 4(-3) + 15$
$= -54 + 27 + 12 + 15$
$= 0$
$\therefore$ **$x + 3$ is a factor of $2x^3 + 3x^2 - 4x + 15$.**

3 i $P(-3) = (-3)^3 + a(-3)^2 + b(-3) - 24 = 0$
$-27 + 9a - 3b - 24 = 0$
$9a - 3b - 51 = 0$
$9a - 3b = 51$
$3a - b = 17 \ldots$ ①
$P(-2) = (-2)^3 + a(-2)^2 + b(-2) - 24 = -8$
$-8 + 4a - 2b - 24 = -8$
$4a - 2b - 32 = -8$
$4a - 2b = 24$
$2a - b = 12 \ldots$ ②
① − ②: $a = 5$
Subs. in ①: $3(5) - b = 17$
$b = -2$
$\therefore$ **$a = 5, b = -2$.**

ii $P(-1) = (-1)^4 - 3(-1)^3 + 2(-1)^2 + p(-1) + q = 9$

$1 + 3 + 2 - p + q = 9$

$-p + q + 6 = 9$

$-p + q = 3 \ \ldots$ ①

$P(0) = 0 + q = -2$

$q = -2$

Subs. in ①: $-p + (-2) = 3$

$p = -5$

$\therefore P(x) = x^4 - 3x^3 + 2x^2 - 5x - 2$

$P(1) = 1^4 - 3(1)^3 + 2(1)^2 - 5(1) - 2$

$= 1 - 3 + 2 - 5 - 2$

$= \mathbf{-7}$

4 i Consider $P(x) = (x + 3)(x - 1).Q(x) + (-2x + 3)$

$P(1) = (1 + 3)(1 - 1).Q(1) + (-2(1) + 3)$

$= 1$

$\therefore$ **the remainder is 1.**

ii Consider $P(x) = A(x).Q(x) + R(x)$

$x^4 + kx^2 - 10x - 5 = (x^2 - 2x).Q(x) + (kx - 5)$

As $x^2 - 2x = x(x - 2)$, then let $x = 2$:

$2^4 + k(2)^2 - 10(2) - 5 = (2^2 - 2(2)).Q(2) + (k(2) - 5)$

$16 + 4k - 20 - 5 = 0 + 2k - 5$

$4k - 9 = 0 + 2k - 5$

$2k = 4$

$\mathbf{k = 2}$

5 i Let $P(x) = x^3 + 3x^2 - 10x - 24$.

Try $P(-2) = (-2)^3 + 3(-2)^2 - 10(-2) - 24$

$= -8 + 12 + 20 - 24$

$= 0$

$\therefore x = -2$ is a root and hence $(x + 2)$ is a factor.

$$\begin{array}{r} x^2+x-12 \\ x+2\overline{\smash{)}x^3+3x^2-10x-24} \\ \underline{x^3+2x^2} \qquad\qquad \\ x^2-10x \qquad \\ \underline{x^2+2x} \qquad \\ -12x-24 \\ \underline{-12x-24} \\ 0 \end{array}$$

$\therefore (x + 2)(x^2 + x - 12) = 0$

$(x + 2)(x + 4)(x - 3) = 0$

$\therefore \mathbf{x = -2, -4, 3}$

ii Let $P(x) = 6x^3 + 7x^2 - x - 2$.

Try $P(-1) = 6(-1)^3 + 7(-1)^2 - (-1) - 2$

$= -6 + 7 + 1 - 2$

$= 0$

$\therefore x = -1$ is a root and hence $(x + 1)$ is a factor.

$$\begin{array}{r} 6x^2+x-2 \\ x+1\overline{\smash{)}6x^3+7x^2-x-2} \\ \underline{6x^3+6x^2} \qquad\qquad \\ x^2-x \qquad \\ \underline{x^2+x} \qquad \\ -2x-2 \\ \underline{-2x-2} \\ 0 \end{array}$$

$\therefore (x + 1)(6x^2 + x - 2) = 0$

$(x + 1)(3x + 2)(2x - 1) = 0$

$\therefore x = -1, -\frac{2}{3}, \frac{1}{2}$

$\therefore$ **the zeros are** $\mathbf{-1, -\frac{2}{3}, \frac{1}{2}}$.

4 Sums and products of roots of polynomials ⊃ p. 31

1 i a $\boldsymbol{\alpha + \beta + \gamma = 1, \alpha\beta + \alpha\gamma + \beta\gamma = -2, \alpha\beta\gamma = -3}$

b $\frac{1}{\alpha} + \frac{1}{\beta} + \frac{1}{\gamma} = \frac{\beta\gamma + \alpha\gamma + \alpha\beta}{\alpha\beta\gamma}$

$= -2 \div -3$

$= \mathbf{\frac{2}{3}}$

c $\alpha^2 + \beta^2 + \gamma^2 = (\alpha + \beta + \gamma)^2 - 2(\alpha\beta + \alpha\gamma + \beta\gamma)$

$= 1^2 - 2(-2)$

$= \mathbf{5}$

d $(\alpha + 1)(\beta + 1)(\gamma + 1) = (\alpha\beta + \alpha + \beta + 1)(\gamma + 1)$

$= \alpha\beta\gamma + \alpha\beta + \alpha\gamma + \alpha + \beta\gamma + \beta + \gamma + 1$

$= \alpha\beta\gamma + (\alpha\beta + \alpha\gamma + \beta\gamma) + (\alpha + \beta + \gamma) + 1$

$= -3 + (-2) + 1 + 1$

$= \mathbf{-3}$

ii a $\frac{\alpha}{\beta} + \frac{\beta}{\alpha} = \frac{\alpha^2 + \beta^2}{\alpha\beta}$

$= \frac{(\alpha+\beta)^2 - 2\alpha\beta}{\alpha\beta}$

$= \frac{\left(\frac{3}{2}\right)^2 - 2(2)}{2}$

$= \mathbf{-\frac{7}{8}}$

b $\frac{1}{\alpha^2} + \frac{1}{\beta^2} = \frac{\beta^2 + \alpha^2}{\alpha^2\beta^2}$

$= \frac{(\alpha+\beta)^2 - 2\alpha\beta}{(\alpha\beta)^2}$

$= \frac{\left(\frac{3}{2}\right)^2 - 2(2)}{2^2}$

$= \mathbf{-\frac{7}{16}}$

c $(3\alpha + 4\beta)(3\beta + 4\alpha) = 9\alpha\beta + 12\alpha^2 + 12\beta^2 + 16\alpha\beta$

$= 12(\alpha^2 + \beta^2) + 25\alpha\beta$

$= 12[(\alpha + \beta)^2 - 2\alpha\beta] + 25\alpha\beta$

$= 12\left[\left(\frac{3}{2}\right)^2 - 2(2)\right] + 25(2)$

$= \mathbf{29}$

d $\left(\frac{1}{\alpha} + \frac{1}{\beta}\right)^2$

$= \left(\frac{\beta + \alpha}{\alpha\beta}\right)^2$

$= \left(\frac{3}{2} \div 2\right)^2$

$= \left(\frac{3}{4}\right)^2$

$= \mathbf{\frac{9}{16}}$

2 i Let the roots be α, β and $\alpha + \beta$.

$\alpha + \beta + \alpha + \beta = 10$

$2(\alpha + \beta) = 10$

$\alpha + \beta = 5$

$\therefore$ third root is 5.

$\therefore x = 5$ is a root and hence $(x - 5)$ is a factor.

$$\begin{array}{r} x^2-5x+4 \\ x-5\overline{)x^3-10x^2+29x-20} \\ \underline{x^3-5x^2} \\ -5x^2+29x \\ \underline{-5x^2+25x} \\ 4x-20 \\ \underline{4x-20} \\ 0 \end{array}$$

$\therefore (x - 5)(x^2 - 5x + 4) = 0$

$(x - 5)(x - 1)(x - 4) = 0$

$\mathbf{x = 5, 1, 4}$

ii Let the roots be α, β and $\alpha\beta$.

$\alpha \times \beta \times \alpha\beta = 36$

$\therefore (\alpha\beta)^2 = 36$

$\alpha\beta = \pm 6$

But roots are positive integers:

$\therefore$ third root is 6.

$\therefore x = 6$ is a root and hence $(x - 6)$ is a factor.

Now, let $P(x) = x^3 + bx^2 + 36x - 36$

$P(6) = 6^3 + b(6)^2 + 36(6) - 36 = 0$

$36b = -396$

$b = -11$

$\therefore P(x) = x^3 - 11x^2 + 36x - 36$

$$\begin{array}{r} x^2-5x+6 \\ x-6\overline{)x^3-11x^2+36x-36} \\ \underline{x^3-6x^2} \\ -5x^2+36x \\ \underline{-5x^2+30x} \\ -6x-36 \\ \underline{-6x-36} \\ 0 \end{array}$$

$\therefore (x - 6)(x^2 - 5x + 6) = 0$

$(x - 6)(x - 2)(x - 3) = 0$

$x = 6, 2, 3$

$\therefore$ **the roots are 6, 2, 3.**

5 Multiplicity of roots ⊃ p. 32

1 i Let $P(x) = (x + 1)^2.Q(x)$

$P'(x) = 2(x + 1).Q(x) + (x + 1)^2.Q'(x)$

$P'(-1) = 2(-1 + 1).Q(-1) + (-1 + 1)^2.Q'(-1)$

$= \mathbf{0}$

ii Let $P(x) = (x - 4)^3.Q(x)$

$P'(x) = 3(x - 4)^2.Q(x) + (x - 4)^3.Q'(x)$

$P''(x) = 6(x - 4).Q(x) + 3(x - 4)^2.Q'(x)$
$+ 3(x - 4)^2.Q'(x) + (x - 4)^3.Q''(x)$

$P''(4) = 6(4 - 4).Q(4) + 3(4 - 4)^2.Q'(4) +$
$3(4 - 4)^2.Q'(4) + (4 - 4)^3.Q''(x)$

$= \mathbf{0}$

2 i $P(-2) = (-2)^3 + a(-2)^2 + b(-2) - 16$

$= -8 + 4a - 2b - 16 = 0$

$4a - 2b = 24$

$2a - b = 12 \ldots$ ①

$P'(x) = 3x^2 + 2ax + b$

$P'(-2) = 3(-2)^2 + 2a(-2) + b = 0$

$12 - 4a + b = 0$

$-4a + b = -12 \ldots$ ②

② + ①: $-2a = 0$

$a = 0$

Subs. in ①: $2(0) - b = 12$

$-b = 12$

$b = -12$

$\therefore \mathbf{a = 0, b = -12}$

ii $P(-1) = (-1)^4 + (-1)^3 + p(-1)^2 + q(-1) + r = 0$

$1 - 1 + p - q + r = 0$

$p - q + r = 0 \ldots$ ①

$P'(x) = 4x^3 + 3x^2 + 2px + q$

$P'(-1) = 4(-1)^3 + 3(-1)^2 + 2p(-1) + q = 0$

$-4 + 3 - 2p + q = 0$

$-2p + q = 1 \ldots$ ②

$P''(x) = 12x^2 + 6x + 2p$

$P''(-1) = 12(-1)^2 + 6(-1) + 2p = 0$

$12 - 6 + 2p = 0$

$2p = -6$

$p = -3$

Subs. into ②: $-2(-3) + q = 1$

$q = -5$

Subs. into ①: $-3 - (-5) + r = 0$

$2 + r = 0$

$r = -2$

$\therefore \mathbf{p = -3, q = -5, r = -2}$

6 Graphs of polynomial functions ⊃ pp. 33–34

1 i $P(x) = (x - 4)^2(x + 2)$

Roots at $x = 4$ (twice) and $x = -2$

y-intercept of $(-4)^2(2) = 32$

Also consider the value of the polynomial for $x > 4$:

e.g. $P(5) = (5 - 4)^2(5 + 2) > 0$

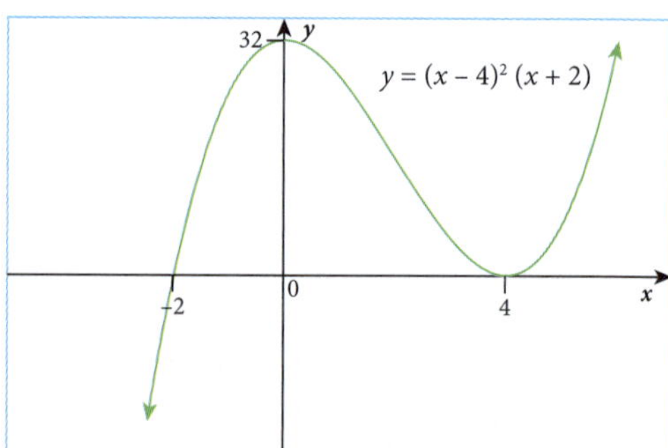

ii $P(x) = x(x + 1)(1 - x)^3$

Roots at $x = 0$, $x = -1$ and $x = 1$ (three times)

y-intercept of 0

Also consider the value of the polynomial for $x > 1$:

e.g. $P(2) = 2(2 + 1)(1 - 2)^3 < 0$

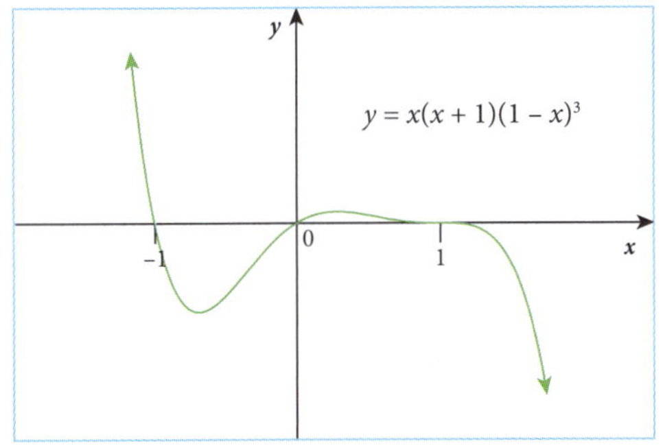

2 i e.g. $P(x) = -(x - 1)(x + 2)^3$

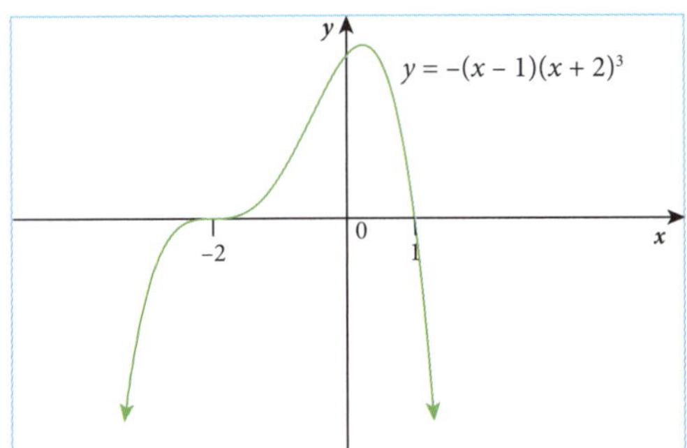

ii e.g. $P(x) = 2x^3(x + 5)(x - 4)^2$

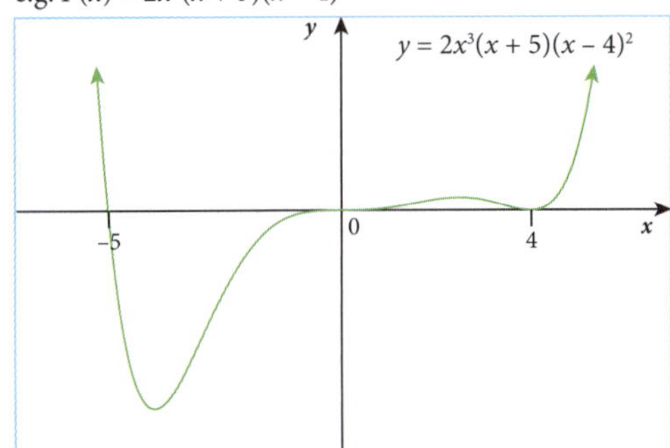

3 i Double root at $x = -2$ and roots at $x = 0$ and $x = 3$. Also $P(4) > 0$.

Possibly $P(x) = x(x + 2)^2(x - 3)$

ii Double root at $x = -3$, root at $x = 0$ and triple root at $x = 2$. Also $P(3) < 0$.

Possibly $P(x) = -x(x + 3)^2(x - 2)^3$

4 i $P(x) = x^3 + x^2 - 5x + 3$

Consider $P(1) = 1^3 + 1^2 - 5(1) + 3$

$= 1 + 1 - 5 + 3$

$= 0$

$\therefore (x - 1)$ is a factor.

$$
\begin{array}{r}
x^2+2x-3 \\
x-1\overline{\smash{)}\,x^3+x^2-5x+3} \\
\underline{x^3-x^2} \qquad\quad\; \\
2x^2-5x \quad\;\; \\
\underline{2x^2-2x} \quad\;\; \\
-3x+3 \\
\underline{-3x+3} \\
0
\end{array}
$$

$\therefore P(x) = (x - 1)(x^2 + 2x - 3)$

$= (x - 1)(x + 3)(x - 1)$

$= (x - 1)^2(x + 3)$

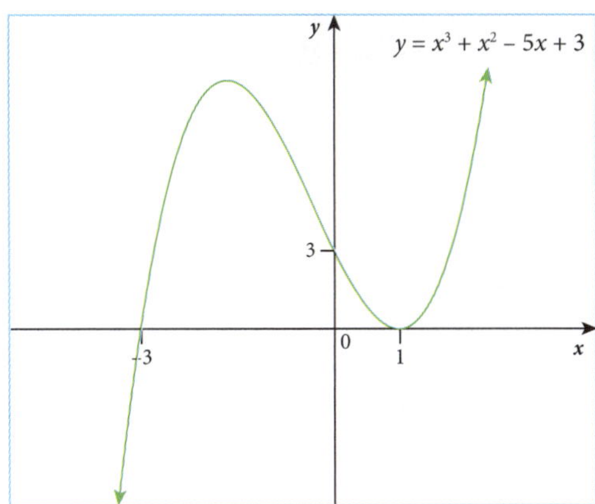

ii $P(x) = -x^4 - 4x^3 + 3x^2 + 18x$

$= -x(x^3 + 4x^2 - 3x - 18)$

Consider $P(-3) = -(-3)[(-3)^3 + 4(-3)^2 - 3(-3) - 18]$

$= 3[-27 + 36 + 9 - 18]$

$= 0$

$\therefore (x + 3)$ is a factor.

$$
\begin{array}{r}
x^2+x-6 \\
x+3\overline{\smash{)}\,x^3+4x^2-3x-18} \\
\underline{x^3+3x^2} \qquad\qquad\; \\
x^2-3x \qquad\; \\
\underline{x^2+3x} \qquad\; \\
-6x-18 \\
\underline{-6x-18} \\
0
\end{array}
$$

$\therefore P(x) = -x(x + 3)(x^2 + x - 6)$

$= -x(x + 3)(x + 3)(x - 2)$

$= -x(x + 3)^2(x - 2)$

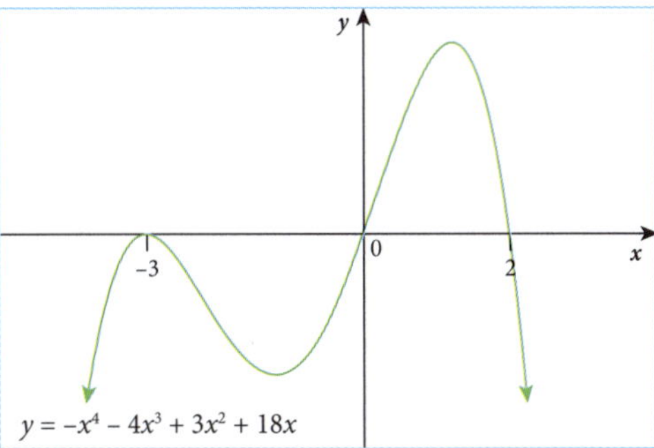

YEAR 11 EXAM-TYPE QUESTIONS

Part A

1 **A.** Let $P(x) = 2x^3 - 5x^2 + 3x - 4$

$P(-1) = 2(-1)^3 - 5(-1)^2 + 3(-1) - 4$

$= -2 - 5 - 3 - 4$

$= -14$

2 **D.** Product of roots = −72

$\therefore$ correct answer is either B or D.

Sum of roots in pairs = −18

$\therefore$ in D: $(3)(-4) + (3)(6) + (-4)(6) = -18$

3 **D.** Let $P(x) = (x + 2)(x - 1)Q(x) + (x + 9)$

$P(1) = (1 + 2)(1 - 1)Q(1) + (1 + 9)$

$= 10$

4 **B.** Subs. $x = 1$ in each $P(x)$.

$\therefore$ correct answer is either A, B or D.

Consider $P'(1)$:

In B, $P'(x) = 5x^4 - 3x^2 - 2x$

$P'(1) = 5(1)^4 - 3(1)^2 - 2(1) = 0$

$\therefore x = 1$ is root of $P(x)$ and $P'(x)$.

5 **C.**

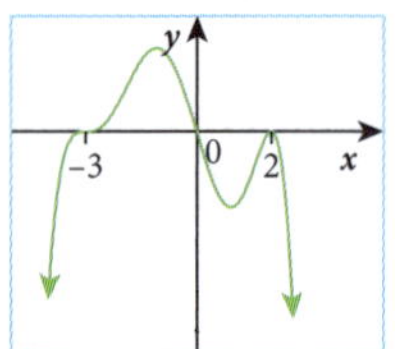

Part B

6 a $\alpha + \beta + \gamma = -\dfrac{b}{a} = \mathbf{2}$ ✓

b $\alpha\beta\gamma = -\dfrac{d}{a} = \mathbf{-32}$ ✓

c $\therefore \alpha + \beta + (-\alpha) = 2$

$\therefore \beta = 2$, i.e. one root is 2. ✓

$\therefore P(2) = 2^3 - 2(2)^2 - k(2) + 32 = 0$

$8 - 8 - 2k + 32 = 0$

$\mathbf{k = 16}$ ✓

7 a $P(-2) = (-2 + 2)(-2 - 1)Q(-2) + a(-2) + b = 0$

$-2a + b = 0 \ldots$ ① ✓

$P(1) = (1 + 2)(1 - 1)Q(1) + a(1) + b = 6$

$a + b = 6 \ldots$ ②

② − ①: $3a = 6$

$a = 2$

Subs. in ②: $2 + b = 6$

$b = 4$

$\therefore \mathbf{a = 2, b = 4}$ ✓

b $\mathbf{2x + 4}$ ✓

8 Sum: $\alpha + 3 + (-4) = \dfrac{4}{a}$

$\alpha - 1 = \dfrac{4}{a}$

$\alpha = 1 + \dfrac{4}{a} \ldots$ ① ✓

Product: $\alpha \times 3 \times (-4) = -\dfrac{60}{a}$

$-12\alpha = -\dfrac{60}{a}$

From ①, $-12\left(1 + \dfrac{4}{a}\right) = -\dfrac{60}{a}$

$-12 - \dfrac{48}{a} = -\dfrac{60}{a}$

$\dfrac{12}{a} = 12$

$a = 1$ ✓

Subs. in ①: $\alpha = 1 + 4$

$= \mathbf{5}$ ✓

9 a Subs. $k = 14$:

$\therefore P(x) = x^3 - x^2 - 14x + 24$

$P(-4) = (-4)^3 - (-4)^2 - 14(-4) + 24 = 0$

$= -64 - 16 + 56 + 24$

$= 0$

$\therefore \mathbf{k = 14}$ ✓

b

$$\begin{array}{r} x^2 - 5x + 6 \\ x + 4 \overline{\smash{)}\, x^3 - x^2 - 14x + 24} \\ \underline{x^3 + 4x^2 \qquad\qquad\quad} \\ -5x^2 - 14x \qquad \\ \underline{-5x^2 - 20x \qquad} \\ 6x + 24 \\ \underline{6x + 24} \\ 0 \end{array}$$

$P(x) = (x + 4)(x^2 - 5x + 6)$ ✓

$= (x + 4)(x - 2)(x - 3) = 0$

$x = -4, 2, 3$

$\therefore$ **the zeros are −4, 2, 3.** ✓

10 $P'(x) = 12x^2 - 3 = 0$

$12x^2 = 3$

$x = \pm\dfrac{1}{2}$ ✓

Now, $P\left(\dfrac{1}{2}\right) = 4\left(\dfrac{1}{2}\right)^3 - 3\left(\dfrac{1}{2}\right) + k = 0$

$\dfrac{1}{2} - \dfrac{3}{2} + k = 0$

$k = 1$

Also, $P\left(-\dfrac{1}{2}\right) = 4\left(-\dfrac{1}{2}\right)^3 - 3\left(-\dfrac{1}{2}\right) + k = 0$

$-\dfrac{1}{2} + \dfrac{3}{2} + k = 0$

$k = -1$

$\therefore$ **possible values of k are ± 1.** ✓

11 Subs. $x = 3$:

$3^3 - 3(3)^2 + 2(3) + d = (3 - 3)Q(3) + 2$ ✓

$27 - 27 + 6 + d = 2$

$\mathbf{d = -4}$ ✓

12 a $\alpha + \beta + \gamma + \delta = -p$ ✓

$\boldsymbol{\alpha\beta\gamma + \alpha\beta\delta + \alpha\gamma\delta + \beta\gamma\delta = -r}$ ✓

b Now, $(\alpha + \beta + \gamma + \delta)^2$

$= (\alpha + \beta + \gamma + \delta)(\alpha + \beta + \gamma + \delta)$

$= \alpha^2 + \alpha\beta + \alpha\gamma + \alpha\delta + \beta\alpha + \beta^2 + \beta\gamma + \beta\delta + \gamma\alpha + \gamma\beta + \gamma^2 + \gamma\delta + \delta\alpha + \delta\beta + \delta\gamma + \delta^2$

$= \alpha^2 + \beta^2 + \gamma^2 + \delta^2 + 2(\alpha\beta + \alpha\gamma + \alpha\delta + \beta\gamma + \beta\delta + \gamma\delta)$ ✓

$\therefore \alpha^2 + \beta^2 + \gamma^2 + \delta^2 = (\alpha + \beta + \gamma + \delta)^2 - 2(\alpha\beta + \alpha\gamma + \alpha\delta + \beta\gamma + \beta\delta + \gamma\delta)$

$= (-p)^2 - 2(q)$

$= \boldsymbol{p^2 - 2q}$ ✓

13 a Sum: $-b = 0 + 2 - 2$

$= 0$

$\therefore b = 0$ ✓

Sum of pairs: $c = (0)(2) + (0)(-2) + (2)(-2)$

$= -4$

$\therefore c = -4$

Product: $-d = (0)(2)(-2)$

$= 0$

$\therefore d = 0$

$\therefore \boldsymbol{b = 0, c = -4, d = 0}$ ✓

Alternatively, as roots are $x = 0, 2, -2$, then $P(x) = x(x - 2)(x + 2)$ ✓

$P(x) = x(x^2 - 4)$

$= x^3 - 4x$

Hence, $b = 0$, $c = -4$, $d = 0$ ✓

b $P(x) = x^3 - 4x$

$= x(x^2 - 4)$

$= x(x - 2)(x + 2)$

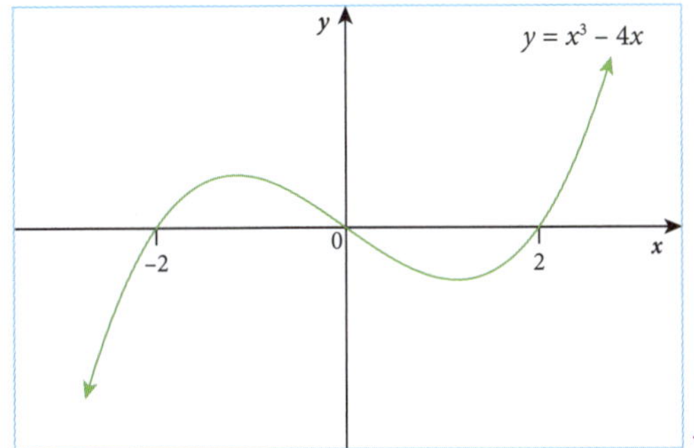

✓

c $x^2 \times \dfrac{x^2 - 4}{x} > 0 \times x^2$

$x(x^2 - 4) > 0$

$\boldsymbol{-2 < x < 0}$ **or** $\boldsymbol{x > 2}$ **(from graph)** ✓

14 Let the roots be α, $-\alpha$ and β.

$\therefore$ sum of roots: $\alpha + (-\alpha) + \beta = -b$

$\beta = -b$

$\therefore$ one root is $-b$. ✓

$\therefore$ Subs. $x = -b$ into polynomial:

$(-b)^3 + b(-b)^2 + c(-b) + d = 0$

$-b^3 + b^3 - cb + d = 0$

$\boldsymbol{d = bc}$ ✓

15 $P(2) = 2^3 + b(2)^2 + c(2) - 12 = 0$

$8 + 4b + 2c - 12 = 0$

$4b + 2c = 4$

$2b + c = 2 \ldots$ ① ✓

Also, $(x - 2)$ is a factor of $P'(x)$:

$P'(x) = 3x^2 + 2bx + c$

$P'(2) = 3(2)^2 + 2b(2) + c = 0$

$12 + 4b + c = 0$

$4b + c = -12 \ldots$ ② ✓

② − ①: $2b = -14$

$b = -7$

Subs. in ①: $2(-7) + c = 2$

$-14 + c = 2$

$c = 16$

$\therefore \boldsymbol{b = -7, c = 16}$ ✓

CHAPTER 3

FURTHER TRIGONOMETRIC IDENTITIES

TOPIC: TRIGONOMETRIC FUNCTIONS

YEAR 11

1 Sums and differences of angles

- **sin (*A* + *B*)** = $\sin A \cos B + \cos A \sin B$
- **sin (*A* – *B*)** = $\sin A \cos B - \cos A \sin B$
- **cos (*A* + *B*)** = $\cos A \cos B - \sin A \sin B$
- **cos (*A* – *B*)** = $\cos A \cos B + \sin A \sin B$
- **tan (*A* + *B*)** = $\dfrac{\tan A + \tan B}{1 - \tan A \tan B}$
- **tan (*A* – *B*)** = $\dfrac{\tan A - \tan B}{1 + \tan A \tan B}$

EXAMPLE 1

Given $\cos(x - y) = \cos x \cos y + \sin x \sin y$, prove $\sin(x + y) = \sin x \cos y + \cos x \sin y$.

Replace x with $90° - x$:

$$\cos(90° - x - y) = \cos(90° - x)\cos y + \sin(90° - x)\sin y$$
$$\cos(90° - (x + y)) = \sin x \cos y + \cos x \sin y$$
$$\sin(x + y) = \sin x \cos y + \cos x \sin y$$

$\cos(90° - \theta) = \sin\theta$

EXAMPLE 2

Find the exact value of sin 42° cos 18° + cos 42° sin 18°.

$$\begin{aligned}\sin 42° \cos 18° + \cos 42° \sin 18° &= \sin(42° + 18°)\\ &= \sin 60°\\ &= \frac{\sqrt{3}}{2}\end{aligned}$$

Use exact trigonometric values.

EXAMPLE 3

Find the exact value of tan 75°. Leave your answer with a rational denominator.

$$\begin{aligned}\tan 75° &= \tan(30° + 45°)\\ &= \frac{\tan 30° + \tan 45°}{1 - \tan 30° \tan 45°}\\ &= \frac{\frac{1}{\sqrt{3}} + 1}{1 - \frac{1}{\sqrt{3}} \times 1} \times \frac{\sqrt{3}}{\sqrt{3}}\\ &= \frac{1 + \sqrt{3}}{\sqrt{3} - 1} \times \frac{\sqrt{3} + 1}{\sqrt{3} + 1}\\ &= \frac{\sqrt{3} + 1 + 3 + \sqrt{3}}{3 - 1}\\ &= \frac{2\sqrt{3} + 4}{2}\\ &= \sqrt{3} + 2\end{aligned}$$

Use exact trigonometric values.

EXAMPLE 4

If $\sin x = \frac{3}{4}$ and $\cos y = \frac{4}{5}$, where x and y are acute, find the exact value of $\cos(x + y)$.

As $\sin x = \frac{3}{4}$, then $\cos x = \frac{\sqrt{7}}{4}$.

Also, as $\cos y = \frac{4}{5}$, then $\sin y = \frac{3}{5}$.

$$\begin{aligned}\cos(x + y) &= \cos x \cos y - \sin x \sin y\\ &= \frac{\sqrt{7}}{4} \times \frac{4}{5} - \frac{3}{4} \times \frac{3}{5}\\ &= \frac{4\sqrt{7} - 9}{20}\end{aligned}$$

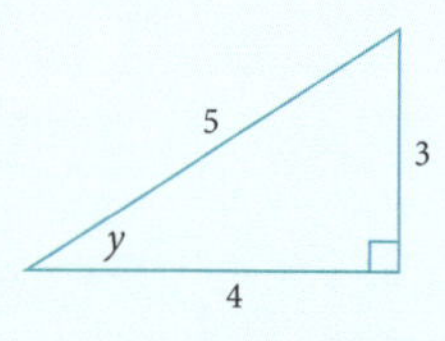

Use right triangles and Pythagoras' theorem.

→ PRACTICE

1 i Given $\sin(x + y) = \sin x \cos y + \cos x \sin y$, replace y with $-y$ to prove $\sin(x - y) = \sin x \cos y - \cos x \sin y$.

ii Use the results for $\sin(x + y)$ and $\cos(x + y)$ to prove $\tan(x + y) = \dfrac{\tan x + \tan y}{1 - \tan x \tan y}$.

2 Find the exact value of:

i $\cos 80° \cos 35° + \sin 80° \sin 35°$

ii $\dfrac{\tan 17° + \tan 13°}{1 - \tan 17° \tan 13°}$

3 Find the exact value, leaving answers with a rational denominator:

i $\sin 15°$

ii $\tan 195°$

4 i If $\sin x = \dfrac{2}{\sqrt{6}}$ and $\cos y = \dfrac{3}{4}$, where x and y are acute, find the exact value of $\cos(x - y)$.

ii If $\cos\theta = \dfrac{\sqrt{5}}{3}$ and $\cos\phi = \dfrac{\sqrt{5}}{2\sqrt{3}}$, where θ and ϕ are acute, show that $\tan(\theta - \phi) = 3\sqrt{35} - 8\sqrt{5}$.

Answers ➲ p. 48

2 Double-angle formulae

➔ $\sin 2A = 2\sin A\cos A$

➔ $\cos 2A = \cos^2 A - \sin^2 A$
$= 1 - 2\sin^2 A$
$= 2\cos^2 A - 1$

➔ $\tan 2A = \dfrac{2\tan A}{1 - \tan^2 A}$

EXAMPLE 1

Derive the result $\sin 2x = 2\sin x\cos x$.

$\sin 2x = \sin(x + x)$
$= \sin x\cos x + \cos x\sin x$
$= 2\sin x\cos x$

Use $\sin(A + B) = \sin A\cos B + \cos A\sin B$

EXAMPLE 2

Simplify:

a $2\cos^2 3\beta - 1$

b $\sin 2\gamma\cos 2\gamma$

a $2\cos^2 3\beta - 1 = \cos 6\beta$

b $\sin 2\gamma\cos 2\gamma = \dfrac{1}{2}\sin 2(2\gamma)$
$= \dfrac{1}{2}\sin 4\gamma$

Use double-angle results.

EXAMPLE 3

If $\tan\alpha = \dfrac{2}{3}$, find the value of $\sin 2\alpha$.

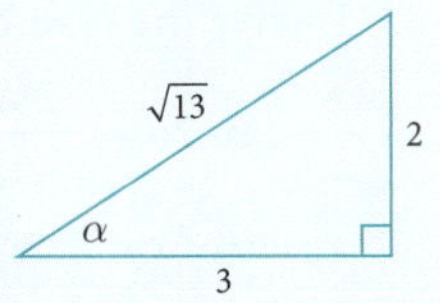

$\therefore \sin\alpha = \dfrac{2}{\sqrt{13}}$ and $\cos\alpha = \dfrac{3}{\sqrt{13}}$

$\sin 2\alpha = 2\sin\alpha\cos\alpha$
$= 2 \times \dfrac{2}{\sqrt{13}} \times \dfrac{3}{\sqrt{13}}$
$= \dfrac{12}{13}$

Use a right triangle and Pythagoras' theorem.

EXAMPLE 4

Find the exact value of $\dfrac{2\tan 15°}{1 - \tan^2 15°}$.

$\dfrac{2\tan 15°}{1 - \tan^2 15°} = \tan(2 \times 15°)$
$= \tan 30°$
$= \dfrac{1}{\sqrt{3}}$

Use exact trig ratios.

EXAMPLE 5

Simplify $\dfrac{\sin 2x\cos x - \cos 2x\sin x}{\cos 2x\cos x + \sin 2x\sin x}$.

$\dfrac{\sin 2x\cos x - \cos 2x\sin x}{\cos 2x\cos x + \sin 2x\sin x} = \dfrac{\sin(2x - x)}{\cos(2x - x)}$
$= \dfrac{\sin x}{\cos x}$
$= \tan x$

Use the sum and difference of trig results.

EXAMPLE 6

Prove $\dfrac{\sin 2A + 1}{\cos 2A} = \dfrac{\cos A + \sin A}{\cos A - \sin A}$.

$\text{LHS} = \dfrac{\sin 2A + 1}{\cos 2A}$

$= \dfrac{2\sin A\cos A + \sin^2 A + \cos^2 A}{\cos^2 A - \sin^2 A}$

(using $\sin^2 A + \cos^2 A = 1$)

$= \dfrac{(\cos A + \sin A)^2}{(\cos A - \sin A)(\cos A + \sin A)}$

$= \dfrac{\cos A + \sin A}{\cos A - \sin A}$

$= \text{RHS}$

Start with the LHS.

→ PRACTICE

1 Derive the result:

i $\tan 2\theta = \dfrac{2\tan\theta}{1-\tan^2\theta}$

ii $\cos 2x = 2\cos^2 x - 1$

2 Simplify:

a $1 - 2\sin^2\dfrac{\theta}{2}$

b $\dfrac{2\tan 3A}{1-\tan^2 3A}$

3 i If $\tan\alpha = \dfrac{3}{5}$, find the value of $\sin 2\alpha$.

ii If $\cos\beta = \dfrac{2}{\sqrt{11}}$, find the value of $\tan 2\beta$.

4 Find the exact value of:

i $\dfrac{2\tan 22.5^o}{1-\tan^2 22.5^o}$

ii $\cos^2 105° - \sin^2 105°$

5 Simplify:

i $(\sin x + \cos x)^2$

ii $\dfrac{1}{1-\tan 2\varphi} - \dfrac{1}{1+\tan 2\varphi}$

6 Prove:

i $\dfrac{\sin 2A}{1+\cos 2A} = \tan A$

ii $\dfrac{\cot x + \tan x}{\cot x - \tan x} = \sec 2x$

Answers ⊃ pp. 48–49

3 The t-formulae $\left(t = \tan\dfrac{A}{2}\right)$

→ For $t = \tan\dfrac{A}{2}$:

- $\sin A = \dfrac{2t}{1+t^2}$
- $\cos A = \dfrac{1-t^2}{1+t^2}$
- $\tan A = \dfrac{2t}{1-t^2}$

EXAMPLE 1

If $t = \tan\dfrac{\theta}{2}$, prove $\sin\theta = \dfrac{2t}{1+t^2}$.

$$\sin\theta = \sin 2\left(\frac{\theta}{2}\right)$$
$$= 2\sin\frac{\theta}{2}\cos\frac{\theta}{2}$$
$$= 2\times\frac{t}{\sqrt{1+t^2}}\times\frac{1}{\sqrt{1+t^2}}$$
$$= \frac{2t}{1+t^2}$$

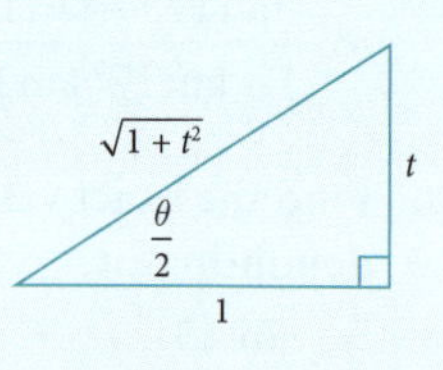

Use $\tan\dfrac{\theta}{2} = \dfrac{t}{1}$

EXAMPLE 2

Express $\tan\theta + \sec\theta$ in terms of t, where $t = \tan\dfrac{\theta}{2}$.

$$\tan\theta + \sec\theta = \tan\theta + \frac{1}{\cos\theta}$$
$$= \frac{2t}{1-t^2} + \frac{1+t^2}{1-t^2}$$
$$= \frac{2t+1+t^2}{1-t^2}$$
$$= \frac{(1+t)^2}{(1-t)(1+t)}$$
$$= \frac{1+t}{1-t}$$

Use $\sec\theta = \dfrac{1}{\cos\theta}$.

EXAMPLE 3

Show that $\dfrac{1-\cos\theta}{\sin\theta} = \tan\dfrac{\theta}{2}$.

$$\text{LHS} = \frac{1-\cos\theta}{\sin\theta}$$
$$= \left(1-\frac{1-t^2}{1+t^2}\right)\div\frac{2t}{1+t^2}$$
$$= \frac{1+t^2-1+t^2}{1+t^2}\times\frac{1+t^2}{2t}$$
$$= \frac{2t^2}{2t}$$
$$= t$$
$$= \tan\frac{\theta}{2}$$
$$= \text{RHS}$$

Rewrite $\sin\theta$ and $\cos\theta$ in terms of $\tan\dfrac{\theta}{2}$.

→ PRACTICE

1 If $t = \tan\frac{\theta}{2}$, prove:

i $\cos\theta = \frac{1-t^2}{1+t^2}$

ii $\tan\theta = \frac{2t}{1-t^2}$

2 Express the following in terms of t, where $t = \tan\frac{\theta}{2}$:

i $1 - \frac{1}{2}\sin\theta\tan\frac{\theta}{2}$

ii $\frac{\tan\theta - \tan\frac{\theta}{2}}{\cot\frac{\theta}{2} + \tan\theta}$

3 Show that:

i $\operatorname{cosec}\theta + \cot\theta = \cot\frac{\theta}{2}$

ii $\cot\frac{\theta}{2} - \tan\frac{\theta}{2} = 2\cot\theta$

Answers ➲ pp. 49-50

4 Trigonometric products as sums and differences

→ $\cos A\cos B = \frac{1}{2}[\cos(A-B) + \cos(A+B)]$

→ $\sin A\sin B = \frac{1}{2}[\cos(A-B) - \cos(A+B)]$

→ $\sin A\cos B = \frac{1}{2}[\sin(A+B) + \sin(A-B)]$

→ $\cos A\sin B = \frac{1}{2}[\sin(A+B) - \sin(A-B)]$

These results lead to:

→ $\cos A + \cos B = 2\cos\frac{A+B}{2}\cos\frac{A-B}{2}$

→ $\cos A - \cos B = -2\sin\frac{A+B}{2}\sin\frac{A-B}{2}$

→ $\sin A + \sin B = 2\sin\frac{A+B}{2}\cos\frac{A-B}{2}$

→ $\sin A - \sin B = 2\cos\frac{A+B}{2}\sin\frac{A-B}{2}$

EXAMPLE 1

Using the results for $\sin(x+y)$ and $\sin(x-y)$, show $\sin x\cos y = \frac{1}{2}[\sin(x+y) + \sin(x-y)]$.

$\sin(x+y) = \sin x\cos y + \cos x\sin y \ldots$ ①

$\sin(x-y) = \sin x\cos y - \cos x\sin y \ldots$ ②

① + ②: $\sin(x+y) + \sin(x-y) = 2\sin x\cos y$

$\therefore \sin x\cos y = \frac{1}{2}[\sin(x+y) + \sin(x-y)]$

Use simultaneous equations.

EXAMPLE 2

Write the product $\cos 4t\sin 3t$ as a sum of trigonometric identities.

$$\cos 4t\sin 3t = \frac{1}{2}[\sin(4t+3t) - \sin(4t-3t)]$$

$$= \frac{1}{2}[\sin 7t - \sin t]$$

Use $A = 4t$, $B = 3t$.

EXAMPLE 3

Find the exact value of $\sin 105°\cos 15°$, using an appropriate product-sum identity.

$$\sin 105°\cos 15° = \frac{1}{2}[\sin(105° + 15°) + \sin(105° - 15°)]$$

$$= \frac{1}{2}[\sin 120° + \sin 90°]$$

$$= \frac{1}{2}\left[\frac{\sqrt{3}}{2} + 1\right]$$

$$= \frac{\sqrt{3}+2}{4}$$

Use $A = 105°$, $B = 15°$.

EXAMPLE 4

Write the sum $\sin 5\theta + \sin 3\theta$ as a product.

$$\sin 5\theta + \sin 3\theta = 2\sin\frac{5\theta+3\theta}{2}\cos\frac{5\theta-3\theta}{2}$$

$$= 2\sin 4\theta\cos\theta$$

Use $A = 5\theta$, $B = 3\theta$.

EXAMPLE 5

Find the exact value of cos 75° – cos 15°, using an appropriate difference-product identity.

$$\cos 75° - \cos 15° = -2\sin\frac{75°+15°}{2}\sin\frac{75°-15°}{2}$$
$$= -2\sin 45°\sin 30°$$
$$= -2\times\frac{1}{\sqrt{2}}\times\frac{1}{2}$$
$$= -\frac{1}{\sqrt{2}}$$

Use $A = 75°$, $B = 15°$.

EXAMPLE 6

Prove $\frac{\cos 3\phi + \cos\phi}{\sin 3\phi - \sin\phi} = \cot\phi$.

$$\text{LHS} = \frac{\cos 3\phi + \cos\phi}{\sin 3\phi - \sin\phi}$$
$$= \frac{2\cos 2\phi\cos\phi}{2\cos 2\phi\sin\phi}$$
$$= \cot\phi$$
$$= \text{RHS}$$

$\frac{\cos\phi}{\sin\phi} = \cot\phi$

→ PRACTICE

1 Using the results for:

i sin $(x + y)$ and sin $(x - y)$, show

$\cos x\sin y = \frac{1}{2}[\sin(x+y) - \sin(x-y)]$

ii cos $(x + y)$ and cos $(x - y)$, show

$\sin x\sin y = \frac{1}{2}[\cos(x-y) - \cos(x+y)]$

2 Write the following products as a sum or a difference of trigonometric identities:

i $\sin 2\theta\sin\theta$

ii $2\cos 3\alpha\cos(-\alpha)$

3 Using an appropriate identity, find the exact value of:

i cos 45° cos 15°

ii cos 75° sin 195°

4 Write the following as products:

i $\cos 6x + \cos 2x$

ii $\sin m - \sin 5m$

5 By expressing as a product, find the exact value of:

i cos 105° + cos 15°

ii sin 75° – sin 165°

6 Prove the following:

i $\frac{\sin 2t + \sin 4t}{\cos 2t - \cos 4t} = \cot t$

ii $\frac{\sin\alpha + \sin\beta}{\cos\alpha + \cos\beta} = \tan\frac{\alpha+\beta}{2}$

Answers ➲ pp. 50–51

YEAR 11 EXAM-TYPE QUESTIONS

Total marks: 35

PART A

(Suggested time: 10 minutes)

(1 mark each)

1 Which of these is equivalent to $\sin 4x \cos 4x$?

A $\frac{1}{4}\sin 8x$ B $\frac{1}{2}\sin 8x$ C $2\sin 2x$ D $4\sin 2x$

2 Which of these is equivalent to $\sin(\alpha + \beta) - \sin(\alpha - \beta)$?

A $2\cos\alpha\sin\beta$ B $2\sin\alpha\cos\beta$
C $-2\sin\alpha\cos\beta$ D $-2\cos\alpha\sin\beta$

3 If the exact value of $\cos x$ is $\frac{2}{\sqrt{5}}$, which of these is the value of $\cos 2x$?

A $-\frac{2}{\sqrt{5}}$ B $\frac{1}{5}$ C $\frac{3}{5}$ D $\frac{4}{\sqrt{5}}$

4 Which of these is equivalent to $\cos 3\alpha + \cos 5\alpha$?

A $-2\cos 4\alpha\cos\alpha$ B $-2\cos 4\alpha\sin\alpha$
C $2\sin 4\alpha\sin\alpha$ D $2\cos 4\alpha\cos\alpha$

5 Which of these is equal to $\left(1-\sqrt{2}\sin\frac{\theta}{2}\right)\left(1+\sqrt{2}\sin\frac{\theta}{2}\right)$?

A $\cos 2\theta$ B $\cos\theta$ C $\cos^2\theta$ D $\cos^2\frac{\theta}{2}$

PART B

(Suggested time: 45 minutes)

Show all working.

6 Prove $\frac{\sin 3A+\sin A}{\cos 3A+\cos A} = \tan 2A$. (3 marks)

7 If θ is acute and $\cos\theta = \frac{3}{4}$, find the exact value of $\tan 2\theta$. (3 marks)

8 By using the substitution $t = \tan\frac{\theta}{2}$, show that $\frac{1+\cos\theta+\sin\theta}{1-\cos\theta+\sin\theta} = \cot\frac{\theta}{2}$. (2 marks)

9 a Prove that $\cos 105° = \frac{\sqrt{2}-\sqrt{6}}{4}$. (2 marks)

b Hence find the exact value of $\cos 52\frac{1}{2}°$. (2 marks)

10 Find the exact value of $\cos\frac{11\pi}{12}$. (3 marks)

11 Find the exact value of $\cos 75° \sin 15°$. (2 marks)

12 Simplify $\frac{\tan 2\alpha-\tan\alpha}{\cot\alpha+\tan 2\alpha}$. (3 marks)

13 If $\sin\phi = 0.6$, and ϕ is acute, find the exact value of $\cos\frac{\phi}{2}$. (3 marks)

14 The diagram shows right triangle ABC, where M is midpoint of AC, and $AC = BC = 2$.

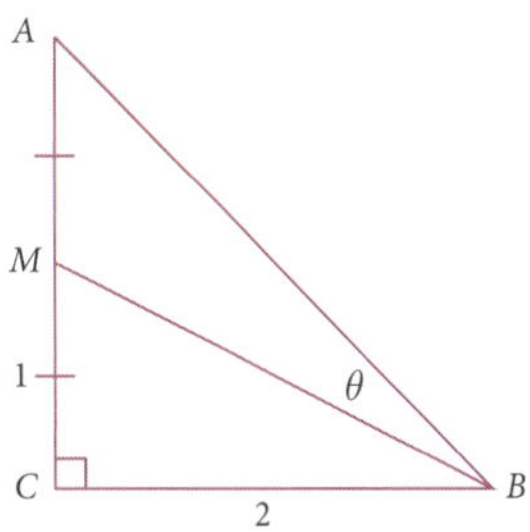

Find the value of $\tan\theta$. (3 marks)

15 Prove the identity $\frac{\cos 4t-\cos 2t}{\sin 4t+\sin 2t} = -\tan t$. (2 marks)

16 If $\tan\alpha$ and $\tan\beta$ are the two values of $\tan\theta$ which satisfy the quadratic equation $a\tan^2\theta + b\tan\theta + c = 0$, show that $\tan(\alpha+\beta) = \frac{b}{c-a}$. (2 marks)

ANSWERS

PRACTICE

1 Sums and differences of angles ➲ pp. 42–43

1 i $\sin(x-y) = \sin x\cos(-y) + \cos x\sin(-y)$
$= \sin x \times \cos y + \cos x \times -\sin y$
$= \mathbf{\sin x \cos y - \cos x \sin y}$

ii $\tan(x+y) = \dfrac{\sin(x+y)}{\cos(x+y)}$

$= \dfrac{\sin x\cos y + \cos x\sin y}{\cos x\cos y - \sin x\sin y}$

$= \dfrac{\dfrac{\sin x\cos y + \cos x\sin y}{\cos x\cos y}}{\dfrac{\cos x\cos y - \sin x\sin y}{\cos x\cos y}}$ (by dividing both top and bottom by $\cos x \cos y$)

$= \dfrac{\dfrac{\sin x}{\cos x} + \dfrac{\sin y}{\cos y}}{1 - \dfrac{\sin x\sin y}{\cos x\cos y}}$

$= \mathbf{\dfrac{\tan x + \tan y}{1 - \tan x\tan y}}$

2 i $\cos 80°\cos 35° + \sin 80°\sin 35° = \cos(80° - 35°)$
$= \cos 45°$
$= \mathbf{\dfrac{1}{\sqrt{2}}}$

ii $\dfrac{\tan 17° + \tan 13°}{1 - \tan 17°\tan 13°} = \tan(17° + 13°)$
$= \tan 30°$
$= \mathbf{\dfrac{1}{\sqrt{3}}}$

3 i $\sin 15° = \sin(45° - 30°)$
$= \sin 45°\cos 30° - \cos 45°\sin 30°$
$= \dfrac{1}{\sqrt{2}} \times \dfrac{\sqrt{3}}{2} - \dfrac{1}{\sqrt{2}} \times \dfrac{1}{2}$
$= \dfrac{\sqrt{3}-1}{2\sqrt{2}} \times \dfrac{\sqrt{2}}{\sqrt{2}}$
$= \mathbf{\dfrac{\sqrt{6}-\sqrt{2}}{4}}$

ii $\tan 195° = \tan(135° + 60°)$
$= \dfrac{\tan 135° + \tan 60°}{1 - \tan 135°\tan 60°}$
$= \dfrac{-1+\sqrt{3}}{1-(-1)(\sqrt{3})}$
$= \dfrac{-1+\sqrt{3}}{1+\sqrt{3}} \times \dfrac{1-\sqrt{3}}{1-\sqrt{3}}$
$= \dfrac{-1+\sqrt{3}+\sqrt{3}-3}{1-3}$
$= \dfrac{-4+2\sqrt{3}}{-2}$
$= \mathbf{2 - \sqrt{3}}$

4 i As $\sin x = \dfrac{2}{\sqrt{6}}$, then $\cos x = \dfrac{\sqrt{2}}{\sqrt{6}}$.

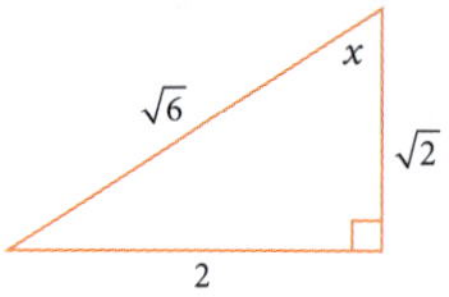

Also, as $\cos y = \dfrac{3}{4}$, then $\sin y = \dfrac{\sqrt{7}}{4}$.

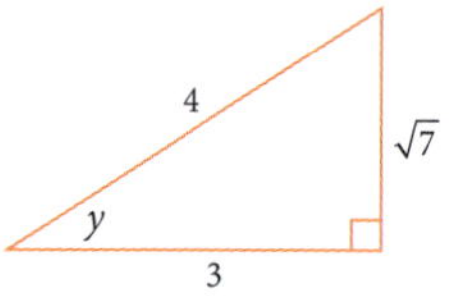

$\cos(x-y) = \cos x\cos y + \sin x\sin y$
$= \dfrac{\sqrt{2}}{\sqrt{6}} \times \dfrac{3}{4} + \dfrac{2}{\sqrt{6}} \times \dfrac{\sqrt{7}}{4}$
$= \dfrac{3\sqrt{2}+2\sqrt{7}}{4\sqrt{6}} \times \dfrac{\sqrt{6}}{\sqrt{6}}$
$= \dfrac{3\sqrt{12}+2\sqrt{42}}{24}$
$= \dfrac{6\sqrt{3}+2\sqrt{42}}{24}$
$= \mathbf{\dfrac{3\sqrt{3}+\sqrt{42}}{12}}$

ii As $\cos\theta = \dfrac{\sqrt{5}}{3}$ then $\tan\theta = \dfrac{2}{\sqrt{5}}$.

Also, as $\cos\phi = \dfrac{\sqrt{5}}{2\sqrt{3}}$, then $\tan\phi = \dfrac{\sqrt{7}}{\sqrt{5}}$.

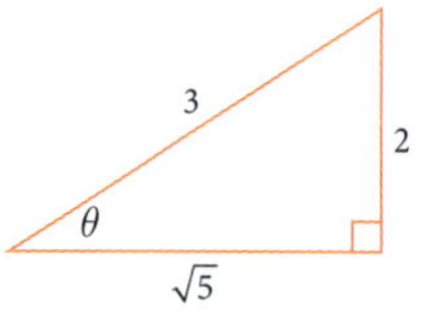

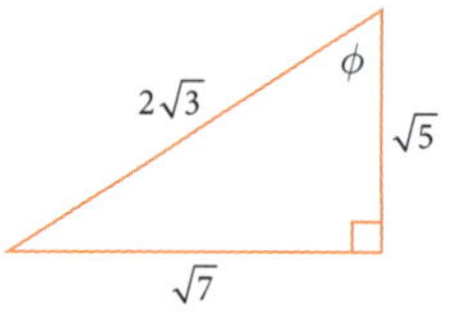

$\tan(\theta - \phi) = \dfrac{\tan\theta - \tan\phi}{1 + \tan\theta\tan\phi}$
$= \dfrac{\dfrac{2}{\sqrt{5}} - \dfrac{\sqrt{7}}{\sqrt{5}}}{1 + \dfrac{2}{\sqrt{5}} \times \dfrac{\sqrt{7}}{\sqrt{5}}}$
$= \dfrac{2-\sqrt{7}}{\sqrt{5}} \div \left(1 + \dfrac{2\sqrt{7}}{5}\right)$
$= \dfrac{2-\sqrt{7}}{\sqrt{5}} \div \dfrac{5+2\sqrt{7}}{5}$
$= \dfrac{2-\sqrt{7}}{\sqrt{5}} \times \dfrac{5}{5+2\sqrt{7}}$
$= \dfrac{2-\sqrt{7}}{1} \times \dfrac{\sqrt{5}}{5+2\sqrt{7}}$
$= \dfrac{2\sqrt{5}-\sqrt{35}}{5+2\sqrt{7}} \times \dfrac{5-2\sqrt{7}}{5-2\sqrt{7}}$
$= \dfrac{10\sqrt{5}-4\sqrt{35}-5\sqrt{35}+2\sqrt{245}}{25-28}$
$= \dfrac{10\sqrt{5}-9\sqrt{35}+14\sqrt{5}}{-3}$
$= \dfrac{9\sqrt{35}-24\sqrt{5}}{3}$
$= \mathbf{3\sqrt{35} - 8\sqrt{5}}$

2 Double-angle formulae ➲ p. 44

1 i $\tan 2\theta = \tan(\theta + \theta)$
$= \dfrac{\tan\theta + \tan\theta}{1 - \tan\theta.\tan\theta}$
$= \mathbf{\dfrac{2\tan\theta}{1 - \tan^2\theta}}$

ii $\cos 2x = \cos(x + x)$
$= \cos x.\cos x - \sin x.\sin x$
$= \cos^2 x - \sin^2 x$
$= \cos^2 x - (1 - \cos^2 x)$
$= \mathbf{2\cos^2 x - 1}$

2 a $1 - 2\sin^2\frac{\theta}{2} = \cos 2\left(\frac{\theta}{2}\right)$
$= \boldsymbol{\cos\theta}$

b $\frac{2\tan 3A}{1-\tan^2 3A} = \tan 2(3A)$
$= \mathbf{\tan 6A}$

3 i As $\sin\alpha = \frac{3}{\sqrt{34}}$ and $\cos\alpha = \frac{5}{\sqrt{34}}$,

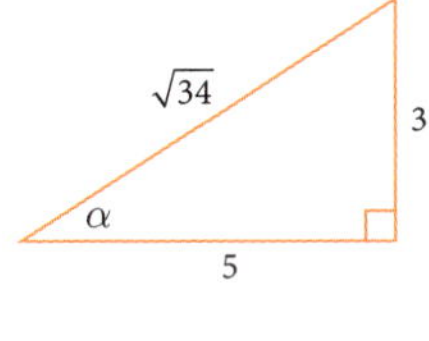

$\sin 2\alpha = 2\sin\alpha\cos\alpha$
$= 2 \times \frac{3}{\sqrt{34}} \times \frac{5}{\sqrt{34}}$
$= \frac{30}{34}$
$= \mathbf{\frac{15}{17}}$

ii As $\tan\beta = \frac{\sqrt{7}}{2}$,

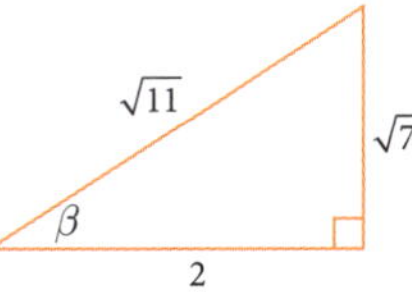

$\tan 2\beta = \frac{2\tan\beta}{1-\tan^2\beta}$
$= \frac{2\left(\frac{\sqrt{7}}{2}\right)}{1-\left(\frac{\sqrt{7}}{2}\right)^2}$
$= \frac{\sqrt{7}}{1-\frac{7}{4}}$
$= \frac{\sqrt{7}}{-\frac{3}{4}}$
$= \mathbf{-\frac{4\sqrt{7}}{3}}$

4 i $\frac{2\tan 22.5^\circ}{1-\tan^2 22.5^\circ} = \tan 2(22.5^\circ)$
$= \tan 45^\circ$
$= \mathbf{1}$

ii $\cos^2 105^\circ - \sin^2 105^\circ = \cos 2(105^\circ)$
$= \cos 210^\circ$
$= \mathbf{-\frac{\sqrt{3}}{2}}$

5 i $(\sin x + \cos x)^2 = \sin^2 x + 2\sin x\cos x + \cos^2 x$
$= \mathbf{1 + \sin 2x}$

ii $\frac{1}{1-\tan 2\varphi} - \frac{1}{1+\tan 2\varphi} = \frac{1+\tan 2\varphi-(1-\tan 2\varphi)}{(1-\tan 2\varphi)(1+\tan 2\varphi)}$
$= \frac{2\tan 2\varphi}{1-\tan^2 2\varphi}$
$= \mathbf{\tan 4\varphi}$

6 i $\text{LHS} = \frac{\sin 2A}{1+\cos 2A}$
$= \frac{2\sin A\cos A}{1+(2\cos^2 A-1)}$
$= \frac{2\sin A\cos A}{2\cos^2 A}$
$= \frac{\sin A}{\cos A}$
$= \tan A$
= **RHS**

ii $\text{LHS} = \frac{\cot x+\tan x}{\cot x-\tan x}$
$= \frac{\frac{1}{\tan x}+\tan x}{\frac{1}{\tan x}-\tan x}$
$= \frac{1+\tan^2 x}{1-\tan^2 x}$ (multiplying top and bottom by $\tan x$)
$= \frac{\sec^2 x}{1-\tan^2 x}$
$= \frac{1}{\cos^2 x} \div \left(1-\frac{\sin^2 x}{\cos^2 x}\right)$
$= \frac{1}{\cos^2 x} \div \frac{\cos^2 x-\sin^2 x}{\cos^2 x}$
$= \frac{1}{\cos^2 x} \times \frac{\cos^2 x}{\cos^2 x-\sin^2 x}$
$= \frac{1}{\cos^2 x-\sin^2 x}$
$= \frac{1}{\cos 2x}$
$= \sec 2x$
= **RHS**

3 The t-formulae $\left(t = \tan\frac{A}{2}\right)$ ➲ p. 45

1 i $\cos\theta = \cos 2\left(\frac{\theta}{2}\right)$
$= \cos^2\frac{\theta}{2} - \sin^2\frac{\theta}{2}$

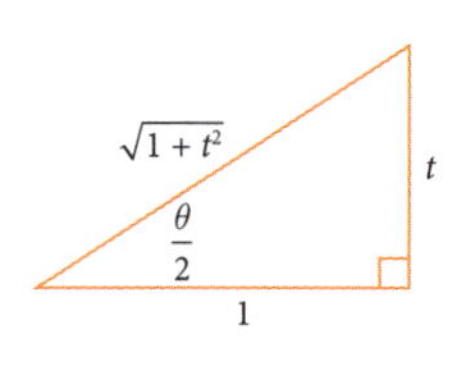

$= \left(\frac{1}{\sqrt{1+t^2}}\right)^2 - \left(\frac{t}{\sqrt{1+t^2}}\right)^2$
$= \mathbf{\frac{1-t^2}{1+t^2}}$

ii $\tan\theta = \tan 2\left(\frac{\theta}{2}\right)$
$= \frac{2\tan\frac{\theta}{2}}{1-\tan^2\frac{\theta}{2}}$
$= \mathbf{\frac{2t}{1-t^2}}$

2 **i** $1-\frac{1}{2}\sin\theta\tan\frac{\theta}{2}=1-\frac{1}{2}\times\frac{2t}{1+t^2}\times t$

$$=1-\frac{t^2}{1+t^2}$$

$$=\frac{1+t^2-t^2}{1+t^2}$$

$$=\mathbf{\frac{1}{1+t^2}}$$

ii $\dfrac{\tan\theta-\tan\frac{\theta}{2}}{\cot\frac{\theta}{2}+\tan\theta}=\dfrac{\frac{2t}{1-t^2}-t}{\frac{1}{t}+\frac{2t}{1-t^2}}$

$$=\frac{2t^2-t^2(1-t^2)}{1-t^2+2t^2}$$ [multiplying both top and bottom by $t(1-t^2)$]

$$=\frac{2t^2-t^2+t^4}{1+t^2}$$

$$=\frac{t^2+t^4}{1+t^2}$$

$$=\frac{t^2(1+t^2)}{1+t^2}$$

$$=\mathbf{t^2}$$

3 **i** LHS = $\operatorname{cosec}\theta+\cot\theta$

$$=\frac{1}{\sin\theta}+\frac{1}{\tan\theta}$$

$$=\frac{1+t^2}{2t}+\frac{1-t^2}{2t}$$

$$=\frac{1+t^2+1-t^2}{2t}$$

$$=\frac{2}{2t}$$

$$=\frac{1}{t}$$

$$=\frac{1}{\tan\frac{\theta}{2}}$$

$$=\cot\frac{\theta}{2}$$

= **RHS**

ii LHS = $\cot\frac{\theta}{2}-\tan\frac{\theta}{2}$

$$=\frac{1}{t}-t$$

$$=\frac{1-t^2}{t}$$

$$=2\times\frac{1-t^2}{2t}$$

$$=2\cot\theta$$

= **RHS**

4 Trigonometric products as sums and differences ➲ p. 46

1 **i** $\sin(x+y)=\sin x\cos y+\cos x\sin y\ \ldots$ ①

$\sin(x-y)=\sin x\cos y-\cos x\sin y\ \ldots$ ②

① – ②: $\sin(x+y)-\sin(x-y)=2\cos x\sin y$

$\therefore \cos x\sin y=\frac{1}{2}[\sin(x+y)-\sin(x-y)]$

ii $\cos(x+y)=\cos x\cos y-\sin x\sin y\ \ldots$ ①

$\cos(x-y)=\cos x\cos y+\sin x\sin y\ \ldots$ ②

② – ①: $\cos(x-y)-\cos(x+y)=2\sin x\sin y$

$\therefore \sin x\sin y=\frac{1}{2}[\cos(x-y)-\cos(x+y)]$

2 **i** $\sin 2\theta\sin\theta=\frac{1}{2}[\cos(2\theta-\theta)-\cos(2\theta+\theta)]$

$$=\mathbf{\frac{1}{2}[\cos\theta-\cos 3\theta]}$$

ii $2\cos 3\alpha\cos(-\alpha)=2\times\frac{1}{2}[\cos(3\alpha-(-\alpha))+\cos(3\alpha+(-\alpha))]$

$$=2\times\frac{1}{2}[\cos 4\alpha+\cos 2\alpha]$$

$$=\mathbf{\cos 4\alpha+\cos 2\alpha}$$

3 **i** $\cos 45°\cos 15°=\frac{1}{2}[\cos(45°-15°)+\cos(45°+15°)]$

$$=\frac{1}{2}[\cos 30°+\cos 60°]$$

$$=\frac{1}{2}\left[\frac{\sqrt{3}}{2}+\frac{1}{2}\right]$$

$$=\mathbf{\frac{\sqrt{3}+1}{4}}$$

ii $\cos 75°\sin 195°=\frac{1}{2}[\sin(75°+195°)-\sin(75°-195°)]$

$$=\frac{1}{2}[\sin 270°-\sin(-120°)]$$

$$=\frac{1}{2}\left[-1-\left(-\frac{\sqrt{3}}{2}\right)\right]$$

$$=\frac{1}{2}\left[\frac{\sqrt{3}}{2}-1\right]$$

$$=\mathbf{\frac{\sqrt{3}-2}{4}}$$

4 **i** $\cos 6x+\cos 2x=2\cos\frac{6x+2x}{2}\cos\frac{6x-2x}{2}$

$$=\mathbf{2\cos 4x\cos 2x}$$

ii $\sin m-\sin 5m=2\cos\frac{m+5m}{2}\sin\frac{m-5m}{2}$

$$=2\cos 3m\sin(-2m)$$

$$=\mathbf{-2\cos 3m\sin 2m}$$

5 **i** $\cos 105°+\cos 15°=2\cos\frac{105°+15°}{2}\cos\frac{105°-15°}{2}$

$$=2\cos 60°\cos 45°$$

$$=2\times\frac{1}{2}\times\frac{1}{\sqrt{2}}$$

$$=\mathbf{\frac{1}{\sqrt{2}}}$$

ii $\sin 75°-\sin 165°=2\cos\frac{75°+165°}{2}\sin\frac{75°-165°}{2}$

$$=2\cos 120°\sin(-45°)$$

$$=2\times-\frac{1}{2}\times-\frac{1}{\sqrt{2}}$$

$$=\mathbf{\frac{1}{\sqrt{2}}}$$

6 i LHS $= \dfrac{\sin 2t + \sin 4t}{\cos 2t - \cos 4t}$

$= \dfrac{2\sin 3t\cos(-t)}{-2\sin 3t\sin(-t)}$

$= \dfrac{\cos t}{-(-\sin t)}$

$= \dfrac{\cos t}{\sin t}$

$= \cot t$

= RHS

ii LHS $= \dfrac{\sin\alpha + \sin\beta}{\cos\alpha + \cos\beta}$

$= \dfrac{2\sin\dfrac{\alpha+\beta}{2}\cos\dfrac{\alpha-\beta}{2}}{2\cos\dfrac{\alpha+\beta}{2}\cos\dfrac{\alpha-\beta}{2}}$

$= \dfrac{\sin\dfrac{\alpha+\beta}{2}}{\cos\dfrac{\alpha+\beta}{2}}$

$= \tan\dfrac{\alpha+\beta}{2}$

= RHS

YEAR 11 EXAM-TYPE QUESTIONS

Part A

1 B. $\sin 4x \cos 4x = \frac{1}{2}\sin 2(4x)$

$= \frac{1}{2}\sin 8x$

2 A. $\sin(\alpha + \beta) - \sin(\alpha - \beta) = \sin\alpha\cos\beta + \cos\alpha\sin\beta - (\sin\alpha\cos\beta - \cos\alpha\sin\beta)$

$= 2\cos\alpha\sin\beta$

3 C. $\cos 2x = 2\cos^2 x - 1$

$= 2\left(\frac{2}{\sqrt{5}}\right)^2 - 1$

$= \frac{8}{5} - 1$

$= \frac{3}{5}$

4 D. $\cos 3\alpha + \cos 5\alpha = 2\cos\dfrac{3\alpha+5\alpha}{2}\cos\dfrac{3\theta-5\theta}{2}$

$= 2\cos 4\alpha\cos(-\alpha)$

$= 2\cos 4\alpha\cos\alpha$

5 B. $\left(1-\sqrt{2}\sin\frac{\theta}{2}\right)\left(1+\sqrt{2}\sin\frac{\theta}{2}\right) = 1 - 2\sin^2\left(\frac{\theta}{2}\right)$

$= \cos 2\left(\frac{\theta}{2}\right)$

$= \cos\theta$

Part B

6 LHS $= \dfrac{\sin 3A + \sin A}{\cos 3A + \cos A}$

$= \dfrac{\sin(2A+A) + \sin(2A-A)}{\cos(2A+A) + \cos(2A-A)}$ ✓

$= \dfrac{2\sin 2A\cos A}{2\cos 2A\cos A}$ ✓

$= \dfrac{\sin 2A}{\cos 2A}$

$= \tan 2A$

= RHS ✓

7 As $\cos\theta = \frac{3}{4}$, $\tan\theta = \frac{\sqrt{7}}{3}$. ✓

$\tan 2\theta = \dfrac{2\tan\theta}{1-\tan^2\theta}$

$= \dfrac{2\left(\frac{\sqrt{7}}{3}\right)}{1-\left(\frac{\sqrt{7}}{3}\right)^2}$ ✓

$= \dfrac{2\sqrt{7}}{3} \div \dfrac{2}{9}$

$= \dfrac{2\sqrt{7}}{3} \times \dfrac{9}{2}$

$= 3\sqrt{7}$ ✓

4
$\sqrt{7}$
θ
3

8 As $\sin\theta = \dfrac{2t}{1+t^2}$, $\cos\theta = \dfrac{1-t^2}{1+t^2}$,

LHS $= \dfrac{1+\cos\theta+\sin\theta}{1-\cos\theta+\sin\theta}$

$= \dfrac{1+\dfrac{1-t^2}{1+t^2}+\dfrac{2t}{1+t^2}}{1-\dfrac{1-t^2}{1+t^2}+\dfrac{2t}{1+t^2}} \times \dfrac{1+t^2}{1+t^2}$ ✓

$= \dfrac{1+t^2+1-t^2+2t}{1+t^2-1+t^2+2t}$

$= \dfrac{2+2t}{2t^2+2t}$

$= \dfrac{2(1+t)}{2t(t+1)}$

$= \dfrac{1}{t}$

$= \cot\dfrac{\theta}{2}$

= RHS ✓

9 a $\cos 105° = \cos(60° + 45°)$

$= \cos 60°\cos 45° - \sin 60°\sin 45°$ ✓

$= \frac{1}{2}\times\frac{1}{\sqrt{2}} - \frac{\sqrt{3}}{2}\times\frac{1}{\sqrt{2}}$

$= \dfrac{1-\sqrt{3}}{2\sqrt{2}}$

$= \dfrac{1-\sqrt{3}}{2\sqrt{2}} \times \dfrac{\sqrt{2}}{\sqrt{2}}$

$= \dfrac{\sqrt{2}-\sqrt{6}}{4}$ ✓

b Let $\theta = 52\frac{1}{2}°$. $\therefore 2\theta = 105°$

$\therefore \cos 2\theta = 2\cos^2\theta - 1 = \dfrac{\sqrt{2}-\sqrt{6}}{4}$ ✓

$2\cos^2\theta = 1 + \dfrac{\sqrt{2}-\sqrt{6}}{4}$

$2\cos^2\theta = \dfrac{4+\sqrt{2}-\sqrt{6}}{4}$

$\cos^2\theta = \dfrac{4+\sqrt{2}-\sqrt{6}}{8}$

$\boldsymbol{\cos\theta = \sqrt{\dfrac{4+\sqrt{2}-\sqrt{6}}{8}}}$ (as $\cos\theta > 0$) ✓

10 $\cos\dfrac{11\pi}{12} = \cos\left(\dfrac{2\pi}{3}+\dfrac{\pi}{4}\right)$ ✓

$= \cos\dfrac{2\pi}{3}\cos\dfrac{\pi}{4} - \sin\dfrac{2\pi}{3}\sin\dfrac{\pi}{4}$

$= -\dfrac{1}{2}\times\dfrac{1}{\sqrt{2}} - \dfrac{\sqrt{3}}{2}\times\dfrac{1}{\sqrt{2}}$ ✓

$= -\dfrac{1}{2\sqrt{2}} - \dfrac{\sqrt{3}}{2\sqrt{2}}$

$= -\left(\dfrac{1+\sqrt{3}}{2\sqrt{2}}\times\dfrac{\sqrt{2}}{\sqrt{2}}\right)$

$= -\sqrt{2}\left(\dfrac{1+\sqrt{3}}{4}\right)$ ✓

11 $\cos 75°\sin 15° = \dfrac{1}{2}[\sin(75°+15°) - \sin(75°-15°)]$ ✓

$= \dfrac{1}{2}[\sin 90° - \sin 60°]$

$= \dfrac{1}{2}\left(1-\dfrac{\sqrt{3}}{2}\right)$

$= \boldsymbol{\dfrac{2-\sqrt{3}}{4}}$ ✓

12 $\dfrac{\tan 2\alpha - \tan\alpha}{\cot\alpha + \tan 2\alpha} = \dfrac{\dfrac{2\tan\alpha}{1-\tan^2\alpha} - \tan\alpha}{\dfrac{1}{\tan\alpha} + \dfrac{2\tan\alpha}{1-\tan^2\alpha}}$ ✓

$= \dfrac{\dfrac{2\tan\alpha}{1-\tan^2\alpha} - \tan\alpha}{\dfrac{1}{\tan\alpha} + \dfrac{2\tan\alpha}{1-\tan^2\alpha}} \times \dfrac{\tan\alpha(1-\tan^2\alpha)}{\tan\alpha(1-\tan^2\alpha)}$

$= \dfrac{2\tan^2\alpha - \tan^2\alpha(1-\tan^2\alpha)}{1-\tan^2\alpha + 2\tan^2\alpha}$ ✓

$= \dfrac{\tan^2\alpha + \tan^4\alpha}{1+\tan^2\alpha}$

$= \dfrac{\tan^2\alpha(1+\tan^2\alpha)}{1+\tan^2\alpha}$

$= \boldsymbol{\tan^2\alpha}$ ✓

13 If $\sin\phi = \dfrac{3}{5}$, then $\cos\phi = \dfrac{4}{5}$. ✓

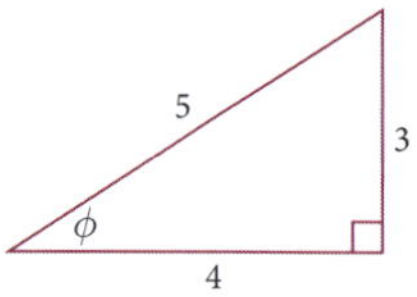

$\therefore \cos 2\left(\dfrac{\phi}{2}\right) = 2\cos^2\dfrac{\phi}{2} - 1$ ✓

$\dfrac{4}{5} = 2\cos^2\dfrac{\phi}{2} - 1$

$2\cos^2\dfrac{\phi}{2} = \dfrac{4}{5} + 1$

$2\cos^2\dfrac{\phi}{2} = \dfrac{9}{5}$

$\cos^2\dfrac{\phi}{2} = \dfrac{9}{10}$

$\boldsymbol{\cos\dfrac{\phi}{2} = \dfrac{3}{\sqrt{10}}}$ $\left(\cos\dfrac{\phi}{2} > 0\right)$ ✓

14 Let $\angle MBC = \alpha$.

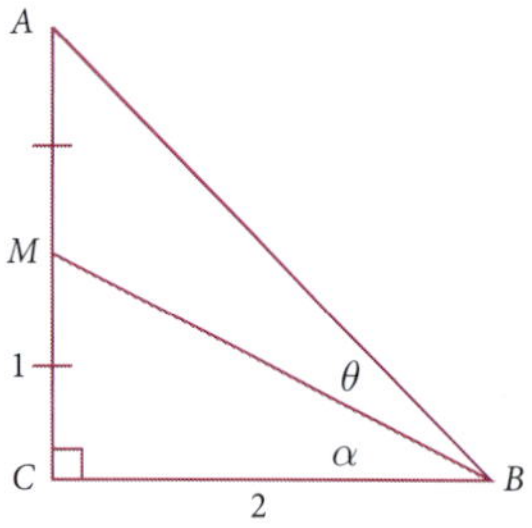

$\tan\alpha = \dfrac{1}{2}$ and $\tan(\alpha+\theta) = \dfrac{2}{2} = 1$. ✓

$\therefore \dfrac{\tan\alpha + \tan\theta}{1-\tan\alpha\tan\theta} = 1$

$\dfrac{\dfrac{1}{2}+\tan\theta}{1-\dfrac{1}{2}\tan\theta} = 1$

$\dfrac{1+2\tan\theta}{2-\tan\theta} = 1$ ✓

$1 + 2\tan\theta = 2 - \tan\theta$

$3\tan\theta = 1$

$\boldsymbol{\tan\theta = \dfrac{1}{3}}$ ✓

15 LHS $= \dfrac{\cos 4t - \cos 2t}{\sin 4t + \sin 2t}$

$= \dfrac{-2\sin\left(\dfrac{4t+2t}{2}\right)\sin\left(\dfrac{4t-2t}{2}\right)}{2\sin\left(\dfrac{4t+2t}{2}\right)\cos\left(\dfrac{4t-2t}{2}\right)}$ ✓

$= \dfrac{-\sin t}{\cos t}$

$= -\tan t$

= **RHS** ✓

16 Sum: $\tan\alpha + \tan\beta = \dfrac{-b}{a}$

Product: $\tan\alpha\tan\beta = \dfrac{c}{a}$ ✓

$\tan(\alpha+\beta) = \dfrac{\tan\alpha+\tan\beta}{1-\tan\alpha\tan\beta}$

$= \dfrac{-\dfrac{b}{a}}{1-\dfrac{c}{a}}$

$= -\dfrac{b}{a} \div \dfrac{a-c}{a}$

$= -\dfrac{b}{a-c}$

$= \boldsymbol{\dfrac{b}{c-a}}$ ✓

CHAPTER 4 INVERSE TRIGONOMETRIC FUNCTIONS

TOPIC: TRIGONOMETRIC FUNCTIONS

YEAR 11

1 Domain, range and graphs

➔ $y = \sin^{-1} x$ (or $\arcsin x$)

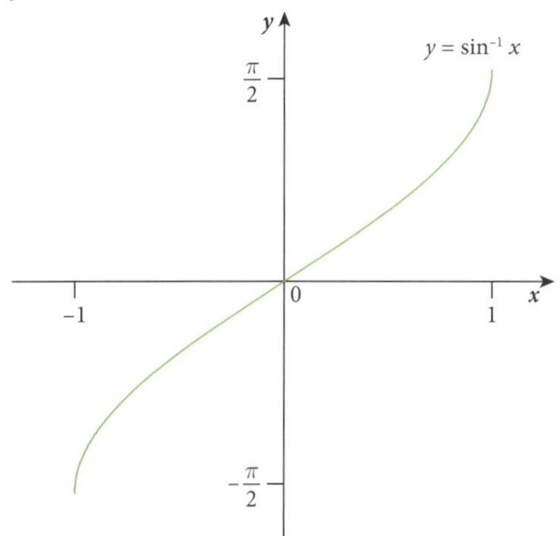

- **Domain**: $-1 \leq x \leq 1$, which can be written as $[-1, 1]$
- **Range**: $-\frac{\pi}{2} \leq y \leq \frac{\pi}{2}$, which can be written as $\left[-\frac{\pi}{2}, \frac{\pi}{2}\right]$

➔ $y = \cos^{-1} x$ (or $\arccos x$)

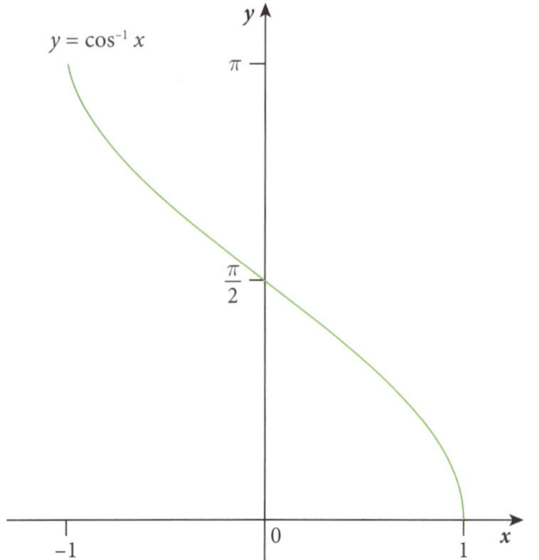

- **Domain**: $-1 \leq x \leq 1$, which can be written as $[-1, 1]$
- **Range**: $0 \leq y \leq \pi$, which can be written as $[0, \pi]$

➔ $y = \tan^{-1} x$ (or $\arctan x$)

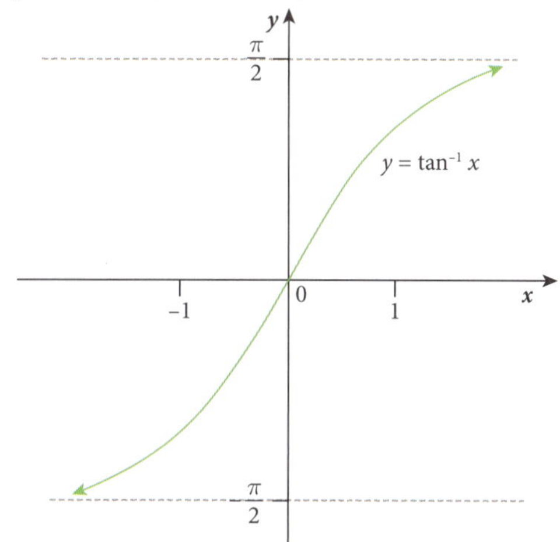

- **Domain**: all real x
- **Range**: $-\frac{\pi}{2} < y < \frac{\pi}{2}$, which can be written as $\left(-\frac{\pi}{2}, \frac{\pi}{2}\right)$

EXAMPLE 1

Find the domain and range of the inverse functions:

a $y = 2\sin^{-1} 3x$

b $y = \frac{1}{2}\cos^{-1}(x+1)$

a Domain: $-1 \leq 3x \leq 1$

$\therefore -\frac{1}{3} \leq x \leq \frac{1}{3}$ or $\left[-\frac{1}{3}, \frac{1}{3}\right]$

Range: Multiply end-values of $-\frac{\pi}{2} \leq y \leq \frac{\pi}{2}$ by 2:

$\therefore -\pi \leq y \leq \pi$

b Domain: $-1 \leq x + 1 \leq 1$

$\therefore -2 \leq x \leq 0$

Range: Multiply end-values of $0 \leq y \leq \pi$ by $\frac{1}{2}$:

$\therefore 0 \leq y \leq \frac{\pi}{2}$ or $\left[0, \frac{\pi}{2}\right]$

Consider the domain and range of $y = \sin^{-1} x$ and $y = \cos^{-1} x$.

EXAMPLE 2

Sketch the following:

a $y = \frac{1}{2}\cos^{-1} 3x$

b $y = 2\tan^{-1}(x+3)$

a

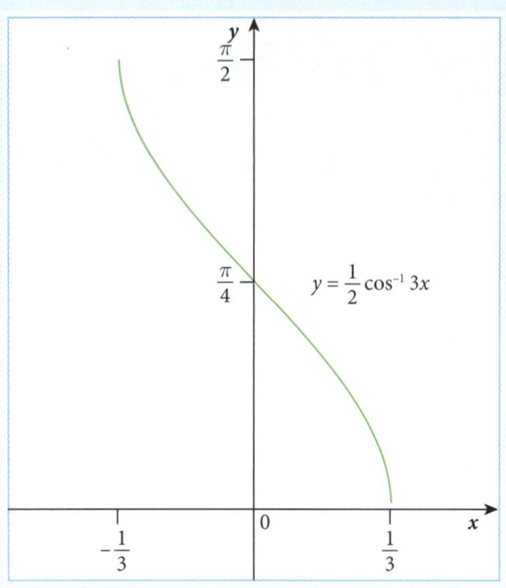

First find the domain and range.

b

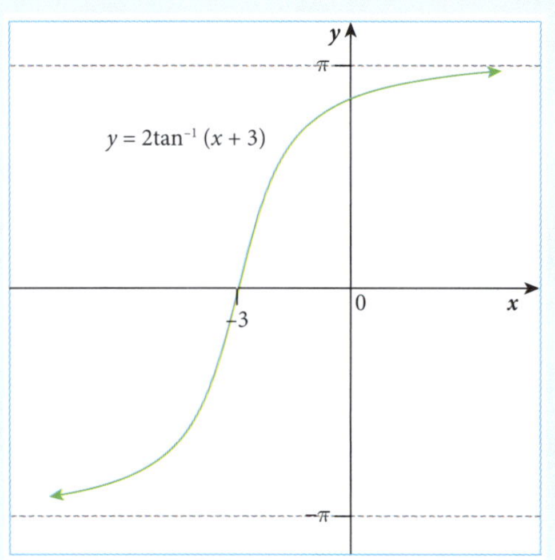

First find the domain and range.

→ PRACTICE

1 Find the domain and range of the inverse functions:

i a $y = 3\sin^{-1} 2x$

b $y = \frac{1}{3}\cos^{-1}(x-1)$

ii $y = 4\tan^{-1}\frac{x}{2}$

iii $y = \frac{3}{2}\sin^{-1}\frac{x+1}{2}$

2 Sketch the following:

i a $y = \frac{1}{2}\sin^{-1} 2x$

b $y = 3\tan^{-1} 5x$

ii $y = 4\cos^{-1}(x-2)$

iii $y = \frac{1}{2}\sin^{-1}\frac{2x+1}{3}$

Answers ➲ p. 59

2 Properties of inverse trigonometric functions

- → $\sin^{-1}[\sin x] = \sin[\sin^{-1} x] = x$
- → $\sin^{-1}(-x) = -\sin^{-1} x$. Hence, $y = \sin^{-1} x$ is an **odd** function.
- → $\cos^{-1}[\cos x] = \cos[\cos^{-1} x] = x$
- → $\cos^{-1}(-x) = \pi - \cos^{-1} x$. Hence, $y = \cos^{-1} x$ is **not** an odd function.
- → $\tan^{-1}[\tan x] = \tan[\tan^{-1} x] = x$
- → $\tan^{-1}(-x) = -\tan^{-1} x$. Hence, $y = \tan^{-1} x$ is an **odd** function.
- → $\sin^{-1} x + \cos^{-1} x = \frac{\pi}{2}$

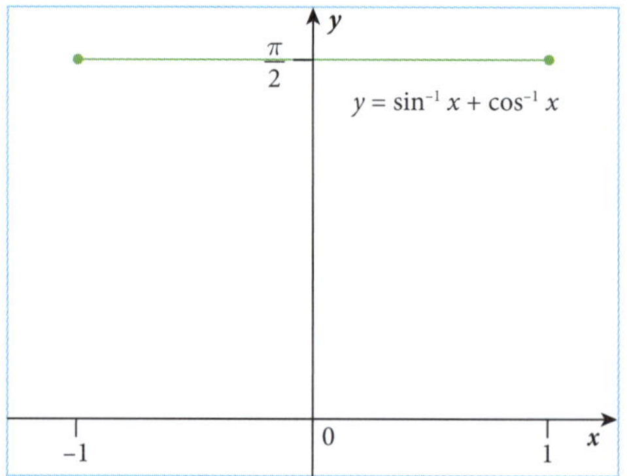

EXAMPLE 1

Prove the result $\sin^{-1}(-x) = -\sin^{-1} x$.

Let $y = \sin^{-1}(-x)$

$\therefore \sin y = -x$, for $-\frac{\pi}{2} \le y \le \frac{\pi}{2}$

$x = -\sin y$

$x = \sin(-y)$

$-y = \sin^{-1} x$

$y = -\sin^{-1} x$

$\therefore \sin^{-1}(-x) = -\sin^{-1} x$

Use $\sin[\sin^{-1}(-x)] = -x$.

EXAMPLE 2

Find:

a $\sin^{-1}\left(-\frac{1}{2}\right)$

b $\tan\left[\cos^{-1}\left(-\frac{1}{2}\right)\right]$

a Let $y = \sin^{-1}\left(-\dfrac{1}{2}\right)$

$\therefore \sin y = -\dfrac{1}{2}$

$y = -\dfrac{\pi}{6}$

$\therefore \sin^{-1}\left(-\dfrac{1}{2}\right) = -\dfrac{\pi}{6}$

Remember the restricted domain.

b Let $y = \cos^{-1}\left(-\dfrac{1}{2}\right)$

$\therefore \cos y = -\dfrac{1}{2}$

$y = \dfrac{2\pi}{3}$

$\therefore \tan\left(\dfrac{2\pi}{3}\right) = -\sqrt{3}$

$\therefore \tan\left[\cos^{-1}\left(-\dfrac{1}{2}\right)\right] = -\sqrt{3}$

Use exact values.

EXAMPLE 3

Find the exact value of $\cos\left[\tan^{-1}\left(\dfrac{4}{3}\right)\right]$.

Let $\alpha = \tan^{-1}\left(\dfrac{4}{3}\right)$

$\tan \alpha = \dfrac{4}{3}$

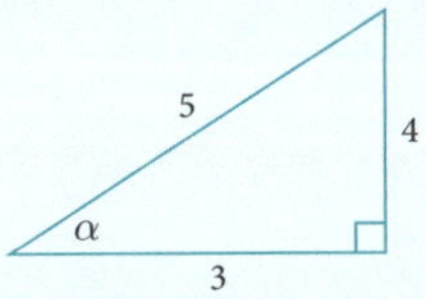

$\therefore \cos \alpha = \dfrac{3}{5}$ (by Pythagoras' theorem)

Use Pythagoras' theorem.

EXAMPLE 4

Show $\sin^{-1}\left(\dfrac{\sqrt{3}}{2}\right) + \cos^{-1}\left(\dfrac{\sqrt{3}}{2}\right) = \dfrac{\pi}{2}$.

$$\begin{aligned} \text{LHS} &= \sin^{-1}\left(\frac{\sqrt{3}}{2}\right) + \cos^{-1}\left(\frac{\sqrt{3}}{2}\right) \\ &= \frac{\pi}{3} + \frac{\pi}{6} \\ &= \frac{\pi}{2} \\ &= \text{RHS} \end{aligned}$$

A 'show' question requires LHS = RHS.

EXAMPLE 5

Find the exact value of $\tan\left[\sin^{-1}\left(\dfrac{2}{\sqrt{13}}\right) + \cos^{-1}\left(\dfrac{1}{\sqrt{3}}\right)\right]$.

Let $\alpha = \sin^{-1}\left(\dfrac{2}{\sqrt{13}}\right)$ and $\beta = \cos^{-1}\left(\dfrac{1}{\sqrt{3}}\right)$.

$\therefore \sin \alpha = \dfrac{2}{\sqrt{13}}$ $\qquad \cos \beta = \dfrac{1}{\sqrt{3}}$

$\therefore \tan \alpha = \dfrac{2}{3}$ $\qquad \tan \beta = \sqrt{2}$

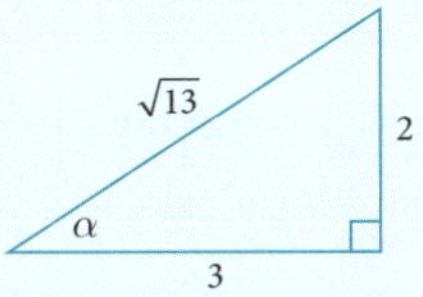

$$\begin{aligned} \tan\left[\sin^{-1}\left(\frac{2}{\sqrt{13}}\right) + \cos^{-1}\left(\frac{1}{\sqrt{3}}\right)\right] &= \tan(\alpha + \beta) \\ &= \frac{\tan\alpha + \tan\beta}{1 - \tan\alpha\tan\beta} \\ &= \frac{\frac{2}{3} + \sqrt{2}}{1 - \frac{2}{3}\times\sqrt{2}} \\ &= \frac{2 + 3\sqrt{2}}{3 - 2\sqrt{2}} \text{ (by multiplying top and bottom by 3)} \end{aligned}$$

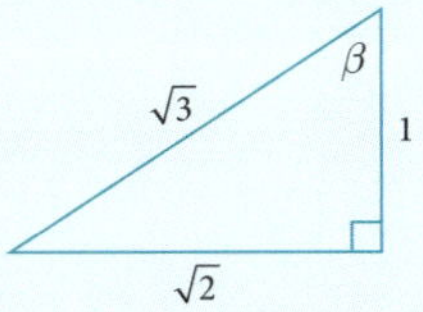

Use the compound angles results from Chapter 3.

➔ PRACTICE

1 Prove the results:

i $\cos^{-1}(-x) = \pi - \cos^{-1} x$

ii $\sin^{-1} x + \cos^{-1} x = \dfrac{\pi}{2}$

2 Find:

i a $\sin^{-1}\left(-\dfrac{\sqrt{3}}{2}\right)$

b $\cos^{-1}\left(-\dfrac{1}{\sqrt{2}}\right)$

ii a $\sin[\tan^{-1}(-\sqrt{3})]$

b $\tan\left[\cos^{-1}\left(-\dfrac{\sqrt{3}}{2}\right)\right]$

3 Find the exact value of:

i $\sin\left[\cos^{-1}\left(\dfrac{3}{\sqrt{13}}\right)\right]$

ii $\tan\left[\sin^{-1}\left(-\dfrac{\sqrt{5}}{3}\right)\right]$

4 Show:

i $\sin^{-1}\left(\dfrac{1}{\sqrt{2}}\right) + \cos^{-1}\left(\dfrac{1}{\sqrt{2}}\right) = \dfrac{\pi}{2}$

ii $\sin^{-1}\left(\dfrac{\sqrt{7}}{5}\right) + \cos^{-1}\left(\dfrac{\sqrt{7}}{5}\right) = \dfrac{\pi}{2}$

5 **Find the exact value of:**

i $\tan\left[\cos^{-1}\left(\frac{3}{5}\right)+\sin^{-1}\left(\frac{\sqrt{5}}{3}\right)\right]$

ii $\cos\left[\tan^{-1}\left(\frac{2}{\sqrt{2}}\right)-\sin^{-1}\left(\frac{1}{2}\right)\right]$

Answers ➲ pp. 59–60

YEAR 11 EXAM-TYPE QUESTIONS

Total marks: 35

PART A

(Suggested time: 10 minutes)

(1 mark each)

1 What is the domain of the function $f(x) = \cos^{-1} 2x$?

A $-1 \leq x \leq 1$ B $-\frac{1}{2} \leq x \leq \frac{1}{2}$

C $-2 \leq x \leq 2$ D $0 \leq x \leq \pi$

2 Which function best describes the following graph?

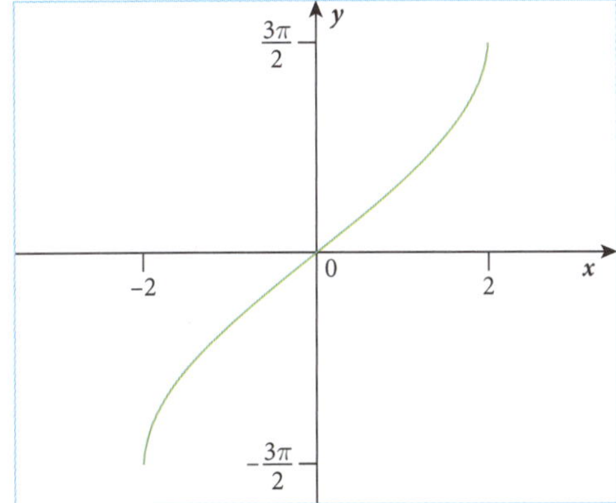

A $y = 2\sin^{-1}\frac{2x}{3}$ B $y = 2\sin^{-1}\frac{3x}{2}$

C $y = \frac{3}{2}\sin^{-1} 2x$ D $y = 3\sin^{-1}\frac{x}{2}$

3 Which of these is the range of $y = \frac{2}{3}\tan^{-1} x^2$?

A $0 \leq y < \frac{\pi}{3}$ B $0 \leq y < \frac{2\pi}{3}$

C $0 \leq y < \frac{4\pi}{3}$ D $-\frac{2\pi}{3} < y < \frac{2\pi}{3}$

4 The graph of the curve $y = \frac{1}{2}\cos^{-1}\frac{x}{3}$ passes through $(p, 0)$ and $(0, q)$. What are the values of p and q?

A $p = \frac{1}{3}, q = \frac{\pi}{2}$ B $p = 3, q = \frac{\pi}{2}$

C $p = \frac{1}{3}, q = \frac{\pi}{4}$ D $p = 3, q = \frac{\pi}{4}$

5 Which of the following represents the graph of $y = \cos^{-1}(x - 1)$?

A

B

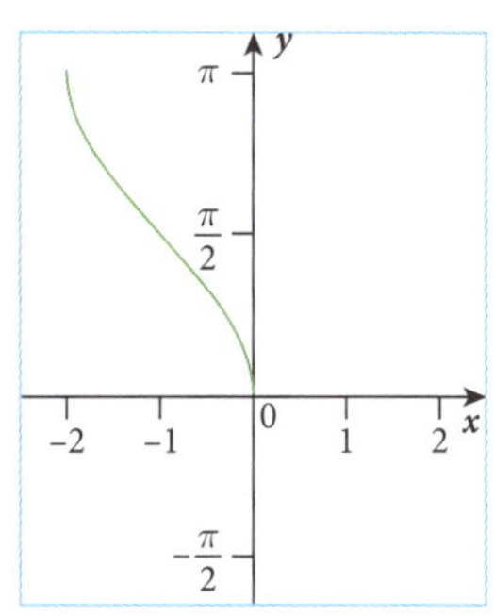

C

D

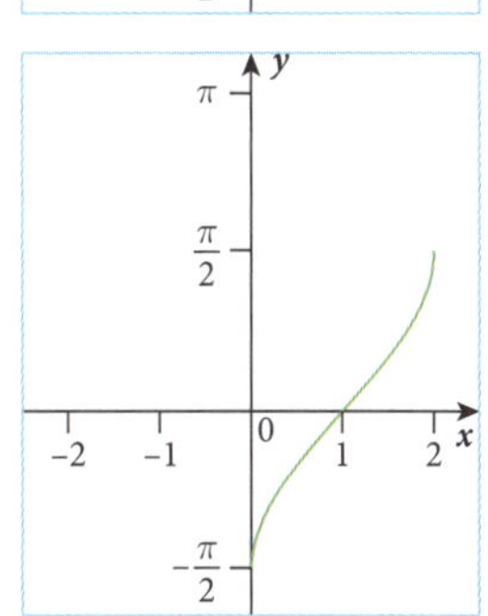

PART B

(Suggested time: 45 minutes)

Show all working.

6 Find the exact value of $\cos^{-1}\left(-\frac{1}{2}\right)$. (1 mark)

7 Sketch the graph of $y = 4\tan^{-1}x$, clearly indicating the range. (2 marks)

8 State the domain and range of $2\cos^{-1}\left(\frac{x}{4}\right)$. (2 marks)

9 Prove that $\cos^{-1}x = \sin^{-1}\sqrt{1-x^2}$, for $0 \leq x \leq 1$. (2 marks)

10 Find the exact values of:

a $\tan\left(\cos^{-1}\frac{2}{\sqrt{7}}\right)$ (2 marks)

b $\sin\left(2\tan^{-1}\frac{\sqrt{3}}{5}\right)$ (3 marks)

11 Evaluate, leaving your answer in exact form, $\sin\left[\cos^{-1}\left(\frac{3}{5}\right) + \tan^{-1}\left(-\frac{2}{3}\right)\right]$. (3 marks)

12 Consider the function $f(x) = \frac{2}{5}\sin^{-1}2x$.

a State the domain and range. (2 marks)

b Hence sketch the graph of $y = f(x)$. (1 mark)

13 The function $f(x) = \sec x$ for $0 \leq x < \frac{\pi}{2}$, and is not defined for other values of x.

a Find the domain of the inverse function $f^{-1}(x)$. (1 mark)

b Show that $f^{-1}(x) = \cos^{-1}\left(\frac{1}{x}\right)$. (2 marks)

14 Use the result $\sin^{-1} x + \cos^{-1} x = \frac{\pi}{2}$ to find the point of intersection of $y = \sin^{-1} x$ and $y = \cos^{-1} x$. (3 marks)

15 Using the result for $\tan(\alpha + \beta)$, prove

$\tan^{-1} 4 - \tan^{-1}\frac{3}{5} = \frac{\pi}{4}$. (3 marks)

16 Find the exact values of x and y which satisfy the simultaneous equations:

$$\sin^{-1} x + \frac{1}{2}\cos^{-1} y = \frac{5\pi}{12}$$

$$2\sin^{-1} x - \cos^{-1} y = \frac{\pi}{6}$$

(3 marks)

ANSWERS

PRACTICE

1 Domain, range and graphs ➲ p. 54

1 i a Domain: $-1 \le 2x \le 1$

$\therefore -\frac{1}{2} \le x \le \frac{1}{2}$, or $\left[-\frac{1}{2}, \frac{1}{2}\right]$

Range: Multiply end-values of $-\frac{\pi}{2} \le y \le \frac{\pi}{2}$ by 3:

$\therefore -\frac{3\pi}{2} \le y \le \frac{3\pi}{2}$ **or** $\left[-\frac{3\pi}{2}, \frac{3\pi}{2}\right]$

b Domain: $-1 \le x - 1 \le 1$

$\therefore 0 \le x \le 2$, or $[0, 2]$

Range: Multiply end-values of $0 \le y \le \pi$ by $\frac{1}{3}$:

$\therefore 0 \le y \le \frac{\pi}{3}$, **or** $\left[0, \frac{\pi}{3}\right]$

ii Domain: all real x or $(-\infty, \infty)$

Range: Multiply end-values of $-\frac{\pi}{2} \le y \le \frac{\pi}{2}$ by 4:

$\therefore -2\pi \le y \le 2\pi$ **or** $[-2\pi, 2\pi]$

iii Domain: $-1 \le \frac{x+1}{2} \le 1$

$-2 \le x + 1 \le 2$

$\therefore -3 \le x \le 1$ or $[-3, 1]$

Range: Multiply end-values of $-\frac{\pi}{2} \le y \le \frac{\pi}{2}$ by $\frac{3}{2}$:

$\therefore -\frac{3\pi}{4} \le y \le \frac{3\pi}{4}$ **or** $\left[-\frac{3\pi}{4}, \frac{3\pi}{4}\right]$

2 i a

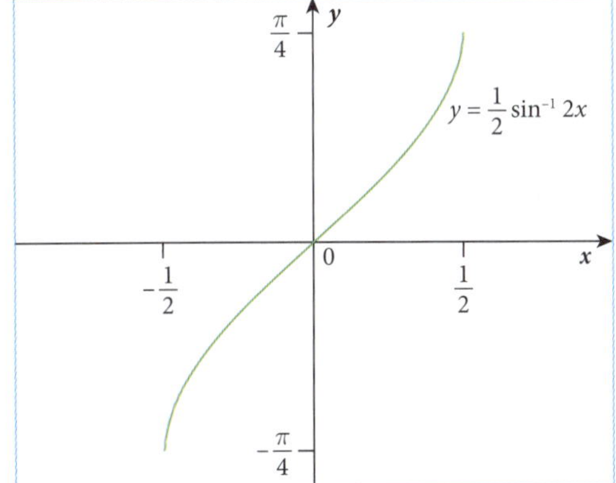

b

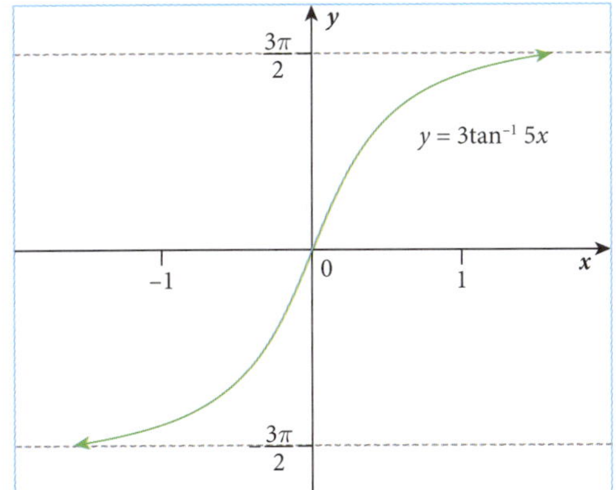

ii

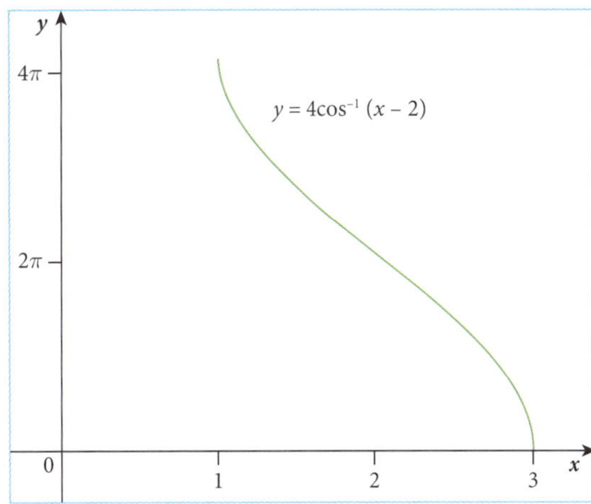

iii

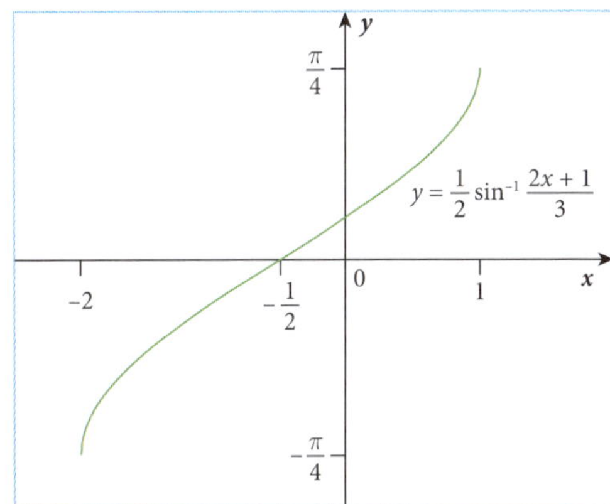

2 Properties of inverse trigonometric functions ➲ pp. 55–56

1 i Let $y = \cos^{-1}(-x)$

$\therefore \cos y = -x$, for $0 \le y \le \pi$

$x = -\cos y$

$x = \cos(\pi - y)$

$\pi - y = \cos^{-1} x$

$y = \pi - \cos^{-1} x$

$\therefore \cos^{-1}(-x) = \pi - \cos^{-1} x$

ii Let $y = \sin^{-1} x$ … ①

$\therefore \sin y = x$

$x = \sin y$

$x = \cos\left(\frac{\pi}{2} - y\right)$

$\therefore \frac{\pi}{2} - y = \cos^{-1} x$ … ②

① + ②: $\sin^{-1} x + \cos^{-1} x = y + \frac{\pi}{2} - y$

$\therefore \sin^{-1} x + \cos^{-1} x = \frac{\pi}{2}$

2 i a Let $y = \sin^{-1}\left(-\frac{\sqrt{3}}{2}\right)$

$\therefore \sin y = -\frac{\sqrt{3}}{2}$

$y = -\frac{\pi}{3}$

$\therefore \mathbf{\sin^{-1}\left(-\frac{1}{2}\right) = -\frac{\pi}{3}}$

b Let $y = \cos^{-1}\left(-\frac{1}{\sqrt{2}}\right)$

$\therefore \cos y = -\frac{1}{\sqrt{2}}$

$y = \frac{3\pi}{4}$

$\therefore \mathbf{\cos^{-1}\left(-\frac{1}{\sqrt{2}}\right) = \frac{3\pi}{4}}$

ii a Let $y = \tan^{-1}(-\sqrt{3})$

$\therefore \tan y = -\sqrt{3}$

$y = -\frac{\pi}{3}$

$\therefore \sin[\tan^{-1}(-\sqrt{3})] = \sin\left[-\frac{\pi}{3}\right]$

$= \mathbf{-\frac{\sqrt{3}}{2}}$

b Let $y = \cos^{-1}\left(-\frac{\sqrt{3}}{2}\right)$

$\therefore \cos y = -\frac{\sqrt{3}}{2}$

$y = \frac{5\pi}{6}$

$\therefore \tan\left[\cos^{-1}\left(-\frac{\sqrt{3}}{2}\right)\right] = \tan\left[\frac{5\pi}{6}\right]$

$= \mathbf{-\frac{1}{\sqrt{3}}}$

3 i Let $\alpha = \cos^{-1}\left(\frac{3}{\sqrt{13}}\right)$

$\cos \alpha = \frac{3}{\sqrt{13}}$

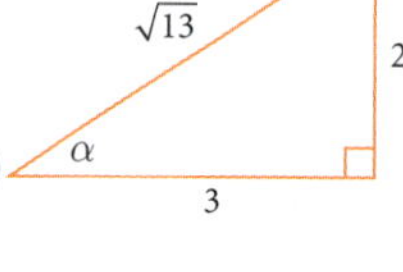

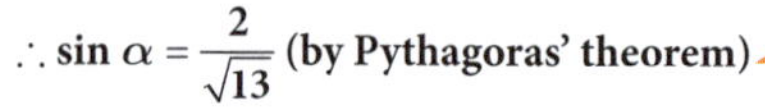

$\therefore \mathbf{\sin \alpha = \frac{2}{\sqrt{13}}}$ **(by Pythagoras' theorem)**

ii Let $\alpha = \sin^{-1}\left(-\frac{\sqrt{5}}{3}\right)$

$\sin \alpha = -\frac{\sqrt{5}}{3}$

Now, α lies in 4th quadrant:

$\therefore \mathbf{\tan \alpha = -\frac{\sqrt{5}}{2}}$ **(by Pythagoras' theorem)**

4 i $\sin^{-1}\left(\frac{1}{\sqrt{2}}\right) + \cos^{-1}\left(\frac{1}{\sqrt{2}}\right) = \frac{\pi}{2}$

$\text{LHS} = \sin^{-1}\left(\frac{1}{\sqrt{2}}\right) + \cos^{-1}\left(\frac{1}{\sqrt{2}}\right)$

$= \frac{\pi}{4} + \frac{\pi}{4}$

$= \frac{\pi}{2}$

= RHS

ii Let $\alpha = \sin^{-1}\left(\frac{\sqrt{7}}{5}\right)$ and let $\beta = \cos^{-1}\left(\frac{\sqrt{7}}{5}\right)$.

$\therefore \sin \alpha = \frac{\sqrt{7}}{5}$ and $\cos \beta = \frac{\sqrt{7}}{5}$

As $\alpha + \beta = \frac{\pi}{2}$, then $\sin^{-1}\left(\frac{\sqrt{7}}{5}\right) + \cos^{-1}\left(\frac{\sqrt{7}}{5}\right) = \frac{\pi}{2}$.

5 i Let $\alpha = \cos^{-1}\left(\frac{3}{5}\right)$ and $\beta = \sin^{-1}\left(\frac{\sqrt{5}}{3}\right)$.

$\therefore \cos \alpha = \frac{3}{5}$ $\quad \sin \beta = \frac{\sqrt{5}}{3}$

$\therefore \tan \alpha = \frac{4}{3}$ $\quad \tan \beta = \frac{\sqrt{5}}{2}$

$\tan\left[\cos^{-1}\left(\frac{3}{5}\right) + \sin^{-1}\left(\frac{\sqrt{5}}{3}\right)\right] = \tan(\alpha + \beta)$

$= \frac{\tan\alpha + \tan\beta}{1 - \tan\alpha\tan\beta}$

$= \frac{\frac{4}{3} + \frac{\sqrt{5}}{2}}{1 - \frac{4}{3} \times \frac{\sqrt{5}}{2}}$

$= \mathbf{\frac{8 + 3\sqrt{5}}{6 - 4\sqrt{5}}}$ (by multiplying top and bottom by 6)

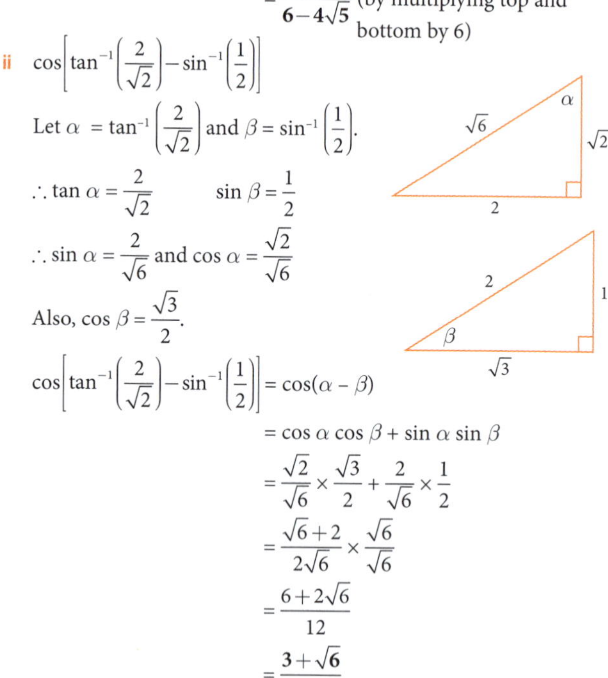

ii $\cos\left[\tan^{-1}\left(\frac{2}{\sqrt{2}}\right) - \sin^{-1}\left(\frac{1}{2}\right)\right]$

Let $\alpha = \tan^{-1}\left(\frac{2}{\sqrt{2}}\right)$ and $\beta = \sin^{-1}\left(\frac{1}{2}\right)$.

$\therefore \tan \alpha = \frac{2}{\sqrt{2}}$ $\quad \sin \beta = \frac{1}{2}$

$\therefore \sin \alpha = \frac{2}{\sqrt{6}}$ and $\cos \alpha = \frac{\sqrt{2}}{\sqrt{6}}$

Also, $\cos \beta = \frac{\sqrt{3}}{2}$.

$\cos\left[\tan^{-1}\left(\frac{2}{\sqrt{2}}\right) - \sin^{-1}\left(\frac{1}{2}\right)\right] = \cos(\alpha - \beta)$

$= \cos \alpha \cos \beta + \sin \alpha \sin \beta$

$= \frac{\sqrt{2}}{\sqrt{6}} \times \frac{\sqrt{3}}{2} + \frac{2}{\sqrt{6}} \times \frac{1}{2}$

$= \frac{\sqrt{6} + 2}{2\sqrt{6}} \times \frac{\sqrt{6}}{\sqrt{6}}$

$= \frac{6 + 2\sqrt{6}}{12}$

$= \mathbf{\frac{3 + \sqrt{6}}{6}}$

YEAR 11 EXAM-TYPE QUESTIONS

Part A

1 B. Domain: $-1 \le 2x \le 1$

$\therefore -\frac{1}{2} \le x \le \frac{1}{2}$

2 D. Domain: $-2 \le x \le 2$ and range: $-\frac{3\pi}{2} \le y \le \frac{3\pi}{2}$

$\therefore$ the graph of $y = 3\sin^{-1}\frac{x}{2}$.

3 A. Range of $y = \tan^{-1} x^2$ is $0 \le y < \frac{\pi}{2}$ or $\left[0, \frac{\pi}{2}\right)$.

Multiply end-values of $0 \le y < \frac{\pi}{2}$ by $\frac{2}{3}$:

$\therefore 0 \le y < \frac{\pi}{3}$ or $\left[0, \frac{\pi}{3}\right)$.

4 D. $y = \frac{1}{2}\cos^{-1}\frac{x}{3}$

Let $y = 0$ $\therefore$ $0 = \frac{1}{2}\cos^{-1}\frac{x}{3}$

$\cos^{-1}\frac{x}{3} = 0$

$\frac{x}{3} = 1$

$x = 3 \quad \therefore p = 3$

Let $x = 0$ $\therefore$ $y = \frac{1}{2}\cos^{-1} 0$

$= \frac{1}{2}\left(\frac{\pi}{2}\right)$

$= \frac{\pi}{4} \quad \therefore q = \frac{\pi}{4}$

5 C. The graph of $y = \cos^{-1} x$ has been translated one unit to the right.

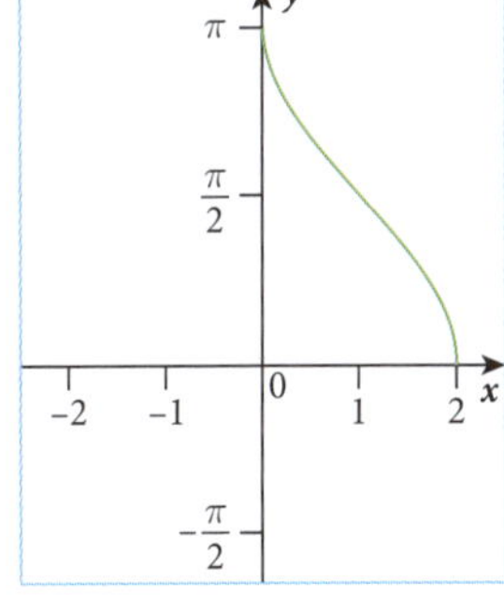

Part B

6 Let $y = \cos^{-1}\left(-\frac{1}{2}\right)$

$\therefore \cos y = -\frac{1}{2}$

$y = \frac{2\pi}{3}$

$\therefore \boldsymbol{\cos^{-1}\left(-\frac{1}{2}\right) = \frac{2\pi}{3}}$ ✓

7 Range is $-2\pi < y < 2\pi$.

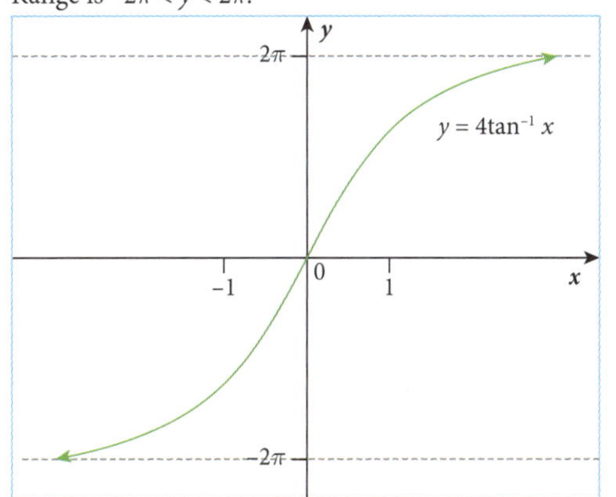

✓

Range is $-2\pi < y < 2\pi$ or **$(-2\pi, 2\pi)$** ✓

8 Domain: $-1 \le \frac{x}{4} \le 1$

$\therefore$ **$-4 \le x \le 4$ or $[-4, 4]$** ✓

Range: Multiply end-values of $0 \le y \le \pi$ by 2:

$\therefore$ **$0 \le y \le 2\pi$ or $[0, 2\pi]$** ✓

9 Let $\cos^{-1}\frac{x}{1} = \alpha$.

$\therefore$ Using Pythagoras' theorem:

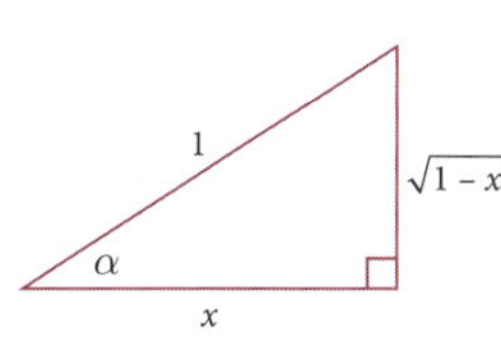

$\sin\alpha = \frac{\sqrt{1-x^2}}{1}$ ✓

$= \sqrt{1-x^2}$

$\alpha = \sin^{-1}\sqrt{1-x^2}$

$\therefore$ **$\cos^{-1} x = \sin^{-1}\sqrt{1-x^2}$** ✓

10 a Let $\alpha = \cos^{-1}\frac{2}{\sqrt{7}}$

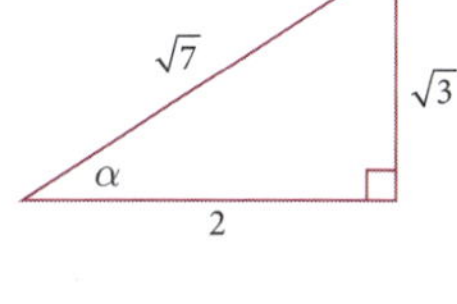

$\cos\alpha = \frac{2}{\sqrt{7}}$ ✓

$\therefore$ **$\tan\alpha = \frac{\sqrt{3}}{2}$** ✓

b 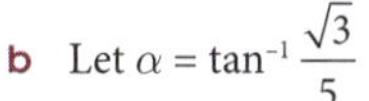 Let $\alpha = \tan^{-1}\frac{\sqrt{3}}{5}$

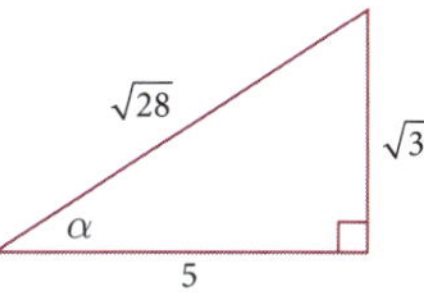

$\tan\alpha = \frac{\sqrt{3}}{5}$ ✓

$\therefore \sin\alpha = \frac{\sqrt{3}}{\sqrt{28}}$ and $\cos\alpha = \frac{5}{\sqrt{28}}$ ✓

$\therefore \sin 2\alpha = 2\sin\alpha\cos\alpha$

$= 2 \times \frac{\sqrt{3}}{\sqrt{28}} \times \frac{5}{\sqrt{28}}$

$= \boldsymbol{\frac{5\sqrt{3}}{14}}$ ✓

11 Let $\alpha = \cos^{-1}\frac{3}{5}$ and $\beta = \tan^{-1}\left(-\frac{2}{3}\right)$.

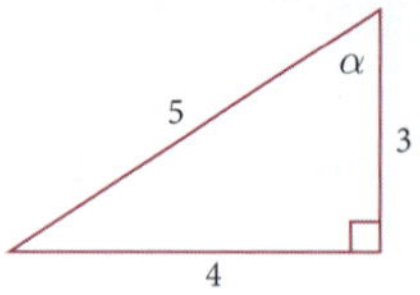

$\therefore \cos\alpha = \frac{3}{5}, \tan\beta = -\frac{2}{3}$. ✓

$\therefore \sin\alpha = \frac{4}{5}, \sin\beta = -\frac{2}{\sqrt{13}}, \cos\beta = \frac{3}{\sqrt{13}}$.

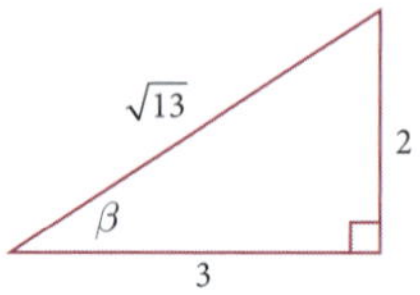

$\therefore \sin(\alpha+\beta) = \sin\alpha\cos\beta + \sin\beta\cos\alpha$ ✓

$= \frac{4}{5}\times\frac{3}{\sqrt{13}} + \left(-\frac{2}{\sqrt{13}}\right)\times\frac{3}{5}$

$= \frac{12}{5\sqrt{13}} - \frac{6}{5\sqrt{13}}$

$= \mathbf{\frac{6}{5\sqrt{13}}}$ ✓

12 a Domain: $-1 \le 2x \le 1$

$\therefore -\frac{1}{2} \le x \le \frac{1}{2}$ ✓

Range: Multiply end-values of $-\frac{\pi}{2} \le y \le \frac{\pi}{2}$ by $\frac{2}{5}$:

$\therefore -\frac{\pi}{5} \le y \le \frac{\pi}{5}$ ✓

b

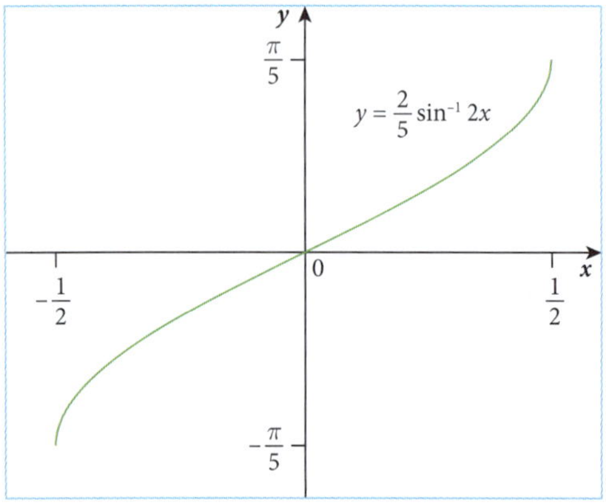

✓

13 a Consider the range of $f(x) = \sec x$ for $0 \le x < \frac{\pi}{2}$: the range is $f(x) \ge 1$.

$\therefore$ **domain of $f^{-1}(x)$ is $x \ge 1$.** ✓

b Let $y = \sec x$

Now, $y \leftrightarrow x$:

$x = \sec y$ ✓

$= \frac{1}{\cos y}$

$\cos y = \frac{1}{x}$

$y = \cos^{-1}\left(\frac{1}{x}\right)$.

$\therefore \mathbf{f^{-1}(x) = \cos^{-1}\left(\frac{1}{x}\right)}$. ✓

14 Using $\cos^{-1}x = \frac{\pi}{2} - \sin^{-1}x$:

Point of intersection when $\sin^{-1}x = \cos^{-1}x$.

$\sin^{-1}x = \frac{\pi}{2} - \sin^{-1}x$ ✓

$2\sin^{-1}x = \frac{\pi}{2}$

$\sin^{-1}x = \frac{\pi}{4}$

$x = \sin\frac{\pi}{4}$

$= \frac{1}{\sqrt{2}}$ ✓

$\therefore$ **point of intersection at $\left(\frac{1}{\sqrt{2}}, \frac{\pi}{4}\right)$** ✓

15 Consider $\tan\left(\tan^{-1}4 - \tan^{-1}\frac{3}{5}\right) = \tan\frac{\pi}{4}$,

i.e. need to show $\tan\left(\tan^{-1}4 - \tan^{-1}\frac{3}{5}\right) = 1$ ✓

Let $\alpha = \tan^{-1}4$ and $\beta = \tan^{-1}\frac{3}{5}$.

$\therefore \tan\alpha = 4$ and $\tan\beta = \frac{3}{5}$ ✓

Now, $\tan(\alpha - \beta) = \frac{\tan\alpha - \tan\beta}{1 + \tan\alpha\tan\beta}$

$= \frac{4 - \frac{3}{5}}{1 + 4\left(\frac{3}{5}\right)}$

$= 1$ ✓

$\therefore \mathbf{\tan^{-1}4 - \tan^{-1}\frac{3}{5} = \frac{\pi}{4}}$

16 $\sin^{-1}x + \frac{1}{2}\cos^{-1}y = \frac{5\pi}{12}$ … ①

$2\sin^{-1}x - \cos^{-1}y = \frac{\pi}{6}$ … ②

2 × ①: $2\sin^{-1}x + \cos^{-1}y = \frac{5\pi}{6}$ … ③

② + ③: $4\sin^{-1}x = \pi$ ✓

$\sin^{-1}x = \frac{\pi}{4}$

$x = \frac{1}{\sqrt{2}}$ ✓

Subs. in ③: $2\left(\frac{\pi}{4}\right) + \cos^{-1}y = \frac{5\pi}{6}$

$\cos^{-1}y = \frac{\pi}{3}$

$y = \frac{1}{2}$

$\therefore \mathbf{x = \frac{1}{\sqrt{2}}}$ **and** $\mathbf{y = \frac{1}{2}}$ ✓

CHAPTER 5
RATES OF CHANGE

TOPIC: CALCULUS

YEAR 11

1 Rates of change with respect to time

- **The rate of change** of a quantity Q with respect to time is measured by $\frac{dQ}{dt}$.
- **Displacement** (x) measures the distance of a particle from a fixed point (origin). Displacement can be positive or negative according to which side of the origin the particle is on.
- **Velocity** (v, or $\dot{x}$), the rate of change of the displacement (x) of a particle moving along the x-axis with respect to time (t), is measured by $\frac{dx}{dt}$.
- **Acceleration** (a, or $\ddot{x}$), the rate of change of the velocity (v) of a particle moving along the x-axis with respect to time (t), is measured by $\frac{dv}{dt}$, or $\frac{d^2x}{dt^2}$.
- **Displacement has magnitude and direction.** Distance has magnitude only and |displacement| = distance.
- **Velocity has magnitude and direction.** Speed has magnitude only and |velocity| = speed.
- **Other terminology**: 'initially': $t = 0$, 'at rest': $v = 0$, 'change of direction': v changes sign, 'velocity increasing at an increasing rate': acceleration is positive.

EXAMPLE 1

Find a formula for the rate of change of P with respect to t if $P = 3t^2 - 4t + 2$.

$\frac{dP}{dt} = 6t - 4$

The derivative describes the rate of change.

EXAMPLE 2

The mass M, in grams, of a melting iceblock is given by the formula $M = 20 + 2t - 4t^2$, where t is the time in minutes. Find the rate at which the iceblock will be melting after 5 minutes.

$\frac{dM}{dt} = 2 - 8t$

Substitute $t = 5$:

Decreasing when $\frac{dM}{dt} < 0$

$\frac{dM}{dt}(5) = 2 - 8(5)$

$= -38$

∴ the volume is decreasing at a rate of 38 g/min.

EXAMPLE 3

The graph shows the displacement x of a particle over time t, where t is in seconds.

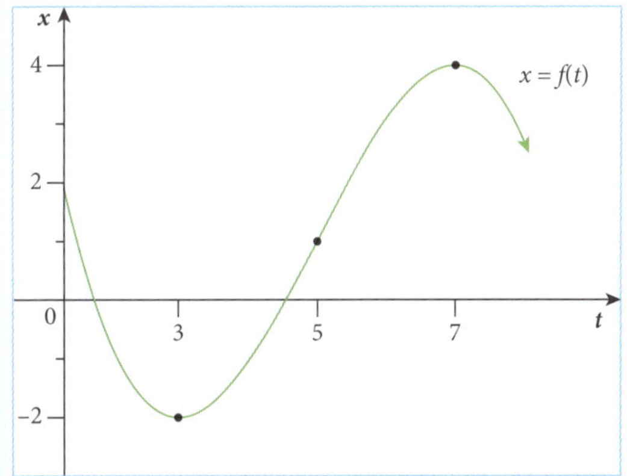

a **When and where is the particle at rest?**

b **How far does the particle travel in the first 5 seconds?**

c **On separate number planes draw graphs of the velocity (v) and acceleration (a) functions.**

a Consider when $v = \frac{dx}{dt} = 0$.

Turning points at $(3, -2)$ and $(7, 4)$.

The particle is at rest when the gradient is zero.

∴ the particle is at rest after 3 seconds at $x = -2$ and after 7 seconds at $x = 4$.

b Initally the particle is at $x = 2$, moves to $x = -2$ and by $t = 5$ is at $x = 1$.

Distance $= 4 + 3$

$= 7$

For distance we ignore the sign of the displacement.

∴ the particle moved 7 units.

c

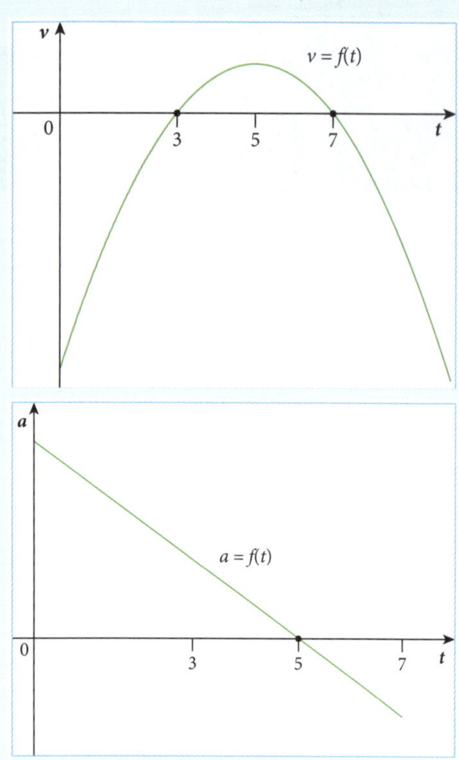

As you differentiate there is one less turning point.

EXAMPLE 4

The displacement x cm of an object over t seconds is given by the equation $x = 12t - t^3$.

a How many times is the object at rest?

b How far does the object move in the first 3 seconds?

a $v = \dfrac{dx}{dt} = 12 - 3t^2 = 0$

$3t^2 = 12$

$t^2 = 4$

$t = 2 \quad (t \geq 0)$

$\therefore$ the particle is at rest once.

At rest means $v = 0$.

b $x(0) = 12(0) - (0)^3$

$= 0$

$x(2) = 12(2) - (2)^3$

$= 16$

$x(3) = 12(3) - (3)^3$

$= 9$

Distance $= 16 + (16 - 9)$

$= 16 + 7$

$= 23$

$\therefore$ the object moves 23 cm.

Find displacement initially, when the particle stops, and at the end.

EXAMPLE 5

The velocity, in cms^{-1}, of a body is given by $v = \sqrt{2t+9}$, where t is time in seconds.

a Find the intial velocity of the body.

b What is the acceleration after 8 seconds?

a Let $t = 0$: $v(0) = \sqrt{2(0)+9}$

$= 3$

$\therefore$ initially the body is moving at 3 cms^{-1}.

b $v = (2t+9)^{\frac{1}{2}}$

$a = \dfrac{dv}{dt} = \dfrac{1}{2}(2t+9)^{-\frac{1}{2}}.2$

$= (2t+9)^{-\frac{1}{2}}$

$= \dfrac{1}{\sqrt{2t+9}}$

$a = \dfrac{dv}{dt} = \dfrac{d^2x}{dt^2}$

Let $t = 8$: $a(8) = \dfrac{1}{\sqrt{2(8)+9}}$

$= \dfrac{1}{5}$

$\therefore$ the acceleration is 0.2 cms^{-2}.

→ PRACTICE

1 Find a formula for the rate of change of:

i Q with respect to t if $Q = 3 - 11t - 4t^2$

ii V with respect to t if $V = \dfrac{3t}{\sqrt{t+2}}$

2 **i** A circular disc expands as it is being heated. The area, in square centimetres, of the disc increases according to the formula $A = 6t^2 + 2t$, where t is in minutes. Find the rate of increase of the area of the disc after 4 minutes.

ii Water is draining from the bottom of a tank. The volume V, in litres, of water in the tank is given by the formula $V = 4800\left(1 - \dfrac{t}{60}\right)^2$, where t is in minutes. At what rate is the water draining from the tank after 15 minutes?

3 i The graph shows the displacement x of a particle over time t.

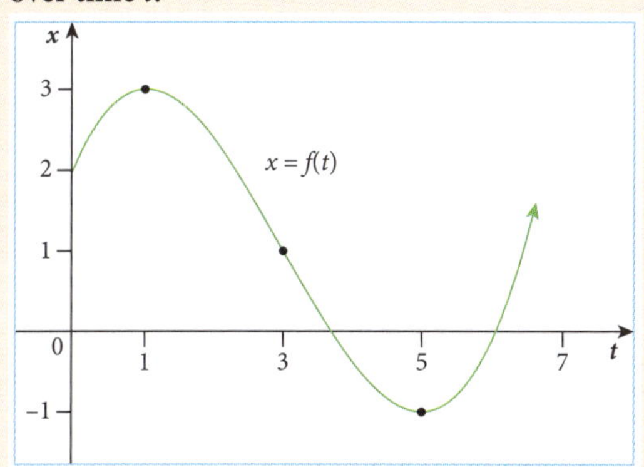

a When and where is the particle at rest?

b How far does the particle travel in the first 5 seconds?

c On separate number planes draw graphs of the velocity (v) and acceleration (a) functions.

ii The graph shows the velocity v of an object over time t.

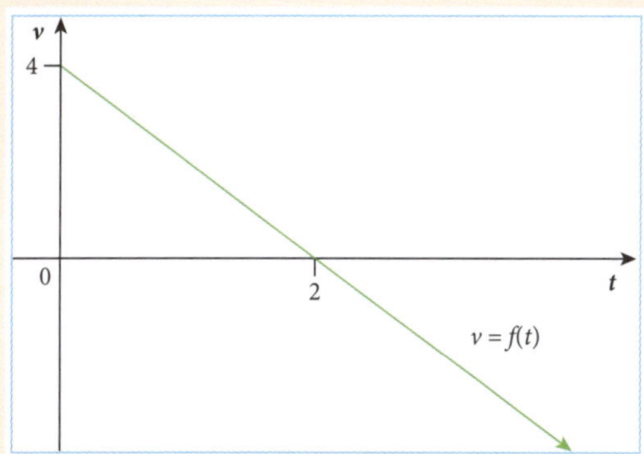

a When is the object stationary?

b Write the equation of the velocity v, in terms of t, and hence graph the acceleration function.

4 i The displacement x cm of a particle after t seconds is given by the equation $x = t^3 - 3t^2 - 9t - 1$, for $t \geq 0$.

a How many times is the object at rest?

b How far does the object move in the first 4 seconds?

ii The displacement, in metres, of a body is given by $x = t^4 - 4t - 3$ where t is time in seconds.

a Find where the body is when it stops.

b What is the exact velocity of the body when the acceleration is 24 ms^{-2}?

5 i The velocity, in cms^{-1}, of a body is given by $v = \sqrt{4t+1}$, where t is time in seconds.

a Find the initial velocity of the body.

b What is the acceleration after 2 seconds?

ii The displacement x cm of a particle after t seconds is given by the equation $x = \dfrac{t-1}{t+1}$.

a Show that the particle passes through O at a speed of 0.5 cms^{-1}.

b Find an expression for the acceleration of the particle.

Answers ➲ pp. 73–74

2 Exponential growth and decay: $N(t) = Ae^{kt}$

➔ **When the rate of change** of a quantity N is proportional to the quantity we say $\dfrac{dN}{dt} = kN$, where k is constant.

➔ **If $N = Ae^{kt}$**, then by differentiation $\dfrac{dN}{dt} = kN$. (A is the initial quantity.)

➔ $\dfrac{dN}{dt}$ **is the rate** at which the quantity is changing.

EXAMPLE 1

Draw a sketch of $f(t) = 4e^{0.2t}$, for $0 \leq t \leq 8$.

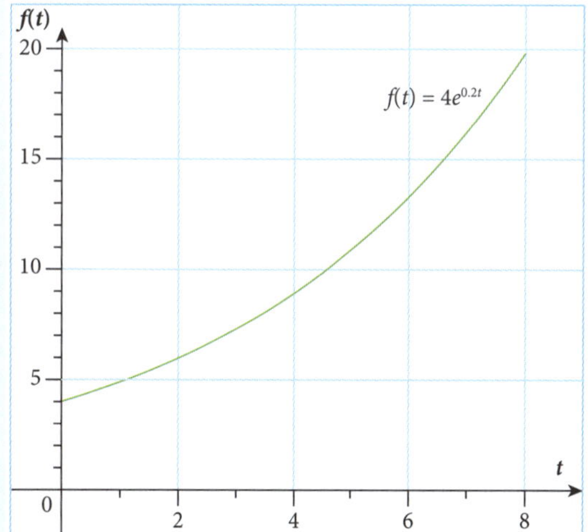

Increasing at an increasing rate.

EXAMPLE 2

A quantity of radioactive material decays according to the equation $\dfrac{dM}{dt} = -kM$, where M is the mass of the material in kilograms, t in years and k is a constant. Show that $M = Ae^{-kt}$ is a solution of the equation, where A is a constant.

$$M = Ae^{-kt}$$

$$\frac{dM}{dt} = -kAe^{-kt}$$

$$= -kM$$

k is the growth/decay constant.

EXAMPLE 3

The population of a colony of rabbits over t months is given by $P = 180e^{0.347t}$. Find:

- **a** **the initial population**
- **b** **the population after 8 months**
- **c** **when the population reaches 5400**

a Let $t = 0$: $P(0) = 180e^{0.347(0)}$

$$= 180$$

Let $t = 0$.

$\therefore$ the initial population was 180 rabbits.

b Let $t = 8$: $P(0) = 180e^{0.347(8)}$

$$= 2889.84\ldots$$

$$= 2890 \text{ (nearest whole)}$$

$\therefore$ the population is 2890 rabbits.

c Let $P = 5400$: $5400 = 180e^{0.347t}$.

$$e^{0.347t} = \frac{5400}{180}$$

$$e^{0.347t} = 30$$

$$\ln e^{0.347t} = \ln 30$$

Take logs of both sides.

$$0.347t = \ln 30$$

$$t = \frac{\ln 30}{0.347}$$

$$= 9.8017\ldots$$

$$= 9.80 \text{ (2 dec. pl.)}$$

$\therefore$ it will take 9.8 months.

EXAMPLE 4

A drug is used to control a medical condition. It is known that the quantity, in millilitres, of the drug remaining in the body after t hours satisfies an equation of the form $Q = 8e^{-kt}$, where k is a constant. After 6 hours the amount remaining in the body is half the initial dose.

- **a** **Find the value of k.**
- **b** **When will one-tenth of the initial dose remain?**
- **c** **Find the rate at which the quantity of drug is reducing after 2 hours.**

a Initial dose = 8 mL

Let $t = 6$, $Q = 4$:

$$4 = 8e^{-k(6)}$$

$$e^{-6k} = 0.5$$

$$-6k = \ln 0.5$$

$$k = \frac{\ln 0.5}{-6}$$

A half-life is the time taken for the quantity to halve in size.

$$= 0.115\,524\,53\ldots$$

$$= 0.1155 \text{ (4 dec. pl.)}$$

b Let $Q = 0.8$:

$$0.8 = 8e^{-kt}$$

$$e^{-kt} = 0.1$$

$$-kt = \ln 0.1$$

$$t = \frac{\ln 0.1}{-k}$$

Use the value of *k* in your calculator's memory.

$$= 19.931\,568\,57\ldots$$

$$= 19.9 \text{ (1 dec. pl.)}$$

$\therefore$ after 19.9 hours

c Let $t = 2$,

$$Q(2) = 8e^{-k(2)}$$

$$= 6.349\,604\,208\ldots$$

$$= 6.35 \text{ (2 dec. pl.)}$$

$$\frac{dQ}{dt} = -kQ$$

Use the value of *k* in your calculator's memory.

$$= -0.1155 \times 6.35$$

$$= -0.733\,535\,042\ldots$$

$$= -0.73 \text{ (2 dec. pl.)}$$

$\therefore$ the quantity is reducing at 0.73 mL/h.

→ PRACTICE

1 Draw a sketch of:

- **i** $f(t) = 2e^{0.4t}$, for $0 \le t \le 5$
- **ii** $Q = 40e^{-0.2t}$, for $0 \le t \le 10$

2 **i** Due to overfishing in a particular bay, the number N of a particular species of fish is dropping according to the formula $\frac{dN}{dt} = -kN$, where t is measured in years and k is a constant. Show that $N = Ae^{-kt}$ is a solution to the equation, where A is a constant.

ii At the beginning of 2010 the population of a town was 8000. Assume the population P is increasing exponentially and satisfies an equation of the form $P = 8000e^{kt}$, where k is a constant and t is measured in years from the beginning of 2010. Show that $P = 8000e^{kt}$ satisfies the differential equation $\frac{dP}{dt} = kP$.

3 i A population of bacteria after t hours is given by the formula $P = 20\,000e^{0.0458t}$. Find:

a the initial population

b the population after 20 hours

c the time it takes for the population to reach half a million

ii A chemical reaction causes the amount of chlorine to be reduced at a rate proportional to the amount of chlorine, in litres, present at any one time. The amount of chlorine is given by the formula $A = A_0e^{-0.0646t}$, where A_0 is constant and t is measured in days. Initially there is 100 L of chlorine present. Find the amount of:

a chlorine present after 15 days

b time it takes for the amount of chlorine to reduce to 500 mL

4 i A cell culture is growing exponentially. It has a doubling time of 4 hours. Initially there are 12 000 cells present. Use the equation $P = P_0e^{kt}$, to find the:

a constant k

b time, to the nearest minute, for the population to triple

c rate at which the population is increasing when it is 10 times its original size

ii The decay of a radium isotope is proportional to its mass. If 60 kg of radium takes 12 years to decay to 50 kg, find the:

a value of the decay constant k, correct to four significant figures

b decay rate when 20 kg of radium remains

c half-life of the radium. Give your answer to the nearest year.

Answers ➲ pp. 74–75

3 Further exponential growth and decay: $N(t) = P + Ae^{kt}$

➔ **When the rate of change** of a quantity N is proportional to the difference between N and P we say $\frac{dN}{dt} = k(N - P)$ where k is constant.

➔ **If $N = P + Ae^{kt}$**, then by differentiation $\frac{dN}{dt} = k(N - P)$, where A is a constant.

EXAMPLE 1

Draw a sketch of $f(t) = 4 + 6e^{-0.5t}$, for $0 \leq t \leq 10$.

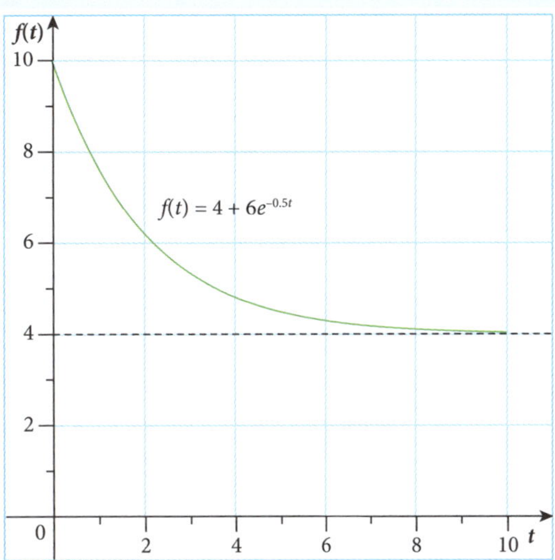

When $t = 0$, $f(0) = 4 + 6$.

EXAMPLE 2

Show that $N = 20 + 60e^{0.05t}$ is a solution to the equation $\frac{dN}{dt} = 0.05(N - 20)$.

$$N = 20 + 60e^{0.05t}$$

$$\frac{dN}{dt} = 0.05 \times 60e^{0.05t}$$

Show by differentiation.

But, $60e^{0.05t} = N - 20$

$$\therefore \frac{dN}{dt} = 0.05(N - 20)$$

EXAMPLE 3

A bottle of soft drink taken out of a refrigerator has a temperature of 3 °C. It is placed in a room with a constant temperatue of 24 °C. After t minutes the temperature, T °C, of the bottle is given by $T = A - Be^{-0.035t}$, where A and B are positive constants. Find the:

a values of A and B

b temperature after 10 minutes

c time taken for the temperature of the bottle to rise to 20 °C

a When $t \to \infty$, $Be^{-0.035t} \to 0$.

$\therefore 24 = A + 0$

$\therefore A = 24$

$\therefore T = 24 - Be^{-0.035t}$

Let $t \to \infty$ to find A, then let $t = 0$ to find B.

When $t = 0$ and $T = 3$:

$3 = 24 - B$

$\therefore B = 21$

$\therefore T = 24 - 21e^{-0.035t}$

b Let $t = 10$: $T(10) = 24 - 21e^{-0.035(10)}$

$= 9.201\,550\,116\ldots$

$= 9.2$ (1 dec. pl.)

$\therefore$ the temperature is 9.2 °C.

c Let $T = 20$: $20 = 24 - 21e^{-0.035t}$

$21e^{-0.035t} = 4$

$e^{-0.035t} = \frac{4}{21}$

$-0.035t = \ln\frac{4}{21}$

Take logs of both sides.

$t = \frac{\ln\frac{4}{21}}{-0.035}$

$= 47.377\,945\,05$

$= 47.38$ (2 dec. pl.)

$\therefore$ it will take 47.38 minutes.

EXAMPLE 4

The rate of change of a certain town's population is proportional to the excess of the population over 6000. The town had a population on 1 January 1990 of 8000 and by 1 January 2005 the population had grown to 10 400. Using the formula $P = 6000 + Be^{kt}$, find the:

a values of B and k

b year in which the population will reach 20 000

a Let $t = 0$ (in 1990), $P = 8000$:

$8000 = 6000 + Be^{k(0)}$

$8000 = 6000 + B$

$B = 2000$

$\therefore P = 6000 + 2000e^{kt}$

Let $t = 15$ (in 2005), $P = 10\,400$:

$10\,400 = 6000 + 2000e^{k(15)}$

$2000e^{15k} = 4400$

$e^{15k} = 2.2$

$15k = \ln 2.2$

$k = \frac{\ln 2.2}{15}$

$= 0.052\,563\,824\ldots$

$= 0.0526$ (4 dec. pl.)

b Let $P = 20\,000$:

$20\,000 = 6000 + 2000e^{kt}$

$2000e^{kt} = 14\,000$

$e^{kt} = 7$

$kt = \ln 7$

Always use the exact value of k in your calculations.

$t = \frac{\ln 7}{k}$

$= 37.019\,950\,23\ldots$

$1990 + 37 = 2027$

$\therefore$ the population will reach 20 000 during 2027.

→ PRACTICE

1 Draw a sketch of:

i $N = 10 + 90e^{-0.4t}$, for $0 \leq t \leq 10$

ii $M = 80 - 75e^{-0.8t}$, for $0 \leq t \leq 5$

2 i Show that $N = 6 + 19e^{0.75t}$ is a solution of the equation $\frac{dN}{dt} = 0.75(N - 6)$.

ii A container is being filled at a variable rate and the volume of liquid in the container at any time t is given by $V = 20(1 - e^{-kt})$. Show that $\frac{dV}{dt} = k(20 - V)$.

3 i A cup of coffee with an initial temperature of 90 °C is placed in a room with a constant temperature of 20 °C. The temperature, T °C, of the coffee after t minutes is given by the formula $T = A + Be^{-0.046t}$, where A and B are constants. The temperature of the coffee drops to 40 °C after 20 minutes. Find the:

a values of A and B

b temperature after 30 minutes

c time taken for the temperature of the coffee to fall to 25 °C

ii The height, H, of an elephant is modelled by $H = A - Be^{-0.1187t}$, where H is measured in metres, t is the age in years and k is a positive constant. It is known that the elephant was 0.9 m high when born and grows to a height of 3.3 m. Find the:

a values of A and B

b height of the 4-year-old elephant

c age of the elephant when it is 3 m tall

4 i A cake was removed from the oven at 10:15 am and has a temperature of 180 °C. It was left to cool in a room that had a temperature of 28 °C. After 30 minutes the temperature of the cake

was 50 °C. Use the equation $T = 28 + Be^{-kt}$, where B and k are constants to find the:

a values of B and k

b time when the temperature of the cake was 35 °C

ii The number (N) of ants in a colony at time t weeks is given by the formula $N = 110\,000 - Be^{-kt}$, where B and k are positive constants. The initial size of the colony, when discovered on 1 March 2019, was 4000 and 6 weeks later the size had increased to 25 000. Find the:

a exact values of B and k

b month when the population exceeded 100 000

Answers ⮌ pp. 75–76

4 Related rates of change

→ **The rate of change of a quantity is related to the rate of change of other quantities.**

→ **The chain rule** is used: $\frac{dy}{dt} = \frac{dy}{dx} \times \frac{dx}{dt}$.

EXAMPLE 1

If $y = x^2 + 2x - 1$ and $\frac{dx}{dt} = -3$, evaluate $\frac{dy}{dt}$ when $x = 4$.

$$\frac{dy}{dx} = 2x + 2$$

$$\frac{dy}{dx}(4) = 2(4) + 2 = 10$$

$$\frac{dy}{dt} = \frac{dy}{dx} \times \frac{dx}{dt} = 10 \times -3 = -30$$

The rate of change of y is related to the other rates of change.

EXAMPLE 2

A spherical balloon is being inflated and its radius is increasing at a constant rate of 2 cm/min. At what rate is the volume increasing when the radius of the balloon is 6 cm?

$$\frac{dr}{dt} = 2$$

$$V = \frac{4}{3}\pi r^3$$

Look at the units: cm/min is $\frac{dr}{dt}$.

$$\frac{dV}{dr} = 4\pi r^2$$

$$\frac{dV}{dr}(6) = 4\pi(6)^2 = 144\pi$$

$$\frac{dV}{dt} = \frac{dV}{dr} \times \frac{dr}{dt} = 144\pi \times 2 = 288\pi$$

∴ the volume is increasing at 288π cm³/min.

EXAMPLE 3

Sand is being poured into a heap in the shape of a right cone at the rate of 6 cm³/min. The height of the heap is always equal to the radius of its circular base. When the heap is 15 cm high, how fast is the area of its base increasing?

$$V = \frac{1}{3}\pi r^2 h$$

As $h = r$, then $V = \frac{1}{3}\pi r^3$:

$$\frac{dV}{dr} = \pi r^2$$

$$\frac{dV}{dr}(15) = \pi(15)^2 = 225\pi$$

This question involves two applications of the chain rule.

$$\frac{dV}{dt} = \frac{dV}{dr} \times \frac{dr}{dt}$$

$$6 = 225\pi \times \frac{dr}{dt}$$

$$\frac{dr}{dt} = \frac{6}{225\pi}$$

Now, $\frac{dA}{dt} = \frac{dA}{dr} \times \frac{dr}{dt}$

As $A = \pi r^2$, then $\frac{dA}{dr} = 2\pi r$

$$\frac{dA}{dr}(15) = 2\pi(15) = 30\pi$$

Now, $\frac{dA}{dt} = 30\pi \times \frac{6}{225\pi} = \frac{180}{225} = \frac{4}{5}$

Use the volume to find $\frac{dr}{dt}$, then find $\frac{dA}{dt}$.

∴ the area is increasing at $\frac{4}{5}$ cm²/min.

→ PRACTICE

1 i If $y = 3x^2 + 5x + 4$ and $\frac{dx}{dt} = -1$, evaluate $\frac{dy}{dt}$ when $x = -2$.

ii If $V = s^3$ and $\frac{dV}{dt} = -1.8$, find the value of $\frac{ds}{dt}$ when $s = 5$.

2 i A circular oil spill lies on the surface of calm water. The radius of the spill is increasing at 2.5 m/min. At what rate is the area increasing when the radius is 12 m?

ii A 3-m long ladder is leaning against a wall. Its base is slipping away from the wall at a constant rate of 1.8 ms^{-1}. At what rate, correct to two decimal places, will the top of the ladder be slipping down the wall at the instant the base is 1.2 m out from the wall?

3 i A metal sphere is dissolving in acid at the rate of 12 cm^3/h. As it dissolves it maintains its spherical shape. What is the rate of decrease of the surface area of the sphere when the radius is 8 cm?

ii The volume of a cube is increasing at the rate of 30 cm^3s^{-1}. Find the rate of increase of its surface area when its volume is 4096 cm^3.

Answers ➲ pp. 77–78

YEAR 11 EXAM-TYPE QUESTIONS

Total marks: 35

PART A

(Suggested time: 10 minutes)

(1 mark each)

1 If $y = x^2$ and $\frac{dx}{dt} = 3$, what is the value of $\frac{dy}{dt}$ when $x = 6$?

A 2 B 6 C 12 D 36

2 As $t \to \infty$, which of these is the limiting value of P if $P = -20 + 60e^{-0.23t}$?

A −80 B −20 C 40 D 41

3 The volume V cm^3 of water flowing through a pipe is given by $V = 2t^3 - 4t^2 + t$, where t is in seconds. Which of these is the rate at which the water is flowing when $t = 2$?

A 2 cm^{3}s^{-1} B 7 cm^{3}s^{-1} C 9 cm^{3}s^{-1} D 48 cm^{3}s^{-1}

4 The graph shows the displacement x of a particle moving along a straight line as a function of time t.

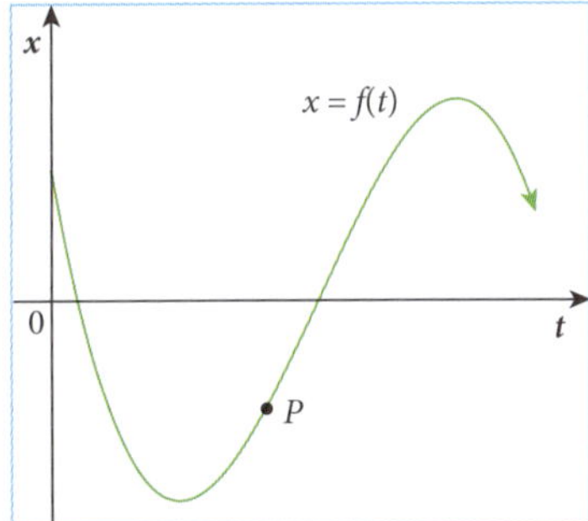

Which of these correctly describes the motion of the particle at P?

A to the right of the origin and moving towards the origin

B to the left of the origin and moving towards the origin

C to the left of the origin and moving away from the origin

D to the right of the origin and moving away from the origin

5 The population P of a city after t years is given by the formula $P = P_0e^{0.02t}$, where P_0 is the initial population. Which of these expressions represents the number of years it takes for the city to double its population?

A 50 ln 0.5 B 25 ln 0.5 C 25 ln 2 D 50 ln 2

PART B

(Suggested time: 45 minutes)

Show all working.

6 The displacement of a particle moving along the x-axis is given by $x = 4 + 8t - t^2$, where x is the displacement from the origin in metres and t is the time in seconds.

a Find the time when the particle is at rest. (1 mark)

b Show that the acceleration of the particle is constant. (1 mark)

c Find the distance travelled by the particle during the first 6 seconds. (2 marks)

7 A council worker accidentally spreads a toxic chemical on a local sports field. The concentration of the chemical in the soil was initially measured at 12 kg/ha. One month later the concentration was found to be 8.6 kg/ha. It is known that the concentration C is given by $C = C_0e^{-kt}$, where C_0 and k are constants, and t is measured in months.

a Evaluate C_0 and k. (2 marks)

b It is safe to use the field when the concentration is below 1.4 kg/ha. After how long must the players wait after the accident before the field can be used? Give your answer to the nearest month. (2 marks)

8 A melting iceberg is decreasing in volume at the rate of $\frac{\pi}{4}$ m^3/h, but it is always conical in shape and its semi-vertical angle remains constant. Initially the iceberg has a base radius 16 m and height 24 m.

a Explain why $r = \frac{2h}{3}$. (1 mark)

b Find the rate at which the height is decreasing when the height of the iceberg is 9 m. (3 marks)

c Find the rate at which the base area is decreasing when the radius is 8 m. (3 marks)

9 According to Boyle's Law, the pressure of a gas is given by the formula $P = \frac{k}{V}$, where k is a constant and V is the volume of the gas. If $k = 60$ for a particular gas, find the rate of change in the pressure, $\frac{dP}{dV}$, when $V = 37.5$. (2 marks)

10 A spherical metal ball is being heated such that the volume increases at a rate of 6π mm^3/min. At what rate is the surface area increasing when the radius is 4 mm? (3 marks)

11 A particle is moving so that its displacement is given by the expression $x = 4t^3 - t^4$, where x is the displacement, in cm, from the origin O and t is time in seconds.

a Show that the particle starts at rest from the origin. (1 mark)

b Find when the particle again returns to O and its speed at that time. (2 marks)

c What is the greatest distance the particle is from O during the time from when it starts to when it returns to O? (3 marks)

12 At any time t minutes, the rate of cooling of a body with temperature T, when the surrounding temperature is S, is given by the differential equation $\frac{dT}{dt} = -k(T - S)$, for some constant k.

a Show that $T = S + Ae^{-kt}$, for some constant A, satisfies this differential equation. (1 mark)

b A metal rod has an initial temperature of 1260 °C and cools to 810 °C in 15 minutes when the surrounding temperature is 18 °C. Find how much longer it will take the rod to cool to 65 °C, giving your answer to the nearest minute. (3 marks)

PRACTICE

1 Rates of change with respect to time ➲ pp. 64–65

1 i $\frac{dQ}{dt} = -11 - 8t$

ii $V = 3t(t+2)^{-\frac{1}{2}}$

$\frac{dV}{dt} = 3.(t+2)^{-\frac{1}{2}} - \frac{1}{2}(t+2)^{-\frac{3}{2}}.1.3t$

$= \frac{3}{2}(t+2)^{-\frac{3}{2}}[2(t+2) - t]$

$= \frac{3}{2}(t+2)^{-\frac{3}{2}}[t+4]$

$= \frac{3(t+4)}{2\sqrt{(t+2)^3}}$

2 i $A = 6t^2 + 2t,$

$\frac{dA}{dt} = 12t + 2$

Let $t = 4$: $\frac{dA}{dt}(4) = 12(4) + 2$

$= 50$

∴ the area is increasing at 50 cm²/min.

ii $\frac{dV}{dt} = 4800 \times 2\left(1 - \frac{t}{60}\right)\cdot\frac{-1}{60}$

$= -160\left(1 - \frac{t}{60}\right)$

Let $t = 15$: $\frac{dV}{dt}(15) = -160\left(1 - \frac{15}{60}\right)$

$= -120$

∴ water is draining at 120 L/min.

3 i a Consider when $v = \frac{dx}{dt} = 0$.

Turning points at (1, 3) and (5, −1).

∴ particle at rest after 1 second at $x = 3$ and after 5 seconds at $x = -1$.

b Initially the particle is at $x = 2$, moves to $x = 3$, then to $x = -1$.

Distance = 3 − 2 + 4

= 5

∴ particle moves 5 units.

c

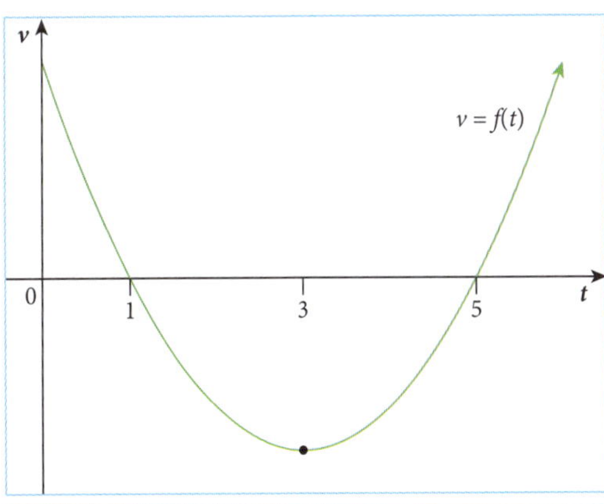

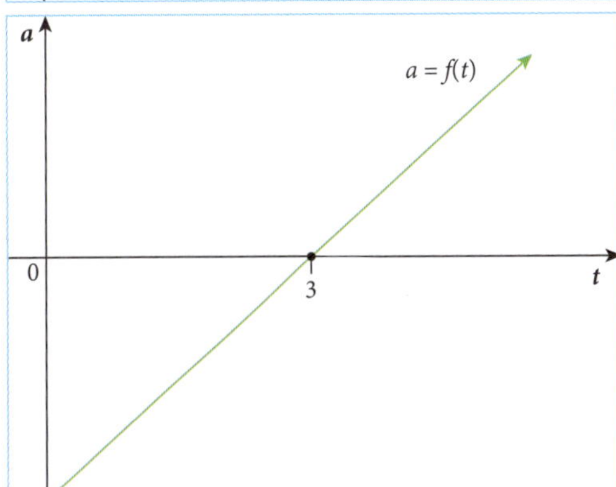

ii a Stationary when $v = 0$: at $t = 2$

∴ object is stationary after 2 seconds.

b $v = 4 - 2t$

$\therefore a = \frac{dv}{dt} = -2$

4 i a $v = 3t^2 - 6t - 9 = 0$

$t^2 - 2t - 3 = 0$

$(t-3)(t+1) = 0$

$t = 3, -1$

But $t \geq 0$, hence, $t = 3$.

∴ at rest on one occasion.

b Let $t = 0$: $x(0) = 0^3 - 3(0)^2 - 9(0) - 1$

$= -1$

Let $t = 3$: $x(3) = 3^3 - 3(3)^2 - 9(3) - 1$

$= -28$

Let $t = 4$: $x(4) = 4^3 - 3(4)^2 - 9(4) - 1$

$= -21$

Total distance = 27 + 7

= 34

$\therefore$ **particle moves 34 cm.**

ii a $v = 4t^3 - 4 = 0$

$4t^3 = 4$

$t = 1$

$\therefore x(1) = 1^4 - 4(1) - 3$

$= -6$

$\therefore$ **object is 6 m to the left of the origin.**

b $a = 12t^2 = 24$

$t^2 = 2$

$t = \sqrt{2}$ (as $t \geq 0$)

$\therefore v(\sqrt{2}) = 4(\sqrt{2})^3 - 4$

$= 8\sqrt{2} - 4$

$= 4(2\sqrt{2} - 1)$

$\therefore$ **object is moving at $4(2\sqrt{2} - 1)$ ms^{-1}.**

5 i a Let $t = 0$: $v(0) = \sqrt{4(0)+1}$

$= 1$

$\therefore$ **the particle is initially moving at 1 cms^{-1}.**

b $v = (4t+1)^{\frac{1}{2}}$

$a = \frac{1}{2}(4t+1)^{-\frac{1}{2}}.4$

$= 2(4t+1)^{-\frac{1}{2}}$

$= \frac{2}{\sqrt{4t+1}}$

Let $t = 2$: $a(2) = \frac{2}{\sqrt{4(2)+1}}$

$= \frac{2}{3}$

$\therefore$ **the particle is accelerating at $\frac{2}{3}$ cms^{-2}.**

ii a Let $x = 0$: $0 = \frac{t-1}{t+1}$

$t - 1 = 0$

$t = 1$

$\therefore$ after 1 second the body passes through the origin.

$v = \frac{dx}{dt}$

$= \frac{(t+1).1-(t-1).1}{(t+1)^2}$ (using quotient rule)

$= \frac{t+1-t+1}{(t+1)^2}$

$= \frac{2}{(t+1)^2}$

Subs. $t = 1$: $v(1) = \frac{2}{(t+1)^2}$

$= 0.5$

$\therefore$ **the body passes through the origin at a speed of 0.5 cms^{-1}.**

b As $v = 2(t + 1)^{-2}$

$a = \frac{dv}{dt}$

$= -4(t + 1)^{-3}$

$= \frac{-4}{(t+1)^3}$

2 Exponential growth and decay: $N(t) = Ae^{kt}$ ➲ pp. 66–67

1 i

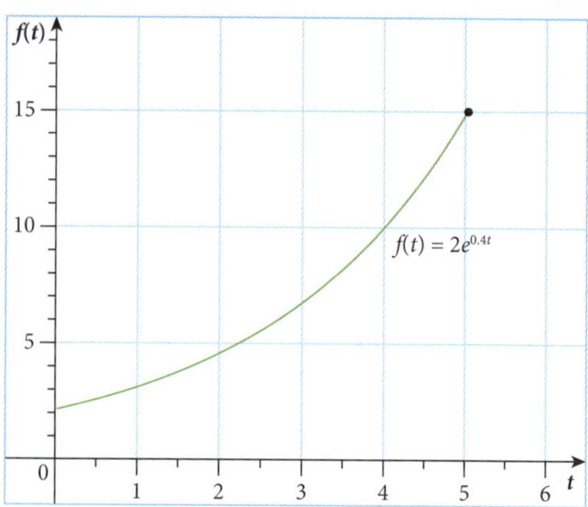

ii

Q

40

30

20

10

0

2 4 6 8 10

t

$Q = 40e^{-0.2t}$

2 i $N = Ae^{-kt}$

$\frac{dN}{dt} = -kAe^{-kt}$

$= \mathbf{-kN}$

ii $P = 8000e^{kt}$

$\frac{dP}{dt} = k \times 8000e^{kt}$

$= \mathbf{kP}$

3 i a Let $t = 0$: $P(0) = 20\,000e^{0.0458(0)}$

$= 20\,000$

$\therefore$ **the initial population is 20 000.**

b Let $t = 20$: $P(20) = 20\,000e^{0.0458(20)}$

$= 49\,985.465\,52\ldots$

$= 49\,985$ (nearest whole)

$\therefore$ **the population is 49 985.**

c Let $P = 500\,000$: $500\,000 = 20\,000e^{0.0458t}$

$e^{0.0458t} = 25$

$0.0458t = \ln 25$

$t = \frac{\ln 25}{0.0458}$

$= 70.281\,131\,55\ldots$

$= 70.3$ (1 dec. pl.)

$\therefore$ **it takes 70.3 hours.**

ii a $A_0 = 100 \quad \therefore A = 100e^{-0.0646t}$

Let $t = 15$: $A(15) = 100e^{-0.0646(15)}$

$= 37.946\,23\ldots$

$= 37.95$ (2 dec. pl.)

$\therefore$ **the amount is 37.95 L.**

b Let $A = 0.5$: $0.5 = 100e^{-0.0646t}$

$e^{-0.0646t} = 0.005$

$-0.0646t = \ln 0.005$

$t = \dfrac{\ln 0.005}{-0.0646}$

$= 82.017\,2967\ldots$

$= 82.02$ (2 dec. pl.)

$\therefore$ **it takes 82.02 days.**

4 i a As $P_0 = 12\,000$, then $P = 12\,000e^{kt}$

Let $P = 24\,000$, $t = 4$:

$24\,000 = 12\,000e^{k(4)}$

$e^{4k} = 2$

$4k = \ln 2$

$\mathbf{k = \dfrac{\ln 2}{4}}$

b Let $P = 36\,000$:

$36\,000 = 12\,000e^{kt}$

$e^{kt} = 3$

$kt = \ln 3$

$t = \dfrac{\ln 3}{k}$

$= 6.339\,850\,003\ldots$

= 6 hours 20 minutes (nearest minute)

c Consider $P = 120\,000$

$\dfrac{dP}{dt} = kP$

$= k(120\,000)$

$= 20\,794.415\,42$

$= 20\,794$ cells/hour

$\therefore$ **population increasing at 20 794 cells/hour.**

ii a Let $M = M_0e^{-kt}$

As $M_0 = 60$, $M = 50$, $t = 12$:

$\therefore 50 = 60e^{-k(12)}$

$e^{-12k} = \dfrac{5}{6}$

$-12k = \ln\dfrac{5}{6}$

$k = \dfrac{\ln\frac{5}{6}}{-12}$

$= 0.015\,193\,463\ldots$

$=$ **0.0152 (4 sig. figs)**

b Let $M = 20$:

$\dfrac{dP}{dt} = -kP$

$= -k(20)$

$= -0.303\,869\,261\ldots$

$= -0.3039$ (4 sig. figs)

$\therefore$ **the mass is decreasing at 0.3039 kg/year.**

c Let $M = 30$:

$\therefore 30 = 60e^{-kt}$

$e^{-kt} = \dfrac{1}{2}$

$-kt = \ln 0.5$

$t = \dfrac{\ln 0.5}{-k}$

$= 45.621\,4082\ldots$

$= 46$ (nearest whole)

$\therefore$ **it will take 46 years.**

3 Further exponential growth and decay: $N(t) = P + Ae^{kt}$ ➲ pp. 68–69

1 i

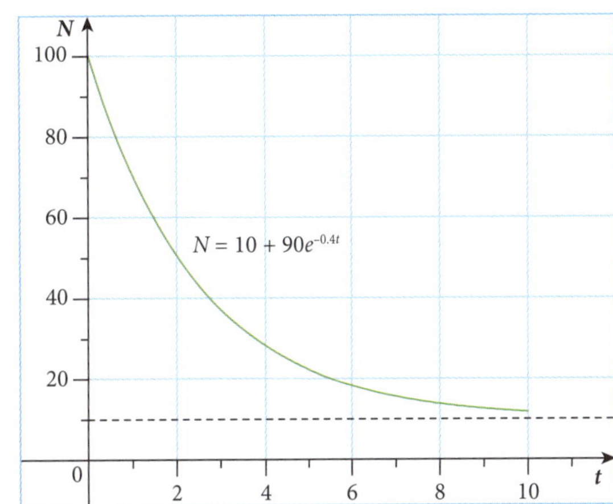

ii

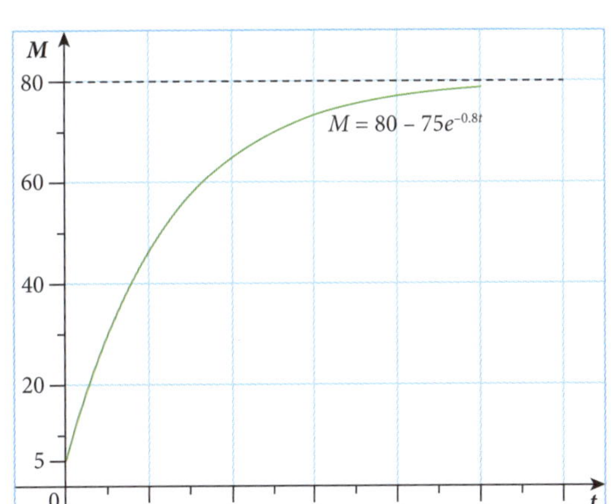

2 i $N = 6 + 19e^{0.75t}$

$\dfrac{dN}{dt} = 0.75 \times 19e^{0.75t}$

But, $19e^{0.75t} = N - 6$

$\therefore \mathbf{\dfrac{dN}{dt} = 0.75(N - 6)}$

ii $V = 20 - 20e^{-kt}$

$\dfrac{dV}{dt} = k.20e^{-kt}$

But, $20e^{-kt} = 20 - V$

$\therefore \mathbf{\dfrac{dV}{dt} = k(20 - V)}$

3 i a When $t \to \infty$, $Be^{-0.046t} \to 0$.

$\therefore 20 = A + 0$

$\therefore A = 20$

$\therefore T = 20 + Be^{-0.046t}$

When $t = 0$ and $T = 90$:

$90 = 20 + B$

$\therefore B = 70$

$\therefore T = 20 + 70e^{-0.046t}$

$\therefore$ **$A = 20$, $B = 70$**

b Let $t = 30$: $T(30) = 20 + 70e^{-0.046(30)}$

$= 37.610\,498\,71\ldots$

$= 37.6$ (1 dec. pl.)

$\therefore$ **the temperature is 37.6 °C.**

c Let $T = 25$: $25 = 20 + 70e^{-0.046t}$

$70e^{-0.046t} = 5$

$e^{-0.046t} = \dfrac{5}{70}$

$-0.046t = \ln\dfrac{1}{14}$

$t = \dfrac{\ln\frac{1}{14}}{-0.046}$

$= 57.370\,811\,51\ldots$

$= 57.37$ (2 dec. pl.)

$\therefore$ **it will take 57.37 minutes.**

ii a When $t \to \infty$, $Be^{-0.1187t} \to 0$.

$\therefore 3.3 = A - 0$

$\therefore A = 3.3$

$\therefore H = 3.3 - Be^{-0.1187t}$

When $t = 0$ and $H = 0.9$:

$0.9 = 3.3 - B$

$\therefore B = 2.4$

$\therefore$ **$H = 3.3 - 2.4e^{-0.1187t}$**

b Let $t = 4$: $H = 3.3 - 2.4e^{-0.1187(4)}$

$= 1.807\,177\,33\ldots$

$= 1.81$ (2 dec. pl.)

$\therefore$ **the height is 1.81 m.**

c Let $H = 3$: $3 = 3.3 - 2.4e^{-0.1187t}$

$2.4e^{-0.1187t} = 0.3$

$e^{-0.1187t} = \dfrac{0.3}{2.4}$

$-0.1187t = \ln 0.125$

$t = \dfrac{\ln 0.125}{-0.1187}$

$= 17.518\,462\,86\ldots$

$= 17.52$ (2 dec. pl.)

$\therefore$ **it will be 17.52 years old.**

4 i a Let $t = 0$, $T = 180$:

$180 = 28 + Be^{-k(0)}$

$180 = 28 + B$

$B = 152$

$\therefore T = 28 + 152e^{-kt}$

Let $t = 30$, $T = 50$:

$50 = 28 + 152e^{-k(30)}$

$152e^{-30k} = 22$

$e^{-30k} = \dfrac{22}{152}$

$-30k = \ln\dfrac{11}{76}$

$k = \dfrac{\ln\frac{11}{76}}{-30}$

$= 0.064\,427\,935\ldots$

$= 0.0644$ (4 dec. pl.)

$\therefore$ **$B = 152$, $k = 0.0644$**

b Let $T = 35$: $35 = 28 + 152e^{-kt}$

$152e^{-kt} = 7$

$-e^{kt} = \dfrac{7}{152}$

$-kt = \ln\dfrac{7}{152}$

$t = \dfrac{\ln\frac{7}{152}}{-k}$

$= 47.773\,847\,54\ldots$

$= 48$ (nearest whole)

10:15 plus 48 minutes is 11:03.

$\therefore$ **the time would be 11:03 am.**

ii a Let $t = 0$, $N = 4000$:

$4000 = 110\,000 - Be^{-k(0)}$

$4000 = 110\,000 - B$

$B = 107\,000$

$\therefore N = 110\,000 - 107\,000e^{-kt}$

Let $t = 6$, $N = 25\,000$:

$25\,000 = 110\,000 - 107\,000e^{-k(6)}$

$107\,000e^{-6k} = 85\,000$

$e^{-6k} = \dfrac{85}{107}$

$-6k = \ln\dfrac{85}{107}$

$k = \dfrac{\ln\frac{85}{107}}{-6}$

$= \dfrac{1}{6}\ln\dfrac{107}{85}$

$\therefore$ **$B = 107\,000$, $k = \dfrac{1}{6}\ln\dfrac{107}{85}$**

b Let $N = 100\,000$: $100\,000 = 110\,000 - 107\,000e^{-kt}$

$107\,000e^{-kt} = 10\,000$

$$-e^{kt} = \frac{10}{107}$$

$$-kt = \ln\frac{10}{107}$$

$$t = \frac{\ln\frac{10}{107}}{-k}$$

$= 61.784\,742\,78\ldots$

$= 1$ year 9.785 weeks (approx.)

1 March 2019 plus 1 year and 9.785 weeks is in May 2020.

$\therefore$ **it would be in May 2020.**

4 Related rates of change ➲ p. 70

1 i $\frac{dy}{dx} = 6x + 5$

$\frac{dy}{dx}(-2) = 6(-2) + 5$

$= -7$

$\frac{dy}{dt} = \frac{dy}{dx} \times \frac{dx}{dt}$

$= -7 \times -1$

$= \mathbf{7}$

ii $\frac{dV}{ds} = 3s^2$

$\frac{dV}{ds}(5) = 3(5)^2$

$= 75$

$\frac{dV}{dt} = \frac{dV}{ds} \times \frac{ds}{dt}$

$-1.8 = 75 \times \frac{ds}{dt}$

$\frac{ds}{dt} = \frac{-1.8}{75}$

$= \mathbf{-0.024}$

2 i $\frac{dr}{dt} = 2.5$

$A = \pi r^2$

$\frac{dA}{dr} = 2\pi r$

$\frac{dA}{dr}(12) = 2\pi\,(12)$

$= 24\pi$

$\frac{dA}{dt} = \frac{dA}{dr} \times \frac{dr}{dt}$

$= 24\pi \times 2.5$

$= 60\pi$

$\therefore$ **the area is increasing at 60π m²/min.**

ii $y = \sqrt{9 - x^2}$

$= (9 - x^2)^{\frac{1}{2}}$

$\frac{dy}{dx} = \frac{1}{2}(9 - x^2)^{-\frac{1}{2}}.-2x$

$= \frac{-x}{\sqrt{9 - x^2}}$

y
3
x

$\frac{dy}{dx}(1.2) = \frac{-1.2}{\sqrt{9 - (1.2)^2}}$

$= -0.436\,435\,78\ldots$

$= -0.436$ (3 dec. pl.)

$\frac{dy}{dt} = \frac{dy}{dx} \times \frac{dx}{dt}$

$= -0.436 \times 1.8$

$= -0.785\,584\,404\ldots$

$= -0.79$ (2 dec. pl.)

$\therefore$ **dropping down the wall at 0.79 ms⁻¹.**

3 i $V = \frac{4}{3}\pi r^3$

$\frac{dV}{dr} = 4\pi r^2$

$\frac{dV}{dr}(8) = 4\pi\,(8)^2$

$= 256\pi$

$\frac{dV}{dt} = \frac{dV}{dr} \times \frac{dr}{dt}$

$-12 = 256\pi \times \frac{dr}{dt}$

$\frac{dr}{dt} = -\frac{3}{64\pi}$

Now, $\frac{dS}{dt} = \frac{dS}{dr} \times \frac{dr}{dt}$

As $S = 4\pi r^2$, then $\frac{dS}{dr} = 8\pi r$

$\frac{dS}{dr}(8) = 8\pi\,(8)$

$= 64\pi$

Now, $\frac{dS}{dt} = 64\pi \times -\frac{3}{64\pi}$

$= -3$

$\therefore$ **the surface area is decreasing at 3 cm²/h.**

ii Let x = side length.

$V = x^3 = 4096$

$x = 16$

$\frac{dV}{dx} = 3x^2$

$\frac{dV}{dx}(16) = 3(16)^2$

$= 768$

$\frac{dV}{dt} = \frac{dV}{dx} \times \frac{dx}{dt}$

$30 = 768 \times \frac{dx}{dt}$

$\frac{dx}{dt} = \frac{5}{128}$

Now, $S = 6x^2$

$\frac{dS}{dx} = 12x$

$\frac{dS}{dx}(16) = 12(16)$

$= 192$

$$\frac{dS}{dt} = \frac{dS}{dx} \times \frac{dx}{dt}$$
$$= 192 \times \frac{5}{128}$$
$$= 7.5$$

$\therefore$ **the surface area is increasing at 7.5 cm²s⁻¹.**

YEAR 11 EXAM-TYPE QUESTIONS

Part A

1 D. $\frac{dy}{dx} = 2x$

$\frac{dy}{dx}(6) = 2(6)$
$= 12$

$\frac{dy}{dt} = \frac{dy}{dx} \times \frac{dx}{dt}$
$= 12 \times 3$
$= 36$

2 B. As $t \to \infty$, $e^{-0.23t} = 0$

$\therefore P = -20 + 60(0)$
$= -20$

3 C. $\frac{dV}{dt} = 6t^2 - 8t + 1$

$\frac{dV}{dt}(2) = 6(2)^2 - 8(2) + 1$
$= 9 \quad \therefore 9 \text{ cm}^3\text{s}^{-1}$

4 B. At P, x is negative, which means it is to the left of the origin. As t increases, it is getting closer to the t-axis and so is moving towards the origin.

5 D. Let $P = 2$, $P_0 = 1$:

$2 = 1e^{0.02t}$

$e^{0.02t} = 2$

$0.02t = \ln 2$

$t = \frac{\ln 2}{0.02}$
$= 50 \ln 2$

Part B

6 a $x = 4 + 8t - t^2$

$v = 8 - 2t = 0$
$2t = 8$
$t = 4$

$\therefore$ **at rest after 4 seconds.** ✓

b $a = \frac{dv}{dt} = -2$

$\therefore$ **acceleration is a constant −2 ms⁻².** ✓

c $x(0) = 4 + 8(0) - (0)^2$
$= 4$

$x(4) = 4 + 8(4) - (4)^2$
$= 20$

$x(6) = 4 + 8(6) - (6)^2$
$= 16$

Distance $= 20 - 4 + 4$
$= 20$

$\therefore$ **the particle travels 20 m.** ✓

7 a When $t = 0$, $C = 12$: $\therefore C_0 = 12$ ✓

$C = 12e^{-kt}$

When $t = 1$, $C = 8.6$:

$8.6 = 12e^{-k(1)}$

$12e^{-k} = 8.6$

$e^{-k} = \frac{8.6}{12}$

$-k = \ln\frac{8.6}{12}$

$k = -\ln\frac{8.6}{12}$
$= 0.333\,144\,446\ldots$
$= $ **0.3331 (4 dec. pl.)** ✓

b Let $C = 1.4$:

$1.4 = 12e^{-kt}$

$e^{-kt} = \frac{1.4}{12}$ ✓

$-kt = \ln\frac{1.4}{12}$

$t = \frac{\ln\frac{1.4}{12}}{-k}$
$= 6.448\,957\,608\ldots$
$= 6.45$ (2 dec. pl.)

$\therefore$ **the field is safe after 6.45 months.** ✓

8 a $\frac{r}{16} = \frac{h}{24}$ (by similar figures)

$\mathbf{r = \frac{2h}{3}}$ ✓

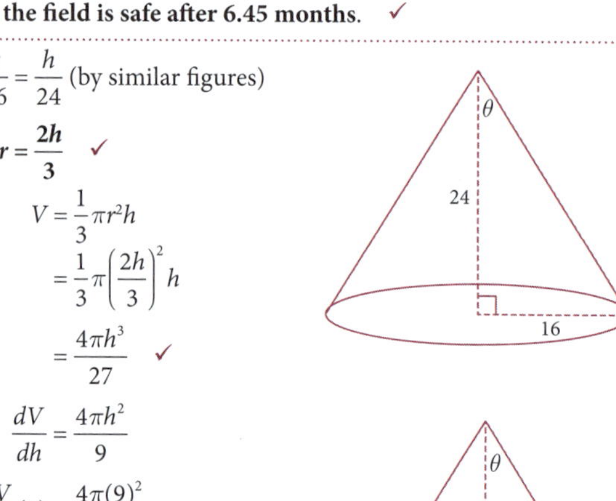

b $V = \frac{1}{3}\pi r^2 h$

$= \frac{1}{3}\pi\left(\frac{2h}{3}\right)^2 h$

$= \frac{4\pi h^3}{27}$ ✓

$\frac{dV}{dh} = \frac{4\pi h^2}{9}$

$\frac{dV}{dh}(9) = \frac{4\pi(9)^2}{9}$
$= 36\pi$ ✓

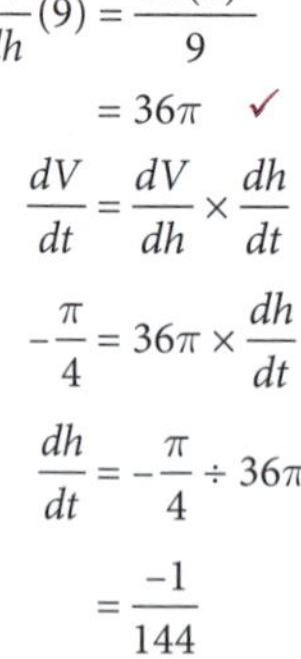

$\frac{dV}{dt} = \frac{dV}{dh} \times \frac{dh}{dt}$

$-\frac{\pi}{4} = 36\pi \times \frac{dh}{dt}$

$\frac{dh}{dt} = -\frac{\pi}{4} \div 36\pi$

$= \frac{-1}{144}$

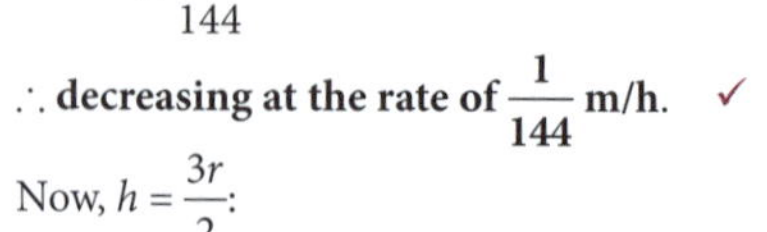

$\therefore$ **decreasing at the rate of $\frac{1}{144}$ m/h.** ✓

c Now, $h = \frac{3r}{2}$:

$V = \frac{1}{3}\pi r^2\left(\frac{3r}{2}\right)$

$= \frac{\pi r^3}{2}$

$\frac{dV}{dr} = \frac{3\pi r^2}{2}$

$$\frac{dV}{dr}(8) = \frac{3\pi(8)^2}{2}$$
$$= 96\pi \checkmark$$
$$\frac{dV}{dt} = \frac{dV}{dr} \times \frac{dr}{dt}$$
$$-\frac{\pi}{4} = 96\pi \times \frac{dr}{dt}$$
$$\frac{dr}{dt} = \frac{-1}{384}$$

Now, $A = \pi r^2$

$$\frac{dA}{dr} = 2\pi r$$
$$\frac{dA}{dr}(8) = 2\pi(8)$$
$$= 16\pi \checkmark$$
$$\frac{dA}{dt} = \frac{dA}{dr} \times \frac{dr}{dt}$$
$$= 16\pi \times \frac{-1}{384}$$
$$= \frac{-\pi}{24}$$

$\therefore$ **decreasing at the rate of $\frac{-\pi}{24}$ m²/h.** ✓

9 $P = 60V^{-1}$

$$\frac{dP}{dV} = -60V^{-2} \checkmark$$
$$= -\frac{60}{V^2}$$
$$= -\frac{60}{37.5^2}$$
$$= -0.0426\,666\,666\ldots$$

$= $ **−0.043 (3 dec. pl.)** ✓

10 $V = \frac{4}{3}\pi r^3$

$$\frac{dV}{dr} = 4\pi r^2$$
$$\frac{dV}{dr}(4) = 4\pi(4)^2$$
$$= 64\pi \checkmark$$
$$\frac{dV}{dt} = \frac{dV}{dr} \times \frac{dr}{dt}$$
$$6\pi = 64\pi \times \frac{dr}{dt}$$
$$\frac{dr}{dt} = \frac{6}{64}$$
$$= \frac{3}{32} \checkmark$$

Now, $S = 4\pi r^2$

$$\frac{dS}{dr} = 8\pi r$$
$$\frac{dS}{dr}(4) = 8\pi(4)$$
$$= 32\pi$$
$$\frac{dS}{dt} = \frac{dS}{dr} \times \frac{dr}{dt}$$
$$= 32\pi \times \frac{3}{32}$$
$$= 3\pi$$

$\therefore$ **increasing at the rate of 3π mm²/min.** ✓

11 a Let $t = 0$: $x(0) = 4(0)^3 - (0)^4$

$$= 0$$
$$v = 12t^2 - 4t^3$$
$$v(0) = 12(0)^2 - 4(0)^3$$
$$= 0 \checkmark$$

$\therefore$ **the particle starts at rest at the origin.**

b $x = 4t^3 - t^4 = 0$

$$t^3(4 - t) = 0$$
$$t = 0, 4$$

$\therefore$ the particle returns to the origin after 4 seconds. ✓

$$v(4) = 12(4)^2 - 4(4)^3$$
$$= -64$$

$\therefore$ **the speed of the particle is 64 cms⁻¹.** ✓

c $v = 12t^2 - 4t^3 = 0$ ✓

$$4t^2(3 - t) = 0$$
$$t = 0, 3$$

$\therefore$ the particle stopped, and turned around after 3 seconds. ✓

Let $t = 3$: $x(3) = 4(3)^3 - (3)^4$

$$= 27$$

$\therefore$ **the greatest distance the particle is from the origin is 27 cm.** ✓

12 a $T = S + Ae^{-kt}$

$$\frac{dT}{dt} = -k.Ae^{-kt}$$

But, $Ae^{-kt} = T - S$

$$\therefore \frac{dT}{dt} = -k(T - S) \checkmark$$

b Let $t \to \infty$, $T = 18$:

$$18 = S + 0 \therefore S = 18$$
$$\therefore T = 18 + Ae^{-kt}$$

Let $t = 0$, $T = 1260$:

$$1260 = 18 + Ae^{-k(0)}$$
$$A = 1242$$
$$\therefore T = 18 + 1242e^{-kt} \checkmark$$

Let $t = 15$, $T = 810$:

$$810 = 18 + 1242e^{-k(15)}$$
$$1242e^{-15k} = 792$$
$$e^{-15k} = \frac{792}{1242}$$
$$-15k = \ln\frac{44}{69}$$
$$k = \frac{\ln\frac{44}{69}}{-15}$$
$$= 0.029\,994\,458\ldots$$
$$= 0.03 \text{ (2 dec. pl.)} \checkmark$$

Let $T = 65$:

$$65 = 18 + 1242e^{-kt}$$
$$1242e^{-kt} = 47$$
$$e^{-kt} = \frac{47}{1242}$$
$$-kt = \ln\frac{47}{1242}$$
$$t = \frac{\ln\frac{47}{1242}}{-k}$$

$= 109.1645215\ldots$

$= 109$ (nearest whole)

Now, $109 - 15 = 94$.

$\therefore$ **another 94 minutes.** ✓

CHAPTER 6 WORKING WITH COMBINATORICS

TOPIC: COMBINATORICS

YEAR 11

1 Fundamental counting principle (multiplication principle)

→ **If an event can happen in *m* different ways and a second event can happen in *n* different ways,** then those successive events can happen in *mn* different ways.

EXAMPLE 1

A Personal Identification Number (PIN) is composed of four digits from 0 to 9. How many different PINs are possible?

$10 \times 10 \times 10 \times 10 = 10^4$
$= 10\,000$

There are 10 digits.

$\therefore$ there are 10 000 different possibilities.

EXAMPLE 2

A restaurant offers a menu consisting of four entrees, six mains and three desserts. What is the probability that two random people have the same three-course meal?

$4 \times 6 \times 3 = 72$

$\therefore$ there are 72 different three-course meals possible.

$P(\text{second person has the same meal as the first}) = \frac{1}{72}$

Imagine the menu.

EXAMPLE 3

The digits 1, 2, 3, 4 and 5 are used to form three-digit numbers where each digit is only used once.

a How many different numbers can be formed?

b What is the probability that a randomly chosen number is even?

a Number $= 5 \times 4 \times 3$
$= 60$

$\therefore$ there are 60 different numbers possible.

b The last digit has to be either a 2 or a 4. This means there are two possibilities for the last digit.

Number $= 4 \times 3 \times 2$
$= 24$

$P(\text{even number}) = \frac{24}{60}$
$= \frac{2}{5}$

Once the last digit is allocated there are only four digits left.

→ PRACTICE

1 i All postcodes in Australia have four digits and the postcodes for South Australia commence with a 5. How many different postcodes are possible in South Australia?

ii A number plate consists of two letters, then two numbers followed by two letters. How many different number-plate combinations are possible? Leave your answer in index form.

2 i In a History test Lexie and Phillie must complete two essays. For the first essay they can choose from options marked as A, B, C or D. Their second essay is to be chosen from three options marked 1, 2 and 3. What is the probability that Lexie and Phillie randomly choose the same two essay options?

ii A security door unlocks when a person presses a certain combination of four digits. I know the first digit is a 7 and the last digit is odd but I have to guess the other digits. What is the probability that I will be able to unlock the door on my second attempt?

3 i The digits 2, 3, 4, 5, 6 and 7 are used to form four-digit numbers where each digit is only used once.

a How many different numbers can be formed?

b What is the probability that a randomly chosen number is greater than 5000?

ii The letters a, b, c, d and e are used to form three-letter 'words', where each letter is only used once.

a How many different words are possible?

b What is the probability that a word starts and ends with a vowel?

Answers ➲ p. 89

2 Factorial notation

➔ $n! = n(n-1)(n-2)(n-3) \ldots 3.2.1$ and, by convention, $0! = 1$.

➔ **If n objects are placed in a row** there are $n!$ different arrangements.

➔ **If n objects are placed around a circle** there are $\frac{n!}{n} = (n-1)!$ different arrangements.

➔ **The number of different ways of arranging n objects** in which x of the objects are of one kind, y of the objects are of another kind, etc is given by $\frac{n!}{x!\,y!\ldots}$.

EXAMPLE 1

Simplify $\frac{(k+1)!}{(k-1)!}$

$$\frac{(k+1)!}{(k-1)!} = \frac{(k+1).k.(k-1)!}{(k-1)!}$$
$$= k(k+1)$$

Write the first few terms in the factorial product.

EXAMPLE 2

Simplify $(k+2)! - (k+1)!$

$$(k+2)! - (k+1)! = (k+1)!(k+2-1)$$
$$= (k+1)(k+1)!$$

Look for a common factor.

EXAMPLE 3

In how many ways can five boys and three girls be arranged in a row if the girls:

a sit together?

b do not sit together?

a Consider the girls as one unit:

B	B	B	B	B	3 GIRLS

The six units can sit in 6! different ways. The girls can sit in 3! different ways.

Number of ways $= 6! \times 3!$
$= 4320$

$\therefore$ they can sit in 4320 different arrangements.

b Number of ways $= 8! - 4320$
$= 36\,000$

$\therefore$ they can sit in 36 000 different arrangements.

Use boxes to show arrangements.

EXAMPLE 4

A father, a mother and their four children sit at a circular table. What is the probability that the parents do not sit together?

Number of arrangements = 5! When parents sit together: Consider parents as one unit, and parents can swap seats: $4! \times 2$

$\therefore$ Arrangements when parents are not together $= 5! - 4! \times 2$

$$\therefore P(\text{parents not together}) = \frac{5! - 4! \times 2}{5!}$$
$$= \frac{72}{120}$$
$$= \frac{3}{5}$$

First find the number of arrangements when the parents are sitting together.

EXAMPLE 5

How many different ordered arrangements can be made from the word CHEESEBALL?

There are 10 letters, but there are 3 Es and 2 Ls.

$$\therefore \text{Total arrangements} = \frac{10!}{3!.2!}$$
$$= 302\,400$$

Carefully count the repeating letters.

➔ PRACTICE

1 Simplify:

i $\frac{(k+2)!}{k!}$

ii $\frac{(n!)^2}{(n+1)!(n-1)!}$

2 Simplify:

i $k! + (k+1)!$

ii $\frac{2m! - (2m-1)!}{4m-2}$

3 i **Six students arrive at the canteen and stand in the same line.**

a **In how many different ways can they arrange themselves?**

b Tennile is last in line and Carmen is first in line. How many different arrangements are possible?

ii Five people are to travel in a car which can seat two people in the front and three people in the back. There are two licensed drivers in the car.

a Caitlyn is not one of the licensed drivers but she wants to sit in the front. How many arrangements are possible?

b What is the probability that Tom drives and the other licensed driver sits behind him?

4 i Four people sit at a round table. What is the probability that two particular people sit together?

ii Four boys and four girls sit around a table. If the seating is arranged randomly, what is the probability that all the boys sit together?

iii Eight people sit around a table. What is the probability that William and Kate do not sit together?

5 How many different ordered arrangements can be made from the word:

i PERMUTATION?

ii MISSISSIPPI?

Answers ⮌ pp. 89–90

3 Pigeonhole principle

→ **If there are n pigeonholes and $n + 1$ pigeons to go into them**, then at least one pigeonhole must hold two or more pigeons.

→ **More generally, if n pigeons are sitting in k pigeonholes, where $n > k$**, then there is at least one pigeonhole with at least $\frac{n}{k}$ pigeons in it.

EXAMPLE 1

What is the smallest group of people that can be assembled where it is guaranteed that at least two people have the same first and last initials in their name?

As 2 students, then $n = 2$.
Number of initial pairings $(k) = 26 \times 26$
$= 676$
Smallest number $= (n - 1)k + 1$
$= 1 \times 676 + 1$
$= 677$
∴ the group must have at least 677 people.

Determine the pigeons (n) and the pigeonholes (k).

EXAMPLE 2

A drawer contains many socks. There are six red socks, eight blue socks, 10 green socks and seven yellow socks. What is the minimum number of randomly chosen socks to be taken from the drawer until it is guaranteed that three of the socks are the same colour?

As 3 socks, then $n = 3$.
Number of colours $(k) = 4$
Minimum number $= (n - 1)k + 1$
$= 2 \times 4 + 1$
$= 9$
∴ at least nine socks are selected.

RRBBGGYY?

→ PRACTICE

1 i What is the smallest group of people that can be assembled where it is guaranteed that at least three people have the same first and last initials in their name?

ii A magician asked his audience to write a two-digit number on a piece of paper. In an audience of 300, is it certain that at least four people have written the same number? Explain, using calculations.

2 i A bag contains eight blue balls, 10 green balls and seven pink balls. How many balls need to be drawn from the bag to ensure there are five balls of the same colour?

ii A bag contains 50 marbles. There is an equal number of white, blue, red, green and yellow marbles in the bag. What is the minimum number of marbles Lara must randomly choose from the bag to ensure that she gets eight marbles of the same colour?

Answers ⮌ p. 90

4 Permutations

→ **Permutation** ${}^nP_r$ is the number of ordered selections of r objects from a total of n different objects.

→ ${}^nP_r = \frac{n!}{(n-r)!}$

EXAMPLE 1

Express $^nP_2$ in factorial form and simplify.

$$^nP_2 = \frac{n!}{(n-2)!}$$
$$= \frac{n(n-1)(n-2)(n-3)...}{(n-2)(n-3)...}$$
$$= n(n-1)$$

$^nP_r = \frac{n!}{(n-r)!}$

EXAMPLE 2

Find the number of arrangements of four digits that can be formed using the digits 0 to 9 if each can only be used once.

$^{10}P_4 = 5040$
This question could also be solved using $10 \times 9 \times 8 \times 7 = 5040$.

This is an ordered selection of four objects from 10.

→ PRACTICE

1 i Express $^{n+1}P_2$ in factorial form and simplify.
ii Prove $^nP_r + r\,^nP_{r-1} = {}^{n+1}P_r$.

2 i A set of 26 cards, each with a different letter of the alphabet, is placed in a hat and four cards are drawn out at random. Find the number of possible words formed.
ii A normal set of 52 cards is shuffled and five randomly chosen cards are placed from left to right on a table. How many different arrangements are possible?

Answers ⊃ p. 90

5 Combinations

→ **Combination ^nC_r, or $\binom{n}{r}$, is the number of unordered selections of r objects from a total of n different objects.**

- ^nC_r is often pronounced 'n choose r'.
- $^nC_r = \frac{n!}{r!(n-r)!}$

EXAMPLE 1

How many ways can a group of four students be chosen from a class of 28 students?

Number of ways $= {}^{28}C_4$
$= 20\,475$

'28 choose 4'

EXAMPLE 2

A team of four men and three women is to be chosen from a group of nine men and seven women.

a How many different teams are possible?
b What is the probability that Rob and Joanne are both chosen?
c Find the probability that Brittany is chosen but Daniel is not.

a Number of possible teams $= {}^9C_4 \times {}^7C_3 = 4410$

b $P(\text{Rob and Joanne chosen}) = \frac{^8C_3 \times {}^6C_2}{^9C_4 \times {}^7C_3}$
$= \frac{840}{4410}$
$= \frac{4}{21}$

If Rob is chosen then there are '8 men choose 3' others.

c $P(\text{Brittany chosen, Daniel not}) = \frac{^6C_2 \times {}^8C_4}{^9C_4 \times {}^7C_3}$
$= \frac{1050}{4410}$
$= \frac{5}{21}$

If Daniel is not chosen then there are '8 men choose 4' others.

→ PRACTICE

1 i There are 28 students in a class. If a group of 20 students is to be selected for a primary school visit, how many different groups are possible?
ii A team of four girls and three boys is to be selected from 12 girls and 15 boys to form a mixed netball team. How many different ways can the team be formed?

2 i A mixed debating team is comprised of two girls and two boys. The team is to be chosen from a group of eight girls and six boys.

a How many different debating teams are possible?

b What is the probability that Lisa and Karl are on the team?

c Find the probability that Fiona is on the team but Harry is not selected.

ii A school committee is to be randomly made up of two teachers, six students and four parents. If five teachers, 14 students and eight parents offer to be part of the committee, what is the probability that:

a two particular students will be on the committee?

b Mrs Rennie (teacher), Jack (student) and his mother will be on the committee?

Answers ➲ p. 90

6 The binomial expansion and Pascal's triangle

➔ **The coefficients** of $(1 + x)^n$ can be determined using Pascal's triangle, or ${}^nC_r$.

➔ $(1 + x)^n = {}^nC_0 + {}^nC_1x + {}^nC_2x^2 + \ldots + {}^nC_kx^k + \ldots + {}^nC_{n-1}x^{n-1} + {}^nC_nx^n$.

➔ $(a + x)^n = {}^nC_0a^n + {}^nC_1\,a^{n-1}x + {}^nC_2\,a^{n-2}x^2 + \ldots + {}^nC_k\,a^{n-k}\,x^k + \ldots + {}^nC_{n-1}\,ax^{n-1} + {}^nC_nx^n$.

➔ **In the expansion** of $(a + x)^n$, the coefficient of x^k is ${}^nC_k$.

EXAMPLE 1

Prove that ${}^nC_{n-r} = {}^nC_r$.

$$\begin{aligned}\text{LHS} &= {}^nC_{n-r}\\ &= \frac{n!}{(n-r)!(n-(n-r))!}\\ &= \frac{n!}{(n-r)!r!}\\ &= {}^nC_r\\ &= \text{RHS}\end{aligned}$$

Show LHS = RHS.

EXAMPLE 2

Find the coefficient of x^4 of the expansion of $(1 + x)^6$.

$(1 + x)^6 = \ldots + {}^6C_4\,1^2x^4 + \ldots$

$\therefore$ the coefficient is ${}^6C_4 = 15$.

Use ${}^nC_k\,1^{6-k}x^k$.

EXAMPLE 3

Find the coefficient of the fourth term in the expansion of $(1 - 2x)^5$.

The fourth term is the term involving x^3.

$(1 - 2x)^5 = \ldots + {}^5C_3\,1^2(-2x)^3 + \ldots$

$\therefore$ the coefficient is ${}^5C_3\,(-2)^3 = -80$

There are $(n + 1)$ terms in the expansion of $(a + x)^n$.

EXAMPLE 4

Find the term independent of x in the expansion of $\left(x - \frac{1}{x^2}\right)^9$.

Consider ${}^9C_kx^{9-k}.(-x^{-2})^k$

$$\begin{aligned}\therefore\quad x^{9-k}.x^{-2k} &= x^0\\ 9 - k - 2k &= 0\\ 9 - 3k &= 0\\ 3k &= 9\\ k &= 3\end{aligned}$$

$\therefore {}^9C_3x^6.\left(\frac{-1}{x^2}\right)^3 = -{}^9C_3$

The term independent of x is the constant term.

EXAMPLE 5

Expand and simplify $(3 - 2x)^4$.

$$\begin{aligned}(3 - 2x)^4 &= {}^4C_0\,(3)^4(-2x)^0 + {}^4C_1\,(3)^3(-2x)^1 + {}^4C_2\,(3)^2(-2x)^2 + {}^4C_3\,(3)^1(-2x)^3 + {}^4C_4\,(3)^0(-2x)^4\\ &= 81 - 216x + 216x^2 - 96x^3 + 16x^4\end{aligned}$$

Use ${}^4C_k$ for coefficients.

➔ PRACTICE

1 i Show $\frac{{}^nP_r}{r!} = {}^nC_r$.

ii Show ${}^nC_r = {}^{n-1}C_{r-1} + {}^{n-1}C_r$.

2 Find the coefficient of x^3 in the expansion of:

i $(2 + x)^7$

ii $(3 - 2x)^5$

3 Find the coefficient of the fifth term of the expansion of:

i $(1 + 3x)^5$

ii $(2a - 3b)^7$

4 **Find the term independent of x in the expansion of:**

i $\left(2x+\dfrac{3}{x^2}\right)^{12}$

ii $\left(\dfrac{2}{x^3}-3x\right)^4(2-x^3)^2$

5 **Expand and simplify:**

i $(2-5x)^4$

ii $(4x^2-3)^3$

Answers ➲ pp. 90–91

YEAR 11 EXAM-TYPE QUESTIONS

Total marks: 35

PART A

(Suggested time: 10 minutes)

(1 mark each)

1 Jordie, Lyndy and four friends sit themselves at random around a circular table. What is the probability that Jordie and Lyndy are not together?

A $\frac{1}{120}$ **B** $\frac{2}{5}$ **C** $\frac{3}{5}$ **D** $\frac{119}{120}$

2 Find the minimum number of students in an assembly hall to be sure that three of them are born in the same month.

A 9 **B** 13 **C** 25 **D** 36

3 How many arrangements of the letters of the word NUMBER are possible if the B and the R are together in any order?

A 5! **B** $2 \times 5!$ **C** 6! **D** $2 \times 6!$

4 Which of these is equivalent to $\frac{{}^nC_{k+1}}{{}^nC_k}$?

A $\frac{n-k}{k+1}$ **B** $\frac{n}{k+1}$ **C** $\frac{n+k}{k+1}$ **D** $\frac{k}{k+1}$

5 Cards with the numbers 1, 2, 3, 4, 5, 6, 7 and 8 are arranged in a circle. How many arrangements are possible if there are at least two odd numbers together?

A 5040 **B** 4896 **C** 4464 **D** 4268

PART B

(Suggested time: 45 minutes)

Show all working.

6 A security lock has 10 buttons labelled as shown.

A B C D E F G H I J

Each person using the lock has been given a four-letter code.

- **a** How many different codes are possible if letters can be repeated and their order is important? (1 mark)
- **b** How many different codes are possible if letters cannot be repeated and their order is important? (1 mark)
- **c** Suppose the lock operates by holding the four buttons down together, so that order is not important. How many different codes are possible? (1 mark)

7 Find the term independent of x in the expansion of $\left(3x^2 + \frac{2}{x}\right)^{12}$. (2 marks)

8 Mr and Mrs Smithers and their five children go to the movies. They are randomly allocated seven adjacent seats in a single row. What is the probability that the five children are allocated seats next to each other? (2 marks)

9 From a group of 10 boys and 10 girls, eight are chosen at random to form a group.

- **a** How many different groups of eight can be formed? (1 mark)
- **b** What is the probability that the group consists of exactly four boys? (1 mark)
- **c** If four boys and four girls are chosen to sit in a circle, what is the probability that the boys and girls alternate? (1 mark)

10 A group of 10 people arrive to eat at a restaurant. The only seating available for them is at two circular tables, one that seats six people and another that seats four.

- **a** Using these tables, how many different seating arangements are there for the 10 people? (2 marks)
- **b** Assuming that the seating arrangement is random, what is the exact probability that a particular couple will be seated at the same table? (3 marks)

11 Solve for n, ${}^nC_2 = 3$. (2 marks)

12 Seventeen students from each of three schools (A, B and C) attend a local student conference. A group of 20 students is then randomly chosen to attend a state conference. Eight students from school A and five students from school B are selected. Find the probability that Jonathon from school B and Anastacia from school C are chosen, but Matea from school A misses out. Leave your answer correct to three decimal places. (2 marks)

13 At an aquarium shop there is a tank of 16 angelfish and another tank of 14 clownfish. A customer chooses six fish at random. What is the probability that the customer will buy at least four angelfish? Leave your answer correct to two decimal places. (2 marks)

14 How many ways can the letters of MATHEMATICS be arranged if:

a there are no restrictions? (1 mark)

b the vowels are grouped together? (1 mark)

15 From a group of six men and eight women a committee of four is to be chosen. Find the probability that the committee contains at least one woman. (2 marks)

16 There are two rows of chairs, with four in the first row and five in the second row. How many ways can nine people be seated if:

a Sonia and Marvin must sit in the second row? (1 mark)

b Grace must sit in the first row and Sarah and Craig must sit in separate rows? (2 marks)

17 The letters of the word SERIES are written in different arrangements. How many possible arrangements are there if:

a the two letter Es are to be together? (1 mark)

b no two consonants are together? (1 mark)

ANSWERS

PRACTICE

1 Fundamental counting principle ➲ pp. 81–82

1 i $1 \times 10 \times 10 \times 10 = 10^3$

$= 1000$

$\therefore$ **there are 1000 different possibilities.**

ii $26 \times 26 \times 10 \times 10 \times 26 \times 26 = 26^4 \times 10^2$

$\therefore$ **there are $26^4 \times 10^2$ different possibilities.**

2 i $4 \times 3 = 12$

$\therefore$ there are 12 different pairs of essays possible.

***P*(second girl chooses the same as the first)** $= \mathbf{\frac{1}{12}}$

ii $10 \times 10 \times 5 = 500$

$\therefore$ there are 500 different possibilities.

$P(\text{fail and then succeed}) = \frac{499}{500} \times \frac{1}{499}$

$= \mathbf{\frac{1}{500}}$

3 i a Number $= 6 \times 5 \times 4 \times 3$

$= 360$

$\therefore$ **there are 360 different numbers possible.**

b The first digit has to be either a 5, 6 or 7. This means there are three possibilities for the first digit.

Number $= 3 \times 5 \times 4 \times 3$

$= 180$

$P(\text{greater than 500}) = \frac{180}{360}$

$= \frac{1}{2}$

ii a Number $= 5 \times 4 \times 3$

$= 60$

$\therefore$ **there are 60 different words possible.**

b There are two vowels, so that two different letters can start the word and the other vowel ends the word. There are three consonants for the middle position.

Number $= 2 \times 3 \times 1$

$= 6$

$P(\text{starts and ends with vowel}) = \frac{6}{60}$

$= \mathbf{\frac{1}{10}}$

2 Factorial notation ➲ pp. 82–83

1 i $\frac{(k+2)!}{k!} = \frac{(k+2)(k+1).k!}{k!}$

$= \mathbf{(k+2)(k+1)}$

ii $\frac{(n!)^2}{(n+1)!(n-1)!} = \frac{n! \times n(n-1)!}{(n+1).n! \times (n-1)!}$

$= \mathbf{\frac{n}{n+1}}$

2 i $k! + (k+1)! = k! + (k+1)k!$

$= k!(1 + k + 1)$

$= \mathbf{k!(k+2)}$

ii $\frac{2m! - (2m-1)!}{4m-2} = \frac{2m(2m-1)! - (2m-1)!}{4m-2}$

$= \frac{(2m-1)!(2m-1)}{2(2m-1)}$

$= \mathbf{\frac{(2m-1)!}{2}}$

3 i a 6 | 5 | 4 | 3 | 2 | 1

Number of ways $= 6!$

$= 720$

$\therefore$ **they can stand in 720 different arrangements.**

b 1 | 4 | 3 | 2 | 1 | 1

Number of ways $= 4!$

$= 24$

$\therefore$ **they can stand in 24 different arrangements.**

ii a
C X
X X X

Number of arrangements $= 2 \times 3!$

$= 12$

$\therefore$ **there are 12 different arrangements.**

b
X T
X X D

Arrangements with Tom driving and other driver behind $= 3!$

Total arrangements $= 2 \times 4!$

$P(\text{event}) = \frac{3!}{2 \times 4!}$

$= \mathbf{\frac{1}{8}}$

4 i
X X
X X

Number of arrangements $= 3!$

Number of arrangements when the two people are together $= 2! \times 2!$

$= 4$

$P(\text{two people together}) = \frac{2! \times 2!}{3!}$

$= \mathbf{\frac{2}{3}}$

ii
X X X X
X X X X

Total arrangements $= 7!$

$P(\text{all boys together}) = \frac{4! \times 4!}{7!}$

$= \mathbf{\frac{4}{35}}$

iii P(William and Kate not together)

$= 1 - P$(William and Kate together)

$= 1 - \frac{2 \times 6!}{7!}$

$= \mathbf{\frac{5}{7}}$

X X X X
X X X X

5 i There are 11 letters, but there are two Ts.

$\therefore$ Total arrangements $= \dfrac{11!}{2!}$

$= \mathbf{19\,958\,400}$

ii There are 11 letters, but there are four Is, four Ss and two Ps.

$\therefore$ Total arrangements $= \dfrac{11!}{4!.4!.2!}$

$= \mathbf{34\,650}$

3 Pigeonhole principle ➲ p. 83

1 i As three students, then $n = 3$.

Number of initial pairings $(k) = 26 \times 26$

$= 676$

Smallest number $= (n - 1)k + 1$

$= 2 \times 676 + 1$

$= 1353$

$\therefore$ **the group must have at least 1353 people.**

ii As four people, then $n = 4$.

Possible numbers $(k) = 10 \times 10$

$= 100$

Smallest audience $= (n - 1)k + 1$

$= 3 \times 100 + 1$

$= 301$

$\therefore$ **no, there must be at least 301 people in the audience.**

2 i As five balls, then $n = 5$.

Number of colours $(k) = 3$

Minimum number $= (n - 1)k + 1$

$= 4 \times 3 + 1$

$= 13$

$\therefore$ **at least 13 balls are drawn.**

ii As eight marbles, then $n = 8$.

Number of colours $(k) = 5$

Minimum number $= (n - 1)k + 1$

$= 7 \times 5 + 1$

$= 36$

$\therefore$ **at least 36 marbles need to be selected.**

4 Permutations ➲ p. 84

1 i $^{n+1}P_2 = \dfrac{(n+1)!}{(n+1-2)!}$

$= \dfrac{(n+1)!}{(n-1)!}$

$= \dfrac{(n+1)n(n-1)(n-2)...}{(n-1)(n-2)...}$

$= \mathbf{n(n+1)}$

ii LHS $= {}^nP_r + r\,{}^nP_{r-1}$

$= \dfrac{n!}{(n-r)!} + r.\dfrac{n!}{(n-r+1)!}$

$= \dfrac{n!(n-r+1)+n!.r}{(n-r+1)!}$

$= \dfrac{n!(n-r+1+r)}{(n-r+1)!}$

$= \dfrac{n!(n+1)}{(n-r+1)!}$

$= \dfrac{(n+1)!}{(n+1-r)!}$

$= {}^{n+1}P_r$

$=$ **RHS**

2 i $\mathbf{^{26}P_4 = 358\,800}$

ii $\mathbf{^{52}P_5 = 311\,875\,200}$

5 Combinations ➲ pp. 84–85

1 i **Number of groups =** $\mathbf{^{28}C_{20}}$

ii **Number of ways =** $\mathbf{^{12}C_4 \times {}^{15}C_3}$

2 i a Possible teams $= {}^8C_2 \times {}^6C_2$

$= \mathbf{420}$

b P(Lisa and Karl on team) $= \dfrac{^7C_1 \times {}^5C_1}{^8C_2 \times {}^6C_2}$

$= \dfrac{35}{420}$

$= \mathbf{\dfrac{1}{12}}$

c P(Fiona on team but Harry not) $= \dfrac{^7C_1 \times {}^5C_2}{^8C_2 \times {}^6C_2}$

$= \dfrac{70}{420}$

$= \mathbf{\dfrac{1}{6}}$

ii a Possible committees $= {}^5C_2 \times {}^{14}C_6 \times {}^8C_4$

P(2 particular students selected) $= \dfrac{^{12}C_4}{^{14}C_6}$

$= \dfrac{495}{3003}$

$= \mathbf{\dfrac{15}{91}}$

b P(Mrs Rennie, Jack and mother on committee)

$= \dfrac{^4C_1 \times {}^{13}C_5 \times {}^7C_3}{^5C_2 \times {}^{14}C_6 \times {}^8C_4}$

$= \mathbf{\dfrac{3}{35}}$

6 The binomial expansion and Pascal's triangle ➲ pp. 85–86

1 i LHS $= \dfrac{^nP_r}{r!}$

$= \dfrac{n!}{(n-r)!} \div r!$

$= \dfrac{n!}{r!(n-r)!}$

$= {}^nC_r$

$=$ **RHS**

ii RHS $= {}^{n-1}C_{r-1} + {}^{n-1}C_r$

$$= \frac{(n-1)!}{(r-1)!(n-1-r+1)!} + \frac{(n-1)!}{r!(n-1-r)!}$$

$$= \frac{(n-1)!}{(r-1)!(n-r)!} + \frac{(n-1)!}{r!(n-r-1)!}$$

$$= \frac{r(n-1)! + (n-r)(n-1)!}{r!(n-r)!}$$

$$= \frac{(n-1)![r+n-r]}{r!(n-r)!}$$

$$= \frac{(n-1)!n}{r!(n-r)!}$$

$$= \frac{n!}{r!(n-r)!}$$

$= {}^nC_r$

= LHS

2 i $(2 + x)^7 = \ldots + {}^7C_3 2^4x^3 + \ldots$

$\therefore$ **the coefficient is $^7C_3 \times 2^4 = 560$.**

ii $(3 - 2x)^5 = \ldots + {}^5C_3 3^2(-2x)^3 + \ldots$

$\therefore$ **the coefficient is $^5C_3 \times 3^2 \times (-2)^3 = -720$.**

3 i The fifth term is the term involving x^4.

$(1 + 3x)^5 = \ldots + {}^5C_4 1^1(3x)^4 + \ldots$

$\therefore$ **the coefficient is $^5C_4 \times 3^4 = 405$.**

ii The fifth term is the term involving b^4.

$(2a - 3b)^7 = \ldots + {}^7C_4(2a)^3(-3b)^4 + \ldots$

$\therefore$ **the coefficient is $^7C_4 \times 2^3 \times (-3)^4 = 22\,680$.**

4 i Consider $^{12}C_4(2x)^{12-k}.(3x^{-2})^k$

$\therefore \quad x^{12-k}.x^{-2k} = x^0$

$12 - k - 2k = 0$

$12 - 3k = 0$

$3k = 12$

$k = 4$

$\therefore$ **$^{12}C_4(2x)^8.(3x^{-2})^4 = {}^{12}C_4.2^8.3^4$**

ii First, the constant term for $(2 - x^3)^2$ is $2^2 = 4$.

Now, consider $^4C_k(2x^{-3})^{4-k}.(-3x)^k$

$x^{-12+3k}.x^k = x^0$

$-12 + 3k + k = 0$

$-12 + 4k = 0$

$4k = 12$

$k = 3$

$\therefore {}^4C_3(2x^{-3}).(-3x)^3 \times 4 = {}^4C_3 \times 2 \times -27 \times 4$

$= -864$

5 i $(2 - 5x)^4 = {}^4C_0(2)^4(-5x)^0 + {}^4C_1(2)^3(-5x)^1 + {}^4C_2(2)^2(-5x)^2 + {}^4C_3(2)^1(-5x)^3 + {}^4C_4(2)^0(-5x)^4$

$= 16 - 160x + 600x^2 - 1000x^3 + 625x^4$

ii $(4x^2 - 3)^3 = {}^3C_0(4x^2)^3(-3)^0 + {}^3C_1(4x^2)^2(-3)^1 + {}^3C_2(4x^2)^1(-3)^2 + {}^3C_3(4x^2)^0(-3)^3$

$= 64x^6 - 144x^4 + 108x^2 - 27$

YEAR 11 EXAM-TYPE QUESTIONS

Part A

1 C. $P(\text{not together}) = 1 - P(\text{together})$

$$= 1 - \frac{2 \times 4!}{5!}$$

$$= \frac{3}{5}$$

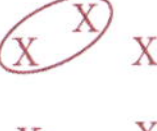

2 C. As three students, then $n = 3$.

Number of months $(k) = 12$

Minimum number $= (n - 1)k + 1$

$= 2 \times 12 + 1$

$= 25$

$\therefore$ There must be at least 25 students in the library.

3 B. Consider B & R as one unit:

$\therefore$ number of arrangements is $5! \times 2$

4 A. $$\frac{^nC_{k+1}}{^nC_k} = \frac{n!}{(k+1)!(n-k-1)!} \div \frac{n!}{k!(n-k)!}$$

$$= \frac{n!}{(k+1)!(n-k-1)!} \times \frac{k!(n-k)!}{n!}$$

$$= \frac{n-k}{k+1}$$

5 B. There are $7! = 5040$ possible arrangements for the 8 cards arranged in a circle. Now the odd and even numbers are alternating in $3! \times 4! = 144$ arrangements. As $5040 - 144$ is 4896, then there are 4896 arrangements.

Part B

6 a Number of codes $= 10^4$

$= 10\,000$ ✓

b Number of codes $= 10 \times 9 \times 8 \times 7$

$= 5040$ ✓

c Number of codes $= {}^{10}C_4$

$= 210$ ✓

7 Consider $^{12}C_k(3x^2)^{12-k}.(2x^{-1})^k$

$\therefore \quad x^{24-2k}.x^{-k} = x^0$ ✓

$24 - 2k - k = 0$

$24 - 3k = 0$

$3k = 24$

$k = 8$

$\therefore {}^{12}C_4(3x^2)^4.(2x^{-1})^8 = {}^{12}C_4.3^4.2^8$ ✓

8 ☐☐ 5 children

As there are 7 seats, the number of arrangements is 7!. ✓

2 parents and the group of children = 3!

Also, the childen can sit in 5! arrangements.

$$P(\text{children sit together}) = \frac{3!.5!}{7!}$$

$$= \frac{\mathbf{1}}{\mathbf{7}}$$ ✓

9 a **Number of groups = $^{20}C_8 = 125\,970$** ✓

b $P(\text{4 boys and 4 girls}) = \dfrac{^{10}C_4 \times {}^{10}C_4}{^{20}C_8}$

$= 0.350\,083\,353\ldots$

= 0.35 (2 dec. pl.) ✓

c **P(4 boys and girls alternating) $= \dfrac{4!.3!}{7!} = \dfrac{1}{35}$** ✓

10 a Number of arrangements $= {}^{10}C_6 \times 5! \times {}^4C_4 \times 3!$ ✓

= 151 200 ✓

b Number of arrangements if both on large table

$= {}^8C_4 \times 5! \times {}^4C_4 \times 3!$ ✓

Number of arrangements if both on small table

$= {}^8C_2 \times 3! \times {}^6C_6 \times 5!$

$P(\text{both on same table}) = \dfrac{^8C_4 \times 5! \times 3! + {}^8C_2 \times 3! \times 5!}{^{10}C_6 \times 5! \times 3!}$ ✓

$= \mathbf{\dfrac{7}{15}}$ ✓

11 $\dfrac{n!}{2!(n-2)!} = 3$

$\dfrac{n(n-1)}{2} = 3$ ✓

$n^2 - n = 6$

$n^2 - n - 6 = 0$

$(n - 3)(n + 2) = 0$

$n = 3, -2$

But $n > 0$, hence $n = 3$. ✓

12 P(Jonathon and Anastacia selected, but Matea not selected)

$= \dfrac{^{16}C_8 \times {}^{16}C_4 \times {}^{16}C_6}{^{17}C_8 \times {}^{17}C_5 \times {}^{17}C_7}$ ✓

$= 0.064\,115\,612\ldots$

= 0.064 (3 dec. pl.) ✓

13 P(at least 4 angelfish) = P(4 angelfish, 2 clownfish) + P(5 angelfish, 1 clownfish) + P(6 angelfish) ✓

$= \dfrac{^{16}C_4 \times {}^{14}C_2 + {}^{16}C_5 \times {}^{14}C_1 + {}^{16}C_6}{^{30}C_6}$

$= 0.395\,402\,298\ldots$

= 0.40 (2 dec. pl.) ✓

14 a Number of ways $= \dfrac{11!}{2!.2!.2!} = 4\,989\,600$ ✓

b [4 vowels][ ][ ][ ][ ][ ][ ][ ]

Consider 4 vowels as a unit, so there are 8 units, with 2 Ms and 2 Ts. Also, the 4 vowels have 2 As.

Number of ways $= \dfrac{8!}{2!2!} \times \dfrac{4!}{2!}$

= 120 960 ✓

15 P(at least one woman) = 1 − P(no woman) ✓

$= 1 - \dfrac{^6C_4}{^{14}C_4}$

$= 1 - 0.014\,985\,014\ldots$

= 0.985 (3 sig. figs) ✓

16 a Sonia can sit on 5 chairs and then Marvin has 4 chairs to choose from. Then the other people have 7! arrangements.

∴ Number of arrangements = 5 × 4 × 7!

= 100 800 ✓

b Grace can sit on 4 chairs, Sarah (front row) has 3 chairs and Craig (second row) has 5 chairs, then other people have 6! chairs. Then Sarah and Craig can swap rows so we multiple by 2. ✓

∴ Number of arrangements = 4 × 3 × 5 × 6! × 2

= 86 400 ✓

17 a [E][E][ ][ ][ ][ ]

The pair of E's can be placed in 5 different places.

The other four letters can be arranged in $\dfrac{4!}{2!}$ ways.

Number of arrangements $= 5 \times \dfrac{4!}{2!}$

= 60 ✓

b [C][ ][C][ ][C][ ]

The consonants can be placed in 2 different ways (1st, 3rd, 5th, or 2nd, 4th, 6th).

The three consonants have 2 S's: $\dfrac{3!}{2!}$, and the vowels have 2 E's: $\dfrac{3!}{2!}$.

Number of arrangements $= 2 \times \dfrac{3!}{2!} \times \dfrac{3!}{2!}$

= 18 ✓

SAMPLE YEAR 11 EXAMINATION

Try to complete this paper as if it is the real thing. These are the instructions you need to follow in the Year 11 exam:

General instructions

- Reading time: 10 minutes
- Working time: 2 hours
- Write using black pen.
- Calculators approved by NESA may be used.
- Attempt all questions.

Total marks: 70

Section I: 10 marks

Section II: 60 marks

Section I: 10 Marks

Allow about 15 minutes for this section.

Objective-response questions

(1 mark each)

Choose the letter A, B, C or D that corresponds to the best answer.

1 What are the number of points of intersection of the graphs $y = |x|$ and $y = |x^2 - 1|$?

A 0 B 2 C 3 D 4

2 The equation $x^3 - x^2 - 8x + 12 = 0$ has a double root at $x = \alpha$. Which of these is the value of α?

A $\alpha = -2$

B $\alpha = -\frac{4}{3}$

C $\alpha = \frac{4}{3}$

D $\alpha = 2$

3 The polynomial $x^3 - x - 4 = 0$ has roots α, β and γ.

Which of these is the value of $\frac{1}{\alpha} + \frac{1}{\beta} + \frac{1}{\gamma}$?

A $-\frac{1}{2}$ B $\frac{1}{4}$ C $\frac{1}{2}$ D $-\frac{1}{4}$

4 Which of the following is the solution of $\frac{x}{x+1} \leq 0$?

A $x \leq 0$

B $-1 < x \leq 0$

C $x > -1$

D $x < -1, x \geq 0$

5 The graph of $y = f(x)$ is shown below.

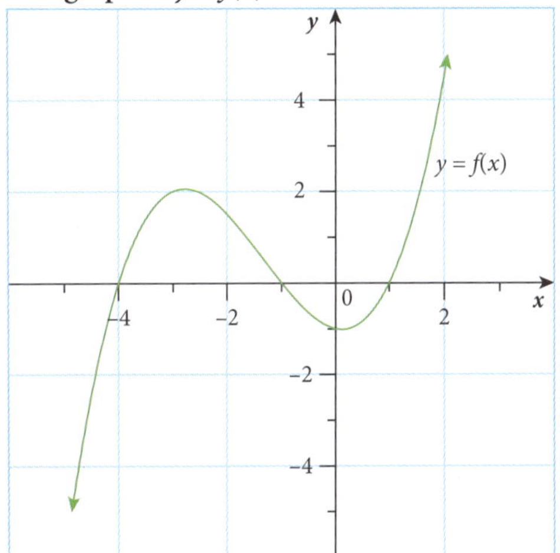

Which of the following could be the graph of $y = f(|x|)$?

A

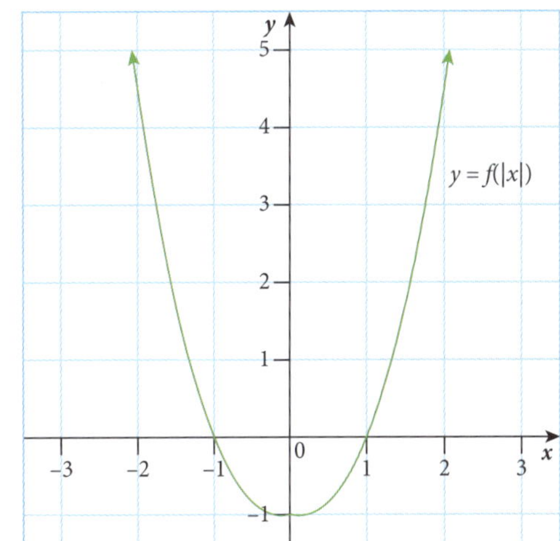

B

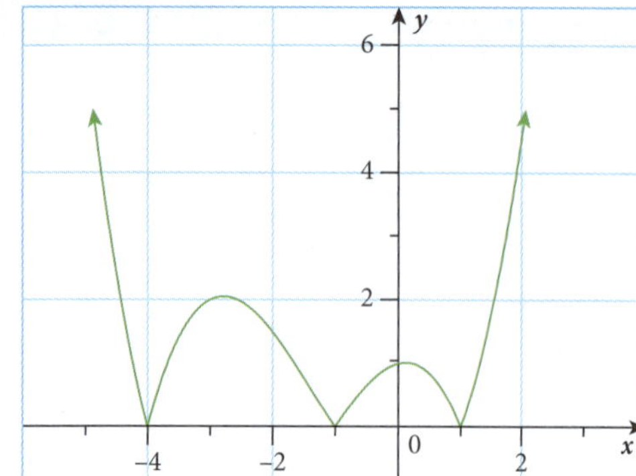

C

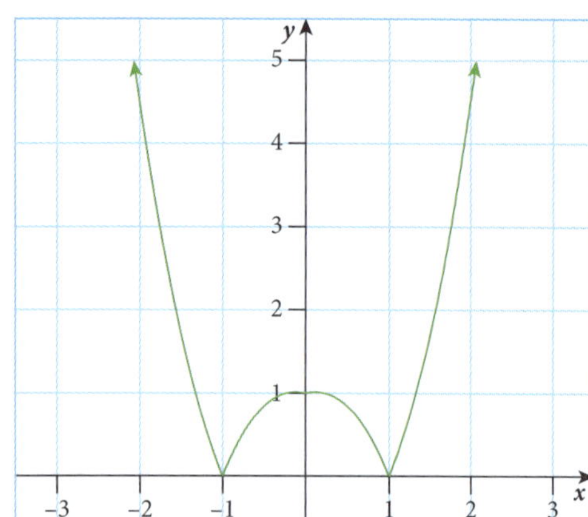

D

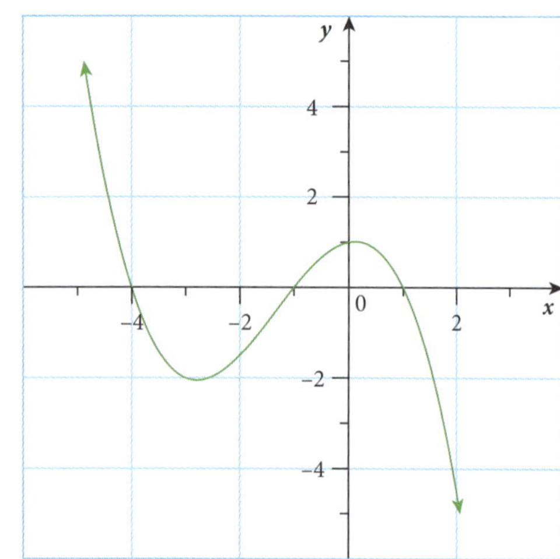

6 Solve $6 - x - 2x^2 \leq 0$.

A $-2 \leq x \leq \frac{3}{2}$ **B** $-2 \leq x \leq \frac{2}{3}$

C $x \leq -2, x \geq \frac{3}{2}$ **D** $x \leq -2, x \geq \frac{2}{3}$

7 Consider the function $f(x) = \sqrt{x+5}$ and its inverse function $f^{-1}(x)$.

Which of these is the value of $f^{-1}(4)$?

A 1 **B** 3 **C** 9 **D** 11

8 Which of these is the exact value of $\cos\frac{5\pi}{12}\sin\frac{\pi}{12}$?

A $\frac{1-\sqrt{3}}{2}$ **B** $\frac{2-\sqrt{3}}{4}$ **C** $\frac{1+\sqrt{3}}{2}$ **D** $\frac{2+\sqrt{3}}{4}$

9 Which of these is the graph of the solution of $|1 - 2x| \leq 5$?

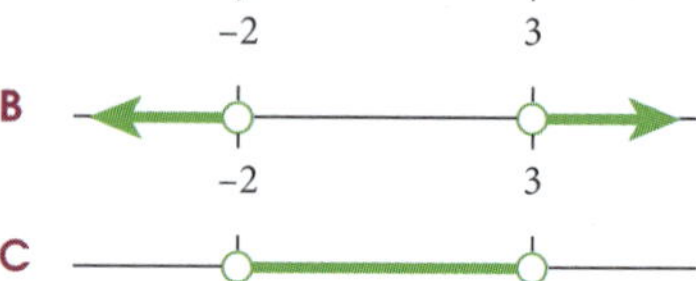

A (number line: -2, 3)

B (number line: -2, 3)

C (number line: -2, 3)

D (number line: -2, 3)

10 When certain crystals are placed in a solution they dissolve in such a way that their volume V mm^3 is given by $V = 20t - \frac{t^2}{4} + 2$, where t is in minutes. After how many minutes are the crystals dissolving at the rate of 10 mm^3/min?

A 8 minutes **B** 15 minutes

C 20 minutes **D** 177 minutes

Section II: 60 marks

Allow about 1 h 45 minutes for this section.

Question 11 (15 marks)

a Find the domain of $f(x) = 4\sin^{-1}\left(\frac{2x-1}{3}\right)$. (2 marks)

b Show that ${}^nC_k + {}^nC_{k-1} = {}^{n+1}C_k$. (2 marks)

c Seven cakes are arranged along a shelf. There are three fairy cakes and four slices. Find the probability that:

- **i** a fairy cake is at each end (1 mark)
- **ii** all the fairy cakes are together (1 mark)

d Find all angles θ, if $0 \leq \theta \leq 2\pi$, for which $\sin 2\theta + 2\sin^2\theta = 0$. (3 marks)

e Solve $\frac{2x+1}{x-2} \leq 1$. (3 marks)

f Find the exact value of $\tan 15°$ in simplified form. (3 marks)

Question 12 (15 marks)

a A particle moving in a straight line is x metres from the origin O at time t seconds, where $x = t^3 - 6t^2 + 9t + 1$.

i Show that the particle is stationary on two occasions. (2 marks)

ii Find the distance travelled in the first 5 seconds. (2 marks)

iii Show that the particle reaches its greatest velocity after 2 seconds. (1 mark)

b Given $P(x) = x^3 - 5x^2 - 8x + 48$.

i Show that $(x + 3)$ is a factor of $P(x)$. (1 mark)

ii Solve $P(x) = 0$. (2 marks)

iii Sketch the graph of $y = x^3 - 5x^2 - 8x + 48$. (1 mark)

iv Hence, or otherwise, find the domain for x for which $x^3 - 5x^2 - 8x + 48 > 0$. (1 mark)

c Functions f and g are defined by $f(x) = 2x + 1$ and $g(x) = 3 - x$.

Find the inverse of the composite function $f \circ g$. (2 marks)

d Given that α is acute and $\sin 2\alpha = \frac{5}{13}$, find $\tan \alpha$. (3 marks)

Question 13 (15 marks)

a A mixed indoor soccer team of six players is to be selected from five girls and four boys. A team must consist of an equal number of boys and girls. Find the number of teams that can be selected such that the team:

i includes Sarah and Mark (1 mark)

ii does not include both Sarah and Mark (1 mark)

b On an island the population at the beginning of 2010 was 5400 and at the beginning of 2016 was 5840. Assume that the population of the island is governed by the equation $P = P_0e^{kt}$, where P_0 and k are constants, and where t is in years.

i Find k, correct to four significant figures. (2 marks)

ii What will the population of the island be at the beginning of 2025? (1 mark)

iii In which year will the population of the island be 9000? (2 marks)

c Show that $\frac{1 - \cos 2x}{\sin 2x} = \tan x$. (2 marks)

d The cubic polynomial $P(x) = x^3 + px^2 + qx + r$, where p, q and r are real numbers, has three zeros 1, α and $-\alpha$. Find the value of:

i p (1 mark)

ii $q + r$ (2 marks)

e A boat is being pulled towards a pier which is 3 m above the surface of the water. The boat is being pulled in at a rate of 1.2 ms^{-1}.

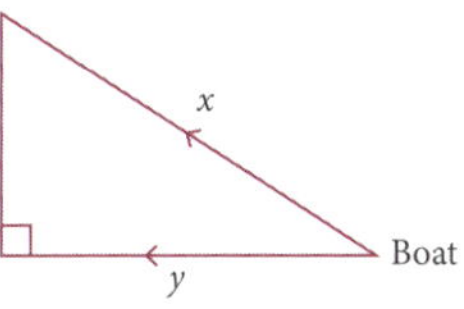

i Show that $\frac{dy}{dx} = \frac{x}{\sqrt{x^2 - 9}}$. (1 mark)

ii How fast is the boat approaching the base of the pier when 5 m of rope remains to be pulled in? (2 marks)

Question 14 (15 marks)

a A function $y = f(x)$ has a double root at $x = -3$ and a single root at $x = 3$. The y-intercept of the curve is 3. Draw separate one-third of a page sketches of:

i $y = f(x)$ (1 mark)

ii $y = \frac{1}{f(x)}$ (2 marks)

iii $y^2 = f(x)$ (2 marks)

b The polynomial $P(x) = x^3 - x^2 - 2x - 3$ has the same remainder when divided by $(x + a)$ and $(x - 2a)$. Find the non-zero values of a. (3 marks)

c When the temperature T of a certain body is 65 °C it is cooling at the rate of 1 °C per minute. Assume Newton's law of cooling, $\frac{dT}{dt} = -k(T - 15)$, where T is the temperature of the body at time t minutes, and k is a constant when the temperature of the surrounding medium is 15 °C.

i Verify that $T = 15 + Ae^{-kt}$ is a solution of the differential equation, where A is a constant. (1 mark)

ii Show that $k = 0.02$. (1 mark)

iii What is the temperature of the body 30 minutes later? Answer to the nearest degree. (2 marks)

d One of the roots of the equation $x^3 + ax^2 + bx + c = 0$ is double the sum of the other two roots. Show that $4a^3 - 18ab + 27c = 0$. (3 marks)

ANSWERS

SAMPLE YEAR 11 EXAMINATION

Section I

1 D.

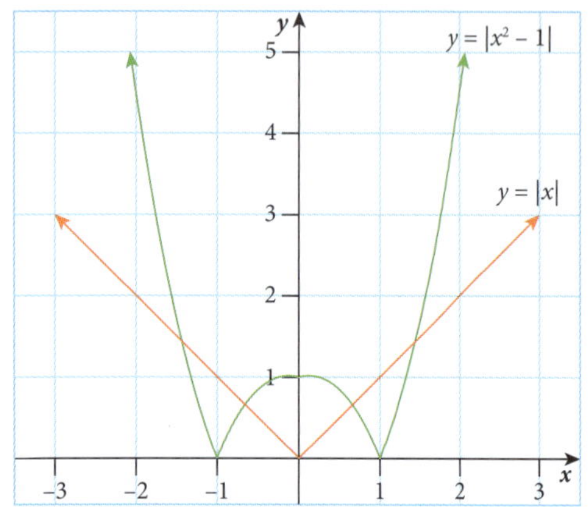

There are 4 points of intersection.

2 D. Let $P(x) = x^3 - x^2 - 8x + 12$

$\therefore P'(x) = 3x^2 - 2x - 8 = 0$

$(3x + 4)(x - 2) = 0$

$x = -\frac{4}{3}, 2$

By inspection, $x = 2$ is also a root of $P(x)$.

3 D. $\frac{1}{\alpha} + \frac{1}{\beta} + \frac{1}{\gamma} = \frac{\beta\gamma + \alpha\gamma + \alpha\beta}{\alpha\beta\gamma}$

$= \frac{-1}{4}$

$= -\frac{1}{4}$

4 B. $(x + 1)^2 \times \frac{x}{x+1} \leq 0$

$x(x + 1) \leq 0$

Let $x(x + 1) = 0$

$x = 0, -1$

Test $x = 1$: $\frac{1}{1+1} \leq 0$ … No!

But $x \neq -1$

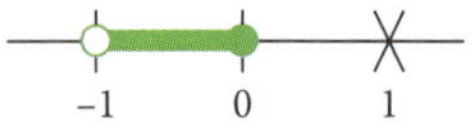

$\therefore$ the solution is $-1 < x \leq 0$.

5 A. The graph on the right side of the y-axis is reflected to the left side.

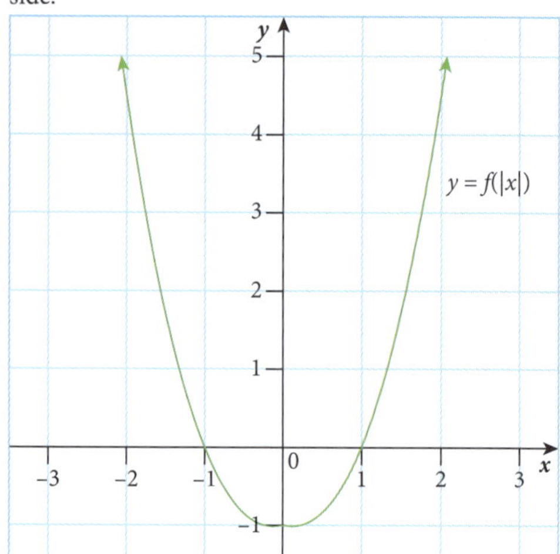

6 C. $6 - x - 2x^2 \leq 0$

$2x^2 + x - 6 \geq 0$

$(2x - 3)(x + 2) \geq 0$

Let $(2x - 3)(x + 2) = 0$

$x = \frac{3}{2}, -2$

Test $x = 0$; $(0 - 3)(0 + 2) \geq 0$? … No!

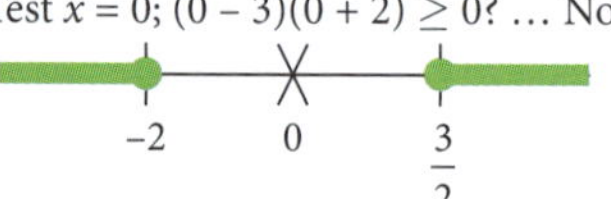

$\therefore$ the solution is $x \leq -2, x \geq \frac{3}{2}$

7 D. For $f^{-1}(4)$, solve $f(x) = 4$:

$\sqrt{x+5} = 4$

$x + 5 = 16$

$x = 11$

8 B. $\cos\frac{5\pi}{12}\sin\frac{\pi}{12} = \frac{1}{2}\left[\sin\left(\frac{5\pi}{12} + \frac{\pi}{12}\right) - \sin\left(\frac{5\pi}{12} - \frac{\pi}{12}\right)\right]$

$= \frac{1}{2}\left[\sin\frac{\pi}{2} - \sin\frac{\pi}{3}\right]$

$= \frac{1}{2}\left[1 - \frac{\sqrt{3}}{2}\right]$

$= \frac{2-\sqrt{3}}{4}$

9 D.

10 C. $\frac{dV}{dt} = 20 - \frac{t}{2} = 10$

$$\frac{t}{2} = 20 - 10$$

$$\frac{t}{2} = 10$$

$$t = 20$$

Section II

11 a Domain of $y = \sin^{-1} x$ is $-1 \leq x \leq 1$.

Hence, domain: $-1 \leq \frac{2x-1}{3} \leq 1$ ✓

$$-3 \leq 2x - 1 \leq 3$$

$$-2 \leq 2x \leq 4$$

$$-1 \leq x \leq 2$$

$\therefore$ **domain is $-1 \leq x \leq 2$.** ✓

b LHS $= {}^nC_k + {}^nC_{k-1}$

$= \frac{n!}{k!.(n-k)!} + \frac{n!}{(k-1)!.(n-k+1)!}$ ✓

$= \frac{n!(n-k+1)+n!.k}{k!.(n-k+1)!}$

$= \frac{n!(n-k+1+k)}{k!.(n-k+1)!}$

$= \frac{n!(n+1)}{k!.(n-k+1)!}$

$= \frac{(n+1)!}{k!.(n+1-k)!}$

$= {}^{n+1}C_k$

$=$ **RHS** ✓

c i [F][ ][ ][ ][ ][ ][F]

$\therefore P(\text{2 fairy cakes on each end}) = \frac{3\times2\times5!}{7!}$

$= \mathbf{\frac{1}{7}}$ ✓

ii [3 fairy cakes][ ][ ][ ][ ]

$\therefore P(\text{fairy cakes together}) = \frac{5!.3!}{7!}$

$= \mathbf{\frac{1}{7}}$ ✓

d $\sin 2\theta + 2\sin^2\theta = 0$

$2\sin\theta\cos\theta + 2\sin^2\theta = 0$

$2\sin\theta(\cos\theta + \sin\theta) = 0$

$2\sin\theta = 0$ $\quad$ $\cos\theta + \sin\theta = 0$

$\sin\theta = 0$ ✓ $\quad$ $1 + \tan\theta = 0$

$\tan\theta = -1$ ✓

$\boldsymbol{\theta = 0, \pi, 2\pi, \frac{3\pi}{4}, \frac{7\pi}{4}}$ ✓

e Multiply both sides by $(x-2)^2$:

$$(x-2)^2 \times \frac{2x+1}{x-2} \leq 1 \times (x-2)^2$$

$$(x-2)(2x+1) \leq (x-2)^2$$

$(x-2)(2x+1) - (x-2)^2 \leq 0$ ✓

$$(x-2)[2x+1-(x-2)] \leq 0$$

$$(x-2)(x+3) \leq 0$$

Let $(x-2)(x+3) = 0$

$x = 2, -3$ ✓

Test $x = 0$: $\frac{0+1}{0-2} \leq 1$? … Yes!

But $x \neq 2$,

$\therefore$ **the solution is $-3 \leq x < 2$.** ✓

f $\tan 15° = \tan(45° - 30°)$

$= \frac{\tan 45° - \tan 30°}{1 + \tan 45° \tan 30°}$ ✓

$= \frac{1 - \frac{1}{\sqrt{3}}}{1 + 1 \times \frac{1}{\sqrt{3}}}$ ✓

$= \frac{\sqrt{3}-1}{\sqrt{3}+1}$

$= \frac{\sqrt{3}-1}{\sqrt{3}+1} \times \frac{\sqrt{3}-1}{\sqrt{3}-1}$

$= \frac{3-2\sqrt{3}+1}{3-1}$

$= \frac{4-2\sqrt{3}}{2}$

$= \mathbf{2 - \sqrt{3}}$ ✓

12 a i $x = t^3 - 6t^2 + 9t + 1$

$v = 3t^2 - 12t + 9 = 0$ ✓

$t^2 - 4t + 3 = 0$

$(t-1)(t-3) = 0$

$t = 1, 3$

$\therefore$ **particle is stationary after 1 second and after 3 seconds.** ✓

ii $x(0) = 0^3 - 6(0)^2 + 9(0) + 1 = 1$ ✓

$x(1) = 1^3 - 6(1)^2 + 9(1) + 1 = 5$

$x(3) = 3^3 - 6(3)^2 + 9(3) + 1 = 1$

$x(5) = 5^3 - 6(5)^2 + 9(5) + 1 = 21$

$\therefore$ particle started at $x = 1$, then to $x = 5$, back to $x = 1$, then to $x = 21$.

$\therefore$ **distance is 4 + 4 + 20 = 28 m.** ✓

iii Greatest velocity when $\frac{dv}{dt} = a = 0$:

$a = 6t - 12 = 0$

$t = 2$

$\therefore$ **greatest velocity after 2 seconds.** ✓

b i $P(-3) = (-3)^3 - 5(-3)^2 - 8(-3) + 48$

$= 0$

$\therefore$ **$(x+3)$ is a factor.** ✓

ii

$$\begin{array}{r} x^2 - 8x + 16 \\ x+3 \overline{\smash{)}\, x^3 - 5x^2 - 8x + 48} \\ \underline{x^3 + 3x^2} \quad\quad\quad\quad \\ -8x^2 - 8x \quad\quad \\ \underline{-8x^2 - 24x} \quad\quad \\ 16x + 48 \\ \underline{16x + 48} \\ 0 \end{array}$$ ✓

$(x+3)(x^2 - 8x + 16) = 0$

$(x+3)(x-4)^2 = 0$

$\mathbf{x = -3, 4}$ ✓

iii $y = (x + 3)(x - 4)^2$:

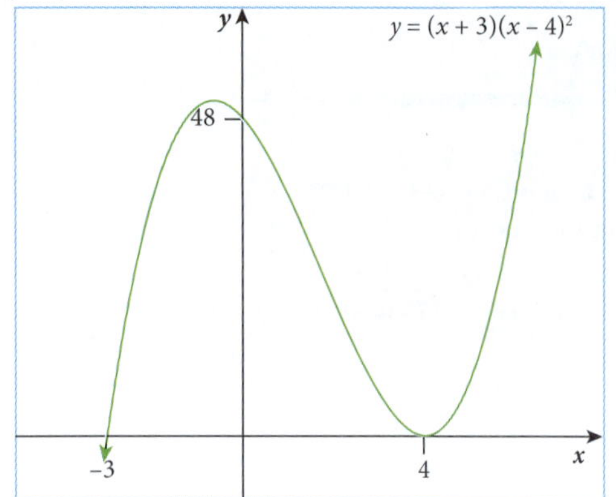

iv **From the graph, $-3 < x < 4$ or $x > 4$.** ✓

c $f \circ g = f(g(x))$

$f \circ g = 2(3 - x) + 1$

$= 6 - 2x + 1$

$= 7 - 2x$ ✓

Let $y = 7 - 2x$

$(f \circ g)^{-1} : x = 7 - 2y$

$2y = 7 - x$

$y = \dfrac{7 - x}{2}$

$\therefore (f \circ g)^{-1} = \dfrac{7 - x}{2}$ ✓

d $\tan 2\alpha = \dfrac{5}{12}$, by Pythagoras' theorem.

$\dfrac{2\tan\alpha}{1 - \tan^2\alpha} = \dfrac{5}{12}$ ✓

$24 \tan \alpha = 5 - 5 \tan^2 \alpha$

$5 \tan^2 \alpha + 24 \tan \alpha - 5 = 0$

$(5 \tan \alpha - 1)(\tan \alpha + 5) = 0$

$\tan \alpha = \dfrac{1}{5}, -5.$ ✓

But $\tan \alpha > 0$, hence $\tan \alpha = \dfrac{1}{5}$. ✓

13 a i ${}^4C_2 \times {}^3C_2 = 18$ ✓

ii ${}^5C_3 \times {}^4C_3 - {}^4C_2 \times {}^3C_2$

$= 40 - 18$

$= 22$ ✓

b i When $t = 0$, $P = 5400$, hence $P_0 = 5400$.

$\therefore P = 5400e^{kt}$

Let $P = 5840$, $t = 6$:

$5840 = 5400e^{k(6)}$ ✓

$e^{6k} = \dfrac{5840}{5400}$

$6k = \ln\dfrac{146}{135}$

$k = \dfrac{\ln\frac{146}{135}}{6}$

$= 0.013\,055\,307\ldots$

$= \mathbf{0.013\,06}$ **(4 sig. figs)** ✓

ii Let $t = 15$:

$P = 5400e^{k(15)}$

$= 6568.126\,038\ldots$

$= 6568$ (nearest whole)

$\therefore$ **the population will be 6568.** ✓

iii Let $P = 9000$:

$9000 = 5400e^{kt}$

$e^{kt} = \dfrac{9000}{5400}$

$kt = \ln\dfrac{5}{3}$ ✓

$k = \dfrac{\ln\frac{5}{3}}{k}$

$= 39.127\,813\,35\ldots$

$\therefore$ **in the 39th year which is in 2049.** ✓

c LHS $= \dfrac{1 - \cos 2x}{\sin 2x}$

$= \dfrac{1 - (1 - 2\sin^2 x)}{2\sin x \cos x}$

$= \dfrac{2\sin^2 x}{2\sin x \cos x}$ ✓

$= \dfrac{\sin x}{\cos x}$

$= \tan x$

$=$ **RHS** ✓

d i Sum: $1 + \alpha + (-\alpha) = -p$

$\therefore 1 = -p$

$\mathbf{p = -1}$ ✓

ii Pairs: $1(\alpha) + 1(-\alpha) + \alpha(-\alpha) = q$

$-\alpha^2 = q$ ✓

Also, product: $1(\alpha)(-\alpha) = -r$

$-\alpha^2 = -r$

$\therefore q = -r$

$\therefore \mathbf{q + r = 0}$ ✓

e i By Pythagoras: $y = \sqrt{x^2 - 9}$

$= (x^2 - 9)^{\frac{1}{2}}$

$\dfrac{dy}{dx} = \dfrac{1}{2}(x^2 - 9)^{-\frac{1}{2}}.2x$

$= \dfrac{\mathbf{x}}{\sqrt{\mathbf{x^2 - 9}}}$ ✓

ii Let $x = 5$: $\dfrac{dy}{dx} = \dfrac{5}{\sqrt{5^2 - 9}}$

$= 1.25$

$\dfrac{dy}{dt} = \dfrac{dy}{dx} \times \dfrac{dx}{dt}$ ✓

$= 1.25 \times 1.2$

$= 1.5$

$\therefore$ **the boat is travelling at 1.5 ms^{-1}.** ✓

14 a i

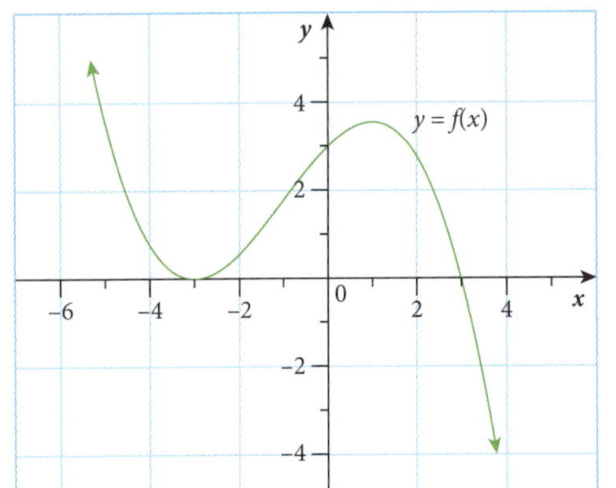

✓

ii

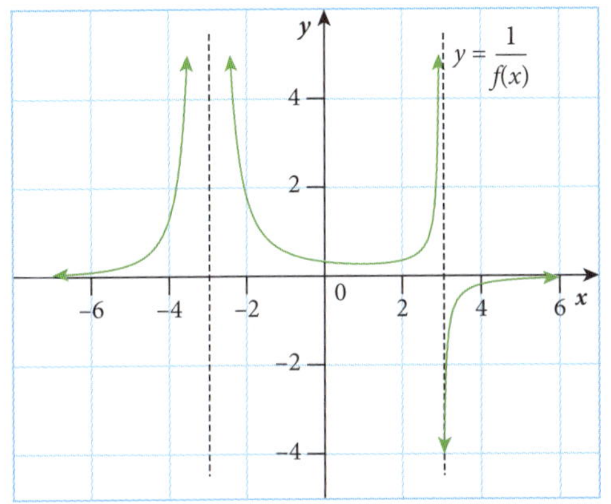

✓✓

iii

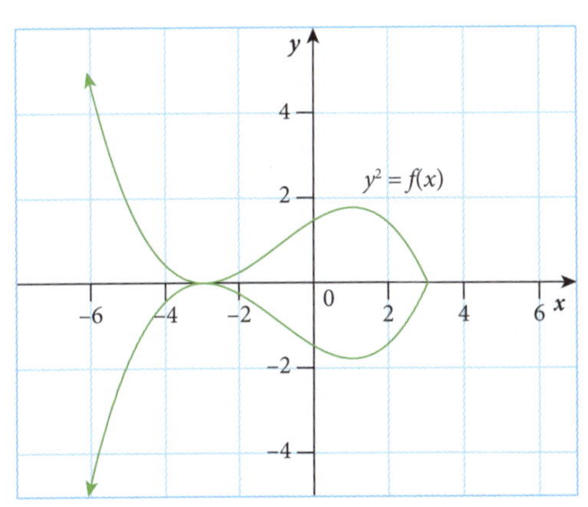

✓✓

b

$$P(-a) = P(2a) \quad ✓$$

$$(-a)^3 - (-a)^2 - 2(-a) - 3 = (2a)^3 - (2a)^2 - 2(2a) - 3$$

$$-a^3 - a^2 + 2a - 3 = 8a^3 - 4a^2 - 4a - 3$$

$$9a^3 - 3a^2 - 6a = 0$$

$$3a(3a^2 - a - 2) = 0$$

$$3a(3a + 2)(a - 1) = 0 \quad ✓$$

$$a = 0, -\frac{2}{3}, 1$$

But $a \neq 0$: $\therefore$ $\boldsymbol{a = -\frac{2}{3}, 1}$ ✓

c i

$$\frac{dT}{dt} = -kAe^{-kt}$$

But $T - 15 = Ae^{-kt}$

$$\therefore \boldsymbol{\frac{dT}{dt} = -k(T - 15)} \quad ✓$$

ii Let $T = 65$, $\frac{dT}{dt} = -1$:

$$-1 = -k(65 - 15)$$

$$-50k = -1$$

$$\boldsymbol{k = 0.02} \quad ✓$$

iii $T = 15 + Ae^{-0.02t}$

Let $T = 65$, $t = 0$:

$$65 = 15 + Ae^{-0.02(0)}$$

$$A = 50$$

$$T = 15 + 50e^{-0.02t} \quad ✓$$

Let $t = 30$:

$$T = 15 + 50e^{-0.02(30)}$$

$$= 42.440\,5818\ldots$$

$$= 42 \text{ (nearest degree)}$$

$\therefore$ **the temperature is 42 °C.** ✓

d Let roots be α, β and $2(\alpha + \beta)$.

Sum: $\alpha + \beta + 2(\alpha + \beta) = -a$

$$3\alpha + 3\beta = -a$$

$$\alpha + \beta = -\frac{a}{3} \ldots \quad ① \quad ✓$$

Also, $(\alpha + \beta)^2 = \frac{a^2}{9}$

$$\alpha^2 + 2\alpha\beta + \beta^2 = \frac{a^2}{9}$$

$$\alpha^2 + \beta^2 = \frac{a^2}{9} - 2\alpha\beta \ldots \quad ②$$

Pairs: $\alpha\beta + 2\alpha^2 + 2\alpha\beta + 2\alpha\beta + 2\beta^2 = b$

$$2\alpha^2 + 2\beta^2 + 5\alpha\beta = b$$

$$2(\alpha^2 + \beta^2) + 5\alpha\beta = b$$

From ②: $2\left(\frac{a^2}{9} - 2\alpha\beta\right) + 5\alpha\beta = b$

$$\frac{2a^2}{9} - 4\alpha\beta + 5\alpha\beta = b$$

$$\alpha\beta = b - \frac{2a^2}{9} \ldots \quad ③ \quad ✓$$

Product: $\alpha\beta \times 2(\alpha + \beta) = -c$

From ② and ③: $\left(b - \frac{2a^2}{9}\right) \times 2 \times -\frac{a}{3} = -c$

$$-\frac{2ab}{3} + \frac{4a^3}{27} = -c$$

$$-18ab + 4a^3 = -27c$$

$$\boldsymbol{4a^3 - 18ab + 27c = 0} \quad ✓$$

CHAPTER 8 PROOF BY MATHEMATICAL INDUCTION

TOPIC: PROOF

YEAR 12

1 Mathematical induction relating to series

- **Mathematical induction is a method** of mathematical proof used to prove propositions (also known as statements) involving the natural numbers.
- **Start with the initial statement**: prove true for the first value of n.
- **Then the inductive statement**: assume true for $n = k$ and then prove it true for $n = k + 1$.
- **You need to establish $S_k + T_{k+1} = S_{k+1}$**, where S_k = sum to k terms, T_{k+1} = (k+1)th term and S_{k+1} = sum to (k+1) terms.

EXAMPLE 1

Use mathematical induction to prove that, for $n \geq 1$, $2 + 6 + 12 + \ldots + n(n + 1) = \frac{n}{3}(n + 1)(n + 2)$.

If $n = 1$, LHS = 2 RHS $= \frac{1}{3}(1 + 1)(1 + 2)$.

$= 2$

$\therefore$ true for $n = 1$.

Assume true for $n = k$:

i.e. $S_k = 2 + 6 + 12 + \ldots + k(k + 1) = \frac{k}{3}(k + 1)(k + 2)$

Then if $n = k + 1$:

i.e. need to prove $S_k + T_{k+1} = S_{k+1}$

i.e. $\frac{k}{3}(k + 1)(k + 2) + (k + 1)(k + 2) = \frac{k+1}{3}(k + 2)(k + 3)$

LHS $= \frac{k}{3}(k + 1)(k + 2) + (k + 1)(k + 2)$

$= \frac{1}{3}(k + 1)(k^2 + 2k + 3k + 6)$

$= \frac{1}{3}(k + 1)(k^2 + 5k + 6)$

$S_k + T_{k+1} = S_{k+1}$

$= \frac{k+1}{3}(k + 2)(k + 3)$

= RHS

$\therefore$ hence the claim is true for $n = k + 1$.

Since shown true for $n = 1$, is true for $n = 2, 3, 4, \ldots$ and so true for all integers $n \geq 1$.

EXAMPLE 2

Use mathematical induction to show that for integer $n \geq 2$, $2 \times 1 + 3 \times 2 + 4 \times 3 + \ldots + n(n - 1) = \frac{n(n^2 - 1)}{3}$

If $n = 2$: LHS $= 2 \times 1$ RHS $= \frac{2(2^2 - 1)}{3}$

$= 2$ $= 2$

To start, prove true for $n = 2$.

$\therefore$ true for $n = 1$.

Assume true for $n = k$:

i.e. $S_k = 2 \times 1 + 3 \times 2 + 4 \times 3 + \ldots + k(k - 1) = \frac{k(k^2 - 1)}{3}$.

Then if $n = k + 1$:

i.e. need to prove $S_k + T_{k+1} = S_{k+1}$

i.e. $\frac{k(k^2 - 1)}{3} + (k + 1)(k) = \frac{(k+1)((k+1)^2 - 1)}{3}$

$= \frac{(k+1)(k^2 + 2k + 1 - 1)}{3}$

$= \frac{(k+1)(k^2 + 2k)}{3}$

$= \frac{k(k+1)(k+2)}{3}$

Now, LHS $= \frac{k(k^2 - 1)}{3} + k(k + 1)$

$= \frac{k(k+1)(k - 1 + 3]}{3}$

$= \frac{k(k+1)(k+2)}{3}$

= RHS

$\therefore$ hence the claim is true for $n = k + 1$.

Since shown true for $n = 2$, is true for $n = 3, 4, \ldots$ and so true for all integers $n \geq 2$.

PRACTICE

1 **Use mathematical induction to prove that, for $n \geq 1$:**

i $1 + 4 + 9 + \ldots + n^2 = \frac{n}{6}(n + 1)(2n + 1)$

ii $\frac{1}{1}+\frac{1}{1+2}+\frac{1}{1+2+3}+\ldots+\frac{1}{1+2+3+\ldots+n}$
$=\frac{2n}{n+1}$

2 **Use mathematical induction to show that for all positive integers $n \geq 1$:**

i $9+14+19+\ldots+(5n+4)=\frac{5n^2+13n}{2}$

ii $1(1!)+2(2!)+3(3!)+\ldots+n(n!)=(n+1)!-1$

Answers ➲ pp. 104–105

2 Mathematical induction relating to divisibility

- **Start with the initial statement**: prove true for the first value of n.
- **Then the inductive statement**: assume true for $n = k$ and then prove true for $n = k + 1$.
- **If M is an integer** then $5M$ is divisible by 5, or a multiple of 5.

EXAMPLE 1

Use mathematical induction to prove that $3^{2n} - 1$ is divisible by 8 for all positive integers n.

If $n = 1$, then $3^2 - 1 = 9 - 1$
$= 8$, which is divisible by 8.
Assume true for $n = k$:
i.e. $3^{2k} - 1 = 8M$, where M is an integer.
Then if $n = k + 1$:
$3^{2(k+1)} - 1 = 3^{2k+2} - 1$
$= 3^{2k}.3^2 - 1$
$= 9.3^{2k} - 1$
$= 8.3^{2k} + 3^{2k} - 1$
$= 8.3^{2k} + 8M$
$= 8(3^{2k} + M)$, which is divisible by 8.
Hence, the claim is true for $n = k + 1$.
Since shown true for $n = 1$, is true for $n = 2, 3, 4, \ldots$ and so true for all integers $n \geq 1$.

Let the assumed statement equal $8M$.

EXAMPLE 2

Use mathematical induction to prove that $7^n + 3^n$ is a multiple of 10 for all odd positive integers n.

If $n = 1$, then $7^1 + 3^1 = 7 + 3$
$= 10$, which is a multiple of 10.
$\therefore$ true for $n = 1$.
Assume true for $n = k$:
i.e. $7^k + 3^k = 10M$, where M is an integer.
Then if $n = k + 2$:
$7^{k+2} + 3^{k+2} = 7^2.7^k + 3^2.3^k$
$= 49.7^k + 9.3^k$
$= 40.7^k + 9.7^k + 9.3^k$
$= 40.7^k + 9(7^k + 3^k)$
$= 40.7^k + 9(10M)$
$= 40.7^k + 90M$
$= 10(4.7^k + 9M)$, which is a multiple of 10.
Hence, the claim is true for $n = k + 2$.
Since shown true for $n = 1$, is true for $n = 3, 5, 7, \ldots$ and so true for all odd integers $n \geq 1$.

Prove true for $n = k + 2$.

➔ PRACTICE

1 **Use mathematical induction to prove that:**

i **$5^{2n} - 1$ is divisible by 3 for all positive integers n**

ii **$4^{3n} + 8$ is divisible by 9 for all integers $n \geq 1$**

2 **Use mathematical induction to show that:**

i **$5^n + 3^n$ is a multiple of 2 for all integers $n > 0$**

ii **$3^{2n+1} + 2^{n+2}$ is a multiple of 7 for all integers $n \geq 1$**

Answers ➲ p. 105

3 False 'proofs by induction'

- **The result has not been proved** if only one of the steps is proved.

EXAMPLE 1

Martha attempted to use mathematical induction to prove that $2 + 4 + \ldots + 2n = n(n - 1) + 2$, for $n \geq 1$. At which step (initial or inductive) did the proof fail? Support your decision with working.

If $n = 1$: LHS $= 2$ RHS $= 1(1 - 1) + 2$
$= 2$ $= 2$
$\therefore$ true for $n = 1$.
Assume true for $n = k$
i.e. $S_k = 2 + 4 + \ldots + 2k = k(k - 1) + 2$.
Then if $n = k + 1$

i.e. need to prove $S_k + T_{k+1} = S_{k+1}$

i.e. $k(k-1) + 2 + 2(k+1) = k(k-1) + 2$

LHS $= k^2 - k + 2 + 2k + 2$ $\quad$ RHS $= k(k-1) + 2$

$= k^2 + k + 4$ $\quad$ $= k^2 - k + 2$

$\therefore$ LHS $\neq$ RHS

Hence, the claim is not true for $n = k + 1$.

$\therefore$ the proof fails the inductive step.

Simplify LHS and RHS separately.

→ PRACTICE

1 **A student attempted to prove the following 'results' using mathematical induction. At which step (initial or inductive) did the proof fail? Support your decision with working.**

i $1 + 4 + 7 + \ldots + (3n - 2) = \frac{n}{3}(n+2)$, **for** $n \geq 1$

ii $5^n + 3$ **is divisible by 8 for all positive integers** n

Answers ➲ pp. 105–106

HSC EXAM-TYPE QUESTIONS

Now for the real thing! The following questions are modelled on the types of questions you will face in the HSC Examination. If you get extensive practice at answering these sorts of questions, you will be more confident answering them in the HSC Examination.

Remember: The Answer section is located straight after this section so that you can check the answers readily. There are worked solutions and also ticks that indicate the mark allocation for all questions. Both these features will help you to strive for full marks.

Make sure also that when you check your work you highlight any questions you found difficult and earmark these areas for extra study.

Total marks: 35

PART A

(Suggested time: 10 minutes)
(1 mark each)

1 In the mathematical induction proof of $\frac{1}{1(2)}+\frac{1}{2(3)}+\frac{1}{3(4)}+\ldots+\frac{1}{n(n+1)}=\frac{n}{n+1}$, for all positive integers n, which of these is an expression for S_{k+1}?

A $S_{k+1}=\frac{1}{k(k+1)}$ **B** $S_{k+1}=\frac{1}{(k+1)(k+2)}$

C $S_{k+1}=\frac{k}{k+1}$ **D** $S_{k+1}=\frac{k+1}{k+2}$

2 For any positive integer n, $2.4^{2n+1}+3^{3n+1}$ is divisible by α. Which of these is the value of α?

A $\alpha=9$ **B** $\alpha=11$ **C** $\alpha=13$ **D** $\alpha=15$

3 If $1+2+3+\ldots+n=\frac{n}{2}(n+1)$, which of these is correct for the equation $S_k+T_{k+1}=S_{k+1}$, for all positive integers n?

A $\frac{k}{2}(k+1)+k=\frac{k}{2}(3k+1)$

B $\frac{k}{2}(k+1)+k+1=\frac{k+1}{2}(2k+1)$

C $\frac{k}{2}(k+1)+k+1=\frac{k+1}{2}(k+2)$

D $\frac{k}{2}(k+1)+k+1=\frac{(k+1)(k+1)}{2}$

4 For the proposition '$S(n)=n^3+(n+1)^3+(n+2)^3$ is divisible by β, for all natural number n where $n>1$', which of these is the value of β?

A $\beta=6$ **B** $\beta=8$ **C** $\beta=9$ **D** $\beta=10$

5 If the sum to n terms is given as $S_n=2^{n+1}-2$, which of these is an expression for S_{k+1}?

A $S_{k+1}=2(2^{k+1}-1)$ **B** $S_{k+1}=2(2^k-1)$

C $S_{k+1}=2(2^{k-1}-1)$ **D** $S_{k+1}=2^k(2^k-1)$

PART B

(Suggested time: 45 minutes)

Show all working.

6 For positive integer n, prove, using mathematical induction, $1+4+7+\ldots+(3n-2)=\frac{n(3n-1)}{2}$. (3 marks)

7 Prove, using mathematical induction, the statement '$2^{3n}-1$ is divisible by 7 for each natural number n'. (3 marks)

8 Use mathematical induction to prove for $n\geq 1$, $1^3+2^3+3^3+\ldots+n^3=\frac{n^2(n+1)^2}{4}$. (3 marks)

9 Use mathematical induction to prove that for $n\geq 1$, $2^{3n}-3^n$ is a multiple of 5. (3 marks)

10 Use the principle of mathematical induction to prove the statement below for integer $n>0$,
$\left(1-\frac{1}{2}\right)\left(1-\frac{1}{3}\right)\ldots\left(1-\frac{1}{n+1}\right)=\frac{1}{n+1}$. (3 marks)

11 Use mathematical induction to prove $1.2+2.3+\ldots+n(n+1)=\frac{n(n+1)(n+2)}{3}$, for all positive integer n. (3 marks)

12 Prove for $n>0$, $\frac{1}{3}+\frac{1}{9}+\frac{1}{27}+\ldots+\frac{1}{3^n}=\frac{3^n-1}{2(3)^n}$ by mathematical induction. (3 marks)

13 Prove, using mathematical induction, $\left(1-\frac{1}{2^2}\right)\left(1-\frac{1}{3^2}\right)\ldots\left(1-\frac{1}{n^2}\right)=\frac{n+1}{2n}$, for all integer $n\geq 2$. (3 marks)

14 Prove, using mathematical induction, that for each positive integer n, n^3+5n+6 is a multiple of 3. (3 marks)

15 Prove, for $n\geq 0$,
$\frac{1}{1.2.3}+\frac{1}{2.3.4}+\frac{1}{3.4.5}+\ldots+\frac{1}{(n+1)(n+2)(n+3)}$
$=\frac{(n+1)(n+4)}{4(n+2)(n+3)}$ by mathematical induction. (3 marks)

PRACTICE

1 Mathematical induction relating to series ➲ pp. 100–101

1 i If $n = 1$, LHS $= 1$ RHS $= \frac{1}{6}(1 + 1)(2(1) + 1)$.
$= 1$

$\therefore$ true for $n = 1$

Assume true for $n = k$:

i.e. $S_k = 1 + 4 + 9 + \ldots + k^2 = \frac{k}{6}(k + 1)(2k + 1)$.

Then if $n = k + 1$:

i.e. need to prove $S_k + T_{k+1} = S_{k+1}$

i.e. $\frac{k}{6}(k + 1)(2k + 1) + (k + 1)^2 = \frac{k+1}{6}(k + 2)(2k + 3)$

$$\begin{aligned}\text{LHS} &= \frac{k}{6}(k + 1)(2k + 1) + (k + 1)^2\\ &= \frac{k+1}{6}[k(2k + 1) + 6(k + 1)]\\ &= \frac{k+1}{6}(2k^2 + k + 6k + 6)\\ &= \frac{k+1}{6}(2k^2 + 7k + 6)\\ &= \frac{k+1}{6}(k + 2)(2k + 3)\\ &= \text{RHS}\end{aligned}$$

$\therefore$ hence the claim is true for $n = k + 1$.

Since shown true for $n = 1$, is true for $n = 2, 3, 4, \ldots$ and so true for all integers $n \geq 1$.

ii If $n = 1$:   LHS $= \frac{1}{1} = 1$ RHS $= \frac{2(1)}{1+1} = 1$

$\therefore$ true for $n = 1$.

Assume true for $n = k$:

i.e. $S_k = \frac{1}{1} + \frac{1}{1+2} + \frac{1}{1+2+3} + \ldots + \frac{1}{1+2+3+\ldots+k} = \frac{2k}{k+1}$

Then if $n = k + 1$:

i.e. need to prove $S_k + T_{k+1} = S_{k+1}$

i.e. $\frac{2k}{k+1} + \frac{1}{1+2+3+\ldots+(k+1)} = \frac{2(k+1)}{k+2}$

Now, $1 + 2 + \ldots + (k + 1)$ is Arithmetic, with $n = k + 1$, $a = 1$, $\ell = k + 1$ and $S_n = \frac{n}{2}[a + \ell]$.

Hence, $1 + 2 + \ldots + (k + 1) = \frac{k+1}{2}[1 + k + 1] = \frac{(k+1)(k+2)}{2}$

$$\begin{aligned}\text{LHS} &= \frac{2k}{k+1} + \frac{1}{1+2+3+\ldots+(k+1)}\\ &= \frac{2k}{k+1} + \frac{2}{(k+1)(k+2)}\\ &= \frac{2k(k+2)+2}{(k+1)(k+2)}\\ &= \frac{2k^2+4k+2}{(k+1)(k+2)}\\ &= \frac{2(k^2+2k+1)}{(k+1)(k+2)}\\ &= \frac{2(k+1)^2}{(k+1)(k+2)}\\ &= \frac{2(k+1)}{(k+2)}\\ &= \text{RHS}\end{aligned}$$

$\therefore$ hence the claim is true for $n = k + 1$.

Since shown true for $n = 1$, is true for $n = 2, 3, 4, \ldots$ and so true for all integers $n \geq 1$.

2 i If $n = 1$: LHS $= 9$ RHS $= \frac{5(1)^2+13(1)}{2} = 9$

$\therefore$ true for $n = 1$.

Assume true for $n = k$:

i.e. $S_k = 9 + 14 + 19 + \ldots + (5k + 4) = \frac{5k^2+13k}{2}$

Then if $n = k + 1$:

i.e. need to prove $S_k + T_{k+1} = S_{k+1}$

i.e. $\frac{5k^2+13k}{2} + 5k + 9 = \frac{5(k+1)^2+13(k+1)}{2}$

$$\begin{aligned}\text{LHS} &= \frac{5k^2+13k}{2} + 5k + 9\\ &= \frac{5k^2+13k+10k+18}{2}\\ &= \frac{5k^2+23k+18}{2}\end{aligned}$$

$$\begin{aligned}\text{RHS} &= \frac{5(k+1)^2+13(k+1)}{2}\\ &= \frac{5k^2+10k+5+13k+13}{2}\\ &= \frac{5k^2+23k+18}{2}\end{aligned}$$

$\therefore$ LHS = RHS

$\therefore$ hence the claim is true for $n = k + 1$.

Since shown true for $n = 1$, is true for $n = 2, 3, 4, \ldots$ and so true for all integers $n \geq 1$.

ii If for $n = 1$:   LHS $= 1 \times 1! = 1 = 1$ RHS $= (1 + 1)! - 1 = 2 - 1$

$\therefore$ true for $n = 1$.

Assume true for $n = k$:

i.e. $S_k = 1(1!) + 2(2!) + \ldots + k(k!) = (k + 1)! - 1$

Then if $n = k + 1$:

i.e. need to prove $S_k + T_{k+1} = S_{k+1}$

i.e. $(k + 1)! - 1 + (k + 1)(k + 1)! = (k + 2)! - 1$

$$\begin{aligned}\text{LHS} &= (k + 1)! - 1 + (k + 1)(k + 1)! \\ &= (k + 1)! + (k + 1)(k + 1)! - 1 \\ &= (k + 1)![1 + (k + 1)] - 1 \\ &= (k + 2)(k + 1)! - 1 \\ &= (k + 2)! - 1 \\ &= \text{RHS}\end{aligned}$$

$\therefore$ hence the claim is true for $n = k + 1$.

Since shown true for $n = 1$, is true for $n = 2, 3, 4, \ldots$ for all integers $n \geq 1$.

2 Mathematical induction relating to divisibility ➲ p. 101

1 i If $n = 1$, $5^2 - 1 = 25 - 1$
$= 24$, which is divisible by 3.

$\therefore$ true for $n = 1$.

Assume true for $n = k$:

i.e. $5^{2k} - 1 = 3M$, where M is an integer.

Then if $n = k + 1$:

$$\begin{aligned}5^{2(k+1)} - 1 &= 5^{2k+2} - 1 \\ &= 5^{2k}.5^2 - 1 \\ &= 25.5^{2k} - 1 \\ &= 24.5^{2k} + 5^{2k} - 1 \\ &= 24.5^{2k} + 3M \\ &= 3(8.5^{2k} + M), \text{ which is divisible by 3.}\end{aligned}$$

Hence, the claim is true for $n = k + 1$.

Since shown true for $n = 1$, is true for $n = 2, 3, 4, \ldots$ and so true for all integers $n \geq 1$.

ii If $n = 1$: $4^3 + 8 = 64 + 8$
$= 72$, which is divisible by 9.

$\therefore$ true for $n = 1$.

Assume true for $n = k$:

i.e. $4^{3k} + 8 = 9M$, where M is an integer.

Then if $n = k + 1$:

$$\begin{aligned}4^{3(k+1)} + 8 &= 4^{3k+3} + 8 \\ &= 4^3.4^{3k} + 8 \\ &= 64.4^{3k} + 8 \\ &= 63.4^{3k} + 4^{3k} + 8 \\ &= 63.4^{3k} + 9M \\ &= 9(7.4^{3k} + M), \text{ which is divisible by 9.}\end{aligned}$$

Hence, the claim is true for $n = k + 1$.

Since shown true for $n = 1$, is true for $n = 2, 3, 4, \ldots$ and so true for all integers $n \geq 1$.

2 i If $n = 1$: $5^1 + 3^1 = 5 + 3$
$= 8$, which is a multiple of 2.

$\therefore$ true for $n = 1$.

Assume true for $n = k$:

i.e. $5^k + 3^k = 2M$, where M is an integer.

Then if $n = k + 1$:

$$\begin{aligned}5^{k+1} + 3^{k+1} &= 5.5^k + 3.3^k \\ &= 2.5^k + 3.5^k + 3.3^k \\ &= 2.5^k + 3(5^k + 3^k) \\ &= 2.5^k + 3(2M) \\ &= 2(5^k + 3M), \text{ which is a multiple of 2.}\end{aligned}$$

Hence, the claim is true for $n = k + 1$.

Since shown true for $n = 1$, is true for $n = 2, 3, 4, \ldots$ and so true for all integers $n \geq 1$.

ii If $n = 1$, then $3^3 + 2^3 = 27 + 8$
$= 35$, which is a multiple of 7.

$\therefore$ true for $n = 1$.

Assume true for $n = k$:

i.e. $3^{2k+1} + 2^{k+2} = 7M$, where M is an integer.

Then if $n = k + 1$:

$$\begin{aligned}3^{2k+3} + 2^{k+3} &= 3^2.3^{2k+1} + 2^1.2^{k+2} \\ &= 9.3^{2k+1} + 2.2^{k+2} \\ &= 7.3^{2k+1} + 2.3^{2k+1} + 2.2^{k+2} \\ &= 7.3^{2k+1} + 2(3^{2k+1} + 2^{k+2}) \\ &= 7.3^{2k+1} + 2(7M) \\ &= 7(3^{2k+1} + 2M), \text{ which is a multiple of 7.}\end{aligned}$$

Hence, the claim is true for $n = k + 1$.

Since shown true for $n = 1$, is true for $n = 2, 3, 4, \ldots$ and so true for all integers $n \geq 1$.

3 False 'proofs by induction' ➲ p. 102

1 i If $n = 1$: LHS $= 3(1) - 2$ RHS $= \frac{1}{3}(1+2)$
$= 1$ $= 1$

$\therefore$ true for $n = 1$.

Assume true for $n = k$:

i.e. $S_k = 1 + 4 + 7 + \ldots + (3k - 2) = \frac{k}{3}(k+2)$.

Then if $n = k + 1$:

i.e. $S_k + T_{k+1} = S_{k+1}$

i.e. $\frac{k}{3}(k+2) + 3(k + 1) - 2 = \frac{k+1}{3}(k+3)$

$$\text{LHS} = \frac{k^2+2k}{3} + 3k + 3 - 2 \qquad \text{RHS} = \frac{(k+1)(k+3)}{3}$$

$$= \frac{k^2+2k}{3} + 3k + 1 \qquad = \frac{k^2+4k+3}{3}$$

$$= \frac{k^2+2k+9k+3}{3}$$

$$= \frac{k^2+11k+3}{3}$$

$\therefore$ LHS $\neq$ RHS

Hence, the claim is not true for $n = k + 1$.

$\therefore$ **the proof fails the inductive step.**

ii $5^n + 3$ is divisible by 8 for all positive integers n.

If $n = 1$: $5^1 + 3 = 5 + 3$
$= 8$, which is divisible by 8.

$\therefore$ true for $n = 1$.

Assume true for $n = k$: $5^k + 3 = 8M$, where M is an integer.

Then if $n = k + 1$:

$5^{k+1} + 3 = 5.5^k + 3$
$= 5.5^k + 15 - 12$
$= 5(5^k + 3) - 12$
$= 5(8M) - 12$
$= 40M - 12$, which is not a multiple of 8.

Hence, the claim is not true for $n = k + 1$.

$\therefore$ **the proof fails the inductive step.**

HSC EXAM-TYPE QUESTIONS

Part A

1 D. If $S_n = \dfrac{n}{n+1}$, then $S_{k+1} = \dfrac{k+1}{k+1+1}$
$= \dfrac{k+1}{k+2}$

2 B. By inspection, $\alpha = 11$, as $2.4^{2(1)+1} + 3^{3(1)+1} = 209$ which is divisible by 11.

3 C. $\dfrac{k}{2}(k+1) + k + 1 = \dfrac{k+1}{2}(k+1+1)$

$S_k = \dfrac{k}{2}(k+1)$

$T_{k+1} = k + 1$

$S_{k+1} = \dfrac{(k+1)(k+2)}{2}$

4 C. By inspection, $\beta = 9$, as $2^3 + (2 + 1)^3 + (2 + 2)^3 = 99$ which is divisible by 9.

5 A. Substitute $n = k + 1$:

$S_{k+1} = 2^{k+1+1} - 2$
$= 2^{k+2} - 2$
$= 2.2^{k+1} - 2$
$= 2(2^{k+1} - 1)$

Part B

6 If $n = 1$: LHS $= 3(1) - 2$ RHS $= \dfrac{1[3(1)-1]}{2}$
$= 1$ $= 1$

$\therefore$ true for $n = 1$.

Assume true for $n = k$:

i.e. $S_k = 1 + 4 + 7 + \ldots + (3k - 2) = \dfrac{k(3k-1)}{2}$ ✓

Then $n = k + 1$:

i.e. need to prove $S_k + T_{k+1} = S_{k+1}$

i.e. $\dfrac{k(3k-1)}{2} + 3k + 1 = \dfrac{(k+1)(3k+2)}{2}$ ✓

LHS $= \dfrac{k(3k-1)}{2} + 3k + 1$
$= \dfrac{3k^2 - k + 6k + 2}{2}$
$= \dfrac{3k^2 + 5k + 2}{2}$
$= \dfrac{(k+1)(3k+2)}{2}$
$=$ RHS

$\therefore$ hence the claim is true for $n = k + 1$

Since shown true for $n = 1$, is true for $n = n = 2, 3, 4, \ldots$ and so true for all integers $n \geq 1$. ✓

7 If $n = 1$: $2^3 - 1 = 8 - 1$
$= 7$, which is divisible by 7. ✓

$\therefore$ true for $n = 1$.

Assume true for $n = k$:

i.e. $2^{3k} - 1 = 7M$, where M is an integer.

Then if $n = k + 1$:

$2^{3(k+1)} - 1 = 2^{3k+3} - 1$ ✓
$= 2^{3k}.2^3 - 1$
$= 8.2^{3k} - 1$
$= 7.2^{3k} + 2^{3k} - 1$
$= 7.2^{3k} + 7M$
$= 7(2^{3k} + M)$, which is divisible by 7.

$\therefore$ hence the claim is true for $n = k + 1$.

Since shown true for $n = 1$, is true for $n = 2, 3, 4, \ldots$ and so true for all integers $n \geq 1$. ✓

8 If $n = 1$: LHS $= 1^3$ RHS $= \dfrac{1^2(1+1)^2}{4}$
$= 1$ $= 1$

$\therefore$ true for $n = 1$.

Assume true for $n = k$:

i.e. $S_k = 1^3 + 2^3 + 3^3 + \ldots + k^3 = \dfrac{k^2(k+1)^2}{4}$. ✓

Then if $n = k + 1$:

i.e. need to prove $S_k + T_{k+1} = S_{k+1}$

i.e. $\dfrac{k^2(k+1)^2}{4} + (k + 1)^3 = \dfrac{(k+1)^2(k+2)^2}{4}$ ✓

LHS $= \dfrac{k^2(k+1)^2}{4} + (k + 1)^3$
$= \dfrac{(k+1)^2[k^2+4k+4]}{4}$
$= \dfrac{(k+1)^2(k+2)^2}{4}$
$=$ RHS

$\therefore$ hence the claim is true for $n = k + 1$.

Since shown true for $n = 1$, is true for $n = 2, 3, 4, \ldots$ and so true for all integers $n \geq 1$. ✓

9 Step 1: Prove true for $n = 1$:

$2^3 - 3^1 = 8 - 3$
$= 5$, which is a multiple of 5.

$\therefore$ true for $n = 1$. ✓

Step 2: Assume true for $n = k$:

$2^{3k} - 3^k = 5M$, where M is an integer.

Now prove true for $n = k + 1$:

$2^{3k+3} - 3^{k+1} = 2^3.2^{3k} - 3.3^k$ ✓
$= 8.2^{3k} - 3.3^k$
$= 5.2^{3k} + 3.2^{3k} - 3.3^k$
$= 5.2^{3k} + 3(2^{3k} - 3^k)$
$= 5.2^{3k} + 3(5M)$
$= 5(2^{3k} + 3M)$, which is a multiple of 5.

$\therefore$ true for $n = k + 1$.

Since shown true for $n = 1$, is true for $n = 2, 3, 4, \ldots$ and so true for all integers $n \geq 1$. ✓

10 If $n = 1$: LHS $= 1 - \dfrac{1}{1+1}$ RHS $= \dfrac{1}{1+1}$

$= \dfrac{1}{2}$ $= \dfrac{1}{2}$

$\therefore$ true for $n = 1$. ✓

Assume true for $n = k$:

i.e. $\left(1-\dfrac{1}{2}\right)\left(1-\dfrac{1}{3}\right)\ldots\left(1-\dfrac{1}{k+1}\right) = \dfrac{1}{k+1}$

Then if $n = k + 1$:

i.e. need to prove $\dfrac{1}{k+1} \times \left(1-\dfrac{1}{k+2}\right) = \dfrac{1}{k+2}$ ✓

$$\begin{aligned}\text{LHS} &= \frac{1}{k+1} \times \left(1-\frac{1}{k+2}\right)\\ &= \frac{1}{k+1} \times \frac{k+2-1}{k+2}\\ &= \frac{1}{k+1} \times \frac{k+1}{k+2}\\ &= \frac{1}{k+2}\\ &= \text{RHS}\end{aligned}$$

$\therefore$ hence the claim is true for $n = k + 1$.

Since shown true for $n = 1$, is true for $n = 2, 3, 4, \ldots$ and so true for all integers $n \geq 1$. ✓

11 If $n = 1$: LHS $= 1(1 + 1)$ RHS $= \dfrac{1(1+1)(1+2)}{3}$

$= 2$ $= 2$

$\therefore$ true for $n = 1$ ✓

Assume true for $n = k$:

i.e. $S_k = 1.2 + 2.3 + \ldots + k(k + 1) = \dfrac{k(k+1)(k+2)}{3}$

Then if $n = k + 1$:

i.e. need to prove $S_k + T_{k+1} = S_{k+1}$

i.e. $\dfrac{k(k+1)(k+2)}{3} + (k + 1)(k + 2) = \dfrac{(k+1)(k+2)(k+3)}{3}$ ✓

$$\begin{aligned}\text{LHS} &= \frac{k(k+1)(k+2)}{3} + (k+1)(k+2)\\ &= \frac{k(k+1)(k+2)+3(k+1)(k+2)}{3}\\ &= \frac{(k+1)(k+2)(k+3)}{3}\\ &= \text{RHS}\end{aligned}$$

$\therefore$ hence the claim is true for $n = k + 1$.

Since shown true for $n = 1$, is true for $n = 2, 3, 4, \ldots$ and so true for all integers $n \geq 1$. ✓

12 If $n = 1$: LHS $= \dfrac{1}{3^1}$ RHS $= \dfrac{3^1-1}{2(3)^1}$

$= \dfrac{1}{3}$ $= \dfrac{1}{3}$

$\therefore$ true for $n = 1$ ✓

Assume true for $n = k$:

i.e. $S_k = \dfrac{1}{3} + \dfrac{1}{9} + \dfrac{1}{27} + \ldots + \dfrac{1}{3^k} = \dfrac{3^k-1}{2(3)^k}$.

Then if $n = k + 1$:

i.e. need to prove $S_k + T_{k+1} = S_{k+1}$

i.e. $\dfrac{3^k-1}{2(3)^k} + \dfrac{1}{3^{k+1}} = \dfrac{3^{k+1}-1}{2(3)^{k+1}}$ ✓

$$\begin{aligned}\text{LHS} &= \frac{3^k-1}{2(3)^k} + \frac{1}{3^{k+1}}\\ &= \frac{3(3^k-1)+2}{2(3)^{k+1}}\\ &= \frac{3.3^k-3+2}{2(3)^{k+1}}\\ &= \frac{3^{k+1}-1}{2(3)^{k+1}}\\ &= \text{RHS}\end{aligned}$$

$\therefore$ hence the claim is true for $n = k + 1$.

Since shown true for $n = 1$, is true for $n = 2, 3, 4, \ldots$ and so true for all integers $n \geq 1$. ✓

13 If $n = 2$: LHS $= 1 - \dfrac{1}{2^2}$ RHS $= \dfrac{2+1}{2(2)}$

$= \dfrac{3}{4}$ $= \dfrac{3}{4}$

$\therefore$ true for $n = 2$ ✓

Assume true for $n = k$:

i.e. $\left(1-\dfrac{1}{2^2}\right)\left(1-\dfrac{1}{3^2}\right)\ldots\left(1-\dfrac{1}{k^2}\right) = \dfrac{k+1}{2k}$.

Then if $n = k + 1$:

i.e. need to prove $\dfrac{k+1}{2k} \times \left(1-\dfrac{1}{(k+1)^2}\right) = \dfrac{k+2}{2(k+1)}$ ✓

$$\begin{aligned}\text{LHS} &= \frac{k+1}{2k} \times \left(1-\frac{1}{(k+1)^2}\right)\\ &= \frac{k+1}{2k} \times \frac{(k+1)^2-1}{(k+1)^2}\\ &= \frac{k^2+2k+1-1}{2k(k+1)}\\ &= \frac{k^2+2k}{2k(k+1)}\\ &= \frac{k(k+2)}{2k(k+1)}\\ &= \frac{k+2}{2(k+1)}\\ &= \text{RHS}\end{aligned}$$

$\therefore$ hence the claim is true for $n = k + 1$.

Since shown true for $n = 2$, is true for $n = 3, 4, \ldots$ and so true for all integers $n \geq 2$. ✓

14 If $n = 1$: $1^3 + 5(1) + 6 = 1 + 5 + 6$

$= 12$, which is a multiple of 3.

$\therefore$ true for $n = 1$. ✓

Assume true for $n = k$:

$k^3 + 5k + 6 = 3M$, where M is an integer.

Then if $n = k + 1$:

$$(k+1)^3 + 5(k+1) + 6 = k^3 + 3k^2 + 3k + 1 + 5k + 5 + 6 \quad ✓$$
$$= k^3 + 5k + 6 + 3k^2 + 3k + 6$$
$$= 3M + 3(k^2 + k + 2)$$
$$= 3(M + k^2 + k + 2)\text{, which is a multiple of 3.}$$

$\therefore$ hence the claim is true for $n = k + 1$.

Since shown true for $n = 1$, is true for $n = 2, 3, 4, \ldots$ and so true for all integers $n \geq 1$. ✓

15 If $n = 0$: LHS $= \dfrac{1}{(0+1)(0+2)(0+3)} = \dfrac{1}{6}$ RHS $= \dfrac{(0+1)(0+4)}{4(0+2)(0+3)} = \dfrac{1}{6}$

$\therefore$ true for $n = 0$ ✓

Assume true for $n = k$:

i.e. $S_k = \dfrac{1}{6} + \dfrac{1}{24} + \dfrac{1}{60} + \ldots + \dfrac{1}{(k+1)(k+2)(k+3)}$

$= \dfrac{(k+1)(k+4)}{4(k+2)(k+3)}$

Then if $n = k + 1$:

i.e. need to prove $S_k + T_{k+1} = S_{k+1}$

i.e. $\dfrac{(k+1)(k+4)}{4(k+2)(k+3)} + \dfrac{1}{(k+2)(k+3)(k+4)} = \dfrac{(k+2)(k+5)}{4(k+3)(k+4)}$ ✓

$$\text{LHS} = \frac{(k+1)(k+4)}{4(k+2)(k+3)} + \frac{1}{(k+2)(k+3)(k+4)}$$
$$= \frac{(k+1)(k+4)^2 + 4}{4(k+2)(k+3)(k+4)}$$
$$= \frac{(k+1)(k^2 + 8k + 16) + 4}{4(k+2)(k+3)(k+4)}$$
$$= \frac{k^3 + 8k^2 + 16k + k^2 + 8k + 16 + 4}{4(k+2)(k+3)(k+4)}$$
$$= \frac{k^3 + 9k^2 + 24k + 20}{4(k+2)(k+3)(k+4)}$$
$$= \frac{(k+2)^2(k+5)}{4(k+2)(k+3)(k+4)} \text{ (as } (k+2)^2(k+5) = k^3 + 9k^2 + 24k + 20)$$
$$= \frac{(k+2)(k+5)}{4(k+3)(k+4)}$$
$$= \text{RHS}$$

$\therefore$ hence the claim is true for $n = k + 1$

Since shown true for $n = 0$, is true for $n = 1, 2, 3, \ldots$ and so true for all integers $n \geq 0$. ✓

CHAPTER 9 INTRODUCTION TO VECTORS

TOPIC: VECTORS

YEAR 12

1 Terminology

- **A vector is a quantity** with magnitude (length) and direction.
- **Vectors can be written** as $\underset{\sim}{a}$, $\boldsymbol{a}$ or $\overrightarrow{OA}$.
- **A position vector** is the distance and direction of a point from the origin, e.g. $\overrightarrow{OA}$.
- **A displacement vector** (or relative vector) is the distance and direction of a point from another, e.g. $\overrightarrow{AB}$.
- **The negative of the vector** from A to B is the vector from B to A. This means $\overrightarrow{AB} = -\overrightarrow{BA}$.

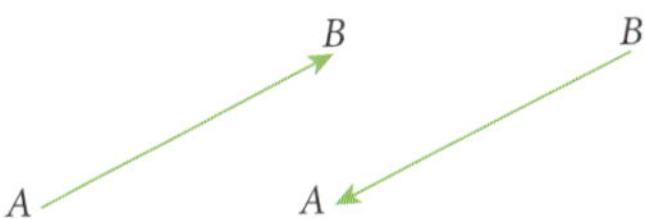

- **The length of vector** AB is written as $|\overrightarrow{AB}|$.
- $\underset{\sim}{i}$ **and** $\underset{\sim}{j}$ **are unit vectors** in the positive directions of the x-axis and y-axis respectively.
- **A scalar, λ, is a real number** and multiplies a vector.
- $\underset{\sim}{u}$ **and** $\underset{\sim}{v}$ **are parallel if** $\underset{\sim}{v} = \lambda \underset{\sim}{u}$, where $\lambda \neq 0$.

EXAMPLE 1

On a number plane draw examples of the vectors $\overrightarrow{OA}$ and $\overrightarrow{AB}$, where $|\overrightarrow{OA}| = |\overrightarrow{AB}|$.

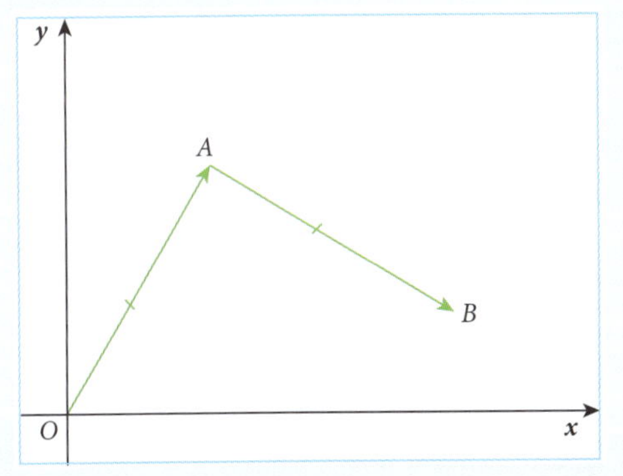

Vectors have same length.

PRACTICE

1 i The vector $\overrightarrow{OA}$ is drawn below. On the same number plane draw examples of $\overrightarrow{AB}$ and $\overrightarrow{BO}$ if $|\overrightarrow{OA}| = |\overrightarrow{AB}| = |\overrightarrow{BO}|$. What shape is formed?

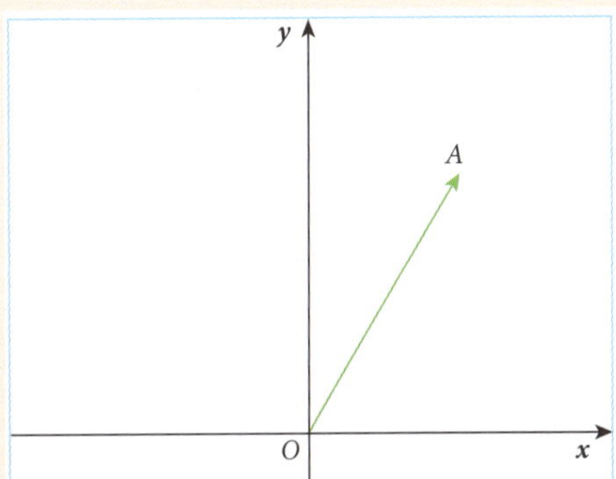

ii On a number plane draw $\overrightarrow{OA}$, $\overrightarrow{AB}$, $\overrightarrow{OC}$ and $\overrightarrow{BC}$ where $|\overrightarrow{OA}| = |\overrightarrow{BC}|$ and $|\overrightarrow{AB}| = |\overrightarrow{OC}|$. What shape is formed?

Answers ➲ p. 116

2 Expressing vectors in different forms

- **A component form of a vector** expresses the vector in terms of unit vectors in the form of $x\underset{\sim}{i} + y\underset{\sim}{j}$, e.g. $4\underset{\sim}{i} + 3\underset{\sim}{j}$ means 4 units in x-direction and 3 units in y-direction.
- **An ordered pair of a vector** is written with round or angle brackets, e.g. $4\underset{\sim}{i} + 3\underset{\sim}{j}$ is (4, 3) or <4, 3>.
- **A column vector** is written in the form $\begin{pmatrix} 4 \\ 3 \end{pmatrix}$.

EXAMPLE 1

Express $\overrightarrow{OA}$ in component form.

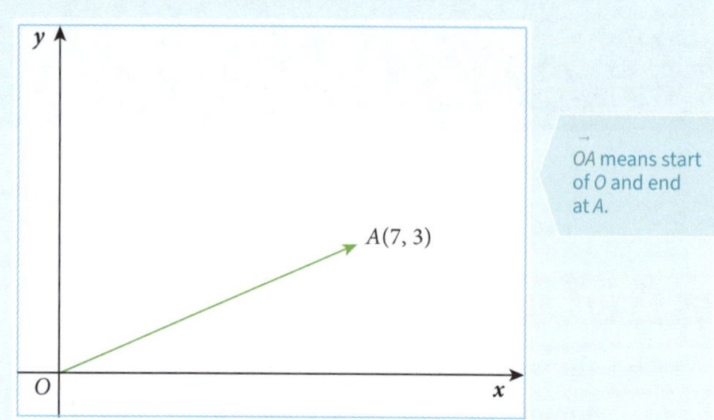

$\overrightarrow{OA}$ means start of O and end at A.

From O to A is 7 units in x-direction and 3 units in y-direction. Hence, $\overrightarrow{OA} = 7\underset{\sim}{i} + 3\underset{\sim}{j}$.

EXAMPLE 2

Express the vector $\underset{\sim}{p}$ in column vector form.

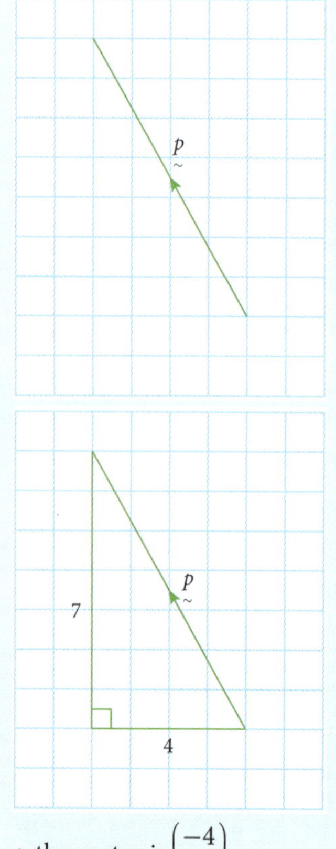

Take note of direction of vector.

$\therefore$ the vector is $\begin{pmatrix} -4 \\ 7 \end{pmatrix}$.

EXAMPLE 3

On the grid below, draw the vectors represented by $\underset{\sim}{a} = 7\underset{\sim}{i} + 3\underset{\sim}{j}$ and $\underset{\sim}{b} = \begin{pmatrix} -4 \\ -5 \end{pmatrix}$.

Vectors can be expressed in a variety of formats.

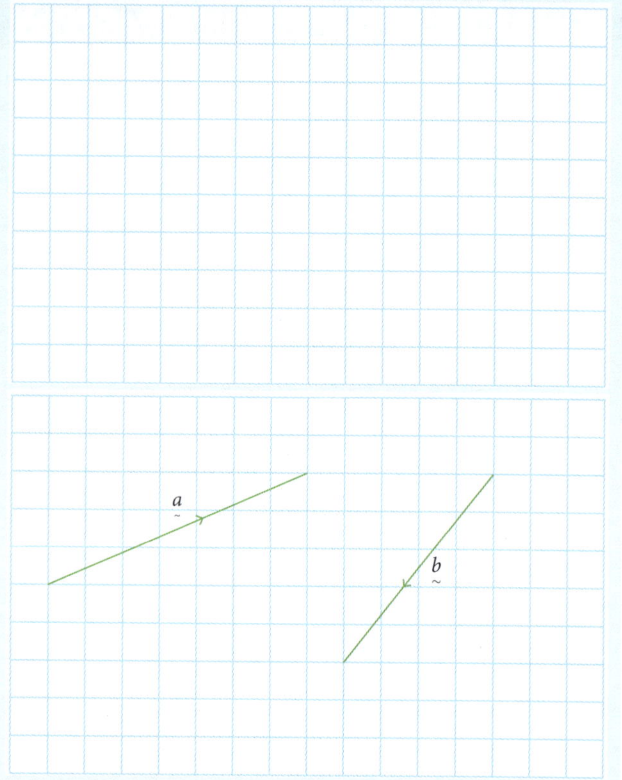

→ PRACTICE

1 Express the following vectors in component form:

i $\overrightarrow{OB}$

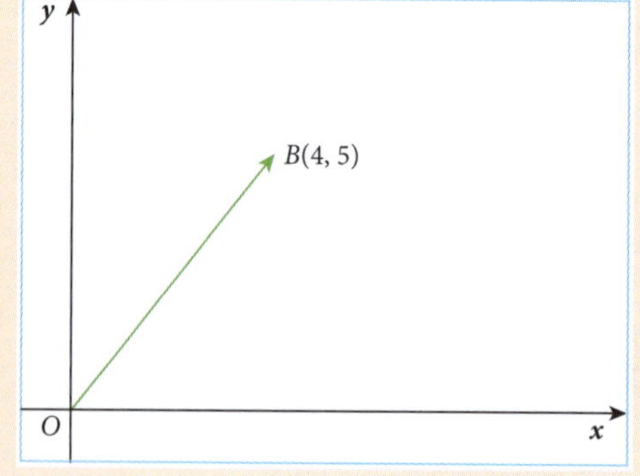

ii Express $\overrightarrow{AB}$ in component form.

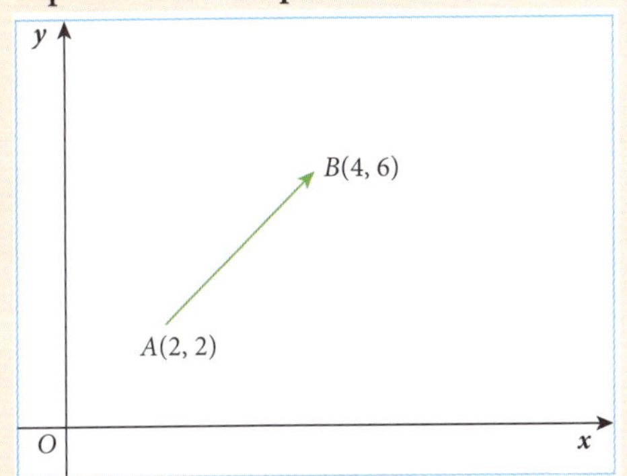

2 i Express $\underset{\sim}{u}$ in column vector form.

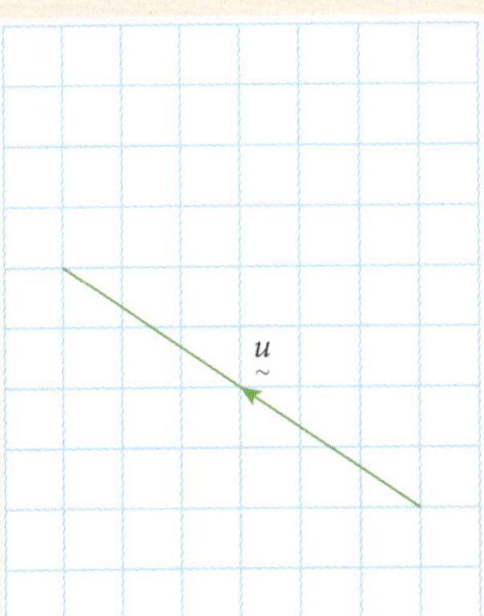

ii Express $\underset{\sim}{v}$ in component form.

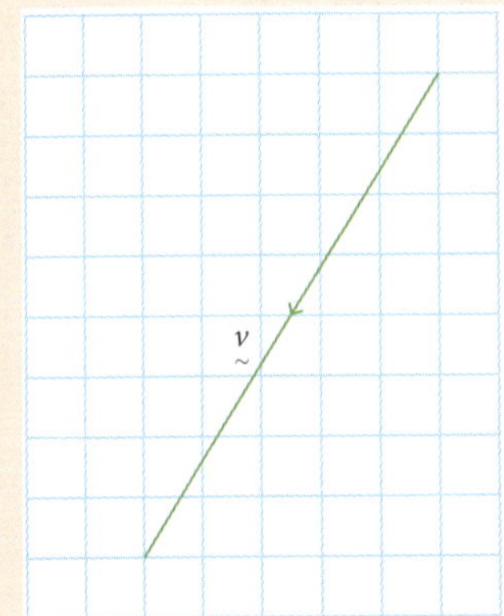

3 On the grid below draw the vectors represented by:

i $\underset{\sim}{u} = 7\underset{\sim}{i} + 3\underset{\sim}{j}$ and $\underset{\sim}{v} = \begin{pmatrix} -5 \\ 3 \end{pmatrix}$

ii $\underset{\sim}{s} = -7\underset{\sim}{i}$ and $\underset{\sim}{t} = \begin{pmatrix} 0 \\ -4 \end{pmatrix}$

Answers ➲ p. 116

3 Operations with vectors

- **Multiplication**: Multiplying a vector by a scalar (real number) changes its length, or direction. The resultant vector is parallel to the original.

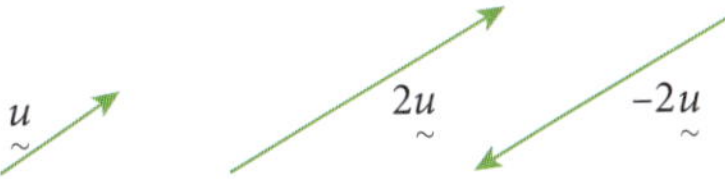

- **Addition**: Two vectors $\underset{\sim}{u}$ and $\underset{\sim}{v}$ can be added geometrically by forming a parallelogram. Also, $\underset{\sim}{u} + \underset{\sim}{v} = \underset{\sim}{v} + \underset{\sim}{u}$.

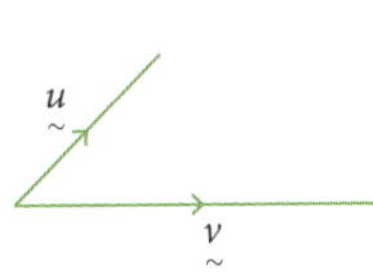

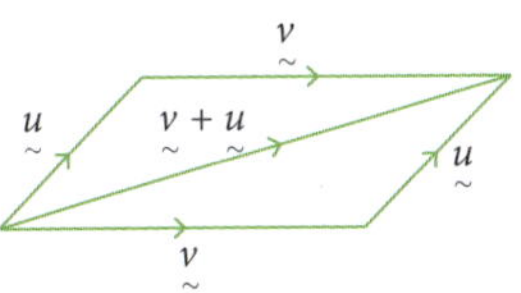

- **Subtraction:** To find $\underset{\sim}{u} - \underset{\sim}{v}$, we add $\underset{\sim}{u}$ and $-\underset{\sim}{v}$.

➔ Using **component** form, if $\underset{\sim}{a} = x_1\underset{\sim}{i} + y_1\underset{\sim}{j}$ and $\underset{\sim}{b} = x_2\underset{\sim}{i} + y_2\underset{\sim}{j}$, then:

- $\underset{\sim}{a} + \underset{\sim}{b} = (x_1 + x_2)\underset{\sim}{i} + (y_1 + y_2)\underset{\sim}{i}$
- $\underset{\sim}{a} - \underset{\sim}{b} = (x_1 - x_2)\underset{\sim}{i} + (y_1 - y_2)\underset{\sim}{j}$
- $k\underset{\sim}{a} = kx_1\underset{\sim}{i} + ky_1\underset{\sim}{j}$ for scalar k.

EXAMPLE 1

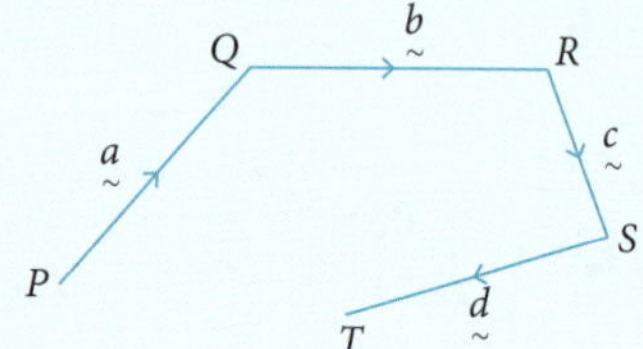

Find in terms of $\underset{\sim}{a}$, $\underset{\sim}{b}$, $\underset{\sim}{c}$ and $\underset{\sim}{d}$:

a $\overrightarrow{PR}$

b $\overrightarrow{TR}$

$\overrightarrow{TR} = -\overrightarrow{AT}$

a $\overrightarrow{PR} = \underset{\sim}{a} + \underset{\sim}{b}$

b $\overrightarrow{TR} = -\underset{\sim}{d} - \underset{\sim}{c} = -(\underset{\sim}{d} + \underset{\sim}{c})$

EXAMPLE 2

If $\underset{\sim}{m} = 3\underset{\sim}{i} + 2\underset{\sim}{j}$ and $\underset{\sim}{n} = 4\underset{\sim}{i} - 3\underset{\sim}{j}$, find $\underset{\sim}{m} + \underset{\sim}{n}$, $\underset{\sim}{m} - \underset{\sim}{n}$ and $2\underset{\sim}{n}$.

$$\underset{\sim}{m} + \underset{\sim}{n} = (3 + 4)\underset{\sim}{i} + (2 - 3)\underset{\sim}{j}$$
$$= 7\underset{\sim}{i} - \underset{\sim}{j}$$
$$\underset{\sim}{m} - \underset{\sim}{n} = (3 - 4)\underset{\sim}{i} + (2 - (-3))\underset{\sim}{j}$$
$$= -\underset{\sim}{i} + 5\underset{\sim}{j}$$
$$2\underset{\sim}{n} = 8\underset{\sim}{i} - 6\underset{\sim}{j}$$

In $2\underset{\sim}{n}$ the scalar is 2.

EXAMPLE 3

If $\underset{\sim}{p} = 5\underset{\sim}{i} + 2\underset{\sim}{j}$ and $\underset{\sim}{q} = \underset{\sim}{i} + 4\underset{\sim}{j}$, show geometrically and algebraically $\underset{\sim}{p} + \underset{\sim}{q}$ and $\underset{\sim}{p} - \underset{\sim}{q}$.

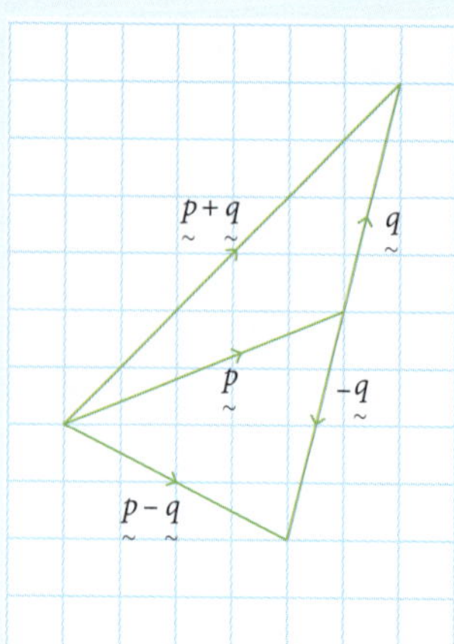

$$\underset{\sim}{p} + \underset{\sim}{q} = (5 + 1)\underset{\sim}{i} + (2 + 4)\underset{\sim}{j}$$
$$= 6\underset{\sim}{i} + 6\underset{\sim}{j}$$
$$\underset{\sim}{p} - \underset{\sim}{q} = (5 - 1)\underset{\sim}{i} + (2 - 4)\underset{\sim}{j}$$
$$= 4\underset{\sim}{i} - 2\underset{\sim}{j}$$

$\underset{\sim}{p} - \underset{\sim}{q}$ is $\underset{\sim}{p} + -\underset{\sim}{q}$.

EXAMPLE 4

Let $O = (0, 0)$, $A = (3, -2)$, $B = (-1, 4)$ and $C = (-5, -3)$ and find in component form:

a $\overrightarrow{OA}$

b $\overrightarrow{BC}$

c $\overrightarrow{OD}$, such that $2(\overrightarrow{OB} - \overrightarrow{OC}) = \overrightarrow{AD}$

a $\overrightarrow{OA} = 3\underset{\sim}{i} - 2\underset{\sim}{j}$

b $\overrightarrow{BC} = \overrightarrow{BO} + \overrightarrow{OC}$
$$= \underset{\sim}{i} - 4\underset{\sim}{j} - 5\underset{\sim}{i} - 3\underset{\sim}{j}$$
$$= -4\underset{\sim}{i} - 7\underset{\sim}{j}$$

c $2(\overrightarrow{OB} - \overrightarrow{OC}) = 2(-\underset{\sim}{i} + 4\underset{\sim}{j} + 5\underset{\sim}{i} + 3\underset{\sim}{j})$
$$= 2(4\underset{\sim}{i} + 7\underset{\sim}{j})$$
$$= 8\underset{\sim}{i} + 14\underset{\sim}{j} = \overrightarrow{AD}$$

Now, $\overrightarrow{AO} + \overrightarrow{OD} = \overrightarrow{AD}$
$$-3\underset{\sim}{i} + 2\underset{\sim}{j} + \overrightarrow{OD} = 8\underset{\sim}{i} + 14\underset{\sim}{j}$$
$$\overrightarrow{OD} = 11\underset{\sim}{i} + 12\underset{\sim}{j}$$

$\overrightarrow{BO} = -\overrightarrow{OB}$

→ PRACTICE

1 i

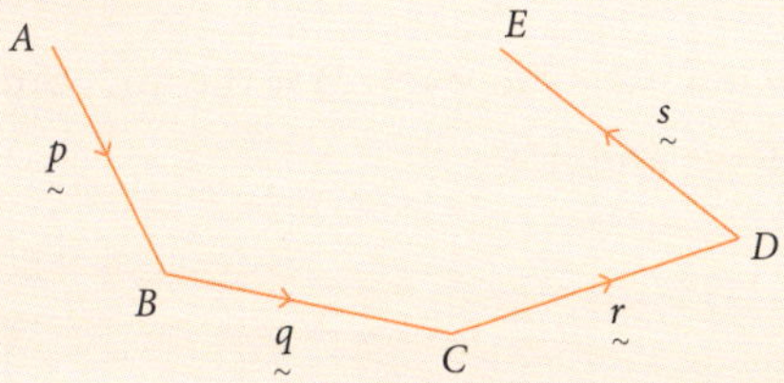

Find in terms of $\underset{\sim}{p}$, $\underset{\sim}{q}$, $\underset{\sim}{r}$ and $\underset{\sim}{s}$:

a $\overrightarrow{AC}$

b $\overrightarrow{EC}$

ii OPQ is an isosceles triangle, where $OP = OQ$. Also, R, S and T are midpoints of OP, OQ and PQ respectively. Let $\overrightarrow{OP} = \underset{\sim}{p}$ and $\overrightarrow{OQ} = \underset{\sim}{q}$.

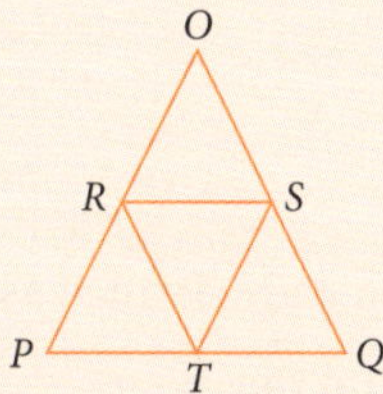

Find in terms of $\underset{\sim}{p}$ and $\underset{\sim}{q}$:

a $\overrightarrow{PQ}$

b $\overrightarrow{OT}$

c $\overrightarrow{TS}$

2 i If $\underset{\sim}{a} = 4\underset{\sim}{i} + \underset{\sim}{j}$ and $\underset{\sim}{b} = 2\underset{\sim}{i} - 5\underset{\sim}{j}$, find $\underset{\sim}{a} + \underset{\sim}{b}$, $\underset{\sim}{a} - \underset{\sim}{b}$ and $3\underset{\sim}{a}$.

ii If $\underset{\sim}{m} = -2\underset{\sim}{i} - 4\underset{\sim}{j}$ and $\underset{\sim}{n} = 3\underset{\sim}{i} - 6\underset{\sim}{j}$, find $-\underset{\sim}{m} + 2\underset{\sim}{n}$ and $2(\underset{\sim}{m} - \underset{\sim}{n})$.

3 i If $\underset{\sim}{a} = 6\underset{\sim}{i} + 2\underset{\sim}{j}$ and $\underset{\sim}{b} = -2\underset{\sim}{i} + 4\underset{\sim}{j}$, show geometrically and algebraically $\underset{\sim}{a} + \underset{\sim}{b}$ and $\underset{\sim}{a} - \underset{\sim}{b}$.

ii If $\underset{\sim}{m} = 2\underset{\sim}{i} + \underset{\sim}{j}$ and $\underset{\sim}{n} = -\underset{\sim}{i} - 2\underset{\sim}{j}$, show geometrically and algebraically $3\underset{\sim}{m} - 2\underset{\sim}{n}$.

4 Find the following vectors in component form if:

i $O = (0, 0)$, $P = (-4, -1)$, $Q = (1, -3)$ and $R = (5, 4)$:

a $\overrightarrow{OQ}$

b $\overrightarrow{QR}$

c $\overrightarrow{OS}$, such that $3(\overrightarrow{OR} - \overrightarrow{OP}) = \overrightarrow{QS}$

ii $O = (0, 0)$, $W = (3, 5)$, $X = (1, -2)$ and $Y = (-2, -3)$:

a $\overrightarrow{YW}$

b $\overrightarrow{ZO}$, such that $\frac{1}{2}(\overrightarrow{WX} + \overrightarrow{YX}) = \overrightarrow{WZ}$

Answers ➲ pp. 116–117

HSC EXAM-TYPE QUESTIONS

Total marks: 35

PART A

(Suggested time: 10 minutes)

(1 mark each)

1

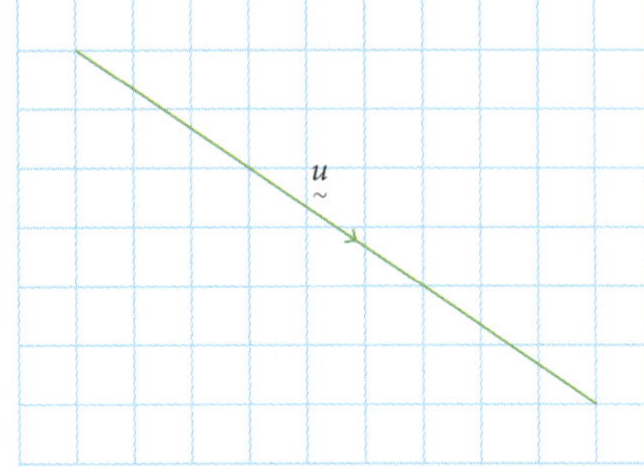

Which of these is the component form of $\underset{\sim}{u}$?

A $-9\underset{\sim}{i}+6\underset{\sim}{j}$ **B** $-6\underset{\sim}{i}+9\underset{\sim}{j}$

C $6\underset{\sim}{i}-9\underset{\sim}{j}$ **D** $9\underset{\sim}{i}-6\underset{\sim}{j}$

2 The diagram shows a pentagon $ABCDE$, where $AB \parallel DC$.

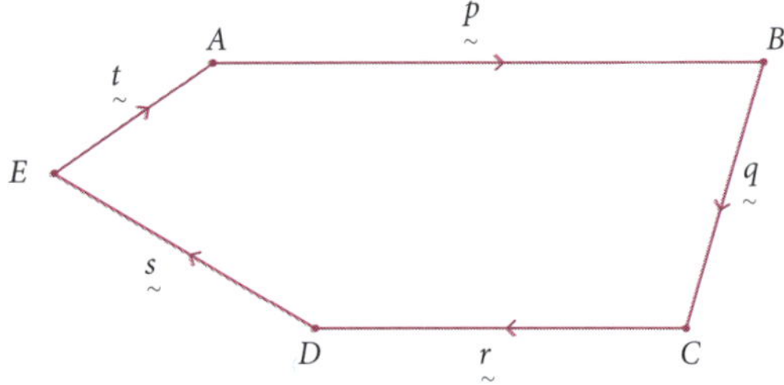

$\overrightarrow{AB}=\underset{\sim}{p}$, $\overrightarrow{BC}=\underset{\sim}{q}$, $\overrightarrow{CD}=\underset{\sim}{r}$, $\overrightarrow{DE}=\underset{\sim}{s}$, $\overrightarrow{EA}=\underset{\sim}{t}$.

Which of these is a true result?

A $\underset{\sim}{p}=-\underset{\sim}{r}$ **B** $\underset{\sim}{p}+\underset{\sim}{q}=\underset{\sim}{r}$

C $\underset{\sim}{p}+\underset{\sim}{q}+\underset{\sim}{r}=\underset{\sim}{s}+\underset{\sim}{t}$ **D** $\underset{\sim}{p}+\underset{\sim}{q}+\underset{\sim}{r}+\underset{\sim}{s}=-\underset{\sim}{t}$

3 Find the values of m and n if $\underset{\sim}{u}=3\underset{\sim}{i}-(m+1)\underset{\sim}{j}$, $\underset{\sim}{v}=5\underset{\sim}{i}+6\underset{\sim}{j}$ and $\underset{\sim}{u}+\underset{\sim}{v}=2n\underset{\sim}{i}+2\underset{\sim}{j}$.

A $m=2$ and $n=4$ **B** $m=2$ and $n=8$

C $m=3$ and $n=4$ **D** $m=3$ and $n=8$

4 Given $\overrightarrow{AB}=2\underset{\sim}{i}+3\underset{\sim}{j}$, $\overrightarrow{BC}=3\underset{\sim}{i}-9\underset{\sim}{j}$ and $\overrightarrow{CD}=-2\underset{\sim}{i}-3\underset{\sim}{j}$, what are the coordinates of A if $D=(2,-5)$?

A $(-1, 4)$ **B** $(-1, -8)$ **C** $(5, -8)$ **D** $(5, -14)$

5 $ABCD$ is a parallelogram. M is the midpoint of BC. Let $\overrightarrow{AB}=\underset{\sim}{u}$ and $\overrightarrow{AD}=\underset{\sim}{v}$.

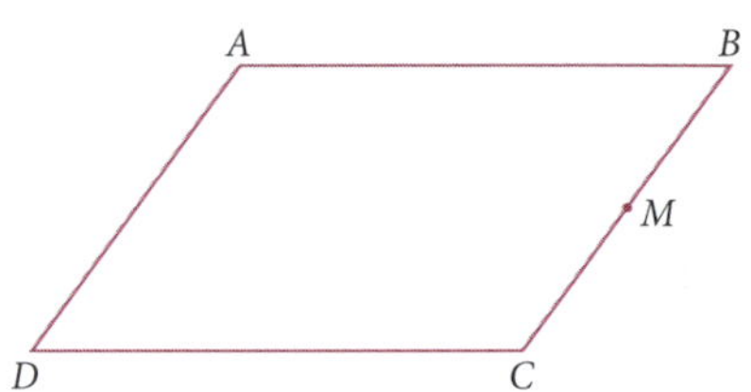

If P is the midpoint of DM, which of these is $\overrightarrow{DP}$?

A $\frac{1}{2}\left(\underset{\sim}{u}+\frac{1}{2}\underset{\sim}{v}\right)$ **B** $\frac{1}{2}\left(\underset{\sim}{u}-\frac{1}{2}\underset{\sim}{v}\right)$

C $\frac{1}{2}\left(\frac{1}{2}\underset{\sim}{v}-\underset{\sim}{u}\right)$ **D** $\frac{1}{2}(\underset{\sim}{u}-\underset{\sim}{v})$

PART B

(Suggested time: 45 minutes)

Show all working.

6 If vectors $\underset{\sim}{p}$ and $\underset{\sim}{q}$ are defined as $5\underset{\sim}{i}-2\underset{\sim}{j}$ and $-4\underset{\sim}{i}+2\underset{\sim}{j}$ respectively, find the:

a vector $\underset{\sim}{p}-\underset{\sim}{q}$ (1 mark)

b values of m and n if

$2\underset{\sim}{p}+\frac{1}{2}\underset{\sim}{q}=(m+2)\underset{\sim}{i}+(1-2n)\underset{\sim}{j}$ (3 marks)

7 Let $\underset{\sim}{a}=\begin{pmatrix}3\\-1\end{pmatrix}$ and $\underset{\sim}{b}=\begin{pmatrix}-2\\3\end{pmatrix}$.

Find the values of p and q if

$(3p-2)(\underset{\sim}{a}+\underset{\sim}{b})=(q-2)(\underset{\sim}{a}-\underset{\sim}{b})$. (3 marks)

8 In shape $ABCD$, M is the midpoint of CA, $AB=DC$ and $AB \parallel DC$. Let $\overrightarrow{BC}=\underset{\sim}{u}$ and $\overrightarrow{BA}=\underset{\sim}{v}$.

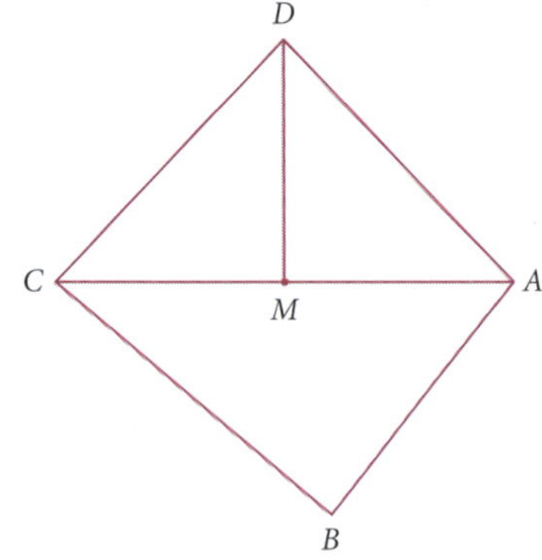

a Explain why $\overrightarrow{CM}=\frac{1}{2}(\underset{\sim}{v}-\underset{\sim}{u})$. (1 mark)

b Find $\overrightarrow{MD}$. (1 mark)

9 The vectors $\underset{\sim}{p}$ and $\underset{\sim}{q}$ are shown on the diagram below.

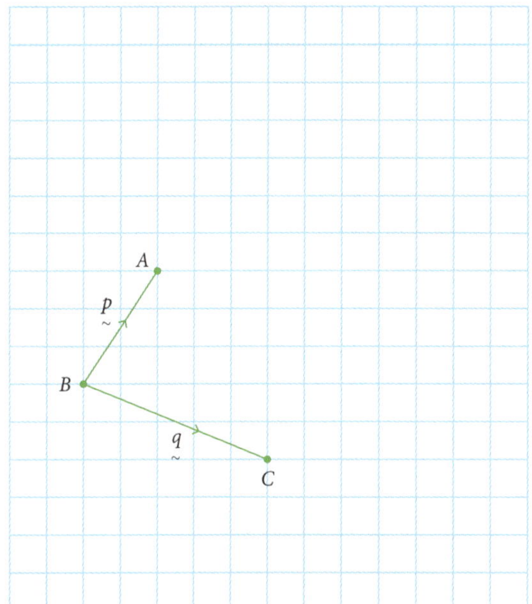

Show graphically the vector $3\underset{\sim}{p} - 2\underset{\sim}{q}$. (2 marks)

10 Find the scalars a and b where

$a\begin{pmatrix} 4 \\ -2 \end{pmatrix} + b\begin{pmatrix} 3 \\ -1 \end{pmatrix} = \begin{pmatrix} 4 \\ 0 \end{pmatrix}$. (2 marks)

11 The diagram shows triangle ABC. M is the midpoint of BC and $BT = 2TA$.

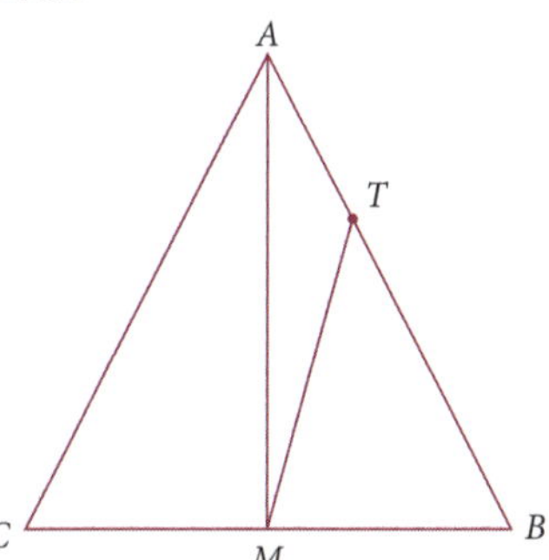

Let $\overrightarrow{AC} = \underset{\sim}{u}$ and $\overrightarrow{AB} = \underset{\sim}{v}$. Find an expression for $\overrightarrow{MT}$ in terms of $\underset{\sim}{u}$ and $\underset{\sim}{v}$. (2 marks)

12 Let $O = (0, 0)$, $A = (-2, 4)$, $B = (3, 5)$ and $C = (5, -4)$. Find in component form:

a $\overrightarrow{BC}$ (1 mark)

b $\overrightarrow{OD}$ such that $2(\overrightarrow{OB} - \overrightarrow{OC}) = \overrightarrow{AD}$ (2 marks)

13 Solve for vector $\underset{\sim}{u}$: $3\underset{\sim}{u} + 2\begin{pmatrix} 4 \\ -1 \end{pmatrix} = \underset{\sim}{u} + 3\begin{pmatrix} 2 \\ 4 \end{pmatrix}$. Leave your answer in component form. (2 marks)

14 $ABCD$ is a parallelogram. M is the midpoint of BC. N is on AB where $AN{:}NB = 2{:}1$.

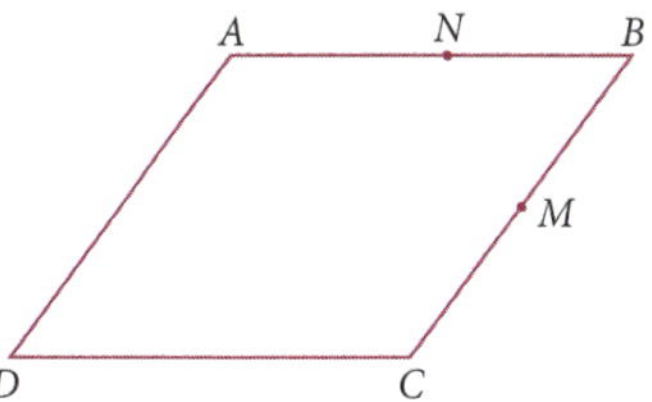

Let $\overrightarrow{DC} = \underset{\sim}{u}$ and $\overrightarrow{AD} = \underset{\sim}{v}$. Express $\overrightarrow{MN}$ in terms of $\underset{\sim}{u}$ and $\underset{\sim}{v}$. (2 marks)

15 PQR is a triangle. T is on PQ such that $PT : TQ = 3 : 2$.

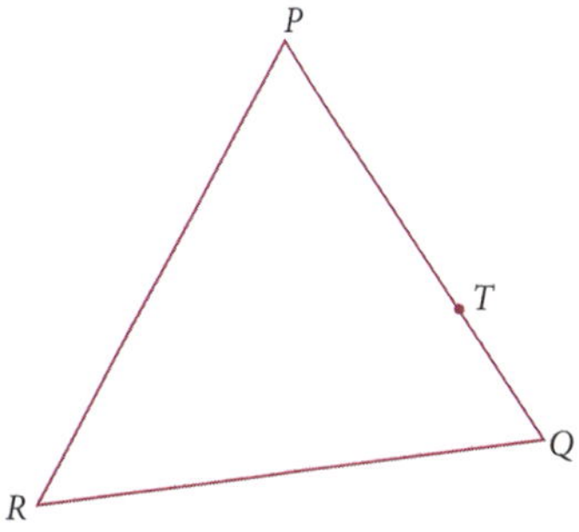

Let $\overrightarrow{RP} = \underset{\sim}{u}$ and $\overrightarrow{RQ} = \underset{\sim}{v}$. Also, $\overrightarrow{RT} = k(2\underset{\sim}{u} + 3\underset{\sim}{v})$, where k is a scalar quantity. Find the value of k. (2 marks)

16 $ABCD$ is drawn below and M, N, P and Q are the midpoints of AB, BC, CD and DA respectively.

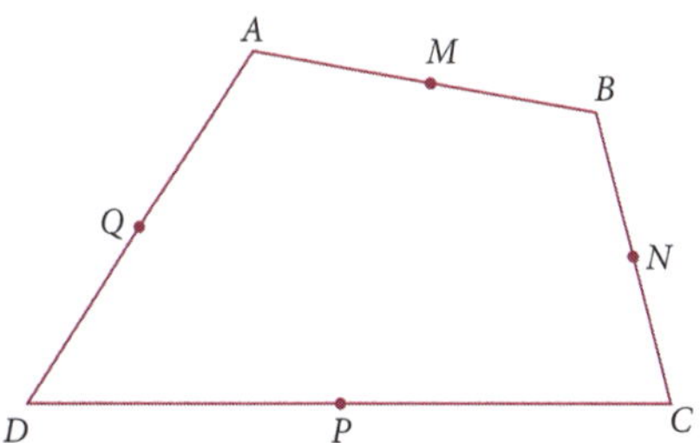

Let $\overrightarrow{AD} = 2\underset{\sim}{u}$, $\overrightarrow{DC} = 2\underset{\sim}{v}$ and $\overrightarrow{CB} = 2\underset{\sim}{w}$. Show that $QM \parallel PN$. (3 marks)

17 Given $\underset{\sim}{a} = 3\underset{\sim}{i} + (p + q)\underset{\sim}{j}$, $\underset{\sim}{b} = (p - q)\underset{\sim}{i} - 5\underset{\sim}{j}$ and $\underset{\sim}{a} - \underset{\sim}{b} = 6\underset{\sim}{i} + 4\underset{\sim}{j}$, write $2\underset{\sim}{a} + 3\underset{\sim}{b}$ in component form. (3 marks)

PRACTICE

1 Terminology ➲ p. 109

1 i

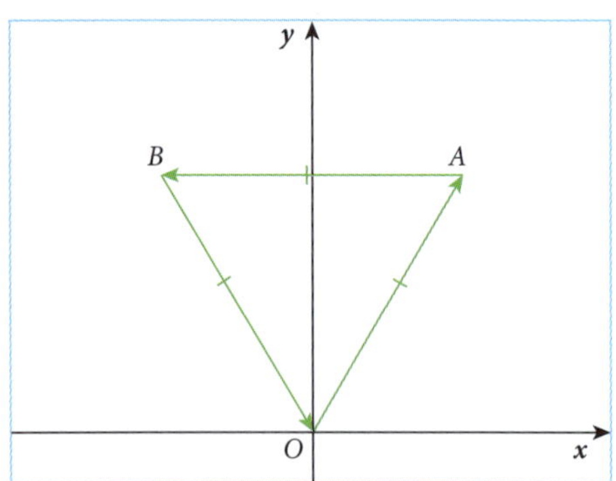

The shape is an equilateral triangle.

ii

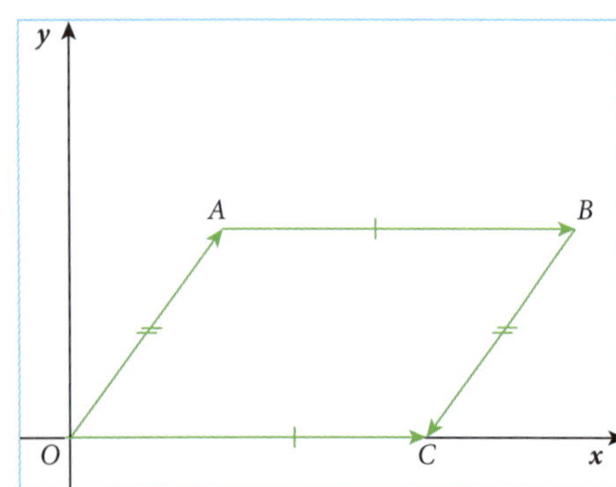

The shape is a parallelogram.

2 Expressing vectors in different forms ➲ pp. 110–111

1 i $\vec{OB} = 4\underset{\sim}{i} + 5\underset{\sim}{j}$

ii As $4 - 2 = 2$ and $6 - 2 = 4$ then $\vec{AB} = 2\underset{\sim}{i} + 4\underset{\sim}{j}$.

2 i

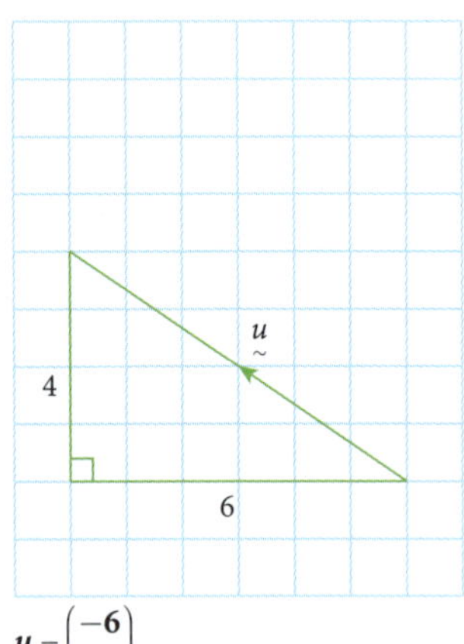

$\underset{\sim}{u} = \begin{pmatrix} -6 \\ 4 \end{pmatrix}$

ii

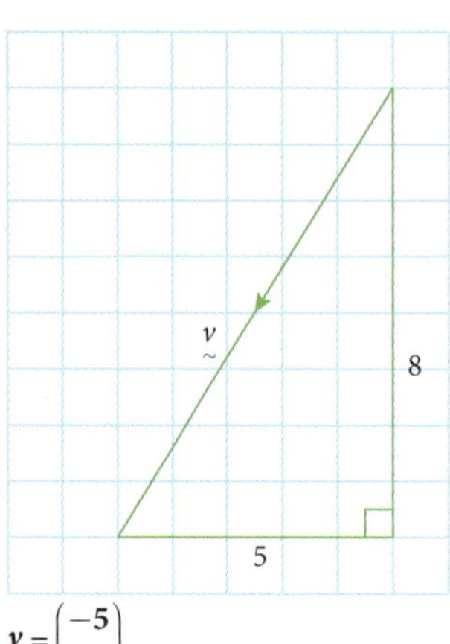

$\underset{\sim}{v} = \begin{pmatrix} -5 \\ -8 \end{pmatrix}$

3 i

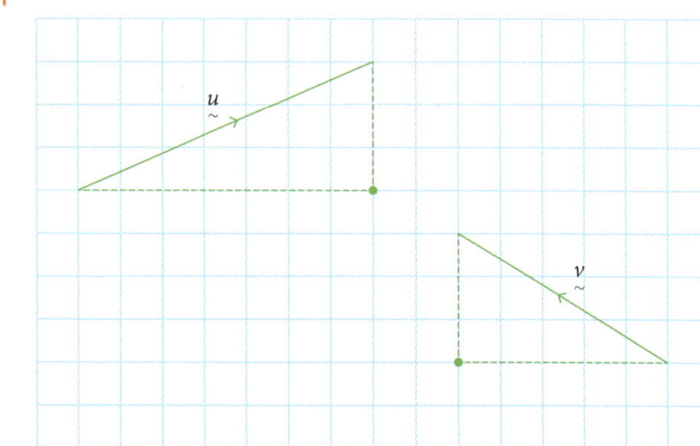

ii

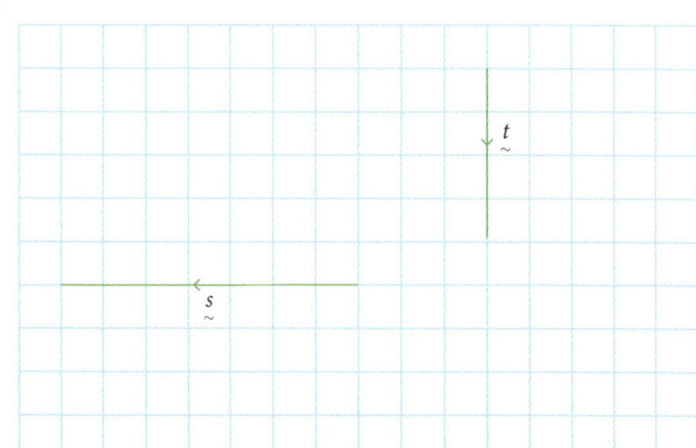

3 Operations with vectors ➲ p. 113

1 i a $\vec{AC} = \underset{\sim}{p} + \underset{\sim}{q}$

b $\vec{EC} = -\underset{\sim}{s} - \underset{\sim}{r} = -(\underset{\sim}{s} + \underset{\sim}{r})$

ii a $\vec{PQ} = \vec{PO} + \vec{OQ}$

$= -\underset{\sim}{p} + \underset{\sim}{q}$

b $\vec{OT} = \vec{OP} + \vec{PT}$

$= \underset{\sim}{p} + \frac{1}{2}(\underset{\sim}{q} - \underset{\sim}{p})$

$= \underset{\sim}{p} + \frac{1}{2}\underset{\sim}{q} - \frac{1}{2}\underset{\sim}{p}$

$= \frac{1}{2}\underset{\sim}{p} + \frac{1}{2}\underset{\sim}{q}$

$= \mathbf{\frac{1}{2}(\underset{\sim}{p} + \underset{\sim}{q})}$

c $\vec{TS} = \vec{TO} + \vec{OS}$

$= -\frac{1}{2}(\underset{\sim}{p} + \underset{\sim}{q}) + \frac{1}{2}\underset{\sim}{q}$

$= -\frac{1}{2}\underset{\sim}{p} - \frac{1}{2}\underset{\sim}{q} + \frac{1}{2}\underset{\sim}{q}$

$= \mathbf{-\frac{1}{2}\underset{\sim}{p}}$

2 i $\underset{\sim}{a} + \underset{\sim}{b} = (4 + 2)\underset{\sim}{i} + (1 - 5)\underset{\sim}{j}$

$= \mathbf{6\underset{\sim}{i} - 4\underset{\sim}{j}}$

$\underset{\sim}{a} - \underset{\sim}{b} = (4 - 2)\underset{\sim}{i} + (1 - (-5))\underset{\sim}{j}$

$= \mathbf{2\underset{\sim}{i} + 6\underset{\sim}{j}}$

$\mathbf{3\underset{\sim}{a} = 12\underset{\sim}{i} + 3\underset{\sim}{j}}$

ii $-\underset{\sim}{m} + 2\underset{\sim}{n} = (2 + 6)\underset{\sim}{i} + (4 - 12)\underset{\sim}{j}$

$= \mathbf{8\underset{\sim}{i} - 8\underset{\sim}{j}}$

$2(\underset{\sim}{m} - \underset{\sim}{n}) = 2[(-2 - 3)\underset{\sim}{i} + (-4 - (-6))\underset{\sim}{j}]$

$= 2[-5\underset{\sim}{i} + 2\underset{\sim}{j}]$

$= \mathbf{-10\underset{\sim}{i} + 4\underset{\sim}{j}}$

3 i

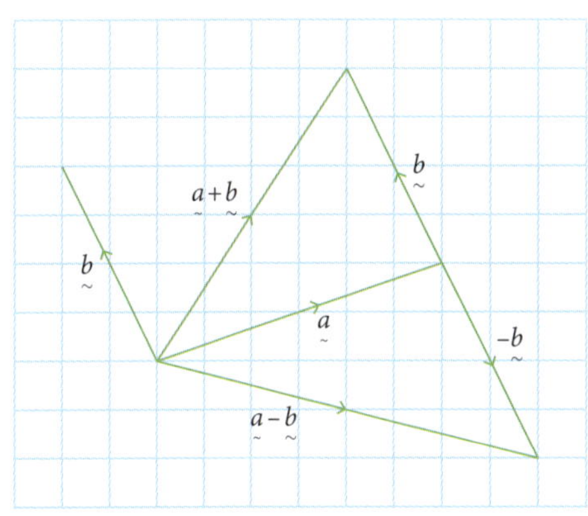

$\underset{\sim}{a} + \underset{\sim}{b} = (6 - 2)\underset{\sim}{i} + (2 + 4)\underset{\sim}{j}$

$= 4\underset{\sim}{i} + 6\underset{\sim}{j}$

$\underset{\sim}{a} - \underset{\sim}{b} = (6 + 2)\underset{\sim}{i} + (2 - 4)\underset{\sim}{j}$

$= \mathbf{8\underset{\sim}{i} - 2\underset{\sim}{j}}$

ii

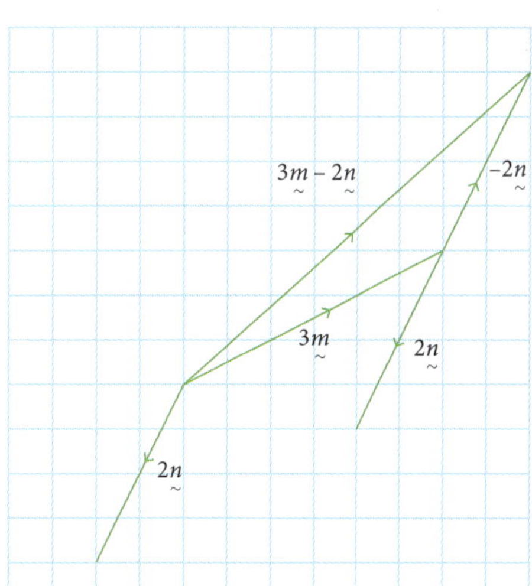

$3\underset{\sim}{m} - 2\underset{\sim}{n} = (6 - 2(-1))\underset{\sim}{i} + (3 - 2(-2))\underset{\sim}{j}$

$= \mathbf{8\underset{\sim}{i} + 7\underset{\sim}{j}}$

4 i a $\mathbf{\vec{OQ} = \underset{\sim}{i} - 3\underset{\sim}{j}}$

b $\vec{QR} = \vec{QO} + \vec{OR}$

$= -\underset{\sim}{i} + 3\underset{\sim}{j} + 5\underset{\sim}{i} + 4\underset{\sim}{j}$

$= \mathbf{4\underset{\sim}{i} + 7\underset{\sim}{j}}$

c $3(\vec{OR} - \vec{OP}) = 3(5\underset{\sim}{i} + 4\underset{\sim}{j} + 4\underset{\sim}{i} + \underset{\sim}{j})$

$= 3(9\underset{\sim}{i} + 5\underset{\sim}{j})$

$= 27\underset{\sim}{i} + 15\underset{\sim}{j} = \vec{QS}$

Now, $\vec{QO} + \vec{OS} = \vec{QS}$

$-\underset{\sim}{i} + 3\underset{\sim}{j} + \vec{OS} = 27\underset{\sim}{i} + 15\underset{\sim}{j}$

$\mathbf{\vec{OS} = 28\underset{\sim}{i} + 12\underset{\sim}{j}}$

ii a $\vec{YW} = \vec{YO} + \vec{OW}$

$= 2\underset{\sim}{i} + 3\underset{\sim}{j} + 3\underset{\sim}{i} + 5\underset{\sim}{j}$

$= \mathbf{5\underset{\sim}{i} + 8\underset{\sim}{j}}$

b $\frac{1}{2}(\vec{WX} + \vec{YX}) = \frac{1}{2}(\vec{WO} + \vec{OX} + \vec{YO} + \vec{OX})$

$= \frac{1}{2}(-3\underset{\sim}{i} - 5\underset{\sim}{j} + \underset{\sim}{i} - 2\underset{\sim}{j} + 2\underset{\sim}{i} + 3\underset{\sim}{j} + \underset{\sim}{i} - 2\underset{\sim}{j})$

$= \frac{1}{2}(\underset{\sim}{i} - 6\underset{\sim}{j}) = \vec{WZ}$

Now, $\vec{WO} - \vec{ZO} = \vec{WZ}$

$-3\underset{\sim}{i} - 5\underset{\sim}{j} - \vec{ZO} = \frac{1}{2}(\underset{\sim}{i} - 6\underset{\sim}{j})$

$\vec{ZO} = -3\underset{\sim}{i} - 5\underset{\sim}{j} - \frac{1}{2}\underset{\sim}{i} + 3\underset{\sim}{j}$

$= \mathbf{\frac{1}{2}(-7\underset{\sim}{i} - 4\underset{\sim}{j})}$

HSC EXAM-TYPE QUESTIONS

Part A

1 D. 9 across and 6 down means $9i - 6j$.

2 D. As $p + q + r + s + t = 0$, then $p + q + r + s = -t$

3 C. $-(m+1) + 6 = 2$

$-m - 1 + 6 = 2$

$-m + 5 = 2$

$m = 3$

$\therefore 3 + 5 = 2n$

$2n = 8$

$n = 4$

4 A. $\vec{AD} = 2i + 3j + 3i - 9j - 2i - 3j$

$= 3i + 9j$

$\vec{AD} = \vec{AO} + \vec{OD}$

$3i - 9j = \vec{AD} + 2i - 5j$

$\therefore \vec{AO} = i - 4j$

$\therefore \vec{DA} = -i + 4j \quad \therefore A(-1, 4)$

5 B.

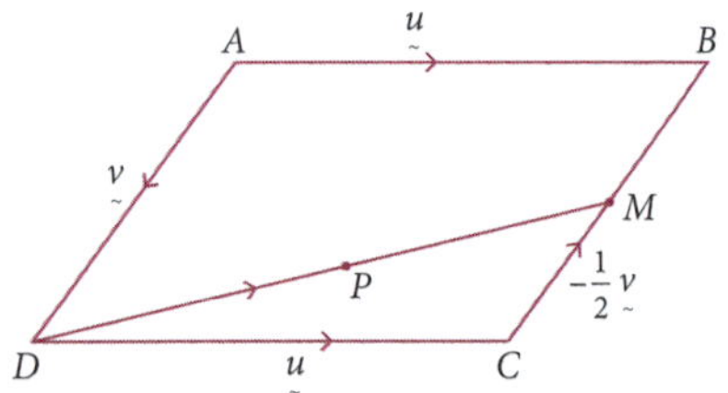

Hence, $\vec{DC} = u$ and $\vec{CM} = -\frac{1}{2}v$.

$\therefore \vec{DM} = u - \frac{1}{2}v$

$\therefore \vec{DP} = \frac{1}{2}(u - \frac{1}{2}v)$

Part B

6 a $p - q = 5i - 2j - (-4i + 2j)$

$= \mathbf{9i - 4j}$

b $2p + \frac{1}{2}q = 2(5i - 2j) + \frac{1}{2}(-4i + 2j)$

$= 10i - 4j - 2i + j$

$= 8i - 3j$

Now, $m + 2 = 8$

$m = 6$

Also, $1 - 2n = -3$

$-2n = -4$

$n = 2$

$\therefore \mathbf{m = 6 \text{ and } n = 2}$

7 $a = \begin{pmatrix} 3 \\ -1 \end{pmatrix}$ and $b = \begin{pmatrix} -2 \\ 3 \end{pmatrix}$

$\therefore a + b = \begin{pmatrix} 1 \\ 2 \end{pmatrix}$ and $a - b = \begin{pmatrix} 5 \\ -4 \end{pmatrix}$

$(3p - 2)(i + 2j) = (q - 2)(5i - 4j)$

$\therefore (3p - 2)i + 2(3p - 2)j = 5(q - 2)i - 4(q - 2)j$

$3p - 2 = 5q - 10$

$3p - 5q = -8 \ldots ①$

$6p - 4 = -4q + 8$

$6p + 4q = 12$

$3p + 2q = 6 \ldots ②$

$① - ②$: $-7q = -14$

$q = 2$

Substitute in ②:

$3p + 2(2) = 6$

$3p = 2$

$p = \frac{2}{3}$

$\therefore \mathbf{p = \frac{2}{3} \text{ and } q = 2}$

8 a $\vec{CA} = v - u$

As M is midpoint of CA, then $\vec{CM} = \frac{1}{2}(v - u)$.

b $\vec{CD} = v$ ($AB = DC$ and $AB \parallel DC$)

$\vec{MD} = v - \frac{1}{2}(v - u)$

$= \mathbf{\frac{1}{2}(u + v)}$

9

A
p
B
q
C
$3p - 2q$

10 $4a + 3b = 4 \ldots ①$

$-2a - b = 0 \ldots ②$

$-4a - 2b = 0 \ldots ③$

$① + ③$: $b = 4$

Subs. in ①: $4a + 3(4) = 4$

$4a = 4 - 12$

$4a = -8$

$a = -2$

$\therefore \mathbf{a = -2, b = 4}$

11 $\vec{CB} = \underset{\sim}{v} - \underset{\sim}{u}$

$\vec{MB} = \frac{1}{2}(\underset{\sim}{v} - \underset{\sim}{u})$

As $\vec{AB} = \underset{\sim}{v}$, then $\vec{TB} = \frac{2}{3}\underset{\sim}{v}$, and hence, $\vec{BT} = -\frac{2}{3}\underset{\sim}{v}$ ✓

$$\begin{aligned}\vec{MT} &= \frac{1}{2}(\underset{\sim}{v} - \underset{\sim}{u}) - \frac{2}{3}\underset{\sim}{v}\\ &= \frac{1}{2}\underset{\sim}{v} - \frac{1}{2}\underset{\sim}{u} - \frac{2}{3}\underset{\sim}{v}\\ &= -\frac{1}{6}\underset{\sim}{v} - \frac{1}{2}\underset{\sim}{u}\\ &= \mathbf{-\frac{1}{6}(3\underset{\sim}{u} + \underset{\sim}{v})}\end{aligned}$$ ✓

12 a $$\begin{aligned}\vec{BC} &= \vec{BO} + \vec{OC}\\ &= -3\underset{\sim}{i} - 5\underset{\sim}{j} + 5\underset{\sim}{i} - 4\underset{\sim}{j}\\ &= \mathbf{2\underset{\sim}{i} - 9\underset{\sim}{j}}\end{aligned}$$ ✓

b $$\begin{aligned}2(\vec{OB} - \vec{OC}) &= 2(3\underset{\sim}{i} + 5\underset{\sim}{j} - 5\underset{\sim}{i} + 4\underset{\sim}{j})\\ &= 2(-2\underset{\sim}{i} + 9\underset{\sim}{j})\\ &= -4\underset{\sim}{i} + 18\underset{\sim}{j} = \vec{AD}\end{aligned}$$ ✓

But $\vec{AD} = \vec{AO} + \vec{OD}$

$\therefore -4\underset{\sim}{i} + 18\underset{\sim}{j} = 2\underset{\sim}{i} - 4\underset{\sim}{j} + \vec{OD}$

$\therefore \mathbf{\vec{OD} = -6\underset{\sim}{i} + 22\underset{\sim}{j}}$ ✓

13 $$\begin{aligned}3\underset{\sim}{u} + 2\begin{pmatrix}4\\-1\end{pmatrix} &= \underset{\sim}{u} + 3\begin{pmatrix}2\\4\end{pmatrix}\\ 3\underset{\sim}{u} - \underset{\sim}{u} &= 3\begin{pmatrix}2\\4\end{pmatrix} - 2\begin{pmatrix}4\\-1\end{pmatrix} \checkmark\\ 2\underset{\sim}{u} &= \begin{pmatrix}6\\12\end{pmatrix} - \begin{pmatrix}8\\-2\end{pmatrix}\\ &= \begin{pmatrix}-2\\14\end{pmatrix}\\ \underset{\sim}{u} &= \begin{pmatrix}-1\\7\end{pmatrix}\\ &= \mathbf{-\underset{\sim}{i} + 7\underset{\sim}{j}}\end{aligned}$$ ✓

14 $\vec{MB} = -\frac{1}{2}\underset{\sim}{v}$ and $\vec{BN} = -\frac{1}{3}\underset{\sim}{u}$ ✓

$$\begin{aligned}\therefore \vec{MN} &= -\frac{1}{2}\underset{\sim}{v} + (-\frac{1}{3}\underset{\sim}{u})\\ &= \mathbf{-\frac{1}{6}(3\underset{\sim}{v} + 2\underset{\sim}{u})}\end{aligned}$$ ✓

15 $$\begin{aligned}\vec{RT} &= \vec{RQ} + \vec{QT}\\ &= \underset{\sim}{v} + \frac{2}{5}(\underset{\sim}{u} - \underset{\sim}{v}) \checkmark\\ &= \underset{\sim}{v} + \frac{2}{5}\underset{\sim}{u} - \frac{2}{5}\underset{\sim}{v}\\ &= \frac{2}{5}\underset{\sim}{u} + \frac{3}{5}\underset{\sim}{v}\\ &= \frac{1}{5}(2\underset{\sim}{u} + 3\underset{\sim}{v})\end{aligned}$$

$\therefore \mathbf{k = \frac{1}{5}}$ ✓

16 $2\underset{\sim}{u} + 2\underset{\sim}{v} + 2\underset{\sim}{w} + \vec{BA} = 0$ ✓

$\therefore \vec{BA} = -(2\underset{\sim}{u} + 2\underset{\sim}{v} + 2\underset{\sim}{w})$

$\therefore \vec{AB} = 2\underset{\sim}{u} + 2\underset{\sim}{v} + 2\underset{\sim}{w}$

$\vec{AM} = \underset{\sim}{u} + \underset{\sim}{v} + \underset{\sim}{w}$

But $\vec{AM} = \underset{\sim}{u} + \vec{QM}$ ✓

$$\begin{aligned}\therefore \vec{QM} &= \underset{\sim}{u} + \underset{\sim}{v} + \underset{\sim}{w} - \underset{\sim}{u}\\ &= \underset{\sim}{v} + \underset{\sim}{w}\end{aligned}$$

But $\vec{PN} = \underset{\sim}{v} + \underset{\sim}{w}$.

$\therefore$ **$\vec{QM} = \vec{PN}$ and hence, $QM \parallel PN$** ✓

17 $$\begin{aligned}3 - p + q &= 6\\ -p + q &= 3 \ldots ①\\ p + q + 5 &= 4\\ p + q &= -1 \ldots ② \checkmark\end{aligned}$$

① + ②: $2q = 2$

$q = 1$

Substitute in ②:

$p + 1 = -1$

$p = -2$ ✓

$\therefore \underset{\sim}{a} = 3\underset{\sim}{i} - \underset{\sim}{j}, \underset{\sim}{b} = -3\underset{\sim}{i} - 5\underset{\sim}{j}$

$\therefore \mathbf{2\underset{\sim}{a} + 3\underset{\sim}{b} = -3\underset{\sim}{i} - 17\underset{\sim}{j}}$ ✓

CHAPTER 10

FURTHER OPERATIONS WITH VECTORS

TOPIC: VECTORS

YEAR 12

1 Magnitude and direction of vectors

- **The magnitude of a vector** $\overrightarrow{AB}$ is the distance between points A and B.
- **If $\underset{\sim}{u} = x\underset{\sim}{i} + y\underset{\sim}{j}$** then the magnitude $\left|\underset{\sim}{u}\right| = \left|x\underset{\sim}{i} + y\underset{\sim}{j}\right| = \sqrt{x^2+y^2}$, using Pythagoras.

  Hence, $\left|\underset{\sim}{u}\right|^2 = x^2 + y^2$.
- **Non-zero vector $\underset{\sim}{u}$** converts to a unit vector $\hat{\underset{\sim}{u}}$ by dividing by its length: $\hat{\underset{\sim}{u}} = \dfrac{\underset{\sim}{u}}{\left|\underset{\sim}{u}\right|}$.
- **If $\underset{\sim}{u} = x\underset{\sim}{i} + y\underset{\sim}{j}$**, then the directional angle of the vector is given as $\theta = \tan^{-1}\left(\dfrac{y}{x}\right)$, which is the angle formed between the vector and the horizontal in the positive direction.
- **Also, for $\underset{\sim}{u} = x\underset{\sim}{i} + y\underset{\sim}{j}$**, $x = \left|\underset{\sim}{u}\right|\cos\theta$ and $y = \left|\underset{\sim}{u}\right|\sin\theta$.

EXAMPLE 1

If $\underset{\sim}{u} = \underset{\sim}{i} + 2\underset{\sim}{j}$ and $\underset{\sim}{v} = 3\underset{\sim}{i} - \underset{\sim}{j}$, find $\left|\underset{\sim}{u}+\underset{\sim}{v}\right|$.

$$\begin{aligned}\left|\underset{\sim}{u}+\underset{\sim}{v}\right| &= \left|\underset{\sim}{i}+2\underset{\sim}{j}+3\underset{\sim}{i}-\underset{\sim}{j}\right| \\ &= \left|4\underset{\sim}{i}+\underset{\sim}{j}\right| \\ &= \sqrt{4^2+1^2} \\ &= \sqrt{17}\end{aligned}$$

Remember: $\left|\underset{\sim}{u}\right| = \sqrt{x^2+y^2}$

EXAMPLE 2

If the vectors $\underset{\sim}{u} = 3\underset{\sim}{i} + m\underset{\sim}{j}$ and $\underset{\sim}{v} = 5\underset{\sim}{i} - 2\underset{\sim}{j}$ have the same magnitude, find the value of positive m.

As $\left|\underset{\sim}{u}\right| = \left|\underset{\sim}{v}\right|$, then $\sqrt{3^2+m^2} = \sqrt{5^2+(-2)^2}$

$$\begin{aligned}\sqrt{m^2+9} &= \sqrt{29} \\ m^2+9 &= 29 \\ m^2 &= 20 \\ m &= \pm\sqrt{20} \\ m &= 2\sqrt{5}, \text{ as } m > 0.\end{aligned}$$

Solve an equation to find m.

EXAMPLE 3

Find the unit vector $\hat{\underset{\sim}{u}}$ if $\underset{\sim}{u} = -3\underset{\sim}{i} + 4\underset{\sim}{j}$.

$$\begin{aligned}\left|\underset{\sim}{u}\right| &= \sqrt{(-3)^2+4^2} \\ &= 5 \\ \hat{\underset{\sim}{u}} &= \frac{\underset{\sim}{u}}{\left|\underset{\sim}{u}\right|} \\ &= \frac{-3\underset{\sim}{i}+4\underset{\sim}{j}}{5} \\ &= -\frac{3}{5}\underset{\sim}{i} + \frac{4}{5}\underset{\sim}{j}\end{aligned}$$

Unit vector is the vector divided by its length.

EXAMPLE 4

Find θ, the direction of $\underset{\sim}{v}$, if $\underset{\sim}{v} = -4\underset{\sim}{i} + 4\underset{\sim}{j}$.

$$\begin{aligned}\theta &= \tan^{-1}\left(\frac{4}{-4}\right) \\ &= 135^\circ\end{aligned}$$

Take care with the correct quadrant.

EXAMPLE 5

A vector $\underset{\sim}{u}$ has a magnitude of 2 and makes an angle of 60°. Express the vector in the form $\underset{\sim}{u} = x\underset{\sim}{i} + y\underset{\sim}{j}$.

$$\begin{aligned}\text{Use } x &= \left|\underset{\sim}{u}\right|\cos\theta \\ &= 2\cos 60^\circ \\ &= 2\times\frac{1}{2} \\ &= 1 \\ \text{Use } y &= \left|\underset{\sim}{u}\right|\sin\theta \\ &= 2\sin 60^\circ \\ &= 2\times\frac{\sqrt{3}}{2} \\ &= \sqrt{3} \\ \therefore \underset{\sim}{u} &= x\underset{\sim}{i} + y\underset{\sim}{j} \\ &= \underset{\sim}{i} + \sqrt{3}\underset{\sim}{j}\end{aligned}$$

Use $\underset{\sim}{u} = \left(\left|\underset{\sim}{u}\right|\cos\theta, \left|\underset{\sim}{u}\right|\sin\theta\right)$.

➔ PRACTICE

1 If $\underset{\sim}{u} = 5\underset{\sim}{i} - 3\underset{\sim}{j}$ and $\underset{\sim}{v} = \underset{\sim}{i} + 2\underset{\sim}{j}$, find:

i $|\underset{\sim}{u}+\underset{\sim}{v}|$

ii $|\underset{\sim}{v}-\underset{\sim}{u}|$

2 If $\underset{\sim}{u} = m\underset{\sim}{i} + 2\underset{\sim}{j}$ and $\underset{\sim}{v} = 2\underset{\sim}{i} - 4\underset{\sim}{j}$, find the value of m when:

i $|\underset{\sim}{u}| = |\underset{\sim}{v}|$

ii $|\underset{\sim}{u}-\underset{\sim}{v}| = 6$

3 i Find the unit vector $\hat{\underset{\sim}{u}}$ if $\underset{\sim}{u} = 6\underset{\sim}{i} - 3\underset{\sim}{j}$.

ii Show that, given $\underset{\sim}{v} = 4\underset{\sim}{i} + 4\underset{\sim}{j}$, then $\hat{\underset{\sim}{v}} = \frac{1}{\sqrt{2}}(\underset{\sim}{i}+\underset{\sim}{j})$.

4 Find θ, the direction of $\underset{\sim}{u}$, if:

i $\underset{\sim}{u} = -\sqrt{3}\,\underset{\sim}{i} - \underset{\sim}{j}$

ii $\underset{\sim}{u} = 2\underset{\sim}{i} - 5\underset{\sim}{j}$, to the nearest degree

5 i Express the vector $\underset{\sim}{u}$ in the form $x\underset{\sim}{i} + y\underset{\sim}{j}$, if the vector has a magnitude 4 and makes an angle of 120°.

ii Use the grid to represent the vector $\underset{\sim}{v}$ with magnitude $3\sqrt{2}$ and making an angle of 315° with the horizontal in the positive direction.

Answers ➲ p. 128

2 Scalar product (or dot product) of vectors

➔ If $\underset{\sim}{u} = x_1\underset{\sim}{i} + y_1\underset{\sim}{j}$, and $\underset{\sim}{v} = x_2\underset{\sim}{i} + y_2\underset{\sim}{j}$, the scalar product of vectors is defined by $\underset{\sim}{u}.\underset{\sim}{v} = x_1x_2 + y_1y_2$.

➔ Also, $\underset{\sim}{u}.\underset{\sim}{v} = |\underset{\sim}{u}||\underset{\sim}{v}|\cos\theta$, where θ is the angle between the vectors $\underset{\sim}{u}$ and $\underset{\sim}{v}$.

➔ To find the angle between two vectors θ, use $\cos\theta = \frac{\underset{\sim}{u}.\underset{\sim}{v}}{|\underset{\sim}{u}||\underset{\sim}{v}|}$, or $\cos\theta = \frac{x_1x_2 + y_1y_2}{|\underset{\sim}{u}||\underset{\sim}{v}|}$.

➔ Also, $\underset{\sim}{u}.\underset{\sim}{v} = \underset{\sim}{v}.\underset{\sim}{u}$; $k(\underset{\sim}{u}.\underset{\sim}{v}) = (k\underset{\sim}{u}).\underset{\sim}{v} = \underset{\sim}{u}.(k\underset{\sim}{v})$; $\underset{\sim}{v}.\underset{\sim}{v} = |\underset{\sim}{v}|^2$.

EXAMPLE 1

If $\underset{\sim}{u} = 3\underset{\sim}{i} + 2\underset{\sim}{j}$ and $\underset{\sim}{v} = 4\underset{\sim}{i} - 5\underset{\sim}{j}$, find:

a $\underset{\sim}{u}.\underset{\sim}{v}$

b $\underset{\sim}{u}.\underset{\sim}{u} + \underset{\sim}{v}.\underset{\sim}{v}$

a $\underset{\sim}{u}.\underset{\sim}{v} = 3 \times 4 + 2 \times (-5)$

$= 2$

b $\underset{\sim}{u}.\underset{\sim}{u} + \underset{\sim}{v}.\underset{\sim}{v} = 3^2 + 2^2 + 4^2 + (-5)^2$

$= 54$

$\underset{\sim}{u}.\underset{\sim}{v} = x_1x_2 + y_1y_2$

EXAMPLE 2

If $|\underset{\sim}{u}| = 5$, $|\underset{\sim}{v}| = 4$ and the angle between $\underset{\sim}{u}$ and $\underset{\sim}{v}$ is 60°, find $\underset{\sim}{u}.\underset{\sim}{v}$.

$\underset{\sim}{u}.\underset{\sim}{v} = |\underset{\sim}{u}||\underset{\sim}{v}|\cos\theta$

$= 5 \times 4 \times \cos 60°$

$= 5 \times 4 \times \frac{1}{2}$

$= 10$

Use exact trig ratios.

EXAMPLE 3

P and Q are defined by the position vectors $\underset{\sim}{u}$ and $\underset{\sim}{v}$ respectively, where $\underset{\sim}{u} = 3\underset{\sim}{i} + 5\underset{\sim}{j}$ and $\underset{\sim}{v} = -4\underset{\sim}{i} + 2\underset{\sim}{j}$. If O is the origin, find the size of $\angle POQ$, to the nearest degree.

$\cos\theta = \frac{\underset{\sim}{u}.\underset{\sim}{v}}{|\underset{\sim}{u}||\underset{\sim}{v}|}$

$= \frac{3\times(-4)+5\times 2}{\sqrt{3^2+5^2}.\sqrt{(-4)^2+2^2}}$

$= \frac{-2}{\sqrt{34}.\sqrt{20}}$

$= -0.076\,696\,498\ldots$

$\theta = 94.398\,705\,36\ldots$

$= 94$ (nearest whole)

$\therefore \angle POQ = 94°$

$0° \le \theta \le 180°$

➔ PRACTICE

1 i If $\underset{\sim}{u} = 5\underset{\sim}{i} - 3\underset{\sim}{j}$ and $\underset{\sim}{v} = 2\underset{\sim}{i} - \underset{\sim}{j}$, find:

a $\underset{\sim}{u}.\underset{\sim}{v}$

b $\underset{\sim}{u}.\underset{\sim}{u} + \underset{\sim}{v}.\underset{\sim}{v}$

ii If $\underset{\sim}{u} = -3\underset{\sim}{j}$ and $\underset{\sim}{v} = -5\underset{\sim}{i} - 2\underset{\sim}{j}$, find:

a $\underset{\sim}{u}.\underset{\sim}{v}$

b $\underset{\sim}{v}.(\underset{\sim}{u} + \underset{\sim}{v})$

2 i If $|\underset{\sim}{u}| = 3\sqrt{2}$, $|\underset{\sim}{v}| = 6$ and the angle between $\underset{\sim}{u}$ and $\underset{\sim}{v}$ is 45°, find $\underset{\sim}{u}.\underset{\sim}{v}$.

ii If $\underset{\sim}{u} = \underset{\sim}{i} - 2\underset{\sim}{j}$ and $\underset{\sim}{v} = 3\underset{\sim}{i} + b\underset{\sim}{j}$, and the angle between the vectors is 60°, find the value of b correct to two decimal places.

3 i A and B are defined by the position vectors $\underset{\sim}{u}$ and $\underset{\sim}{v}$ respectively, where $\underset{\sim}{u} = -2\underset{\sim}{i} + 5\underset{\sim}{j}$ and $\underset{\sim}{v} = 6\underset{\sim}{i} + 5\underset{\sim}{j}$. If O is the origin, find the size of $\angle AOB$, to the nearest degree.

ii Find the acute angle between the vectors $\underset{\sim}{i} + 3\underset{\sim}{j}$ and $-3\underset{\sim}{i} + 2\underset{\sim}{j}$.

Answers ➲ pp. 127–128

3 Parallel and perpendicular vectors

→ **Vectors $\underset{\sim}{u}$ and $\underset{\sim}{v}$ are parallel** if $\underset{\sim}{u}.\underset{\sim}{v} = \pm|\underset{\sim}{u}||\underset{\sim}{v}|$, or $\underset{\sim}{u} = \lambda\underset{\sim}{v}$.

→ **Vectors $\underset{\sim}{u}$ and $\underset{\sim}{v}$ are perpendicular (or orthogonal)** if $\underset{\sim}{u}.\underset{\sim}{v} = 0$.

EXAMPLE 1

Use the dot product to determine whether the vectors $\underset{\sim}{u} = 2\underset{\sim}{i} + 4\underset{\sim}{j}$ and $\underset{\sim}{v} = -6\underset{\sim}{i} + 3\underset{\sim}{j}$ are parallel, perpendicular or neither.

$$\begin{aligned}\underset{\sim}{u}.\underset{\sim}{v} &= x_1x_2 + y_1y_2\\ &= 2 \times (-6) + 4 \times 3\\ &= 0\end{aligned}$$

Use the dot product.

$\therefore$ the vectors are perpendicular.

EXAMPLE 2

Find the value of m if the vectors:

a $\underset{\sim}{u} = 3\underset{\sim}{i} + m\underset{\sim}{j}$ and $\underset{\sim}{v} = \underset{\sim}{i} - 2\underset{\sim}{j}$ are parallel

b $\underset{\sim}{u} = 4\underset{\sim}{i} - 3\underset{\sim}{j}$ and $\underset{\sim}{v} = (m + 2)\underset{\sim}{i} + 8\underset{\sim}{j}$ are perpendicular

a If parallel, then $\underset{\sim}{u} = \lambda\underset{\sim}{v}$.

$$3\underset{\sim}{i} + m\underset{\sim}{j} = \lambda(\underset{\sim}{i} - 2\underset{\sim}{j})$$
$$\therefore 3\underset{\sim}{i} = \lambda\underset{\sim}{i}$$
$$\therefore \lambda = 3$$
$$\therefore m\underset{\sim}{j} = 3(-2\underset{\sim}{j})$$
$$\therefore m = -6$$

Parallel if $\underset{\sim}{u} = \lambda\underset{\sim}{v}$; perpendicular if $\underset{\sim}{u}.\underset{\sim}{v} = 0$

b
$$\begin{aligned}x_1x_2 + y_1y_2 &= 0\\ 4(m + 2) + (-3) \times 8 &= 0\\ 4m + 8 - 24 &= 0\\ 4m &= 16\\ m &= 4\end{aligned}$$

→ PRACTICE

1 Use the dot product to determine whether these vectors are parallel, perpendicular or neither:

i $\underset{\sim}{u} = -6\underset{\sim}{i} + 3\underset{\sim}{j}$ and $\underset{\sim}{v} = -4\underset{\sim}{i} - 8\underset{\sim}{j}$

ii $\underset{\sim}{u} = 3\underset{\sim}{i} - 6\underset{\sim}{j}$ and $\underset{\sim}{v} = -\underset{\sim}{i} + 2\underset{\sim}{j}$

2 Find the value of m if the vectors:

i a $\underset{\sim}{u} = 5\underset{\sim}{i} + m\underset{\sim}{j}$ and $\underset{\sim}{v} = -3\underset{\sim}{i} - 4\underset{\sim}{j}$ are parallel

b $\underset{\sim}{u} = -2\underset{\sim}{i} - 6\underset{\sim}{j}$ and $\underset{\sim}{v} = 3\underset{\sim}{i} + m\underset{\sim}{j}$ are perpendicular

ii a $\underset{\sim}{u} = 4\underset{\sim}{i} + (2m + 3)\underset{\sim}{j}$ and $\underset{\sim}{v} = 3\underset{\sim}{i} - \underset{\sim}{j}$ are parallel

b $\underset{\sim}{u} = -5\underset{\sim}{i}$ and $\underset{\sim}{v} = (2 - 3m)\underset{\sim}{i} + \underset{\sim}{j}$ are orthogonal

Answers ➲ p. 129

4 Vector projection

→ **The vector projection** of $\underset{\sim}{u}$ in the direction of vector $\underset{\sim}{v}$ can be expressed as $\text{proj}_{\underset{\sim}{v}}\,\underset{\sim}{u} = \underset{\sim}{u}_v = \dfrac{\underset{\sim}{u}.\underset{\sim}{v}}{|\underset{\sim}{v}|^2}\underset{\sim}{v}$

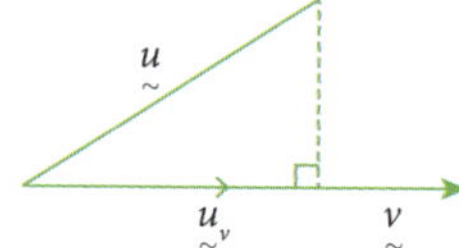

EXAMPLE 1

If $\underset{\sim}{u} = 2\underset{\sim}{i} + 3\underset{\sim}{j}$ and $\underset{\sim}{v} = \underset{\sim}{i} + 4\underset{\sim}{j}$, find the vector projection of $\underset{\sim}{u}$ in the direction of $\underset{\sim}{v}$.

$$\begin{aligned}\underset{\sim}{u}.\underset{\sim}{v} &= 2 \times 1 + 3 \times 4\\ &= 14\\ |\underset{\sim}{v}|^2 &= 1^2 + 4^2\\ &= 17\\ \therefore \quad \underset{\sim}{u}_v &= \frac{14}{17}(\underset{\sim}{i} + 4\underset{\sim}{j})\end{aligned}$$

Use $\underset{\sim}{u}_v = \dfrac{\underset{\sim}{u}.\underset{\sim}{v}}{|\underset{\sim}{v}|^2}\underset{\sim}{v}$

→ PRACTICE

1 i If $\underset{\sim}{u} = 2\underset{\sim}{i} - 4\underset{\sim}{j}$ and $\underset{\sim}{v} = 5\underset{\sim}{i} - 3\underset{\sim}{j}$, find the vector projection of $\underset{\sim}{u}$ in the direction of $\underset{\sim}{v}$.

ii If $\underset{\sim}{u} = -3\underset{\sim}{i} + \underset{\sim}{j}$ and $\underset{\sim}{v} = -\underset{\sim}{i} + 2\underset{\sim}{j}$, find the vector projection of $\underset{\sim}{v}$ in the direction of $\underset{\sim}{u}$.

Answers ➲ p. 129

5 Displacement, force and velocity as vectors

- **The displacement is a vector** which points from the initial position of an object to its final position.
- **The force of an object** is a vector as it has a size, measured in Newtons, N, and a direction.
- **The magnitude of the velocity vector** gives the speed of an object and the vector direction gives its direction.

EXAMPLE 1

Michael cycles 45 km in a northerly direction. He then travels 65 km in an easterly direction. Calculate the magnitude and direction of his displacement ($\underset{\sim}{r}$) from his starting point.

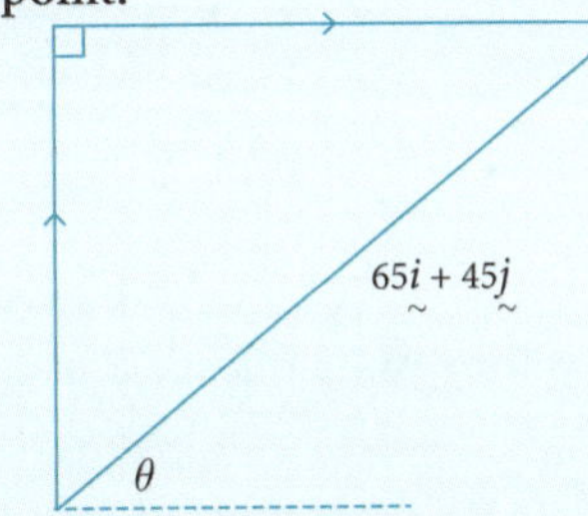

$$\text{Magnitude} = \sqrt{45^2 + 65^2}$$
$$= 79.0569415\ldots$$
$$= 79 \text{ (nearest whole)}$$
$$\tan\theta = \frac{45}{65}$$
$$\theta = 34.69515353\ldots$$
$$= 35 \text{ (nearest whole)}$$

Write $\underset{\sim}{r}$ in component form.

∴ Michael's displacement is 79 km in a direction of 35° above the horizontal.

EXAMPLE 2

Four forces are acting on the object O, as shown.

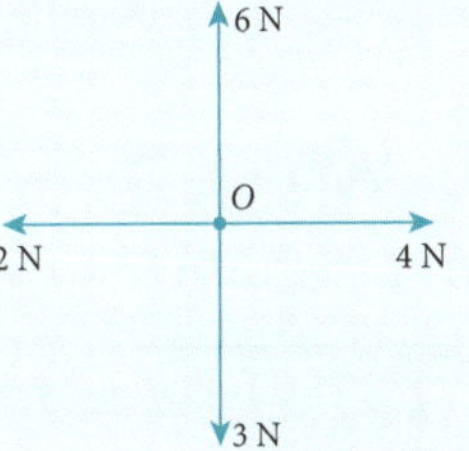

a Express the resultant force $\underset{\sim}{R}$ in component form.

b Calculate the magnitude of $\underset{\sim}{R}$ and the angle that it makes with the horizontal.

a $\underset{\sim}{R} = (4-2)\underset{\sim}{i} + (6-3)\underset{\sim}{j}$
$= 2\underset{\sim}{i} + 3\underset{\sim}{j}$

Use subtraction.

b $|\underset{\sim}{R}| = \sqrt{2^2 + 3^2}$
$= \sqrt{13}$

$\tan\theta = \frac{3}{2}$
$= 56.30993247\ldots$
$= 56$ (nearest whole)

∴ magnitude of $\sqrt{13}$ N at an angle of 56°.

EXAMPLE 3

Find the magnitude and direction of the resultant force of the forces 4 N and 6 N on a particle as shown in the diagram below.

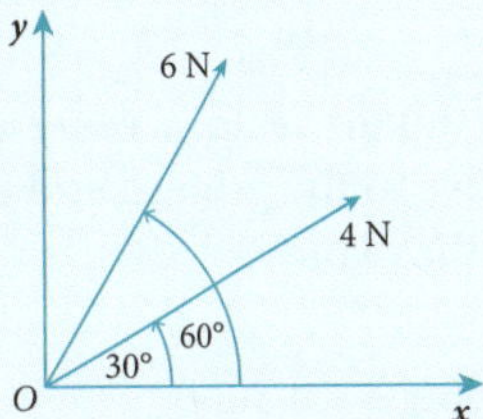

Let force vectors be $\underset{\sim}{u}$ and $\underset{\sim}{v}$.

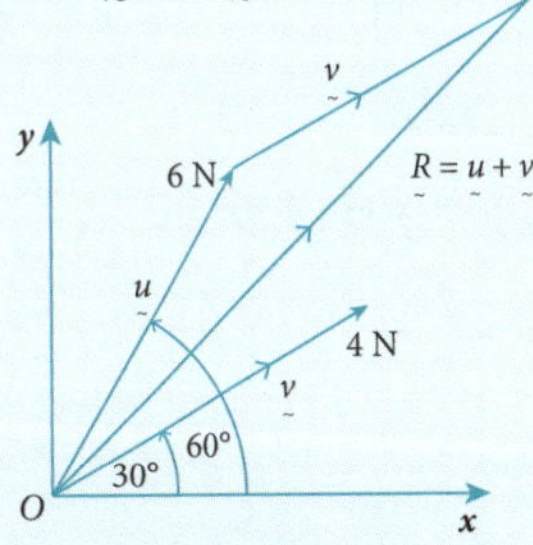

$\underset{\sim}{u} = 6\cos 60°\underset{\sim}{i} + 6\sin 60°\underset{\sim}{j}$
$= 3\underset{\sim}{i} + 3\sqrt{3}\underset{\sim}{j}$

$$\underset{\sim}{v} = 4\cos 30°\underset{\sim}{i} + 4\sin 30°\underset{\sim}{j}$$
$$= 2\sqrt{3}\underset{\sim}{i} + 2\underset{\sim}{j}$$
$$\underset{\sim}{R} = \underset{\sim}{u} + \underset{\sim}{v}$$
$$= 3\underset{\sim}{i} + 3\sqrt{3}\underset{\sim}{j} + 2\sqrt{3}\underset{\sim}{i} + 2\underset{\sim}{j}$$
$$= (3 + 2\sqrt{3})\underset{\sim}{i} + (3\sqrt{3} + 2)\underset{\sim}{j}$$
$$|\underset{\sim}{R}| = \sqrt{(3+2\sqrt{3})^2 + (3\sqrt{3}+2)^2}$$
$$= 9.673\,118\,39\ldots$$
$$= 9.67 \text{ (2 dec. pl.)}$$
$$\tan\theta = \frac{3\sqrt{3}+2}{3+2\sqrt{3}}$$
$$= 1.113\,248\,654\ldots$$
$$\theta = 48.067\,537\,29\ldots$$
$$= 48 \text{ (nearest whole)}$$

$\therefore$ the resultant force is 9.67 N at an angle of 48°.

The resultant vector is $\underset{\sim}{u} + \underset{\sim}{v}$.

EXAMPLE 4

A plane is flying on a bearing of 060°T at a speed of 240 km/h. It is met by a cross wind at a bearing of 330° at a speed of 60 km/h. In the diagram the plane's path is represented by vector $\underset{\sim}{v}$ and the wind's path by vector $\underset{\sim}{w}$.

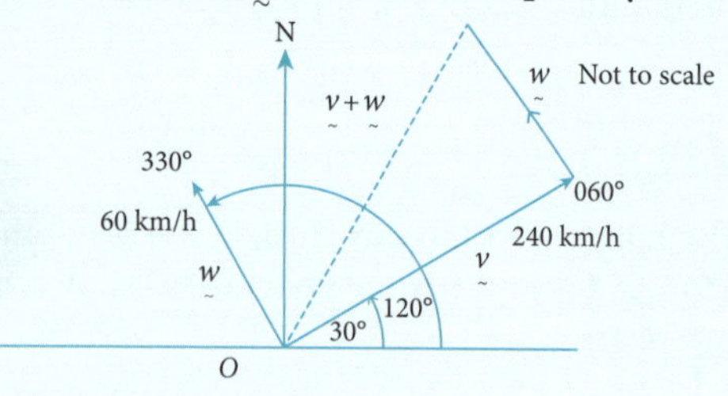

Determine the resultant vector r representing the path of the plane relative to the ground and use it to find the ground speed of the plane.

$$\underset{\sim}{v} = 240(\cos 30°\underset{\sim}{i} + \sin 30°\underset{\sim}{j})$$
$$= 240\left(\frac{\sqrt{3}}{2}\underset{\sim}{i} + \frac{1}{2}\underset{\sim}{j}\right)$$
$$= 120\sqrt{3}\underset{\sim}{i} + 120\underset{\sim}{j}$$
$$\underset{\sim}{w} = 60(\cos 120°\underset{\sim}{i} + \sin 120°\underset{\sim}{j})$$
$$= 60\left(-\frac{1}{2}\underset{\sim}{i} + \frac{\sqrt{3}}{2}\underset{\sim}{j}\right)$$
$$= -30\underset{\sim}{i} + 30\sqrt{3}\underset{\sim}{j}$$
$$\underset{\sim}{r} = \underset{\sim}{v} + \underset{\sim}{w} = (120\sqrt{3} - 30)\underset{\sim}{i} + (120 + 30\sqrt{3})\underset{\sim}{j}$$
$$|\underset{\sim}{r}| = |\underset{\sim}{v} + \underset{\sim}{w}| = \sqrt{(120\sqrt{3}-30)^2 + (120+30\sqrt{3})^2}$$
$$= 247.386\,3375\ldots$$
$$= 247 \text{ (nearest whole)}$$

$\therefore$ ground speed of 247 km/h.

Take care with bearings and angles on the diagram.

→ PRACTICE

1 i Simone sails 3.6 km in an easterly direction and then turns to sail 2.8 km in a northerly direction. Calculate the magnitude and direction of her displacement from her starting point.

ii Ajit runs 4 km due north then walks 6 km on a bearing of 150°. Use vectors to find the magnitude and direction of his displacement from his starting point.

2 i Four forces are acting on the object O, as shown.

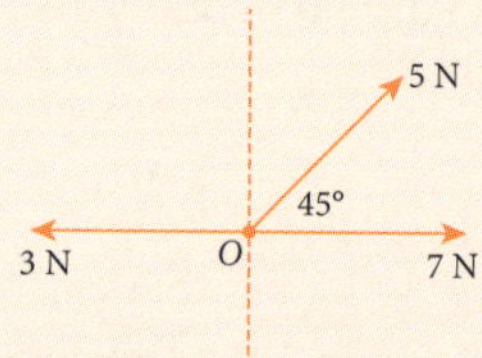

a Express the resultant force $\underset{\sim}{R}$ in component form.

b Calculate the magnitude of $\underset{\sim}{R}$ and the angle that it makes with the horizontal.

ii Three forces are acting on the object O, as shown.

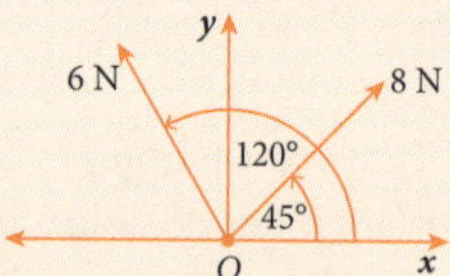

a Express the resultant force $\underset{\sim}{R}$ in component form.

b Calculate the magnitude of $\underset{\sim}{R}$ and the angle that it makes with the horizontal.

3 i Find the magnitude and direction of the resultant force of the forces 6 N and 8 N on a particle as shown in the diagram below.

ii Find the magnitude and direction of the resultant force of the forces $4\sqrt{3}$ N and $3\sqrt{2}$ N on a particle as shown in the diagram below.

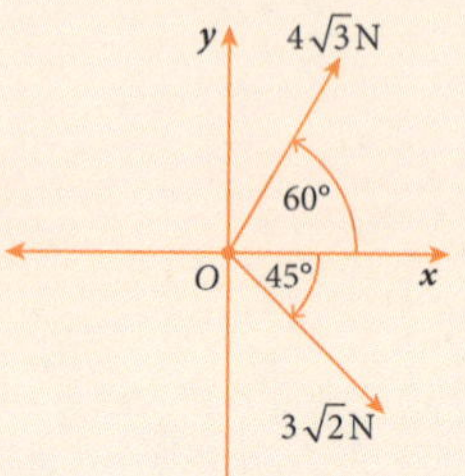

4 **i** **A plane is flying on a bearing of 225°T at a speed of 280 km/h. It is met by a cross wind at a bearing of 150° at a speed of 48 km/h. In the diagram the plane's path is represented by vector $\underset{\sim}{v}$ and the wind's path by vector $\underset{\sim}{w}$.**

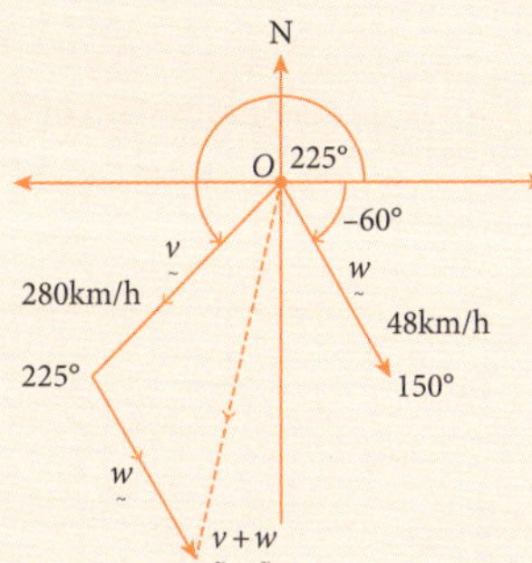

Use the resultant vector $\underset{\sim}{r}$ representing the path of the plane relative to the ground to find the ground speed of the plane.

ii **A cruise ship's captain sets the ship's velocity to be 38 km/h on a bearing of 210°. An ocean current is flowing toward a bearing of 330° at a speed of 12 km/h. Draw a vector diagram of this situation and state the ship's resultant speed and bearing.**

Answers ➲ pp. 130–131

6 Geometric proofs

➔ **You may be required to prove** the following:

- **The diagonals of a parallelogram** meet at right angles if and only if it is a rhombus.
- **The midpoints of the sides of a quadrilateral** join to form a parallelogram.
- **The sum of the squares of the lengths of the diagonals of a parallelogram** is equal to the sum of the squares of the lengths of the sides.

EXAMPLE 1

Using vectors, prove the following result: The diagonals of a rhombus are perpendicular.

Let $ABCD$ be a rhombus.

Let $\overrightarrow{BC} = \underset{\sim}{u}$ and $\overrightarrow{AB} = \underset{\sim}{v}$.

Diagonals are AC and BD.

$\therefore \overrightarrow{AC} = \underset{\sim}{v} + \underset{\sim}{u} = \underset{\sim}{u} + \underset{\sim}{v}$

$\therefore \overrightarrow{BD} = \underset{\sim}{u} - \underset{\sim}{v}$

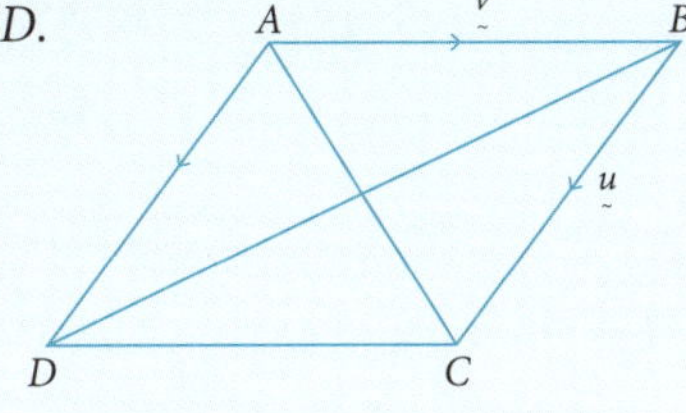

$\therefore \overrightarrow{AC}.\overrightarrow{BD} = (\underset{\sim}{u} + \underset{\sim}{v})(\underset{\sim}{u} - \underset{\sim}{v})$

$= \underset{\sim}{u}.\underset{\sim}{u} - \underset{\sim}{v}.\underset{\sim}{v}$

$= |\underset{\sim}{u}|^2 - |\underset{\sim}{v}|^2$

To show perpendicular, use dot product.

But $|\underset{\sim}{u}|^2 = |\underset{\sim}{v}|^2$ (equal sides of rhombus)

$\therefore \overrightarrow{AC}.\overrightarrow{BD} = 0$

$\therefore AC \perp BD$

EXAMPLE 2

Using vectors, prove the following result: The midpoints of the sides of a quadrilateral join to form a parallelogram.

Let $ABCD$ be a quadrilateral.

Let $\overrightarrow{DC} = \underset{\sim}{a}$, $\overrightarrow{CB} = \underset{\sim}{b}$, $\overrightarrow{DA} = \underset{\sim}{c}$, $\overrightarrow{AB} = \underset{\sim}{d}$.

Record your working on your diagram.

$\overrightarrow{YX} = \frac{1}{2}\underset{\sim}{a} + \frac{1}{2}\underset{\sim}{b}$

$= \frac{1}{2}(\underset{\sim}{a} + \underset{\sim}{b})$

$\overrightarrow{ZW} = \frac{1}{2}\underset{\sim}{c} + \frac{1}{2}\underset{\sim}{d}$

$= \frac{1}{2}(\underset{\sim}{c} + \underset{\sim}{d})$

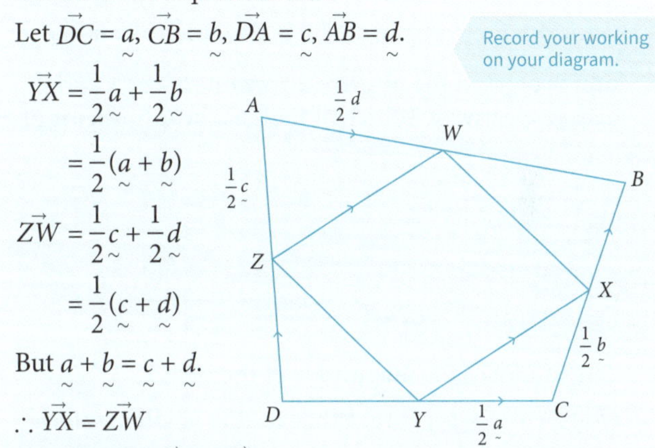

But $\underset{\sim}{a} + \underset{\sim}{b} = \underset{\sim}{c} + \underset{\sim}{d}$.

$\therefore \overrightarrow{YX} = \overrightarrow{ZW}$

$\therefore$ Similarly, $\overrightarrow{WX} = \overrightarrow{ZY}$

$\therefore WXYZ$ is a parallelogram (opposite sides equal and parallel).

➔ PRACTICE

1 **Using vectors, prove the following results.**

i **If the diagonals of a parallelogram are perpendicular, the parallelogram is a rhombus.**

ii **The segment that joins the midpoints of two sides of a triangle is parallel to the third side and has a length equal to half the length of the third side.**

2 **Using vectors, prove the following results.**

i **The sum of the squares of the lengths of the diagonals of a parallelogram is equal to the sum of the squares of the lengths of the sides.**

ii **If the diagonals of a quadrilateral bisect each other, then the quadrilateral is a parallelogram.**

Answers ➲ p. 131

HSC EXAM-TYPE QUESTIONS

Total marks: 35

PART A

(Suggested time: 10 minutes)

(1 mark each)

1 Given $\underset{\sim}{u} = 2\underset{\sim}{i} + m\underset{\sim}{j}$ and $\underset{\sim}{v} = 4\underset{\sim}{i} + 2\underset{\sim}{j}$, for what value of m is the vector $\underset{\sim}{u} - \underset{\sim}{v}$ perpendicular to the vector $\underset{\sim}{v}$?

A $m = -6$ **B** $m = -2$ **C** $m = 2$ **D** $m = 6$

2 Find the exact value of the cosine of the angle between $\underset{\sim}{u} = \begin{pmatrix} 2 \\ 4 \end{pmatrix}$ and $\underset{\sim}{v} = \begin{pmatrix} 4 \\ 2 \end{pmatrix}$.

A 0 **B** $\frac{1}{2}$ **C** $\frac{4}{5}$ **D** 1

3 Consider the vector $\underset{\sim}{v} = 2\sqrt{3}\underset{\sim}{i} + \sqrt{2}\underset{\sim}{j}$, where $\underset{\sim}{i}$ and $\underset{\sim}{j}$ are unit vectors in the positive directions of the x and y axes respectively. Which of these is the unit vector of $\underset{\sim}{v}$?

A $\dfrac{2\sqrt{3}\underset{\sim}{i}+\sqrt{2}\underset{\sim}{j}}{\sqrt{14}}$ **B** $\dfrac{2\sqrt{3}\underset{\sim}{i}+\sqrt{2}\underset{\sim}{j}}{4\sqrt{3}}$

C $\dfrac{2\sqrt{3}\underset{\sim}{i}+\sqrt{2}\underset{\sim}{j}}{12}$ **D** $\dfrac{2\sqrt{3}\underset{\sim}{i}+\sqrt{2}\underset{\sim}{j}}{4\sqrt{6}}$

4 The square of the magnitude of the vector $(p + 1)\underset{\sim}{i} + (3 - p)\underset{\sim}{j}$ is 16. What are the possible values of p?

A $p = 3, -2$ **B** $p = 2, -3$

C $p = -1, -2$ **D** $p = 3, -1$

5 A force $\underset{\sim}{F}$ of magnitude 12 N acts on a body at an angle of 150° in the clockwise direction to the vector $\underset{\sim}{i}$. Which of these is a vector representation of this force?

A $-6\underset{\sim}{i} + 6\sqrt{3}\underset{\sim}{j}$ **B** $-6\underset{\sim}{i} - 6\sqrt{3}\underset{\sim}{j}$

C $-6\sqrt{3}\underset{\sim}{i} - 6\underset{\sim}{j}$ **D** $-6\sqrt{3}\underset{\sim}{i} + 6\underset{\sim}{j}$

PART B

(Suggested time: 45 minutes)

Show all working.

6 If $\underset{\sim}{v}$ and $\underset{\sim}{w}$ are orthogonal vectors, where $\underset{\sim}{v} = 5\underset{\sim}{i} + 3\underset{\sim}{j}$, $\underset{\sim}{w} = (m + 4)\underset{\sim}{i} - 2\underset{\sim}{j}$, find the value of m. (2 marks)

7 The vectors $\underset{\sim}{i}$ and $\underset{\sim}{j}$ are unit vectors along the x-axis and y-axis respectively. The vectors $\underset{\sim}{a} = 2\underset{\sim}{i} + 5\underset{\sim}{j}$ and $\underset{\sim}{b} = \underset{\sim}{i} + 6\underset{\sim}{j}$ are given.

a Find $3\underset{\sim}{a} - 2\underset{\sim}{b}$ in terms of $\underset{\sim}{i}$ and $\underset{\sim}{j}$. (1 mark)

b Another vector $\underset{\sim}{c}$ has the same direction as $3\underset{\sim}{a} - 2\underset{\sim}{b}$ and has a magnitude of 15. Find $\underset{\sim}{c}$ in terms of $\underset{\sim}{i}$ and $\underset{\sim}{j}$. (2 marks)

8 The diagram shows a triangle with vertices A, B and O. The midpoints of line segments AB, BO and OA are P, Q and R respectively.

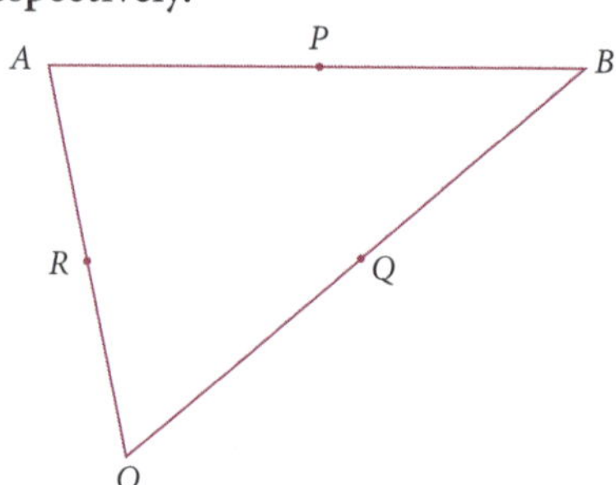

Let O be the origin with vectors $\overrightarrow{OA} = \underset{\sim}{u}$ and $\overrightarrow{OB} = \underset{\sim}{v}$. Use a vector method to prove that $RPBQ$ is a parallelogram. (2 marks)

9 The points A, B and C have position vectors $\underset{\sim}{a} = 3\underset{\sim}{i} + 2\underset{\sim}{j}$, $\underset{\sim}{b} = 5\underset{\sim}{i} + 7\underset{\sim}{j}$ and $\underset{\sim}{c} = 9\underset{\sim}{i} - 2\underset{\sim}{j}$ respectively.

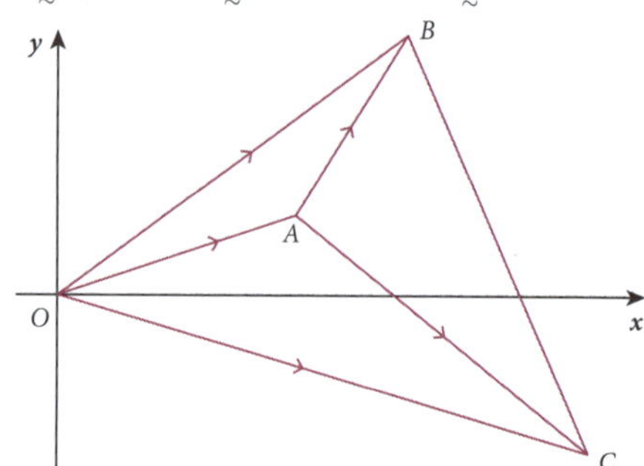

What is the size of the cosine of angle BAC, to the nearest degree? (3 marks)

10 What is the vector projection of $\underset{\sim}{b}$ in the direction of $\underset{\sim}{a}$ if $\underset{\sim}{a} = 3\underset{\sim}{i} + 4\underset{\sim}{j}$ and $\underset{\sim}{b} = -10\underset{\sim}{i} + 15\underset{\sim}{j}$? (2 marks)

11 $OABC$ is a parallelogram. M is the midpoint of BC and N lies on OB such that $ON = 2NB$.

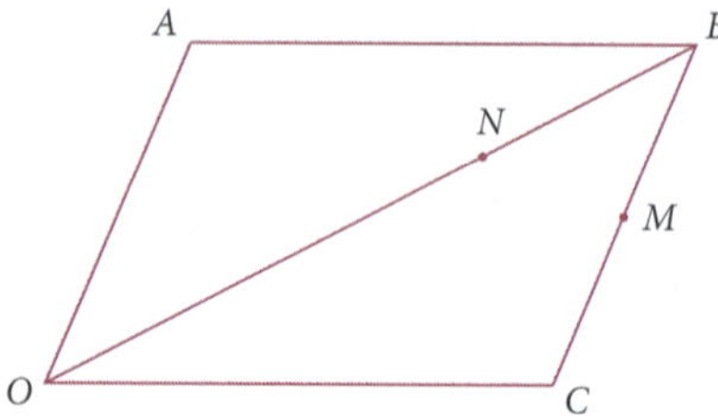

Let $\overrightarrow{OA} = \underset{\sim}{a}$ and $\overrightarrow{OC} = \underset{\sim}{c}$. Use a vector method to show that A, N and M are collinear. (3 marks)

12 Two vectors are given as $\underset{\sim}{s} = 3\underset{\sim}{i} - m\underset{\sim}{j}$ and $\underset{\sim}{t} = -5\underset{\sim}{j} - n\underset{\sim}{j}$, where $m > 0$ and $n > 0$. If $|\underset{\sim}{s}| = 5$, and $\underset{\sim}{s}$ and $\underset{\sim}{t}$ are perpendicular, find the values of m and n. (2 marks)

13 If θ is the angle between $\underset{\sim}{a} = -2\sqrt{3}\underset{\sim}{i} - 2\underset{\sim}{j}$ and $\underset{\sim}{b} = \underset{\sim}{i} + \sqrt{3}\underset{\sim}{j}$, show that $\cos 2\theta = \frac{1}{2}$. (2 marks)

14 A container ship's captain sets the ship's velocity to be $20\sqrt{2}$ km/h on a bearing of 225°. An ocean current to the ship is flowing toward a bearing of 120° at a speed of 16 km/h.

a Draw a vector diagram of this situation. (1 mark)

b What is the resultant velocity and bearing of the ship's path? (3 marks)

15 $ABCD$ is a square where M is a midpoint of AO. Let $\overrightarrow{OC} = a\underset{\sim}{i}$ and $\overrightarrow{OA} = a\underset{\sim}{j}$. P is a point on MB such that $\overrightarrow{MP} = \lambda\overrightarrow{MB}$.

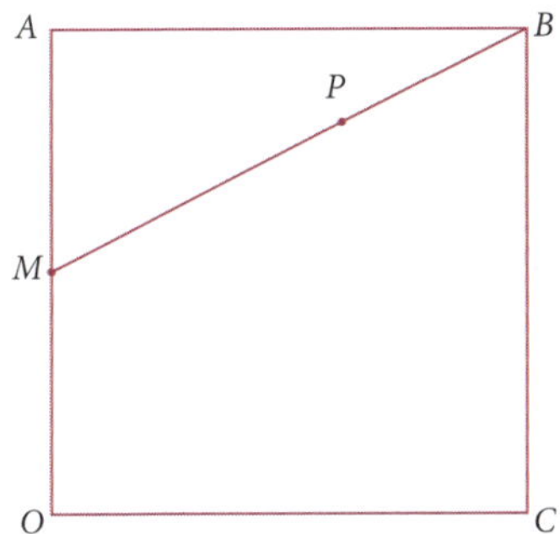

a Show $\overrightarrow{MP} = \lambda\left(a\underset{\sim}{i} + \frac{1}{2}a\underset{\sim}{j}\right)$. (1 mark)

b If MB is perpendicular to CP find the value of λ. (3 marks)

c If $\theta = \angle OCP$, show that $\cos\theta = \frac{\sqrt{5}}{5}$. (3 marks)

ANSWERS

PRACTICE

1 Magnitude and direction of vectors ➲ p. 121

1 i $\left|\underset{\sim}{u}+\underset{\sim}{v}\right| = \left|5\underset{\sim}{i}-3\underset{\sim}{j}+\underset{\sim}{i}+2\underset{\sim}{j}\right|$

$= \left|6\underset{\sim}{i}-\underset{\sim}{j}\right|$

$= \sqrt{6^2+(-1)^2}$

$= \mathbf{\sqrt{37}}$

ii $\left|\underset{\sim}{v}-\underset{\sim}{u}\right| = \left|\underset{\sim}{i}+2\underset{\sim}{j}-5\underset{\sim}{i}+3\underset{\sim}{j}\right|$

$= \left|-4\underset{\sim}{i}+5\underset{\sim}{j}\right|$

$= \sqrt{(-4)^2+5^2}$

$= \mathbf{\sqrt{41}}$

2 i $\left|\underset{\sim}{u}\right| = \left|\underset{\sim}{v}\right|$

$\sqrt{m^2+2^2} = \sqrt{2^2+(-4)^2}$

$\sqrt{m^2+4} = \sqrt{20}$

$m^2 + 4 = 20$

$m^2 = 16$

$\mathbf{m = \pm 4}$

ii $\left|\underset{\sim}{u}-\underset{\sim}{v}\right| = 6$

$\left|m\underset{\sim}{i}+2\underset{\sim}{j}-2\underset{\sim}{i}+4\underset{\sim}{j}\right| = 6$

$\sqrt{(m-2)^2+(2+4)^2} = 6$

$m^2 - 4m + 4 + 36 = 36$

$m^2 - 4m + 4 = 0$

$(m - 2)^2 = 0$

$\mathbf{m = 2}$

3 i $\left|\underset{\sim}{u}\right| = \sqrt{6^2+(-3)^2}$

$= \sqrt{45}$

$= 3\sqrt{5}$

$\hat{\underset{\sim}{u}} = \dfrac{\underset{\sim}{u}}{\left|\underset{\sim}{u}\right|}$

$= \dfrac{6\underset{\sim}{i}-3\underset{\sim}{j}}{3\sqrt{5}}$

$= \dfrac{2\underset{\sim}{i}-\underset{\sim}{j}}{\sqrt{5}}$

$= \mathbf{\dfrac{1}{\sqrt{5}}(2\underset{\sim}{i}-\underset{\sim}{j})}$

ii $\left|\underset{\sim}{v}\right| = \sqrt{4^2+4^2}$

$= \sqrt{32}$

$= 4\sqrt{2}$

$\hat{\underset{\sim}{v}} = \dfrac{\underset{\sim}{v}}{\left|\underset{\sim}{v}\right|}$

$= \dfrac{4\underset{\sim}{i}+4\underset{\sim}{j}}{4\sqrt{2}}$

$= \mathbf{\dfrac{\underset{\sim}{i}+\underset{\sim}{j}}{\sqrt{2}}}$

$= \mathbf{\dfrac{1}{\sqrt{2}}(\underset{\sim}{i}+\underset{\sim}{j})}$

4 i $\theta = \tan^{-1}\left(\dfrac{-1}{-\sqrt{3}}\right)$

$= \tan^{-1}\left(\dfrac{1}{\sqrt{3}}\right)$

$= \mathbf{210°}$

ii $\theta = \tan^{-1}\left(\dfrac{-5}{2}\right)$

$= \mathbf{292°}$ **(nearest degree)**

5 i $x = \left|\underset{\sim}{u}\right|\cos\theta$

$= 4\cos 120°$

$= 4 \times -\dfrac{1}{2}$

$= -2$

$y = \left|\underset{\sim}{u}\right|\sin\theta$

$= 4\sin 120°$

$= 4 \times \dfrac{\sqrt{3}}{2}$

$= 2\sqrt{3}$

$\therefore \underset{\sim}{u} = x\underset{\sim}{i} + y\underset{\sim}{j}$

$= \mathbf{-2\underset{\sim}{i} + 2\sqrt{3}\underset{\sim}{j}}$

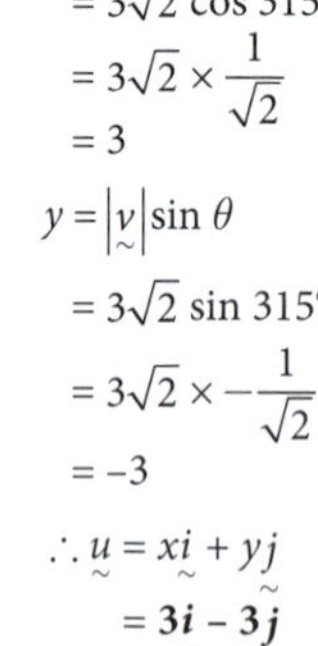

ii $x = \left|\underset{\sim}{v}\right|\cos\theta$

$= 3\sqrt{2}\cos 315°$

$= 3\sqrt{2} \times \dfrac{1}{\sqrt{2}}$

$= 3$

$y = \left|\underset{\sim}{v}\right|\sin\theta$

$= 3\sqrt{2}\sin 315°$

$= 3\sqrt{2} \times -\dfrac{1}{\sqrt{2}}$

$= -3$

$\therefore \underset{\sim}{u} = x\underset{\sim}{i} + y\underset{\sim}{j}$

$= \mathbf{3\underset{\sim}{i} - 3\underset{\sim}{j}}$

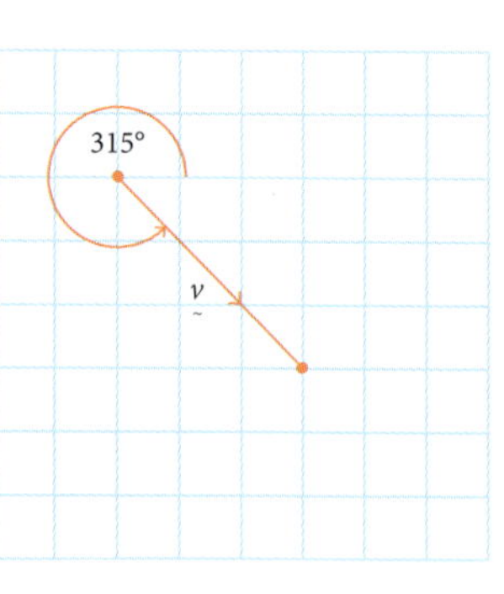

2 Scalar product (or dot product) of vectors ➲ pp. 121–122

1 i a $\underset{\sim}{u}.\underset{\sim}{v} = 5 \times 2 + (-3) \times (-1)$

$= \mathbf{13}$

b $\underset{\sim}{u}.\underset{\sim}{u} + \underset{\sim}{v}.\underset{\sim}{v} = 5^2 + (-3)^2 + 2^2 + (-1)^2$

$= \mathbf{39}$

ii a $\underset{\sim}{u}.\underset{\sim}{v} = 0 \times (-5) + (-3) \times (-2)$

$= \mathbf{6}$

b $\underset{\sim}{v}(\underset{\sim}{u} + \underset{\sim}{v}) = \underset{\sim}{v}.\underset{\sim}{u} + \underset{\sim}{v}.\underset{\sim}{v}$

$= 6 + (-5)^2 + (-2)^2$ using result from a.

$= \mathbf{35}$

2 i $\underset{\sim}{u}.\underset{\sim}{v} = |\underset{\sim}{u}||\underset{\sim}{v}|\cos\theta$

$= 3\sqrt{2} \times 6 \times \cos 45°$

$= 3\sqrt{2} \times 6 \times \dfrac{1}{\sqrt{2}}$

$= \mathbf{18}$

ii $\underset{\sim}{u}.\underset{\sim}{v} = 1 \times 3 + (-2)b$

$= 3 - 2b$

$|\underset{\sim}{u}| = \sqrt{1^2+(-2)^2} = \sqrt{5}$

$|\underset{\sim}{u}| = \sqrt{3^2+b^2} = \sqrt{9+b^2}$

$\underset{\sim}{u}.\underset{\sim}{v} = |\underset{\sim}{u}||\underset{\sim}{v}|\cos\theta$

$3 - 2b = \sqrt{5} \times \sqrt{9+b^2} \times \cos 60°$

$3 - 2b = \sqrt{45+5b^2} \times \dfrac{1}{2}$

$6 - 4b = \sqrt{45+5b^2} \quad \ldots *$

$(6 - 4b)^2 = 45 + 5b^2$

$36 - 48b + 16b^2 = 45 + 5b^2$

$11b^2 - 48b - 9 = 0$

$b = \dfrac{48 \pm \sqrt{(-48)^2 - 4(11)(-9)}}{2(11)}$

$= 4.5437\ldots, -0.1800\ldots$

Now, checking in line *:

$\therefore b = \mathbf{-0.18}$ (2 dec. pl.)

3 i $\cos\theta = \dfrac{\underset{\sim}{u}.\underset{\sim}{v}}{|\underset{\sim}{u}||\underset{\sim}{v}|}$

$= \dfrac{(-2)\times 6 + 5 \times 5}{\sqrt{(-2)^2+5^2}.\sqrt{6^2+5^2}}$

$= \dfrac{13}{\sqrt{29}.\sqrt{61}}$

$= 0.309\,086\,072\ldots$

$\theta = 71.995\,838\,39\ldots$

$= 72$ (nearest whole)

$\therefore \angle \mathbf{AOB = 72°}$

ii $\cos\theta = \dfrac{\underset{\sim}{u}.\underset{\sim}{v}}{|\underset{\sim}{u}||\underset{\sim}{v}|}$

$= \dfrac{1\times(-3)+3\times 2}{\sqrt{1^2+3^2}.\sqrt{(-3)^2+2^2}}$

$= \dfrac{3}{\sqrt{10}.\sqrt{13}}$

$= 0.263\,117\,405\ldots$

$\theta = 74.744\,8813\ldots$

$= 75°$ (nearest degree)

$\therefore$ **the angle is 75°**

3 Parallel and perpendicular vectors ➲ p. 122

1 i $\underset{\sim}{u}.\underset{\sim}{v} = x_1x_2 + y_1y_2$

$= (-6) \times (-4) + 3 \times (-8)$

$= 0$

$\therefore$ **the vectors are perpendicular.**

ii $\underset{\sim}{u}.\underset{\sim}{v} = 3(-1) + (-6)(2)$

$= -15$

$|\underset{\sim}{u}| = \sqrt{3^2+(-6)^2} = \sqrt{45} = 3\sqrt{5}$

$|\underset{\sim}{v}| = \sqrt{(-1)^2+2^2} = \sqrt{5}$

As $3\sqrt{5} \times \sqrt{5} = 15$, then $\underset{\sim}{u}.\underset{\sim}{v} = -|\underset{\sim}{u}||\underset{\sim}{v}|$, and hence vectors are parallel.

2 i a If parallel, then $\underset{\sim}{u} = k\underset{\sim}{v}$.

$5\underset{\sim}{i} + m\underset{\sim}{j} = k(-3\underset{\sim}{i} - 4\underset{\sim}{j})$

$\therefore 5\underset{\sim}{i} = -3k\underset{\sim}{i}$

$\therefore k = -\dfrac{5}{3}$

$\therefore m\underset{\sim}{j} = -\dfrac{5}{3}(-4\underset{\sim}{j})$

$\therefore m = \dfrac{20}{3}$

b $(-2) \times 3 + (-6) \times m = 0$

$-6 - 6m = 0$

$-6m = 6$

$\boldsymbol{m = -1}$

ii a If parallel, then $\underset{\sim}{u} = k\underset{\sim}{v}$.

$4\underset{\sim}{i} + (2m + 3)\underset{\sim}{j} = k(3\underset{\sim}{i} - \underset{\sim}{j})$

$\therefore 4\underset{\sim}{i} = 3k\underset{\sim}{i}$

$\therefore k = \dfrac{4}{3}$

$\therefore (2m + 3)\underset{\sim}{j} = \dfrac{4}{3}(-\underset{\sim}{j})$

$2m + 3 = -\dfrac{4}{3}$

$6m + 9 = -4$

$6m = -13$

$m = -\dfrac{13}{6}$

b $-5(2 - 3m) + 0 \times 1 = 0$

$-10 + 15m = 0$

$15m = 10$

$\boldsymbol{m = \dfrac{2}{3}}$

4 Vector projection ➲ p. 123

1 i $\underset{\sim}{u}.\underset{\sim}{v} = 2 \times 5 + (-4) \times (-3)$

$= 22$

$|\underset{\sim}{v}|^2 = 5^2 + (-3)^2$

$= 34$

$\therefore \underset{\sim}{u}_v = \dfrac{22}{34}(5\underset{\sim}{i} - 3\underset{\sim}{j})$

$= \boldsymbol{\dfrac{11}{17}(5\underset{\sim}{i} - 3\underset{\sim}{j})}$

ii $\underset{\sim}{v}.\underset{\sim}{u} = (-1) \times (-3) + 2 \times 1$

$= 5$

$|\underset{\sim}{u}|^2 = (-3)^2 + 1^2$

$= 10$

$\therefore \underset{\sim}{v}_u = \dfrac{5}{10}(-3\underset{\sim}{i} + \underset{\sim}{j})$

$= \boldsymbol{\dfrac{1}{2}(-3\underset{\sim}{i} + \underset{\sim}{j})}$

5 Displacement, force and velocity as vectors ⊃ pp. 124–125

1 i Magnitude $= \sqrt{3.6^2 + 2.8^2}$

$= 4.5607017\ldots$

$= 4.56$ (2 dec. pl.)

$\tan\theta = \dfrac{2.8}{3.6}$

$\theta = 37.874\,983\,65\ldots$

$= 38$ (nearest whole)

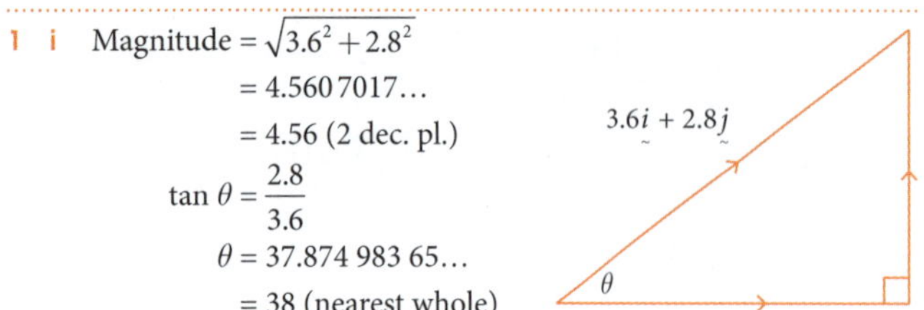

∴ **Simone's displacement is 4.56 km in a direction of 38° above the horizontal.**

ii Let $\underset{\sim}{u}$ = run and $\underset{\sim}{v}$ = walk.

$\therefore \underset{\sim}{u} = 4\underset{\sim}{j}$

$\underset{\sim}{v} = 6(\cos(-60°)\underset{\sim}{i} + \sin(-60°)\underset{\sim}{j})$

$= 6\left(\frac{1}{2}\underset{\sim}{i} - \frac{\sqrt{3}}{2}\right)\underset{\sim}{j}$

$= 3\underset{\sim}{i} - 3\sqrt{3}\underset{\sim}{j}$

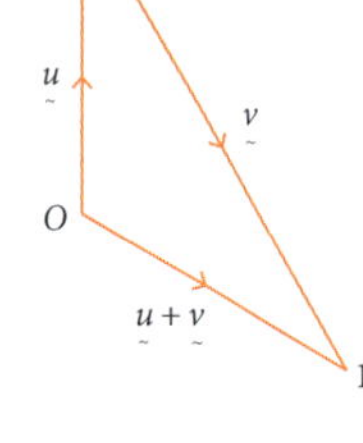

$\underset{\sim}{u} + \underset{\sim}{v} = 4\underset{\sim}{j} + 3\underset{\sim}{i} - 3\sqrt{3}\underset{\sim}{j}$

$= 3\underset{\sim}{i} + (4 - 3\sqrt{3})\underset{\sim}{j}$

$|\underset{\sim}{u} + \underset{\sim}{v}| = \sqrt{3^2 + (4 - 3\sqrt{3})^2}$

$= 3.229\,671\,906\ldots$

$= 3.23$ (2 dec. pl.)

$\tan\theta = \dfrac{4 - 3\sqrt{3}}{3}$

$= -0.398\,717\,474\ldots$

$\theta = 338.261\,9662\ldots$

$= 338°$ (nearest degree)

∴ **Ajit's displacement is 3.23 km in a direction of 338° (or a bearing of 112°).**

2 i a $\underset{\sim}{R} = (6 - 9)\underset{\sim}{i} + (3 - 5)\underset{\sim}{j}$

$= \mathbf{-3\underset{\sim}{i} - 2\underset{\sim}{j}}$

b $|\underset{\sim}{R}| = \sqrt{(-3)^2 + (-2)^2}$

$= \sqrt{13}$

$\tan\theta = \dfrac{-2}{-3}$

$\theta = 33.690\,067\,53\ldots$

$= 34$ (nearest whole)

As force in 3rd quadrant use 180 + 34 = 214.

∴ **magnitude of $\sqrt{13}$ N at an angle of 214°.**

ii a $\underset{\sim}{R} = (7 - 3)\underset{\sim}{i} + 5(\cos 45°\underset{\sim}{i} + \sin 45°\underset{\sim}{j})$

$= 4\underset{\sim}{i} + 5\left(\frac{1}{\sqrt{2}}\underset{\sim}{i} + \frac{1}{\sqrt{2}}\underset{\sim}{j}\right)$

$= \mathbf{\left(4 + \frac{5}{\sqrt{2}}\right)\underset{\sim}{i} + \frac{5}{\sqrt{2}}\underset{\sim}{j}}$

b $|\underset{\sim}{R}| = \sqrt{\left(4 + \frac{5}{\sqrt{2}}\right)^2 + \left(\frac{5}{\sqrt{2}}\right)^2}$

$= 8.323\,717\,394\ldots$

$= 8.32$ (2 dec. pl.)

$\tan\theta = \dfrac{\frac{5}{\sqrt{2}}}{4 + \frac{5}{\sqrt{2}}}$

$\theta = 25.135\,105\,94\ldots$

$= 25$ (nearest whole)

∴ **magnitude of 8.32 N at an angle of 25°.**

3 i Let force vectors be $\underset{\sim}{u}$ and $\underset{\sim}{v}$.

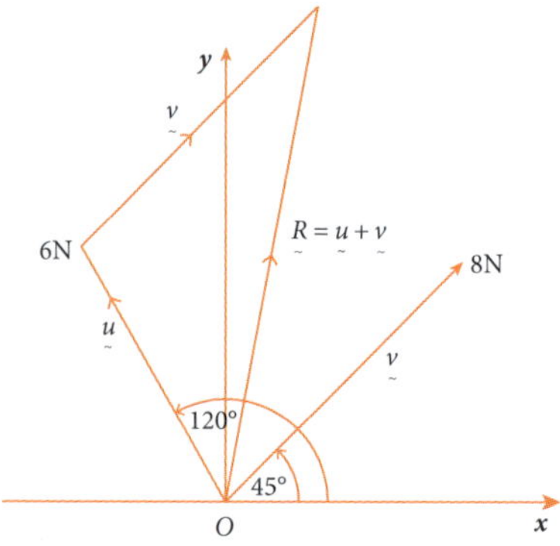

$\underset{\sim}{u} = 6\cos 120°\underset{\sim}{i} + 6\sin 120°\underset{\sim}{j}$

$= -3\underset{\sim}{i} + 3\sqrt{3}\underset{\sim}{j}$

$\underset{\sim}{v} = 8\cos 45°\underset{\sim}{i} + 8\sin 45°\underset{\sim}{j}$

$= \frac{8}{\sqrt{2}}\underset{\sim}{i} + \frac{8}{\sqrt{2}}\underset{\sim}{j}$

$\underset{\sim}{R} = \underset{\sim}{u} + \underset{\sim}{v}$

$= -3\underset{\sim}{i} + 3\sqrt{3}\underset{\sim}{j} + \frac{8}{\sqrt{2}}\underset{\sim}{i} + \frac{8}{\sqrt{2}}\underset{\sim}{j}$

$= \left(-3 + \frac{8}{\sqrt{2}}\right)\underset{\sim}{i} + \left(3\sqrt{3} + \frac{8}{\sqrt{2}}\right)\underset{\sim}{j}$

$|\underset{\sim}{R}| = \sqrt{\left(-3 + \frac{8}{\sqrt{2}}\right)^2 + \left(3\sqrt{3} + \frac{8}{\sqrt{2}}\right)^2}$

$= 11.173\,478\,79\ldots$

$= 11.17$ (2 dec. pl.)

$\tan\theta = \dfrac{3\sqrt{3} + \frac{8}{\sqrt{2}}}{-3 + \frac{8}{\sqrt{2}}}$

$= 4.084\,908\,562\ldots$

$\theta = 76.244\,319\,74\ldots$

$= 76$ (nearest whole)

∴ **the resultant force is 11.17 N at an angle of 76°.**

ii Let force vectors be $\underset{\sim}{u}$ and $\underset{\sim}{v}$.

$\underset{\sim}{u} = 4\sqrt{3}\cos 60°\underset{\sim}{i} + 4\sqrt{3}\sin 60°\underset{\sim}{j}$

$= 2\sqrt{3}\underset{\sim}{i} + 6\underset{\sim}{j}$

$\underset{\sim}{v} = 3\sqrt{2}\cos(-45)°\underset{\sim}{i} + 3\sqrt{2}\sin(-45)°\underset{\sim}{j}$

$= 3\underset{\sim}{i} - 3\underset{\sim}{j}$

$\underset{\sim}{R} = \underset{\sim}{u} + \underset{\sim}{v}$

$= 2\sqrt{3}\underset{\sim}{i} + 6\underset{\sim}{j} + 3\underset{\sim}{i} - 3\underset{\sim}{j}$

$= (2\sqrt{3} + 3)\underset{\sim}{i} + (6 - 3)\underset{\sim}{j}$

$= (2\sqrt{3} + 3)\underset{\sim}{i} + 3\underset{\sim}{j}$

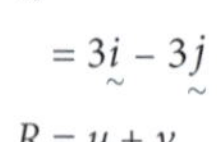

$|\underset{\sim}{R}| = \sqrt{(2\sqrt{3} + 3)^2 + 3^2}$

$= 7.126\,332\,134\ldots$

$= 7.13$ (2 dec. pl.)

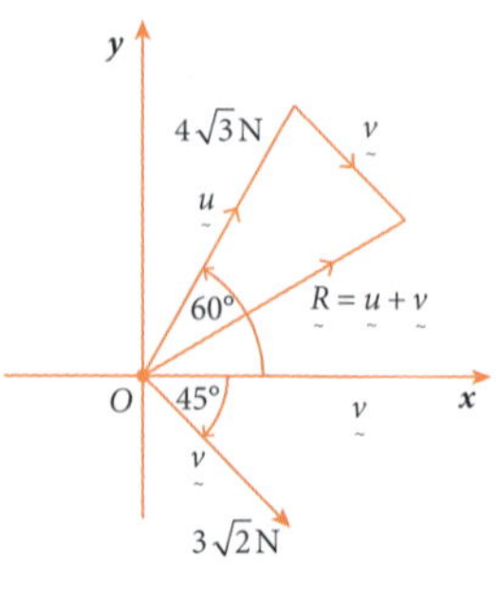

$$\tan\theta = \frac{3}{2\sqrt{3}+3}$$
$= 0.464\,101\,615\ldots$
$\theta = 24.896\,090\,64\ldots$
$= 25$ (nearest whole)

$\therefore$ **the resultant force is 7.13 N at an angle of 25°.**

4 i $\underset{\sim}{v} = 280(\cos 225°\underset{\sim}{i} + \sin 225°\underset{\sim}{j})$

$= \frac{-280}{\sqrt{2}}\underset{\sim}{i} - \frac{280}{\sqrt{2}}\underset{\sim}{j}$

$\underset{\sim}{w} = 48(\cos(-60°)\underset{\sim}{i} + \sin(-60°)\underset{\sim}{j})$

$= 24\underset{\sim}{i} - 24\sqrt{3}\underset{\sim}{j}$

$$\underset{\sim}{r} = \underset{\sim}{v} + \underset{\sim}{w} = \left(\frac{-280}{\sqrt{2}} + 24\right)\underset{\sim}{i} + \left(\frac{-280}{\sqrt{2}} - 24\sqrt{3}\right)\underset{\sim}{j}$$

$$\left|\underset{\sim}{r}\right| = \left|\underset{\sim}{v}+\underset{\sim}{w}\right| = \sqrt{\left(\frac{-280}{\sqrt{2}}+24\right)^2 + \left(\frac{-280}{\sqrt{2}} - 24\sqrt{3}\right)^2}$$

$= 296.076\,0982\ldots$
$= 296$ (2 dec. pl.)

$\therefore$ **ground speed of 296 km/h**

ii Let $\underset{\sim}{s}$ = ship's steering,
$\underset{\sim}{c}$ = the wind's path.
Bearing of 210° means $\theta = 240°$ and bearing of 330° means $\theta = 120°$.

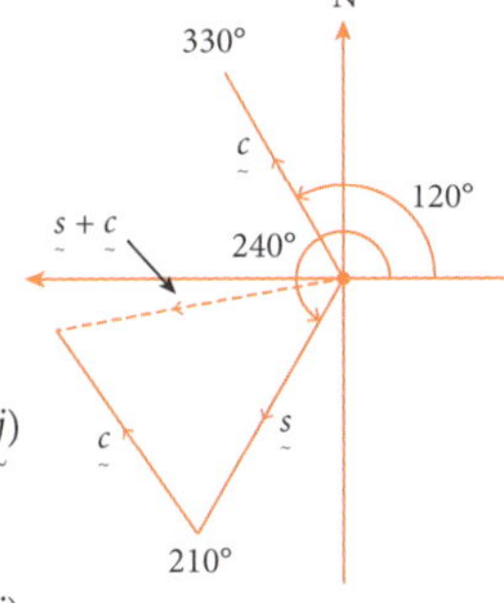

$\underset{\sim}{s} = 38(\cos 240°\underset{\sim}{i} + \sin 240°\underset{\sim}{j})$

$= -19\underset{\sim}{i} - 19\sqrt{3}\underset{\sim}{j}$

$\underset{\sim}{w} = 12(\cos 120°\underset{\sim}{i} + \sin 120°\underset{\sim}{j})$

$= -6\underset{\sim}{i} + 6\sqrt{3}\underset{\sim}{j}$

$\underset{\sim}{r} = \underset{\sim}{s} + \underset{\sim}{w} = -25\underset{\sim}{i} - 13\sqrt{3}\underset{\sim}{j}$

$\left|\underset{\sim}{r}\right| = \left|\underset{\sim}{s}+\underset{\sim}{w}\right| = \sqrt{(-25)^2 + (-13\sqrt{3})^2}$

$= 33.645\,207\,68\ldots$
$= 34$ (2 dec. pl.)

$\therefore$ speed of 34 km/h

$$\tan\theta = \frac{-13\sqrt{3}}{-25}$$
$= 0.900\,666\,42\ldots$
$\theta = 222°$ (3rd quadrant) (nearest whole)

$\therefore$ bearing of $270° - 042° = 228°$

$\therefore$ **speed of 34 km/h on a bearing of 228°**

6 Geometric proofs ➲ p. 125

1 i Let $ABCD$ be a parallelogram.

Let $\overrightarrow{BC} = \underset{\sim}{u}$ and $\overrightarrow{AB} = \underset{\sim}{v}$.
Diagonals are AC and BD, where $AC \perp BD$.

$\therefore \overrightarrow{AC}.\overrightarrow{BD} = 0$

$(\underset{\sim}{u} + \underset{\sim}{v}).(\underset{\sim}{u} - \underset{\sim}{v}) = 0$

$\underset{\sim}{u}.\underset{\sim}{u} - \underset{\sim}{v}.\underset{\sim}{v} = 0$

$\underset{\sim}{u}.\underset{\sim}{u} = \underset{\sim}{v}.\underset{\sim}{v}$

$\therefore |\underset{\sim}{u}| = |\underset{\sim}{v}|$

$\therefore$ **shape is a rhombus (adjacent sides equal)**

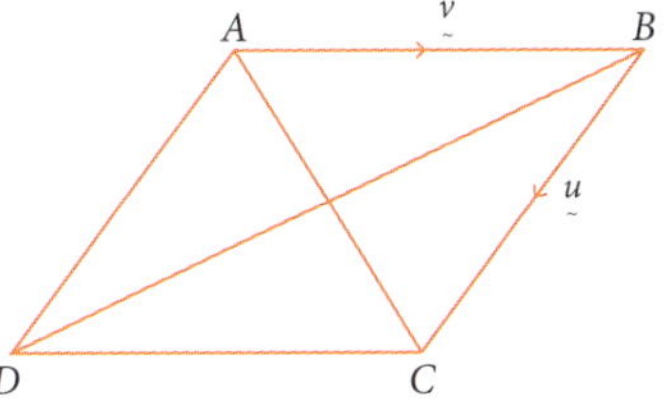

ii ABC is a triangle.

E and D are the midpoints of AC and AB respectively.

Hence, $\overrightarrow{AE} = \frac{1}{2}\overrightarrow{AC}$

Also, $\overrightarrow{AD} = \frac{1}{2}\overrightarrow{AB}$.

Now, $\overrightarrow{CB} = \overrightarrow{AB} - \overrightarrow{AC}$.

Hence, $\overrightarrow{ED} = \frac{1}{2}\overrightarrow{AB} - \frac{1}{2}\overrightarrow{AC}$

$= \frac{1}{2}(\overrightarrow{AB} - \overrightarrow{AC})$

$= \frac{1}{2}\overrightarrow{CB}$

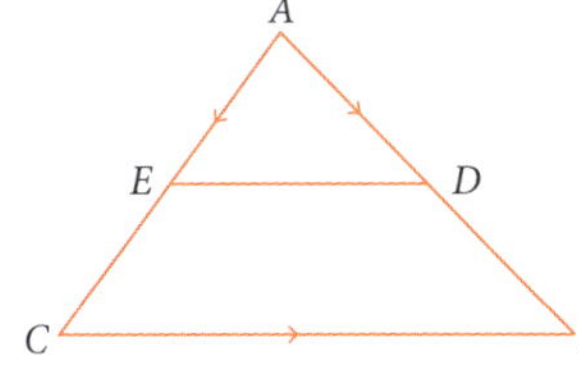

$\therefore \overrightarrow{ED} = \frac{1}{2}\overrightarrow{CB}$

Hence, $CB \parallel ED$.

Also, $|\overrightarrow{ED}| = |\frac{1}{2}\overrightarrow{CB}| = \frac{1}{2}|\overrightarrow{CB}|$ and so CB is half the length of ED.

$\therefore$ CB is parallel to ED and half its length.

$\therefore$ result proven.

2 i Let $ABCD$ be a parallelogram. Let $\overrightarrow{DC} = \underset{\sim}{u}$ and $\overrightarrow{DA} = \underset{\sim}{v}$.

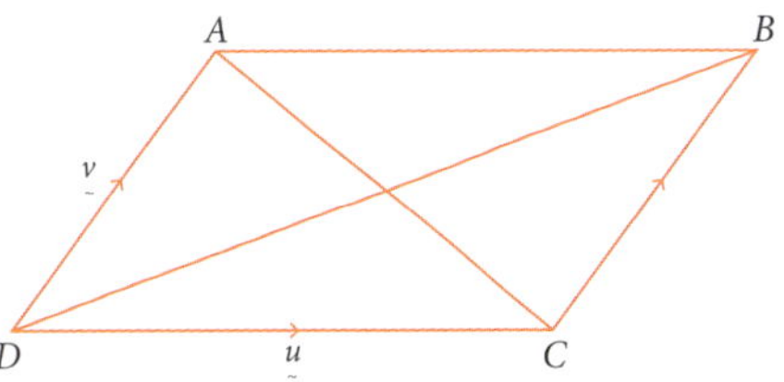

$\therefore$ diagonals are $\underset{\sim}{u} + \underset{\sim}{v}$ and $\underset{\sim}{v} - \underset{\sim}{u}$.

$\therefore |\underset{\sim}{u} + \underset{\sim}{v}|^2 + |\underset{\sim}{v} - \underset{\sim}{u}|^2 = (\underset{\sim}{u} + \underset{\sim}{v})^2 + (\underset{\sim}{v} - \underset{\sim}{u})^2$

$= (\underset{\sim}{u}.\underset{\sim}{u} + 2\underset{\sim}{u}.\underset{\sim}{v} + \underset{\sim}{v}.\underset{\sim}{v}) + (\underset{\sim}{u}.\underset{\sim}{u} - 2\underset{\sim}{u}.\underset{\sim}{v} + \underset{\sim}{v}.\underset{\sim}{v})$

$= 2\left|\underset{\sim}{u}\right|^2 + 2\left|\underset{\sim}{v}\right|^2$

$\therefore |\underset{\sim}{u} + \underset{\sim}{v}|^2 + |\underset{\sim}{v} - \underset{\sim}{u}|^2 = 2\left|\underset{\sim}{u}\right|^2 + 2\left|\underset{\sim}{v}\right|^2$

$\therefore$ result proven.

ii Let $ABCD$ be a quadrilateral.

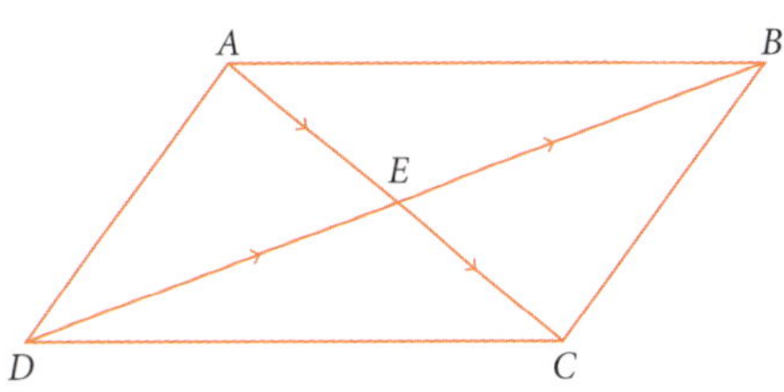

$\therefore \overrightarrow{AE} = \overrightarrow{EC}$ and $\overrightarrow{DE} = \overrightarrow{EB}$ … ①

Also, $\overrightarrow{AB} = \overrightarrow{AE} + \overrightarrow{EB}$.

Also, $\overrightarrow{DC} = \overrightarrow{DE} + \overrightarrow{EC}$

$\therefore$ Using ①: $\overrightarrow{DC} = \overrightarrow{EB} + \overrightarrow{AE}$

$\therefore \overrightarrow{DC} = \overrightarrow{AB}$

Hence, as $\overrightarrow{AB} = \overrightarrow{DC}$ and $\overrightarrow{AB} \parallel \overrightarrow{DC}$, then $ABCD$ is a parallelogram (opposite sides equal and parallel).

HSC EXAM-TYPE QUESTIONS

Part A

1 D. $\underset{\sim}{u} - \underset{\sim}{v} = -2\underset{\sim}{i} + (m-2)\underset{\sim}{j}$ and $\underset{\sim}{v} = 4\underset{\sim}{i} + 2\underset{\sim}{j}$

If perpendicular then $(\underset{\sim}{u} - \underset{\sim}{v}).\underset{\sim}{v} = 0$.

$\therefore (-2) \times 4 + (m-2) \times 2 = 0$

$-8 + 2m - 4 = 0$

$2m = 12$

$m = 6$

2 C. $\cos\theta = \dfrac{\underset{\sim}{u}.\underset{\sim}{v}}{|\underset{\sim}{u}||\underset{\sim}{v}|}$

$= \dfrac{2\times4+4\times2}{\sqrt{2^2+4^2}.\sqrt{4^2+2^2}}$

$= \dfrac{16}{\sqrt{20}.\sqrt{20}}$

$= \dfrac{4}{5}$

3 A. $\hat{\underset{\sim}{v}} = \dfrac{\underset{\sim}{v}}{|\underset{\sim}{v}|}$

$= \dfrac{2\sqrt{3}\underset{\sim}{i}+\sqrt{2}\underset{\sim}{j}}{\sqrt{(2\sqrt{3})^2+(\sqrt{2})^2}}$

$= \dfrac{2\sqrt{3}\underset{\sim}{i}+\sqrt{2}\underset{\sim}{j}}{\sqrt{14}}$

4 D. $|\underset{\sim}{v}|^2 = (p+1)^2 + (3-p)^2 = 16$

$p^2 + 2p + 1 + 9 - 6p + p^2 = 16$

$2p^2 - 4p - 6 = 0$

$p^2 - 2p - 3 = 0$

$(p-3)(p+1) = 0$

$p = 3, -1$

5 C. For 150° in the clockwise direction, use $\theta = 210°$.

$\underset{\sim}{F} = 12(\cos 210°\underset{\sim}{i} + \sin 210°\underset{\sim}{j})$

$= 12 \times -\dfrac{\sqrt{3}}{2}\underset{\sim}{i} + 12 \times -\dfrac{1}{2}\underset{\sim}{j}$

$= -6\sqrt{3}\underset{\sim}{i} - 6\underset{\sim}{j}$

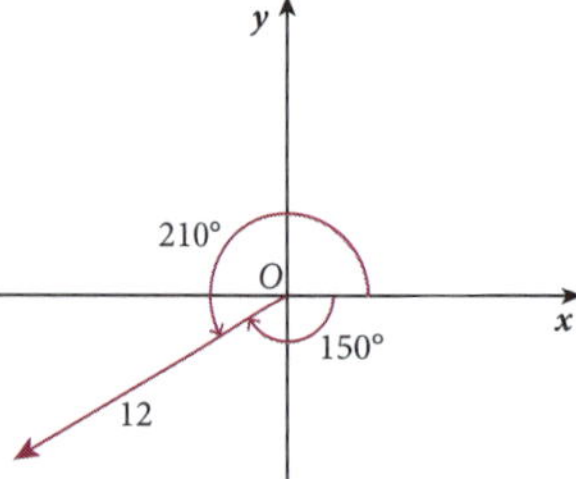

Part B

6 Orthogonal if $\underset{\sim}{v}.\underset{\sim}{w} = 0$. ✓

$5(m+4) + 3(-2) = 0$

$5m + 20 - 6 = 0$

$5m = -14$

$\boldsymbol{m = -\dfrac{14}{5}}$ ✓

7 a $3\underset{\sim}{a} - 2\underset{\sim}{b} = 3(2\underset{\sim}{i} + 5\underset{\sim}{j}) - 2(\underset{\sim}{i} + 6\underset{\sim}{j})$

$= \boldsymbol{4\underset{\sim}{i} + 3\underset{\sim}{j}}$ ✓

b $|4\underset{\sim}{i}+3\underset{\sim}{j}| = \sqrt{4^2+3^2}$

$= 5$ ✓

But $|\underset{\sim}{c}| = 15$.

Hence $\underset{\sim}{c} = 3(4\underset{\sim}{i} + 3\underset{\sim}{j})$. ✓

8 $\overrightarrow{AB} = \underset{\sim}{v} - \underset{\sim}{u}$

$\therefore \overrightarrow{PB} = \dfrac{1}{2}(\underset{\sim}{v} - \underset{\sim}{u})$ (P is midpoint of AB) ✓

Also, $\overrightarrow{OR} = \dfrac{1}{2}\underset{\sim}{u}$ (R is midpoint of OR)

$\overrightarrow{OQ} = \dfrac{1}{2}\underset{\sim}{v}$ (Q is midpoint of OB)

$\overrightarrow{RQ} = \dfrac{1}{2}(\underset{\sim}{v} - \underset{\sim}{u})$

As $\overrightarrow{PB} = \overrightarrow{RQ}$ then $PB = RQ$ and $PB||RQ$.

$\therefore$ ***RPBQ*** **is a parallelogram (opposite sides equal and parallel).** ✓

9 Let $\underset{\sim}{u} = \underset{\sim}{b} - \underset{\sim}{a} = 2\underset{\sim}{i} + 5\underset{\sim}{j}$ ✓

Let $\underset{\sim}{v} = \underset{\sim}{c} - \underset{\sim}{b} = 6\underset{\sim}{i} - 4\underset{\sim}{j}$

$\cos\theta = \dfrac{\underset{\sim}{u}.\underset{\sim}{v}}{|\underset{\sim}{u}||\underset{\sim}{v}|}$

$= \dfrac{2\times6+5\times(-4)}{\sqrt{2^2+5^2}.\sqrt{6^2+(-4)^2}}$ ✓

$= \dfrac{-8}{\sqrt{29}.\sqrt{52}}$

$\theta = 101.888\,658\ldots$

$= 102$ (nearest whole)

$\therefore$ **the angle is 102°.** ✓

10 $\underset{\sim}{a}.\underset{\sim}{b} = 3 \times -10 + 4 \times 15$

$= 30$

$|\underset{\sim}{a}| = \sqrt{3^2+4^2}$

$= 5$ ✓

$\dfrac{\underset{\sim}{a}.\underset{\sim}{b}}{|\underset{\sim}{a}|^2}\underset{\sim}{a} = \dfrac{30}{25}\underset{\sim}{a}$

$= \dfrac{6}{5}(3\underset{\sim}{i} + 4\underset{\sim}{j})$

$= \boldsymbol{\dfrac{18}{5}\underset{\sim}{i} + \dfrac{24}{5}\underset{\sim}{j}}$ ✓

11 $\overrightarrow{ON} = \dfrac{2}{3}(\underset{\sim}{a} + \underset{\sim}{c})$

$\overrightarrow{OM} = \underset{\sim}{c} + \dfrac{1}{2}\underset{\sim}{a}$ ✓

$\overrightarrow{AN} = \dfrac{2}{3}(\underset{\sim}{a} + \underset{\sim}{c}) - \underset{\sim}{a}$

$= \dfrac{2}{3}\underset{\sim}{c} - \dfrac{1}{3}\underset{\sim}{a}$

$= \dfrac{2}{3}(\underset{\sim}{c} - \dfrac{1}{2}\underset{\sim}{a})$ ✓

$\overrightarrow{AM} = -\underset{\sim}{a} + \underset{\sim}{c} + \dfrac{1}{2}\underset{\sim}{a}$

$= \underset{\sim}{c} - \dfrac{1}{2}\underset{\sim}{a}$

As $\overrightarrow{AN} = \dfrac{2}{3}\overrightarrow{AM}$, then $AN||$AM and hence A, N, M lie on same line.

$\therefore$ ***A*, *N* and *M* are collinear.** ✓

12 $|\underset{\sim}{s}| = \sqrt{3^2 + (-m)^2} = 5$

$9 + m^2 = 25$

$m^2 = 16$

$m = 4\ (m > 0)$ ✓

$\therefore \underset{\sim}{s} = 3\underset{\sim}{i} - 4\underset{\sim}{j}$

$\underset{\sim}{s}.\underset{\sim}{t} = 3 \times (-5) + (-4)(-n) = 0$

$-15 + 4n = 0$

$4n = 15$

$n = \dfrac{15}{4}$ ✓

$\therefore$ **$m = 4$ and $n = \dfrac{15}{4}$**

---

13 $\cos\theta = \dfrac{(-2\sqrt{3})(1) + (-2)(\sqrt{3})}{\sqrt{(-2\sqrt{3})^2 + (-2)^2}\sqrt{1^2 + (\sqrt{3})^2}}$

$= \dfrac{-2\sqrt{3} - 2\sqrt{3}}{\sqrt{16}\sqrt{4}}$

$= \dfrac{-4\sqrt{3}}{8} = \dfrac{-\sqrt{3}}{2}$ ✓

$\cos 2\theta = 2\cos^2\theta - 1$

$= 2\left(\dfrac{\sqrt{3}}{2}\right)^2 - 1$

$= 2\left(\dfrac{3}{4}\right) - 1$

$= \dfrac{\mathbf{1}}{\mathbf{2}}$ ✓

---

14 a

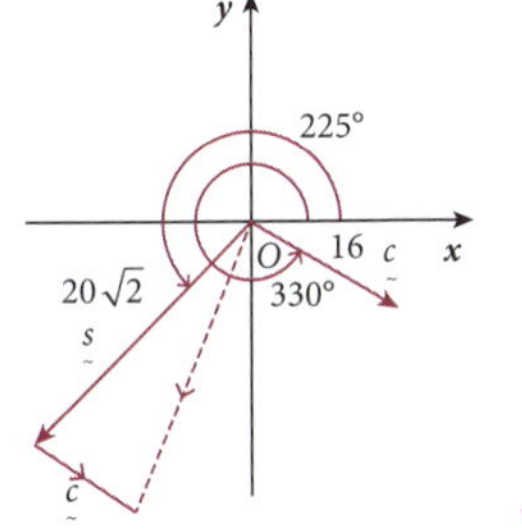

✓

b $\underset{\sim}{s} = 20\sqrt{2}(\cos 225°\underset{\sim}{i} + \sin 225°\underset{\sim}{j})$

$= -20\underset{\sim}{i} - 20\underset{\sim}{j}$ ✓

$\underset{\sim}{w} = 16(\cos 330°\underset{\sim}{i} + \sin 330°\underset{\sim}{j})$

$= 8\sqrt{3}\underset{\sim}{i} - 8\underset{\sim}{j}$

$\underset{\sim}{s} + \underset{\sim}{w} = (-20 + 8\sqrt{3})\underset{\sim}{i} - 28\underset{\sim}{j}$

$|\underset{\sim}{s} + \underset{\sim}{w}| = \sqrt{(-20 + 8\sqrt{3})^2 + (-28)^2}$

$= 28.666\,073\,01\ldots$

$= 29$ (nearest whole)

$\therefore$ speed of 29 km/h ✓

$\tan\theta = \dfrac{-28}{-20 + 8\sqrt{3}}$

$= 4.557\,593\,177\ldots$

$\theta = 258°$ (3rd quadrant) (nearest whole)

$\therefore$ bearing of $180° + (270° - 258°) = 192°$ ✓

$\therefore$ **speed of 29 km/h on a bearing of 192°**

---

15 a $\overrightarrow{MA} = \dfrac{1}{2}a\underset{\sim}{j}$ and $\overrightarrow{AB} = a\underset{\sim}{i}$.

$\therefore \overrightarrow{MB} = a\underset{\sim}{i} + \dfrac{1}{2}a\underset{\sim}{j}$

$\therefore \overrightarrow{\mathbf{MP}} = \boldsymbol{\lambda}\left(\mathbf{a}\underset{\sim}{\mathbf{i}} + \dfrac{\mathbf{1}}{\mathbf{2}}\mathbf{a}\underset{\sim}{\mathbf{j}}\right)$ ✓

b $\overrightarrow{PB} = (1 - \lambda)(a\underset{\sim}{i} + \dfrac{1}{2}a\underset{\sim}{j})$

$\overrightarrow{CP} = a\underset{\sim}{j} - (1 - \lambda)\left(a\underset{\sim}{i} + \dfrac{1}{2}a\underset{\sim}{j}\right)$

$= a\underset{\sim}{j} + (\lambda - 1)\left(a\underset{\sim}{i} + \dfrac{1}{2}a\underset{\sim}{j}\right)$

$= (\lambda - 1)a\underset{\sim}{i} + \left(1 + \dfrac{\lambda - 1}{2}\right)a\underset{\sim}{j}$ ✓

As $MB \perp CP$,

$[a\underset{\sim}{i} + \dfrac{1}{2}a\underset{\sim}{j}].[(\lambda - 1)a\underset{\sim}{i} + \left(1 + \dfrac{\lambda - 1}{2}\right)a\underset{\sim}{j}] = 0$ ✓

$a(\lambda - 1)a + \dfrac{1}{2}a\left(1 + \dfrac{\lambda - 1}{2}\right)a = 0$

$a^2(\lambda - 1) + \dfrac{1}{2}a^2\left(1 + \dfrac{\lambda - 1}{2}\right) = 0$

$\lambda - 1 + \dfrac{1}{2} + \dfrac{\lambda - 1}{4} = 0$

$4\lambda - 4 + 2 + \lambda - 1 = 0$

$5\lambda - 3 = 0$

$\boldsymbol{\lambda} = \dfrac{\mathbf{3}}{\mathbf{5}}$ ✓

c As $\lambda = \dfrac{3}{5}$,

$\overrightarrow{CP} = \left(\dfrac{3}{5} - 1\right)a\underset{\sim}{i} + \left(1 + \dfrac{\frac{3}{5} - 1}{2}\right)a\underset{\sim}{j}$

$= -\dfrac{2}{5}a\underset{\sim}{i} + \dfrac{4}{5}a\underset{\sim}{j}$ ✓

$|\overrightarrow{CP}| = \sqrt{\left(-\dfrac{2}{5}a\right)^2 + \left(\dfrac{4}{5}a\right)^2}$

$= \sqrt{\dfrac{20}{25}}a$

$= \dfrac{2\sqrt{5}}{5}a$ ✓

Also, $\overrightarrow{CO} = -a\underset{\sim}{i}$ and $|\overrightarrow{CO}| = a$

$\cos\theta = \dfrac{-a \times -\frac{2}{5}a + 0 \times \frac{4}{5}a}{\sqrt{a^2} \times \frac{2\sqrt{5}}{5}a}$

$= \dfrac{\frac{2a^2}{5}}{\frac{2a^2\sqrt{5}}{5}}$

$= \dfrac{2a^2}{5} \times \dfrac{5}{2a^2\sqrt{5}}$

$= \dfrac{1}{\sqrt{5}}$

$= \dfrac{\sqrt{\mathbf{5}}}{\mathbf{5}}$ ✓

CHAPTER 11 PROJECTILE MOTION

TOPIC: VECTORS

YEAR 12

1 Position vector as a function of time

- **The location of a particle is** given by its position vector $\underset{\sim}{r}$, where $\underset{\sim}{r} = x\underset{\sim}{i} + y\underset{\sim}{j}$.
- $\underset{\sim}{r}(t) = x(t)\underset{\sim}{i} + y(t)\underset{\sim}{j}$ is a vector function where r, x and y are expressed in terms of time t.

EXAMPLE 1

Find the corresponding cartesian equation for $\underset{\sim}{r}(t) = (2t+3)\underset{\sim}{i} + (t^2+1)\underset{\sim}{j}$.

$$x = 2t + 3$$
$$\therefore t = \frac{x-3}{2}$$

Substitute into $y = t^2 + 1$:

Eliminate the parameter t.

$$y = \left(\frac{x-3}{2}\right)^2 + 1$$
$$\therefore y = \frac{(x-3)^2}{4} + 1$$

EXAMPLE 2

A particle moves so that its position is given by $\underset{\sim}{r}(t) = (2t-1)\underset{\sim}{i} + (t+3)\underset{\sim}{j}$. Find the distance the particle is from the origin after 2 seconds.

$$\underset{\sim}{r}(2) = (2(2)-1)\underset{\sim}{i} + (2+3)\underset{\sim}{j}$$
$$= 3\underset{\sim}{i} + 5\underset{\sim}{j}$$

Evaluate $|\underset{\sim}{r}(2)|$.

$$|\underset{\sim}{r}(2)| = \sqrt{3^2+5^2}$$
$$= \sqrt{34}$$

$\therefore$ the particle is $\sqrt{34}$ units from origin.

EXAMPLE 3

Two particles A and B move so that their vector equations are given by $\underset{\sim}{r}_A(t) = (5t-4)\underset{\sim}{i} + (2t-1)\underset{\sim}{j}$ and $\underset{\sim}{r}_B(t) = t^2\underset{\sim}{i} + (t+3)\underset{\sim}{j}$. Determine whether particles A and B collide.

Equate the components of $\underset{\sim}{j}$:

Equate components of x and y in the vectors $x\underset{\sim}{i} + y\underset{\sim}{j}$.

$$2t - 1 = t + 3$$
$$t = 4$$

Now substitute into components of $\underset{\sim}{i}$ in both equations:

$$\underset{\sim}{r}_A(4) = 5(4) - 4 = 16$$
$$\underset{\sim}{r}_B(4) = 4^2 = 16$$

As $\underset{\sim}{r}_A(4) = \underset{\sim}{r}_B(4)$ for components of $\underset{\sim}{i}$, the particles intersect after 4 seconds.

EXAMPLE 4

Two particles P and Q move so that their vector equations are given by $\underset{\sim}{r}_P(t) = (2t-1)\underset{\sim}{i} + (t+2)\underset{\sim}{j}$ and $\underset{\sim}{r}_Q(t) = t^2\underset{\sim}{i} + (4-3t)\underset{\sim}{j}$. Find the distance between the particles when $t = 2$.

$$\underset{\sim}{r}_P(2) = (2(2)-1)\underset{\sim}{i} + (2+2)\underset{\sim}{j}$$
$$= 3\underset{\sim}{i} + 4\underset{\sim}{j}$$
$$\underset{\sim}{r}_Q(2) = 2^2\underset{\sim}{i} + (4-3(2))\underset{\sim}{j}$$
$$= 4\underset{\sim}{i} - 2\underset{\sim}{j}$$
$$\underset{\sim}{r}_P(2) - \underset{\sim}{r}_Q(2) = (3\underset{\sim}{i} + 4\underset{\sim}{j}) - (4\underset{\sim}{i} - 2\underset{\sim}{j})$$
$$= -\underset{\sim}{i} + 6\underset{\sim}{j}$$

Use $|x\underset{\sim}{i} + y\underset{\sim}{j}| = \sqrt{x^2+y^2}$.

$$\text{Distance} = \sqrt{(-1)^2 + 6^2}$$
$$= \sqrt{37}$$

$\therefore$ the particles are $\sqrt{37}$ units apart.

→ PRACTICE

1 Find the corresponding cartesian equation for:

i $\underset{\sim}{r}(t) = (3 - t)\underset{\sim}{i} + (t^2 + 4t - 1)\underset{\sim}{j}$

ii $\underset{\sim}{r}(t) = 2 \sec t\underset{\sim}{i} + (1 + 4\tan^2 t)\underset{\sim}{j}$

2 i A particle moves so that its vector equation is given by $\underset{\sim}{r}(t) = (2t - 5)\underset{\sim}{i} + (1 + 3t - 2t^2)\underset{\sim}{j}$.

Find the distance the particle is from the origin after 3 seconds.

ii A particle moves so that its vector equation is given by $\underset{\sim}{r}(t) = \cos\frac{t}{6}\underset{\sim}{i} + \left(1 + \sin\frac{t}{4}\right)\underset{\sim}{j}$.

Find the distance the particle is from the origin after 2π seconds.

3 Determine whether particles A and B collide if their vector equations are given by:

i $\underset{\sim}{r}_A(t) = (t^2 - 6)\underset{\sim}{i} + (2t + 2)\underset{\sim}{j}$ and $\underset{\sim}{r}_B(t) = (2t - 3)\underset{\sim}{i} + (3t - 1)\underset{\sim}{j}$

ii $\underset{\sim}{r}_A(t) = (3 + 4t)\underset{\sim}{i} + (t^2 + 3)\underset{\sim}{j}$ and $\underset{\sim}{r}_B(t) = (t^2 + 6)\underset{\sim}{i} + (13 - 3t)\underset{\sim}{j}$.

4 i Two particles P and Q move so that their vector equations are given by $\underset{\sim}{r}_P(t) = (4t - 5)\underset{\sim}{i} + (2t + 3)\underset{\sim}{j}$ and $\underset{\sim}{r}_Q(t) = 2t^2\underset{\sim}{i} + (1 - t)\underset{\sim}{j}$. Find the distance between the particles when $t = 3$.

ii Two particles P and Q move so that their vector equations are given by $\underset{\sim}{r}_P(t) = (2t - 1)\underset{\sim}{i} + (1 - 2t)\underset{\sim}{j}$ and $\underset{\sim}{r}_Q(t) = (t - 2)\underset{\sim}{i} + (3 - t)\underset{\sim}{j}$. When are the particles 5 units apart?

Answers ➲ p. 143

2 Velocity and acceleration vectors as functions of time

→ **If the position vector** is given as $\underset{\sim}{r}(t) = x(t)\underset{\sim}{i} + y(t)\underset{\sim}{j}$, where $x(t)$ and $y(t)$ are differentiable, then:

- **velocity vector** is $\dot{\underset{\sim}{r}}(t) = \underset{\sim}{v}(t) = \dot{x}(t)\underset{\sim}{i} + \dot{y}(t)\underset{\sim}{j}$
- **acceleration vector** is $\ddot{\underset{\sim}{r}}(t) = \underset{\sim}{a}(t) = \ddot{x}(t)\underset{\sim}{i} + \ddot{y}(t)\underset{\sim}{j}$

→ **Speed at time t** is given as $\left|\underset{\sim}{v}(t)\right| = \sqrt{\left(\underset{\sim}{x}(t)\right)^2 + \left(\underset{\sim}{y}(t)\right)^2}$.

→ **Given $\underset{\sim}{v}(t) = \dot{x}(t)\underset{\sim}{i} + \dot{y}(t)\underset{\sim}{j}$,** then $\underset{\sim}{r}(t) = x(t)\underset{\sim}{i} + y(t)\underset{\sim}{j}$, where $x(t) = \int \dot{x}(t)\,dt$ and $y(t) = \int \dot{y}(t)\,dt$.

EXAMPLE 1

A particle moves so that its position vector is given by $\underset{\sim}{r}(t) = 4t^2\underset{\sim}{i} + (2t^3 + 3t + 1)\underset{\sim}{j}$. Find the:

a velocity at time t

b acceleration at time t

a $\underset{\sim}{v}(t) = 8t\underset{\sim}{i} + (6t^2 + 3)\underset{\sim}{j}$ — Differentiate with respect to t.

b $\underset{\sim}{a}(t) = 8\underset{\sim}{i} + 12t\underset{\sim}{j}$

EXAMPLE 2

A particle moves so that its position vector is given by $\underset{\sim}{r}(t) = 3 \sin 2t\underset{\sim}{i} + 4 \cos 2t\underset{\sim}{j}$. Find its speed after $\frac{3\pi}{8}$ seconds, in ms^{-1}.

$$\underset{\sim}{v}(t) = 6 \cos 2t\underset{\sim}{i} - 8 \sin 2t\underset{\sim}{j}$$

$$\underset{\sim}{v}\left(\frac{3\pi}{8}\right) = 6\cos 2\left(\frac{3\pi}{8}\right)\underset{\sim}{i} - 8\sin 2\left(\frac{3\pi}{8}\right)\underset{\sim}{j}$$

$$= 6\cos\frac{3\pi}{4}\underset{\sim}{i} - 8\sin\frac{3\pi}{4}\underset{\sim}{j}$$

$$= -\frac{6}{\sqrt{2}}\underset{\sim}{i} - \frac{8}{\sqrt{2}}\underset{\sim}{j}$$

$$\left|\underset{\sim}{v}\left(\frac{3\pi}{8}\right)\right| = \sqrt{18 + 32}$$

$$= \sqrt{50}$$

$$= 5\sqrt{2}$$

Find $\underset{\sim}{v}(t)$.

$\therefore$ speed is $5\sqrt{2}$ ms^{-1}.

EXAMPLE 3

A particle moves so that its position vector is given by $\underset{\sim}{r}(t) = 2 \sin 3t\underset{\sim}{i} + 2 \cos 3t\underset{\sim}{j}$. Show that the acceleration vector is perpendicular to the velocity vector.

$$\underset{\sim}{r}(t) = 2 \sin 3t\underset{\sim}{i} + 2 \cos 3t\underset{\sim}{j}$$

$$\underset{\sim}{v}(t) = 6 \cos 3t\underset{\sim}{i} - 6 \sin 3t\underset{\sim}{j}$$

$$\underset{\sim}{a}(t) = -18 \sin 3t\underset{\sim}{i} - 18 \cos 3t\underset{\sim}{j}$$

$$\underset{\sim}{a}(t).\underset{\sim}{v}(t) = (-18 \sin 3t) \times 6 \cos 3t + (-18 \cos 3t) \times (-6 \sin 3t)$$

$$= 0$$

$\therefore$ vectors are perpendicular. — Two vectors are perpendicular if their dot product is zero.

EXAMPLE 4

The velocity of an object is given by $\underset{\sim}{v}(t) = 6t^2\underset{\sim}{i} - \sin t\underset{\sim}{j}$ and initially $\underset{\sim}{r} = 2\underset{\sim}{i} + \underset{\sim}{j}$. Determine $\underset{\sim}{r}(t)$.

Consider $x(t) = \int 6t^2\,dt$
$= 2t^3 + c_1$
As $x(0) = 2$,
$2(0)^3 + c_1 = 2$
$\therefore c_1 = 2$
$\therefore x(t) = 2t^3 + 2$

Integrate components separately.

Consider $y(t) = \int -\sin t\,dt$
$= \cos t + c_2$
As $y(0) = 1$,
$\cos(0) + c_2 = 1$
$\therefore c_2 = 0$
$\therefore y(t) = \cos t$
$\therefore \underset{\sim}{r}(t) = (2t^3 + 2)\underset{\sim}{i} + \cos t\underset{\sim}{j}$

➔ PRACTICE

1 i A particle moves so that its position vector is given by $\underset{\sim}{r}(t) = (2t^2 + 1)\underset{\sim}{i} + (4t^2 - 5t + 6)\underset{\sim}{j}$. Find the:

a velocity at time t

b acceleration at time t

ii A particle moves so that its position vector is given by $\underset{\sim}{r}(t) = te^{-t}\underset{\sim}{i} + 3te^{2t}\underset{\sim}{j}$. Find the acceleration vector in terms of t.

2 i A particle moves so that its position vector is given by $\underset{\sim}{r}(t) = 4\cos 3t\underset{\sim}{i} + 2\sin 3t\underset{\sim}{j}$. Find its speed after $\frac{\pi}{6}$ seconds.

ii A particle moves so that its position vector is given by $\underset{\sim}{r}(t) = (t - \sin t)\underset{\sim}{i} + (1 + \cos t)\underset{\sim}{j}$. Show that $|\underset{\sim}{v}(t)| = \sqrt{2(1-\cos t)}$, and hence find the speed of the particle after $\frac{\pi}{3}$ seconds.

3 i A particle moves so that its position vector is given by $\underset{\sim}{r}(t) = 4\cos\frac{t}{2}\underset{\sim}{i} + 4\sin\frac{t}{2}\underset{\sim}{j}$. Show that the acceleration vector is perpendicular to the velocity vector.

ii A particle moves so that its position vector is given by $\underset{\sim}{r}(t) = t\cos t\underset{\sim}{i} + t\sin t\underset{\sim}{j}$.
Show that $\underset{\sim}{v}(t).\underset{\sim}{a}(t) = t$.

4 i The velocity of a car is given by the vector $\underset{\sim}{v}(t) = 4t\underset{\sim}{i} + (e^t + 2)\underset{\sim}{j}$, where velocity is measured in ms^{-1}. If the initial position of the car is $\underset{\sim}{r} = \underset{\sim}{i} + 2\underset{\sim}{j}$, find the position of the car after 3 seconds.

ii The acceleration of an object is given by the vector $\underset{\sim}{a}(t) = -10\underset{\sim}{j}$, where acceleration is in ms^{-2}. Initially, the velocity is given by the vector $\underset{\sim}{v} = \frac{5\sqrt{3}}{2}\underset{\sim}{i} + \frac{5}{2}\underset{\sim}{j}$, and its position is given as $6\underset{\sim}{j}$. Find the position vector in terms of t.

Answers ➲ pp. 144–145

3 Projectile motion

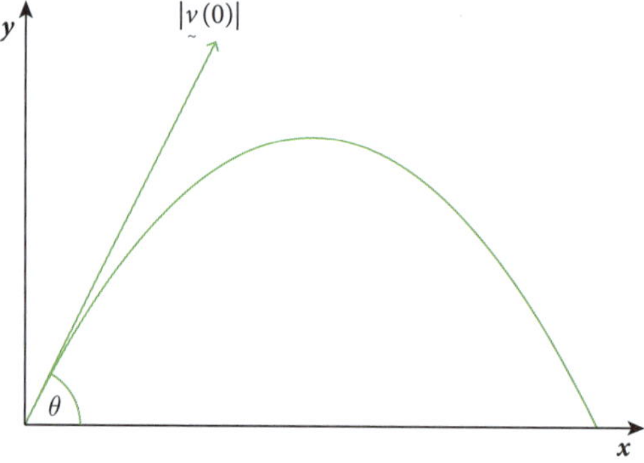

➔ **As the only force is downwards**, then $\ddot{\underset{\sim}{r}}(t) = \underset{\sim}{a}(t) = -g\underset{\sim}{j}$.

➔ **Using integration and $\dot{x}(0) = |\underset{\sim}{v}(0)|\cos\theta$, $\dot{y}(0) = |\underset{\sim}{v}(0)|\sin\theta$**, then $\underset{\sim}{v}(t) = |\underset{\sim}{v}(0)|\cos\theta\,\underset{\sim}{i} + (|\underset{\sim}{v}(0)|\sin\theta - gt)\underset{\sim}{j}$.

➔ **Using integration and $\underset{\sim}{r}(0) = 0$**, then $\underset{\sim}{r}(t) = |\underset{\sim}{v}(0)|t\cos\theta\,\underset{\sim}{i} + \left(|\underset{\sim}{v}(0)|t\sin\theta - \frac{1}{2}gt^2\right)\underset{\sim}{j}$.

➔ **To find maximum height** let $\dot{y}(t) = 0$ to find t, and then substitute into y.

➔ **To find range** let $y(t) = 0$ to find t, and then substitute into x. Greatest range occurs when $\theta = 45°$.

EXAMPLE 1

A person standing on the top of a 40 m–high cliff throws a stone into the water at an angle of 30° with the horizontal and at a velocity of 15 ms^{-1}.

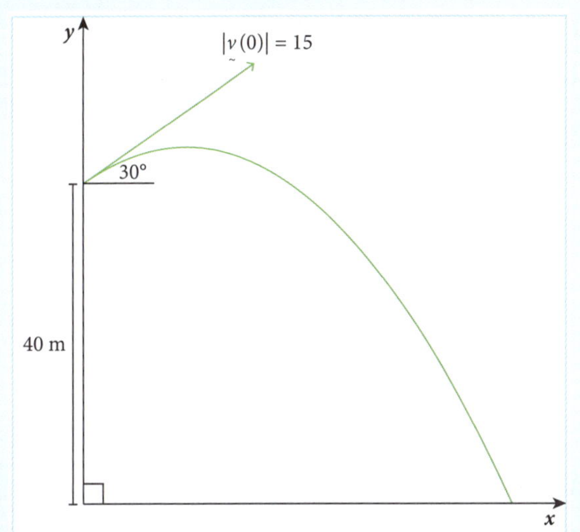

Using the origin as the base of the cliff, derive the velocity vector $\underset{\sim}{v}(t)$ and the position vector $\underset{\sim}{r}(t)$.
(Let $g = -10$ ms^{-2} and neglect air resistance.)

Consider $\underset{\sim}{a}(t) = \ddot{x}(t)\underset{\sim}{i} + \ddot{y}(t)\underset{\sim}{j} = 0\underset{\sim}{i} - 10\underset{\sim}{j}$.

When $t = 0, |\underset{\sim}{v}(0)| = 15, \theta = 30°$.

When $\dfrac{\dot{x}}{15} = \cos 30°$,

then $\dot{x} = 15 \times \cos 30° = \dfrac{15\sqrt{3}}{2}$.

15, $\dot{y}$, 30°, $\dot{x}$

When $\dfrac{\dot{y}}{15} = \sin 30°$,

then $\dot{y} = 15 \times \sin 30° = \dfrac{15}{2}$.

$\underset{\sim}{v}(t) = \int \ddot{x}(t)\, dt\underset{\sim}{i} + \int \ddot{y}(t)\, dt\underset{\sim}{j}$

Consider $\dot{x}(t) = \int 0\, dt$

$= c_1$

As $\dot{x}(0) = \dfrac{15\sqrt{3}}{2}$,

$c_1 = \dfrac{15\sqrt{3}}{2}$

$\therefore \dot{x}(t) = \dfrac{15\sqrt{3}}{2}$

Consider the initial conditions for v.

Consider $\dot{y}(t) = \int -10\, dt$

$= -10t + c_2$

As $\dot{y}(0) = \dfrac{15}{2}$,

$-10(0) + c_2 = \dfrac{15}{2}$

$\therefore c_2 = \dfrac{15}{2}$

$\therefore \dot{y}(t) = \dfrac{15}{2} - 10t$

$\therefore \underset{\sim}{v}(t) = \dfrac{15\sqrt{3}}{2}\underset{\sim}{i} + \left(\dfrac{15}{2} - 10t\right)\underset{\sim}{j}$

Consider $x(t) = \int \dfrac{15\sqrt{3}}{2}\, dt$

$= \dfrac{15\sqrt{3}}{2}t + c_3$

As $x(0) = 0$,

$\dfrac{15\sqrt{3}}{2}(0) + c_3 = 0$

Use different numbered subscripts for constants.

$\therefore c_3 = 0$

$\therefore x(t) = \dfrac{15\sqrt{3}}{2}t$

Consider $y(t) = \int \left(\dfrac{15}{2} - 10t\right)dt$

$= \dfrac{15t}{2} - 5t^2 + c_4$

As $y(0) = 40$,

$\dfrac{5(0)}{2} - 5(0)^2 + c_4 = 40$

$\therefore c_4 = 40$

$\therefore y(t) = \dfrac{15t}{2} - 5t^2 + 40$

$\therefore \underset{\sim}{r}(t) = \dfrac{15\sqrt{3}}{2}t\underset{\sim}{i} + \left(\dfrac{15t}{2} - 5t^2 + 40\right)\underset{\sim}{j}$

EXAMPLE 2

A particle is projected from O at a speed of 60 ms^{-1} at an angle of 45° from the horizontal plane.
Given $\underset{\sim}{r}(t) = |\underset{\sim}{v}(0)|t \cos\theta\, \underset{\sim}{i} + (|\underset{\sim}{v}(0)|t \sin\theta - 5t^2)\underset{\sim}{j}$:

a **find the greatest height of the ball**

b **find the time of flight and the range of the ball when the particle hits the ground**

a Use $|\underset{\sim}{v}(0)| = 60$ and $\theta = 45°$.

$\underset{\sim}{r}(t) = 60t\cos 45°\underset{\sim}{i} + (60t \sin 45° - 5t^2)\underset{\sim}{j}$

$= \dfrac{60t}{\sqrt{2}}\underset{\sim}{i} + \left(\dfrac{60t}{\sqrt{2}} - 5t^2\right)\underset{\sim}{j}$

$= 30\sqrt{2}t\underset{\sim}{i} + (30\sqrt{2}t - 5t^2)\underset{\sim}{j}$

As $y(t) = 30\sqrt{2}t - 5t^2$, then

$\dot{y}(t) = 30\sqrt{2} - 10t = 0$

$10t = 30\sqrt{2}$

$t = 3\sqrt{2}$

Substitute into $y(t) = 30\sqrt{2}t - 5t^2$:

$y(3\sqrt{2}) = 30\sqrt{2}(3\sqrt{2}) - 5(3\sqrt{2})^2$

$= 90$

$\therefore$ greatest height of 90 m.

b Time of flight is $2(3\sqrt{2}) = 6\sqrt{2}$ seconds.

Substitute into $x(t) = 30\sqrt{2}t$:

$x(6\sqrt{2}) = 30\sqrt{2}(6\sqrt{2})$

$= 360$

When projected from *O*, the time of flight is twice the time taken to reach maximum height.

$\therefore$ the time of flight is $6\sqrt{2}$ s and the range is 360 m.

EXAMPLE 3

A golfer hits a golf ball from a point *O* on a flat course at an initial velocity of 50 ms^{-1}. The ball just clears a 15 m tall tree at a distance of 60 m away.

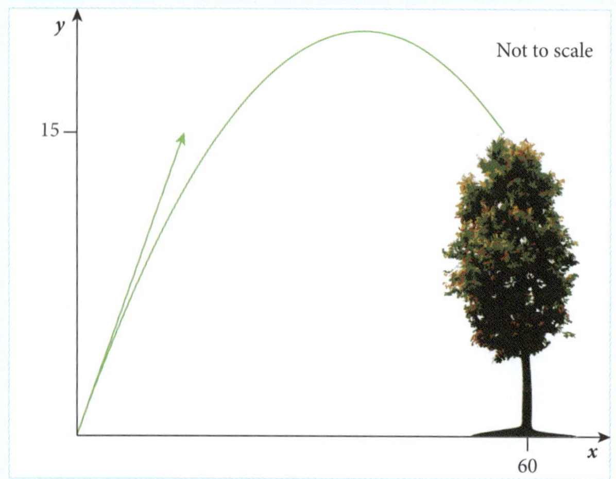

Using $g = 10$ ms^{-2}, the position vector has equation $\underset{\sim}{r}(t) = 50t\cos\theta\,\underset{\sim}{i} + (50t\sin\theta - 5t^2)\underset{\sim}{j}$, where θ is the angle of projection.

a **Show that the cartesian equation of the path of the ball is given by $y = x\tan\theta - \dfrac{x^2(1+\tan^2\theta)}{500}$.**

b **Find the two possible initial angles of projection, to the nearest degree.**

a Let $x = 50t\cos\theta$ and $y = 50t\sin\theta - 5t^2$.

Substitute $t = \dfrac{x}{50\cos\theta}$ in y:

$$y = 50\left(\frac{x}{50\cos\theta}\right)\sin\theta - 5\left(\frac{x}{50\cos\theta}\right)^2$$

$$= x\tan\theta - \frac{x^2}{500\cos^2\theta}$$

$$= x\tan\theta - \frac{x^2\sec^2\theta}{500}$$

$\sec^2\theta = 1 + \tan^2\theta$

$$= x\tan\theta - \frac{x^2(1+\tan^2\theta)}{500}$$

$\therefore$ the trajectory is $y = x\tan\theta - \dfrac{x^2(1+\tan^2\theta)}{500}$.

b Substitute $x = 60$ and $y = 15$ into $y = x\tan\theta - \dfrac{x^2(1+\tan^2\theta)}{500}$:

$$15 = 60\tan\theta - \frac{60^2(1+\tan^2\theta)}{500}$$

$\therefore 3600(1 + \tan^2\theta) - 30\,000\tan\theta + 7500 = 0$

$3600 + 3600\tan^2\theta - 30\,000\tan\theta + 7500 = 0$

$3600\tan^2\theta - 30\,000\tan\theta + 11\,100 = 0$

$36\tan^2\theta - 300\tan\theta + 111 = 0$

$$\tan\theta = \frac{300 \pm \sqrt{(-300)^2 - 4(36)(111)}}{72}$$

$= 0.388\,071\,983\ldots, 7.945\,261\,35\ldots$

$\theta = 21°, 83°$

EXAMPLE 4

An object is projected horizontally from the top of a 34.3 metre cliff into water with a velocity of 42 ms^{-1}.

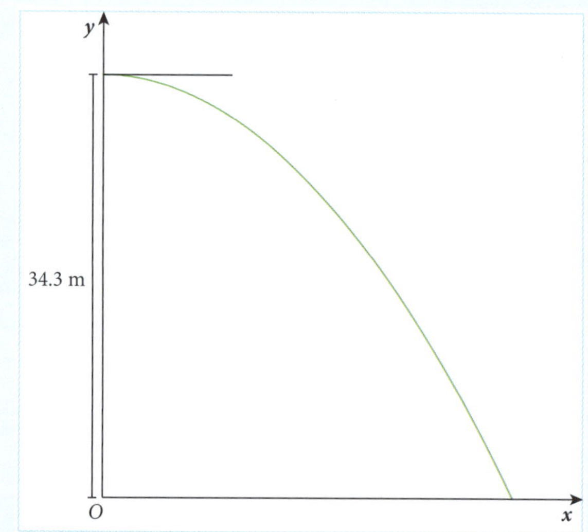

Using $g = 9.8$ ms^{-2}:

a **write $\underset{\sim}{r}(t)$ and $\underset{\sim}{v}(t)$**

b **how far from the base of the cliff does the object land in the water?**

a Use $|\underset{\sim}{v}(0)| = 42$ and $\theta = 0°$,

$\underset{\sim}{r}(t) = 42t\cos 0°\underset{\sim}{i} + (42t\sin 0° - 4.9t^2 + 34.3)\underset{\sim}{j}$

$= 42t\underset{\sim}{i} + (34.3 - 4.9t^2)\underset{\sim}{j}$

$\underset{\sim}{v}(t) = 42\underset{\sim}{i} - 9.8t\underset{\sim}{j}$

$\therefore\ \underset{\sim}{r}(t) = 42t\underset{\sim}{i} + (34.3 - 4.9t^2)\underset{\sim}{j}$ and $\underset{\sim}{v}(t) = 42\underset{\sim}{i} - 9.8t\underset{\sim}{j}$.

b Let $y(t) = 0$:

$34.3 - 4.9t^2 = 0$

$4.9t^2 = 34.3$

$t^2 = 7$

$t = \sqrt{7}\ (t > 0)$

Substitute into $x(t)$:

$x(\sqrt{7}) = 42\sqrt{7}$

$\therefore$ the object lands $42\sqrt{7}$ metres from the cliff.

$v(t) = \dot{r}(t)$

➔ PRACTICE

1 i An object is projected from the top of a 12 m–high tower at an angle of 60° with the horizontal and at a velocity of 28 ms^{-1}.

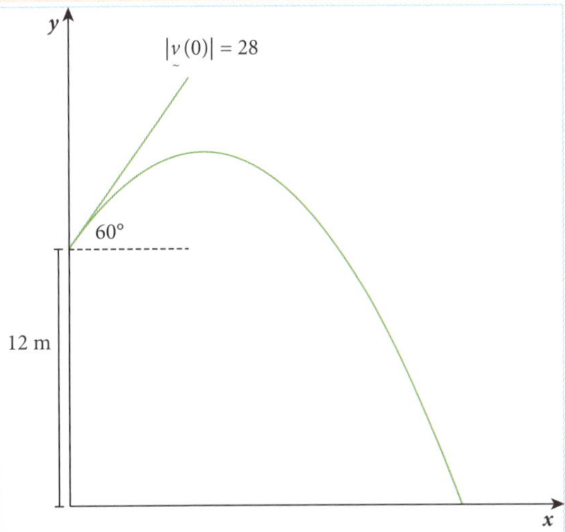

Derive the velocity vector $\underset{\sim}{v}(t)$ and the position vector $\underset{\sim}{r}(t)$. (Let $g = -10$ ms^{-2} and neglect air resistance.)

ii A cricket ball leaves a bowler's hand 2 m from the ground with a velocity of 25 ms^{-1} at an angle of 15° below the horizontal. If the acceleration vector is given as $\underset{\sim}{a}(t) = 0\underset{\sim}{i} - g\underset{\sim}{j}$, show that the position vector is

$\underset{\sim}{r}(t) = 25t\cos 15°\underset{\sim}{i} - \left(25t\sin 15° + \frac{1}{2}gt^2 - 2\right)\underset{\sim}{j}$

2 i An arrow is fired at an angle of 30° and with an initial velocity of 20 ms^{-1} and hits a target. The arrow and the target are at the same height.

You may use $\underset{\sim}{r}(t) = \left|\underset{\sim}{v}(0)\right| t\cos\theta\,\underset{\sim}{i} + \left(\left|\underset{\sim}{v}(0)\right| t\sin\theta - 5t^2\right)\underset{\sim}{j}$). Do NOT prove this.

Find:

a the maximum height of the arrow

b the distance flown by the arrow

ii A particle is projected at an angle of 45° where $\left|\underset{\sim}{v}(0)\right| = V$. Show that the range is four times the maximum height of the particle.

(You may assume the position vector is $\underset{\sim}{r}(t) = \left|\underset{\sim}{v}(0)\right| t\cos\theta\,\underset{\sim}{i} + \left(\left|\underset{\sim}{v}(0)\right| t\sin\theta - \frac{gt^2}{2}\right)\underset{\sim}{j}$.)

3 i a A rugby league football is kicked at 18 m/s from ground level and just passes over a goalpost's crossbar 3 m high and 27 m away.

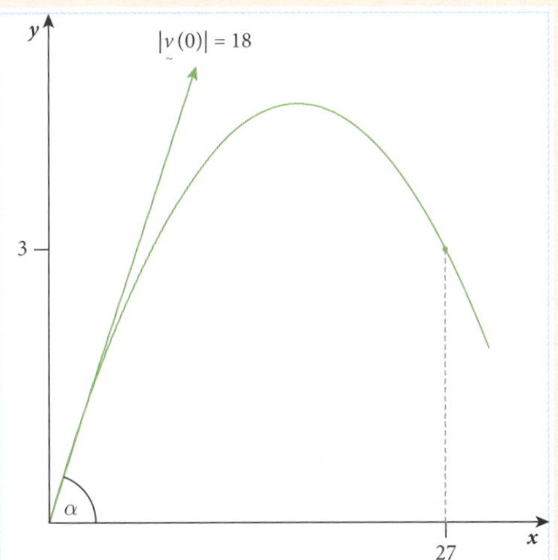

If α is the angle of projection, show that the equation of the trajectory is

$y = x\tan\alpha - \dfrac{5x^2(1+\tan^2\alpha)}{324}$.

(Using $g = 10$ ms^{-2}, the position vector has equation

$\underset{\sim}{r}(t) = \left|\underset{\sim}{v}(0)\right| t\cos\alpha\,\underset{\sim}{i} + \left(\left|\underset{\sim}{v}(0)\right| t\sin\alpha - 5t^2\right)\underset{\sim}{j}$.)

b Find the two possible values of α, to the nearest degree.

c The ball is caught on the full by a ballboy standing 5 m behind the goalpost. Assuming $\alpha < 45°$, at what height was the ball caught?

ii Two objects are set to be launched where object B is positioned 20 m above object A. Object A is projected from the ground at an angle of 60° with a velocity of 20 ms^{-1}. At the same instant object B is projected at an angle of 30° with a velocity of 10 ms^{-1}. Both objects are aimed at the same target.

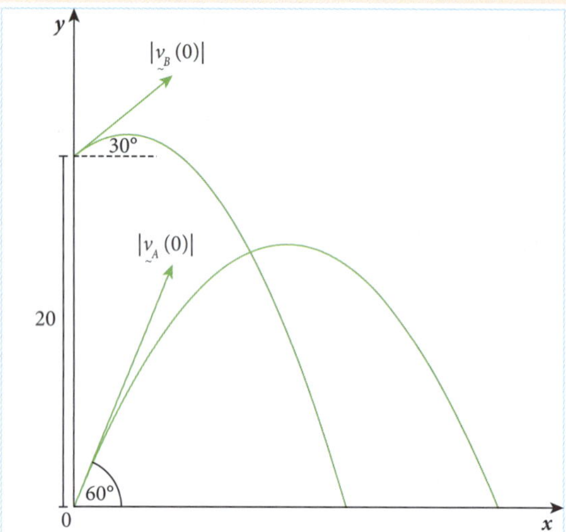

(Using $g = 10$ ms^{-2}, the position vector has equation $\underset{\sim}{r}(t) = \left|\underset{\sim}{v}(0)\right| t \cos \alpha \, \underset{\sim}{i} + \left(\left|\underset{\sim}{v}(0)\right| t \sin \alpha - 5t^2 + h\right)\underset{\sim}{j}$), for initial velocity $\left|\underset{\sim}{v}(0)\right|$, angle of projection α and height above the ground h.)

a Show that the equations of the trajectories of the two objects are given as:

Object A: $y = x\sqrt{3} - \dfrac{x^2}{20}$ and

Object B: $y = \dfrac{x\sqrt{3}}{3} - \dfrac{x^2}{15} + 20$

b Find the point at which the trajectories intersect. Give your answer correct to four decimal places.

c It is known that the objects will not collide if projected at the same instant. Which object should be fired first and how much time should elapse before the second object is projected to ensure collision? Give your answer correct to two decimal places.

4 i From a height of 24 m, an object is projected horizontally from the balcony of an apartment building. If the object has an initial velocity of 18 ms^{-1}, and using $g = 10$ ms^{-2}:

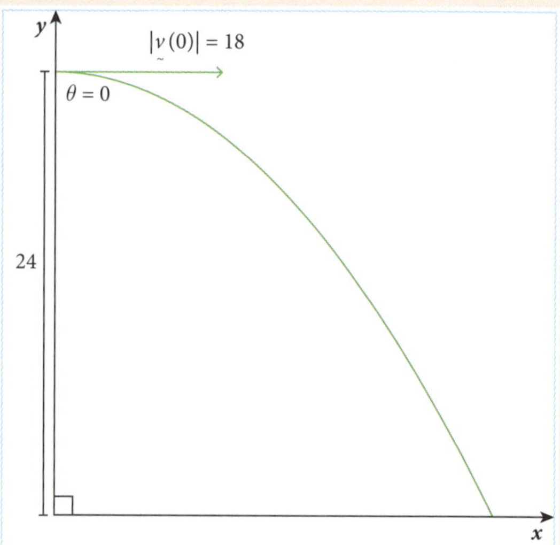

a write $\underset{\sim}{r}(t)$ and $\underset{\sim}{v}(t)$

b how far from the base of the building does the object hit the ground?

ii An object is projected horizontally from the top of a cliff of height 15 m. The object hits the ground 24 m from the base of the cliff.

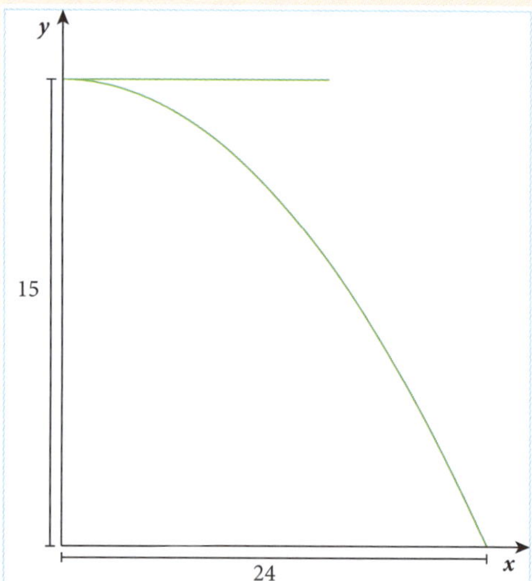

(Using $g = 10$ ms^{-2}, the position vector has equation $\underset{\sim}{r}(t) = \left|\underset{\sim}{v}(0)\right| t \cos \alpha \, \underset{\sim}{i} + \left(\left|\underset{\sim}{v}(0)\right| t \sin \alpha - 5t^2 + 15\right)\underset{\sim}{j}$.)

At what speed and at what acute angle does the object hit the ground?

Answers ➲ pp. 145–147

HSC EXAM-TYPE QUESTIONS

Total marks: 35

PART A

(Suggested time: 10 minutes)

(1 mark each)

1 Which of these is the parametric equation for the position vector $\underset{\sim}{r}(t) = \frac{4-t}{3}\underset{\sim}{i} + (1 - 2t)\underset{\sim}{j}$?

A $y = 6x + 7$ B $y = 3x + 3$

C $y = 3x - 3$ D $y = 6x - 7$

2 A particle moves so that its position is given by $\underset{\sim}{r}(t) = (4t - 5)\underset{\sim}{i} + (t^2 - 12)\underset{\sim}{j}$. Find the distance the particle is from the origin after 3 seconds.

A $\sqrt{45}$ units B $\sqrt{58}$ units

C $\sqrt{74}$ units D $\sqrt{168}$ units

3 A particle moves so that its position vector is given by $\underset{\sim}{r}(t) = (t - e^{-t})\underset{\sim}{i} + te^t\underset{\sim}{j}$. Find the acceleration at time t.

A $\underset{\sim}{a}(t) = e^{-t}\underset{\sim}{i} + (2e^t + te^t)\underset{\sim}{j}$

B $\underset{\sim}{a}(t) = -e^{-t}\underset{\sim}{i} + 2(e^t + te^t)\underset{\sim}{j}$

C $\underset{\sim}{a}(t) = -e^{-t}\underset{\sim}{i} + (2e^t + te^t)\underset{\sim}{j}$

D $\underset{\sim}{a}(t) = e^{-t}\underset{\sim}{i} + 2(e^t + te^t)\underset{\sim}{j}$

4 An object is projected so that its position vector is given by $\underset{\sim}{r}(t) = \frac{\sqrt{2}}{4}t\,\underset{\sim}{i} + \left(\frac{\sqrt{6}}{12}t - 5t^2\right)\underset{\sim}{j}$. What are the values of $\left|\underset{\sim}{v}(0)\right|$ and θ?

A $\left|\underset{\sim}{v}(0)\right| = \frac{1}{\sqrt{6}}$ and $\theta = \frac{\pi}{6}$ B $\left|\underset{\sim}{v}(0)\right| = \frac{1}{\sqrt{2}}$ and $\theta = \frac{\pi}{3}$

C $\left|\underset{\sim}{v}(0)\right| = \frac{1}{\sqrt{2}}$ and $\theta = \frac{\pi}{6}$ D $\left|\underset{\sim}{v}(0)\right| = \frac{1}{\sqrt{6}}$ and $\theta = \frac{\pi}{3}$

5 A golf ball is hit from the ground at an angle of 60° to the horizontal with a velocity of 30 ms^{-1}. The ball is subject only to gravity g and air resistance is negligible. Given that the golf fairway is level, which of these is the distance travelled, in metres, to the landing place of the ball?

A $\frac{15\sqrt{3}}{2g}$ B $\frac{15\sqrt{3}}{g}$ C $\frac{225\sqrt{3}}{g}$ D $\frac{450\sqrt{3}}{g}$

PART B

(Suggested time: 45 minutes)

Show all working.

6 A particle moves so that its position vector is given by $\underset{\sim}{r}(t) = \left(2 - 4\sin\frac{t}{2}\right)\underset{\sim}{i} + \left(3 - 4\cos\frac{t}{2}\right)\underset{\sim}{j}$. Show that the acceleration vector is perpendicular to the velocity vector. (2 marks)

7 A ball is thrown from a window 8 m from the ground with an initial velocity of 25 ms^{-1} at an angle α where $\tan\alpha = \frac{3}{4}$ to the horizontal through O.

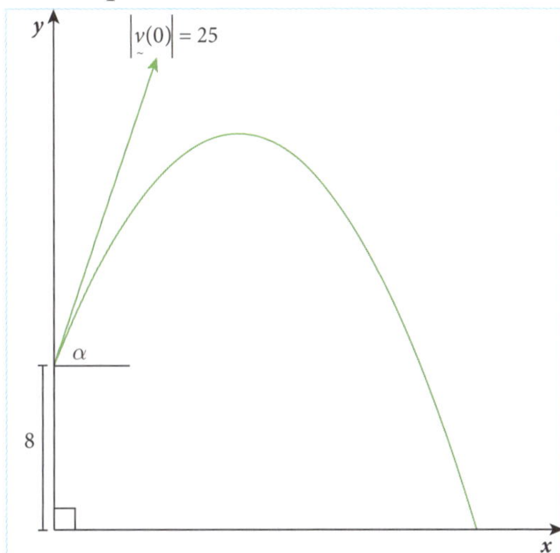

If air resistance is neglected and $g = 10$ ms^{-2}:

a show the position vector is given by the equation $\underset{\sim}{r}(t) = 12t\underset{\sim}{i} + (15t - 5t^2 + 8)\underset{\sim}{j}$ (3 marks)

b find the greatest height attained by the ball (2 marks)

c determine a cartesian equation to represent the trajectory of the ball (1 mark)

8 The velocity of an object is given by $\underset{\sim}{v}(t) = 6e^{2t}\underset{\sim}{i} + 3e^{3t}\underset{\sim}{j}$ and initally $\underset{\sim}{r} = 4\underset{\sim}{i} - 2\underset{\sim}{j}$. Determine $\underset{\sim}{r}(t)$. (2 marks)

9 A projectile leaves a point O on level ground and after 2 seconds just clears a vertical wall whose height above the ground is 6 m and which is 8 m distance from the point of projection. The wall is perpendicular to the level ground.

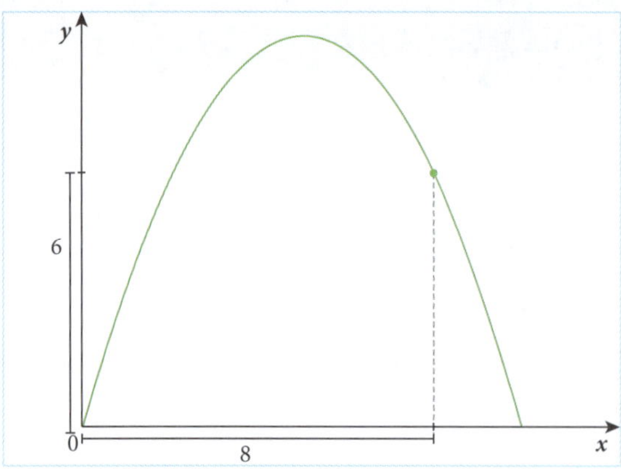

(Using $g = 10$ ms^{-2}, the position vector has equation $\underset{\sim}{r}(t) = |\underset{\sim}{v}(0)|t\cos\alpha\,\underset{\sim}{i} + (|\underset{\sim}{v}(0)|t\sin\alpha - 5t^2)\underset{\sim}{j}$.)

a Find the value of $|\underset{\sim}{v}(0)|$ and $\tan\alpha$. (3 marks)

b Find the velocity vector at the moment the projectile clears the wall. (2 marks)

c How much further past the wall does the projectile hit the ground? (3 marks)

10 Two particles A and B move so that their vector equations are given by $\underset{\sim A}{r}(t) = (3t - 2)\underset{\sim}{i} + (2t + 1)\underset{\sim}{j}$ and $\underset{\sim B}{r}(t) = t^2\underset{\sim}{i} + (5 - 2t)\underset{\sim}{j}$. Find the distance between the particles when $t = 3$. (2 marks)

11 A ball is thrown from a height of 2 m towards the wall of a building. It is thrown with speed of 20 ms^{-1} at an angle of 30° to the horizontal. At its highest point the ball hits the wall and rebounds horizontally from the wall at a speed of 8 ms^{-1}, as shown in the diagram.

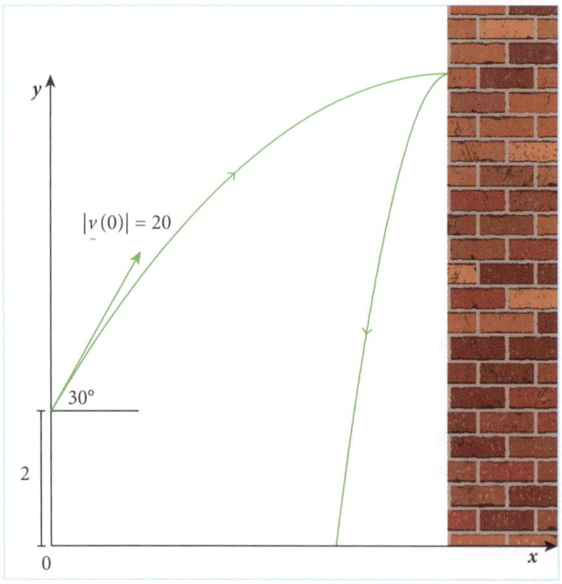

The velocity vector of the ball when it is thrown is given as $\underset{\sim}{v}(t) = |\underset{\sim}{v}(0)|\cos\alpha\,\underset{\sim}{i} + (|\underset{\sim}{v}(0)|\sin\alpha - 10t)\underset{\sim}{j}$.

a Explain why the equation of the position vector is $\underset{\sim}{r}(t) = 10\sqrt{3}t\,\underset{\sim}{i} + (10t - 5t^2 + 2)\underset{\sim}{j}$. (2 marks)

b Show that the ball hits the wall at a height of 7 m. (1 mark)

c How long after being thrown does the ball hit the ground? (2 marks)

d Show that the ball lands approximately 12.26 m from the point of release. (2 marks)

12 A 12 m–tall pyramid is constructed using sandstone blocks as shown in the diagram. An object is projected from the top of the pyramid at an angle of 60° to the horizontal. The object lands on the side of the pyramid at a point P which is three-quarters down the edge of the pyramid.

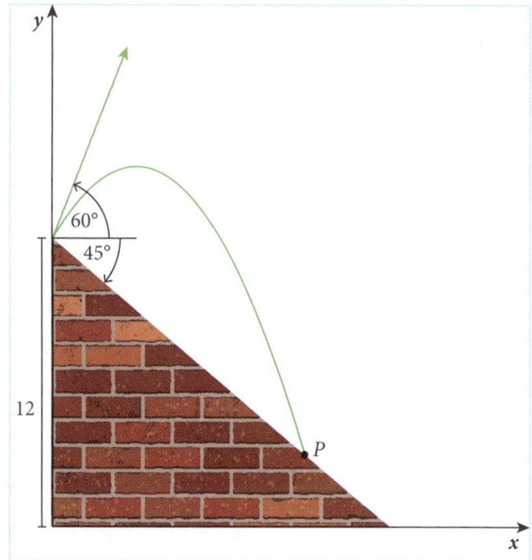

The position vector of the object is defined as $\underset{\sim}{r}(t) = \left(\frac{Vt}{2}\right)\underset{\sim}{i} + \left(\frac{Vt\sqrt{3}}{2} - 5t^2 + 12\right)\underset{\sim}{j}$, where V = initial speed of the object. Show that the initial speed of the object is approximately 8.12 ms^{-1}. (3 marks)

ANSWERS

PRACTICE

1 Position vector as a function of time ➲ p. 135

1 i $\underset{\sim}{r}(t) = (3 - t)\underset{\sim}{i} + (t^2 + 4t - 1)\underset{\sim}{j}$

$x = 3 - t$

$\therefore t = 3 - x$

Substitute into $y = t^2 + 4t - 1$:

$y = (3 - x)^2 + 4(3 - x) - 1$

$= 9 - 6x + x^2 + 12 - 4x - 1$

$= x^2 - 10x + 20$

$\therefore \mathbf{y = x^2 - 10x + 20}$

ii $\underset{\sim}{r}(t) = 2\sec t\underset{\sim}{i} + (1 + 4\tan^2 t)\underset{\sim}{j}$

$x = 2\sec t$

$\therefore \sec t = \dfrac{x}{2}$

$\sec^2 t = \dfrac{x^2}{4}$

$1 + \tan^2 t = \dfrac{x^2}{4}$

$\tan^2 t = \dfrac{x^2 - 4}{4}$

Substitute into $y = 1 + 4\tan^2 t$:

$y = 1 + 4\left(\dfrac{x^2 - 4}{4}\right)$

$= 1 + x^2 - 4$

$= x^2 - 3$

$\therefore \mathbf{y = x^2 - 3}$

2 i $\underset{\sim}{r}(3) = (2(3) - 5)\underset{\sim}{i} + (1 + 3(3) - 2(3)^2)\underset{\sim}{j}$

$= \underset{\sim}{i} - 8\underset{\sim}{j}$

$|\underset{\sim}{r}(3)| = \sqrt{1^2 + (-8)^2}$

$= \sqrt{65}$

$\therefore$ **the particle is $\sqrt{65}$ units from origin.**

ii $\underset{\sim}{r}(2\pi) = \cos\dfrac{2\pi}{6}\underset{\sim}{i} + \left(1 + \sin\dfrac{2\pi}{4}\right)\underset{\sim}{j}$

$= \dfrac{1}{2}\underset{\sim}{i} + (1 + 1)\underset{\sim}{j}$

$= \dfrac{1}{2}\underset{\sim}{i} + 2\underset{\sim}{j}$

$|\underset{\sim}{r}(2\pi)| = \sqrt{\left(\dfrac{1}{2}\right)^2 + 2^2}$

$= \sqrt{\dfrac{1}{4} + 4}$

$= \sqrt{\dfrac{17}{4}}$

$= \dfrac{\sqrt{17}}{2}$

$\therefore$ **the particle is $\dfrac{\sqrt{17}}{2}$ units from origin.**

3 i Equate the components of $\underset{\sim}{i}$: $t^2 - 6 = 2t - 3$

$t^2 - 2t - 3 = 0$

$(t - 3)(t + 1) = 0$

$t = 3$ (as $t \geq 0$)

Equate the components of $\underset{\sim}{j}$: $2t + 2 = 3t - 1$

$3t - 2t = 2 + 1$

$t = 3$

$\therefore$ the common solution is $t = 3$.

$\therefore$ **the particles collide after 3 seconds.**

ii Equate the components of $\underset{\sim}{i}$: $3 + 4t = t^2 + 6$

$t^2 - 4t + 3 = 0$

$(t - 1)(t - 3) = 0$

$t = 1, 3$

Equate the components of $\underset{\sim}{j}$: $t^2 + 3 = 13 - 3t$

$t^2 + 3t - 10 = 0$

$(t + 5)(t - 2) = 0$

$t = -5, 2$

$t = 2$ (as $t \geq 0$)

$\therefore$ there is no common solution of t.

$\therefore$ **the particles do not collide.**

4 i $\underset{\sim}{r}_P(3) = (4(3) - 5)\underset{\sim}{i} + (2(3) + 3)\underset{\sim}{j}$

$= 7\underset{\sim}{i} + 9\underset{\sim}{j}$

$\underset{\sim}{r}_Q(3) = 2(3)^2\underset{\sim}{i} + (1 - 3)\underset{\sim}{j}$

$= 18\underset{\sim}{i} - 2\underset{\sim}{j}$

$\underset{\sim}{r}_P(3) - \underset{\sim}{r}_Q(3) = (7\underset{\sim}{i} + 9\underset{\sim}{j}) - (18\underset{\sim}{i} - 2)$

$= -11\underset{\sim}{i} + 11\underset{\sim}{j}$

Distance $= \sqrt{(-11)^2 + 11^2}$

$= 11\sqrt{2}$

$\therefore$ **the particles are $11\sqrt{2}$ units apart.**

ii $\underset{\sim}{r}_P(t) - \underset{\sim}{r}_Q(t) = (2t - 1 - (t - 2))\underset{\sim}{i} + (1 - 2t - (3 - t))\underset{\sim}{j}$

$= (t + 1)\underset{\sim}{i} + (-2 - t)\underset{\sim}{j}$

Distance $= \sqrt{(t + 1)^2 + (-2 - t)^2}$

$= \sqrt{t^2 + 2t + 1 + 4 + 4t + t^2}$

$= \sqrt{2t^2 + 6t + 5} = 5$

$2t^2 + 6t + 5 = 25$

$2t^2 + 6t - 20 = 0$

$t^2 + 3t - 10 = 0$

$(t + 5)(t - 2) = 0$

$\mathbf{t = 2}$ **(as $t \geq 0$)**

2 Velocity and acceleration vectors as functions of time ➲ p. 136

1 i a $\underset{\sim}{v}(t) = 4t\underset{\sim}{i} + (8t - 5)\underset{\sim}{j}$

b $\underset{\sim}{a}(t) = 4\underset{\sim}{i} + 8\underset{\sim}{j}$

ii $\underset{\sim}{r}(t) = te^{-t}\underset{\sim}{i} + 3te^{2t}\underset{\sim}{j}$

$\underset{\sim}{v}(t) = (e^{-t} - te^{-t})\underset{\sim}{i} + (3e^{2t} + 6te^{2t})\underset{\sim}{j}$

$\underset{\sim}{a}(t) = (-e^{-t} - e^{-t} + te^{-t})\underset{\sim}{i} + (6e^{2t} + 6e^{2t} + 12te^{2t})\underset{\sim}{j}$

$= \mathbf{(-2e^{-t} + te^{-t})\underset{\sim}{i} + (12e^{2t} + 12te^{2t})\underset{\sim}{j}}$

2 i $\underset{\sim}{r}(t) = 4\cos 3t\underset{\sim}{i} + 2\sin 3t\underset{\sim}{j}$

$\underset{\sim}{v}(t) = -12\sin 3t\underset{\sim}{i} + 6\cos 3t\underset{\sim}{j}$

$\underset{\sim}{v}\left(\frac{\pi}{6}\right) = -12\sin 3\left(\frac{\pi}{6}\right)\underset{\sim}{i} + 6\cos 3\left(\frac{\pi}{6}\right)\underset{\sim}{j}$

$= -12\sin\frac{\pi}{2}\underset{\sim}{i} + 6\cos\frac{\pi}{2}\underset{\sim}{j}$

$= -12\underset{\sim}{i}$

$\left|\underset{\sim}{v}\left(\frac{\pi}{6}\right)\right| = \sqrt{(-12)^2 + 0^2}$

$= \mathbf{12}$

ii $\underset{\sim}{r}(t) = (t - \sin t)\underset{\sim}{i} + (1 + \cos t)\underset{\sim}{j}$

$\underset{\sim}{v}(t) = (1 - \cos t)\underset{\sim}{i} - \sin t\underset{\sim}{j}$

$\underset{\sim}{v}\left(\frac{\pi}{6}\right) = \sqrt{(1-\cos t)^2 + (-\sin t)^2}$

$= \sqrt{1 - 2\cos t + \cos^2 t + \sin^2 t}$

$= \sqrt{1 - 2\cos t + 1}$

$= \sqrt{2 - 2\cos t}$

$= \sqrt{2(1 - \cos t)}$

$\left|\underset{\sim}{v}\left(\frac{\pi}{3}\right)\right| = \sqrt{2\left(1 - \cos\frac{\pi}{3}\right)}$

$= \sqrt{2\left(1 - \frac{1}{2}\right)}$

$= \mathbf{1}$

3 i $\underset{\sim}{r}(t) = 4\cos\frac{t}{2}\underset{\sim}{i} + 4\sin\frac{t}{2}\underset{\sim}{j}$

$\underset{\sim}{v}(t) = -2\sin\frac{t}{2}\underset{\sim}{i} + 2\cos\frac{t}{2}\underset{\sim}{j}$

$\underset{\sim}{a}(t) = -\cos\frac{t}{2}\underset{\sim}{i} - \sin\frac{t}{2}\underset{\sim}{j}$

$\underset{\sim}{v}(t).\underset{\sim}{a}(t) = \left(-2\sin\frac{t}{2}\right) \times \left(-\cos\frac{t}{2}\right) + 2\cos\frac{t}{2} \times \left(-\sin\frac{t}{2}\right)$

$= 2\sin\frac{t}{2}\cos\frac{t}{2} - 2\sin\frac{t}{2}\cos\frac{t}{2}$

$= 0$

$\therefore$ **vectors are perpendicular.**

ii $\underset{\sim}{r}(t) = t\cos t\underset{\sim}{i} + t\sin t\underset{\sim}{j}$

$\underset{\sim}{v}(t) = (\cos t - t\sin t)\underset{\sim}{i} + (\sin t + t\cos t)\underset{\sim}{j}$

$\underset{\sim}{a}(t) = (-\sin t - \sin t - t\cos t)\underset{\sim}{i} + (\cos t + \cos t - t\sin t)\underset{\sim}{j}$

$= (-2\sin t - t\cos t)\underset{\sim}{i} + (2\cos t - t\sin t)\underset{\sim}{j}$

$\underset{\sim}{v}(t).\underset{\sim}{a}(t) = (\cos t - t\sin t)(-2\sin t - t\cos t) + (\sin t + t\cos t)(2\cos t - t\sin t)$

$= -2\sin t\cos t - t\cos^2 t + 2t\sin^2 t + t^2\sin t\cos t + 2\sin t\cos t - t\sin^2 t + 2t\cos^2 t - t^2\sin t\cos t$

$= t\sin^2 t + t\cos^2 t$

$= t(\sin^2 t + \cos^2 t)$

$= \mathbf{t}$

4 i $\underset{\sim}{v}(t) = 4t\underset{\sim}{i} + (e^t + 2)\underset{\sim}{j}$

Consider $x(t) = \int 4t\, dt$

$= 2t^2 + c_1$

As $x(0) = 1$,

$2(0)^2 + c_1 = 1$

$\therefore c_1 = 1$

$\therefore x(t) = 2t^2 + 1$

Consider $y(t) = \int (e^t + 2)\, dt$

$= e^t + 2t + c_2$

As $y(0) = 2$,

$e^0 + 2(0) + c_2 = 2$

$\therefore c_2 = 1$

$\therefore y(t) = e^t + 2t + 1$

$\therefore \underset{\sim}{r}(t) = (2t^2 + 1)\underset{\sim}{i} + (e^t + 2t + 1)\underset{\sim}{j}$

$\therefore \underset{\sim}{r}(3) = (2(3)^2 + 1)\underset{\sim}{i} + (e^3 + 2(3) + 1)\underset{\sim}{j}$

$= \mathbf{19\underset{\sim}{i} + (e^3 + 7)\underset{\sim}{j}}$

ii Consider $\underset{\sim}{v}(t) = \dot{x}(t)\underset{\sim}{i} + \dot{y}(t)\underset{\sim}{j}$

Consider $\dot{x}(t) = \int 0\, dt$

$= c_1$

As $\dot{x}(0) = \frac{5\sqrt{3}}{2}$,

$c_1 = \frac{5\sqrt{3}}{2}$

$\therefore \dot{x}(t) = \frac{5\sqrt{3}}{2}$

Consider $\dot{y}(t) = \int -10\, dt$

$= -10t + c_2$

As $\dot{y}(0) = \frac{5}{2}$,

$-10(0) + c_2 = \frac{5}{2}$

$\therefore c_2 = \frac{5}{2}$

$\therefore \dot{y}(t) = \frac{5}{2} - 10t$

$\therefore \underset{\sim}{v}(t) = \frac{5\sqrt{3}}{2}\underset{\sim}{i} + \left(\frac{5}{2} - 10t\right)\underset{\sim}{j}$

Consider $x(t) = \int \frac{5\sqrt{3}}{2}\, dt$

$= \frac{5\sqrt{3}}{2}t + c_3$

As $x(0) = 0$,

$\frac{5\sqrt{3}}{2}(0) + c_3 = 0$

$\therefore c_3 = 0$

$\therefore x(t) = \frac{5\sqrt{3}}{2}t$

Consider $y(t) = \int\left(\frac{5}{2} - 10t\right) dt$

$= \frac{5t}{2} - 5t^2 + c_4$

As $y(0) = 6$,

$\frac{5(0)}{2} - 5(0)^2 + c_4 = 6$

$\therefore c_4 = 6$

$\therefore y(t) = \frac{5t}{2} - 5t^2 + 6$

$\therefore \boldsymbol{\underset{\sim}{r}(t) = \frac{5\sqrt{3}}{2}t\,\underset{\sim}{i} + \left(\frac{5t}{2} - 5t^2 + 6\right)\underset{\sim}{j}}$

3 Projectile motion ⊃ pp. 139–140

1 i Consider $\underset{\sim}{a}(t) = \ddot{x}(t)\underset{\sim}{i} + \ddot{y}(t)\underset{\sim}{j} = 0\underset{\sim}{i} - 10\underset{\sim}{j}$.

When $t = 0$, $|\underset{\sim}{v}(0)| = 28$, $\theta = 60°$.

When $\frac{\dot{x}}{28} = \cos 60°$, then $\dot{x} = 28 \times \cos 60° = 14$.

When $\frac{\dot{y}}{28} = \sin 60°$, then $\dot{y} = 28 \times \sin 60° = 14\sqrt{3}$.

$\underset{\sim}{v}(t) = \int\ddot{x}(t)\, dt\underset{\sim}{i} + \int\ddot{y}(t)\, dt\underset{\sim}{j}$

Consider $\dot{x}(t) = \int 0\, dt$

$= c_1$

As $\dot{x}(0) = 14$,

$c_1 = 14$

$\therefore \dot{x}(t) = 14$

$|\underset{\sim}{v}(0)| = 28$, $\dot{y}$, $60°$, $\dot{x}$

Consider $\dot{y}(t) = \int -10\, dt$

$= -10t + c_2$

As $\dot{y}(0) = 14\sqrt{3}$,

$-10(0) + c_2 = 14\sqrt{3}$

$\therefore c_2 = 14\sqrt{3}$

$\therefore \dot{y}(t) = 14\sqrt{3} - 10t$

$\therefore \underset{\sim}{v}(t) = 14\underset{\sim}{i} + (14\sqrt{3} - 10t)\underset{\sim}{j}$

Consider $x(t) = \int 14\, dt$

$= 14t + c_3$

As $x(0) = 0$,

$14(0) + c_3 = 0$

$\therefore c_3 = 0$

$\therefore x(t) = 14t$

Consider $y(t) = \int(14\sqrt{3} - 10t)\, dt$

$= 14\sqrt{3}t - 5t^2 + c_4$

As $y(0) = 12$,

$\frac{5(0)}{2} - 5(0)^2 + c_4 = 12$

$\therefore c_4 = 12$

$\therefore y(t) = 14\sqrt{3}t - 5t^2 + 12$

$\therefore \boldsymbol{\underset{\sim}{r}(t) = 14t\,\underset{\sim}{i} + (14\sqrt{3}t - 5t^2 + 12)\underset{\sim}{j}}$

ii Consider $\underset{\sim}{a}(t) = \ddot{x}(t)\underset{\sim}{i} + \ddot{y}(t)\underset{\sim}{j} = 0\underset{\sim}{i} - g\underset{\sim}{j}$.

When $t = 0$, $|\underset{\sim}{v}(0)| = 25$, $\theta = -15°$.

When $\frac{\dot{x}}{25} = \cos(-15°)$, then $\dot{x} = 25 \times \cos(-15°) = 25\cos 15°$.

When $\frac{\dot{y}}{25} = \sin(-15°)$, then $\dot{y} = 25 \times \sin(-15°) = -25\sin 15°$.

$\underset{\sim}{v}(t) = \int\ddot{x}(t)\, dt\underset{\sim}{i} + \int\ddot{y}(t)\, dt\underset{\sim}{j}$

Consider $\dot{x}(t) = \int 0\, dt$

$= c_1$

As $\dot{x}(0) = 25\cos 15°$

$c_1 = 25\cos 15°$

$\therefore \dot{x}(t) = 25\cos 15°$

Consider $\dot{y}(t) = \int -g\, dt$

$= -gt + c_2$

As $\dot{y}(0) = -25\sin 15°$,

$-g(0) + c_2 = -25\sin 15°$

$\therefore c_2 = -25\sin 15°$

$\therefore \dot{y}(t) = -25\sin 15° - gt$

$\therefore \underset{\sim}{v}(t) = 25\cos 15°\underset{\sim}{i} + (-25\sin 15° - gt)\underset{\sim}{j}$

Consider $x(t) = \int 25\cos 15°\, dt$

$= 25t\cos 15° + c_3$

As $x(0) = 0$,

$25(0)\cos 15° + c_3 = 0$

$\therefore c_3 = 0$

$\therefore x(t) = 25t\cos 15°$

Consider $y(t) = \int(-25\sin 15° - gt)\, dt$

$= -25t\sin 15° - \frac{1}{2}gt^2 + c_4$

As $y(0) = 2$,

$y(0) = -25(0)\sin 15° - \frac{1}{2}g(0)^2 + c_4 = 2$

$\therefore c_4 = 2$

$\therefore y(t) = -25t\sin 15° - \frac{1}{2}gt^2 + 2$

$\therefore \boldsymbol{\underset{\sim}{r}(t) = 25t\cos 15°\underset{\sim}{i} - (25t\sin 15° + \frac{1}{2}gt^2 - 2)\underset{\sim}{j}}$

2 i a Use $|\underset{\sim}{v}(0)| = 20$ and $\theta = 30°$.

$\underset{\sim}{r}(t) = 20t\cos 30°\underset{\sim}{i} + (20t\sin 30° - 5t^2)\underset{\sim}{j}$

$= 10\sqrt{3}t\underset{\sim}{i} + (10t - 5t^2)\underset{\sim}{j}$

As $y(t) = 10t - 5t^2$, then

$\dot{y}(t) = 10 - 10t = 0$

$10t = 10$

$t = 1$

Substitute into $y(t) = 10t - 5t^2$:

$y(1) = 10(1) - 5(1)^2$

$= 5$

$\therefore$ **greatest height is 5 m.**

b Time of flight is $2(1) = 2$ seconds.

Substitute into $x(t) = 10\sqrt{3}t$:

$x(2) = 10\sqrt{3}(2)$

$= 20\sqrt{3}$

$\therefore$ **the distance is $20\sqrt{3}$ m.**

ii $\underset{\sim}{r}(t) = Vt\cos 45°\underset{\sim}{i} + \left(Vt\sin 45° - \frac{gt^2}{2}\right)\underset{\sim}{j}$

$= \frac{Vt}{\sqrt{2}}\underset{\sim}{i} + \left(\frac{Vt}{\sqrt{2}} - \frac{gt^2}{2}\right)\underset{\sim}{j}$

As $y(t) = \frac{Vt}{\sqrt{2}} - \frac{gt^2}{2}$, then

$\dot{y}(t) = \frac{V}{\sqrt{2}} - gt = 0$

$gt = \frac{V}{\sqrt{2}}$

$t = \frac{V}{g\sqrt{2}}$

Substitute into $y(t) = \frac{Vt}{\sqrt{2}} - \frac{gt^2}{2}$:

$y\left(\frac{V\sqrt{2}}{g}\right) = \frac{V\left(\frac{V}{g\sqrt{2}}\right)}{\sqrt{2}} - \frac{g\left(\frac{V}{g\sqrt{2}}\right)^2}{2}$:

$= \frac{V^2}{2g} - \frac{V^2}{4g}$

$= \frac{V^2}{4g}$

Time of flight is $2\left(\frac{V}{g\sqrt{2}}\right) = \frac{V\sqrt{2}}{g}$ seconds.

Substitute into $x(t) = \frac{Vt}{\sqrt{2}}$:

$x(2) = \frac{V}{\sqrt{2}}\left(\frac{V\sqrt{2}}{g}\right)$

$= \frac{V^2}{g}$

$\therefore$ range $= \frac{V^2}{g}$ and maximum height $= \frac{V^2}{4g}$.

$\therefore$ **the range is four times the maximum height.**

3 i a Let $x = 18t\cos\alpha$ and $y = 18t\sin\alpha - 5t^2$.

Substitute $t = \frac{x}{18\cos\alpha}$ in y:

$y = 18\left(\frac{x}{18\cos\alpha}\right)\sin\alpha - 5\left(\frac{x}{18\cos\alpha}\right)^2$

$= x\tan\alpha - \frac{5x^2}{324\cos^2\alpha}$

$= x\tan\alpha - \frac{5x^2\sec^2\alpha}{324}$

$= x\tan\alpha - \frac{5x^2(1+\tan^2\alpha)}{324}$

$\therefore$ **the trajectory is $y = x\tan\alpha - \frac{5x^2(1+\tan^2\alpha)}{324}$.**

b Substitute $x = 27$ and $y = 3$:

$3 = 27\tan\alpha - \frac{5(27)^2(1+\tan^2\alpha)}{324}$

$972 = 8748\tan\alpha - 3645(1 + \tan^2\alpha)$

$972 = 8748\tan\alpha - 3645 - 3645\tan^2\alpha$

$3645\tan^2\alpha - 8748\tan\alpha + 4617 = 0$

$\tan\alpha = \frac{8748 \pm \sqrt{(-8748)^2 - 4(3645)(4617)}}{2(3645)}$

$= 1.6163332\ldots, 0.7836668\ldots$

$\alpha = 58°, 38°$ (nearest degree)

c Substitute $\alpha = 38°$ and $x = 27 + 3 = 30$:

$y = 30\tan 38° - \frac{5(30)^2(1+\tan^2 38°)}{324}$

$= 1.071\,801\,709\ldots$

$= 1.07$ (2 dec. pl.)

$\therefore$ **the ball is caught at a height of 1.07 m.**

ii a Object A: Let $x = 20t\cos 60°$ and $y = 20t\sin 60° - 5t^2$.

$\therefore x = 10t$ and $y = 10t\sqrt{3} - 5t^2$

As $t = \frac{x}{10}$, then $y = 10\left(\frac{x}{10}\right)\sqrt{3} - 5\left(\frac{x}{10}\right)^2$

$\therefore \boldsymbol{y = x\sqrt{3} - \frac{x^2}{20}}$

Object B: Let $x = 10t\cos 30°$ and $y = 10t\sin 30° - 5t^2 + 20$

$\therefore x = 5t\sqrt{3}$ and $y = 5t - 5t^2 + 20$

As $t = \frac{x}{5\sqrt{3}}$, then $y = 5\left(\frac{x}{5\sqrt{3}}\right) - 5\left(\frac{x}{5\sqrt{3}}\right)^2 + 20$

$= \frac{x}{\sqrt{3}} - \frac{x^2}{15} + 20$

$\therefore \boldsymbol{y = \frac{x\sqrt{3}}{3} - \frac{x^2}{15} + 20}$

b Solve equations simultaneously:

$x\sqrt{3} - \frac{x^2}{20} = \frac{x\sqrt{3}}{3} - \frac{x^2}{15} + 20$

$60x\sqrt{3} - 3x^2 = 20x\sqrt{3} - 4x^2 + 1200$

$x^2 + 40x\sqrt{3} - 1200 = 0$

$x = \frac{-40\sqrt{3} \pm \sqrt{(40\sqrt{3})^2 - 4(1)(-1200)}}{2(1)}$

$= \frac{-40\sqrt{3} \pm \sqrt{9600}}{2}$

$= 14.348\,7787\ldots\ (x > 0)$

$= 14.3488$ (4 dec. pl.) $(x > 0)$

Substitute in y:

$y(14.3488) = (14.3488)\sqrt{3} - \frac{14.3488^2}{20}$

$= 14.558\,447\,56\ldots$

$= 14.5584$ (4 dec. pl.)

$\therefore$ **trajectories intersect at (14.3488, 14.5584).**

c Object A: $x = 10t$:

$14.3488 = 10t$

$t = 1.434\,88$

Object B: $x = 5t\sqrt{3}$:

$14.3488 = 5t\sqrt{3}$

$t = 1.656\,867\,09\ldots$

Difference = 0.22

$\therefore$ **Object A should be projected 0.22 seconds after object B.**

4 i a Use $|\underset{\sim}{v}(0)| = 18$ and $\theta = 0°$,

$\underset{\sim}{r}(t) = 18t\cos 0°\underset{\sim}{i} + (18t\sin 0° - 5t^2 + 24)\underset{\sim}{j}$

$= 18t\underset{\sim}{i} + (24 - 5t^2)\underset{\sim}{j}$

$\underset{\sim}{v}(t) = 18\underset{\sim}{i} - 10t\underset{\sim}{j}$

$\therefore$ **$\underset{\sim}{r}(t) = 18t\underset{\sim}{i} + (24 - 5t^2)\underset{\sim}{j}$ and $\underset{\sim}{v}(t) = 18\underset{\sim}{i} - 10t\underset{\sim}{j}$**

b Let $y(t) = 0$:

$24 - 5t^2 = 0$

$5t^2 = 24$

$t^2 = 4.8$

$t = \sqrt{4.8}\ (t > 0)$

Substitute into $x(t)$:

$x(\sqrt{4.8}) = 18\sqrt{4.8}$

$= 39.436\,024\,14\ldots$

$= 39.44$ (2 dec. pl.)

$\therefore$ **39.44 m from the building.**

ii $\left|\underset{\sim}{v}(0)\right| = V, \alpha = 0°.$

$y(t) = Vt \sin 0 - 5t^2 + 15 = 0$

$5t^2 = 15$

$t^2 = 3$

$t = \sqrt{3}\ (t > 0)$

Now, $x(t) = Vt \cos 0$

$\therefore x(t) = Vt$

Substitute $t = \sqrt{3}$ and $x = 24$:

$V \times \sqrt{3} = 24$

$V = \dfrac{24}{\sqrt{3}} \times \dfrac{\sqrt{3}}{\sqrt{3}}$

$= 8\sqrt{3}$

$\therefore$ object has an initial velocity of $8\sqrt{3}$ ms^{-1} and a time of flight of $\sqrt{3}$ seconds.

Now, $\underset{\sim}{v}(t) = \left|\underset{\sim}{v}(0)\right| \cos 0\ \underset{\sim}{i} + \left(\left|\underset{\sim}{v}(0)\right| \sin 0 - 10t\right) \underset{\sim}{j})$

$= 8\sqrt{3} \cos 0\ \underset{\sim}{i} + (8\sqrt{3} \sin 0 - 10\sqrt{3})\ \underset{\sim}{j})$

$= 8\sqrt{3}\underset{\sim}{i} + (-10\sqrt{3})\ \underset{\sim}{j})$

Let $\left|\underset{\sim}{v}(\sqrt{3})\right|$ = speed the object hits the ground, θ = angle at which the object hits the ground:

$\therefore \left|\underset{\sim}{v}(\sqrt{3})\right| = \sqrt{(8\sqrt{3})^2 + (-10\sqrt{3})^2}$

$= \sqrt{492}$

$= 2\sqrt{123}$

$\theta = \tan^{-1}\left|\dfrac{-10\sqrt{3}}{8\sqrt{3}}\right|$

$= 51.340\,191\,75\ldots$

$= 51$ (nearest whole)

$\therefore$ **the object hits at $2\sqrt{123}$ ms^{-1} at 51° to the horizontal.**

HSC EXAM-TYPE QUESTIONS

Part A

1 D. $x = \dfrac{4 - t}{3}$

$\therefore 3x = 4 - t$

$t = 4 - 3x$

Substitute into $y = 1 - 2t$:

$y = 1 - 2(4 - 3x)$

$= 1 - 8 + 6x$

$= 6x - 7$

2 B. $\underset{\sim}{r}(3) = (4(3) - 5)\underset{\sim}{i} + (3^2 - 12)\ \underset{\sim}{j}$

$= 7\underset{\sim}{i} - 3\underset{\sim}{j}$

$|\underset{\sim}{r}(3)| = \sqrt{7^2 + (-3)^2}$

$= \sqrt{58}$

$\therefore$ the particle is $\sqrt{58}$ units from origin.

3 C. $\underset{\sim}{v}(t) = (1 + e^{-t})\underset{\sim}{i} + (e^{t} + te^{t})\ \underset{\sim}{j}$

$\underset{\sim}{a}(t) = -e^{-t}\underset{\sim}{i} + (e^{t} + e^{t} + te^{t})\ \underset{\sim}{j}$

$= -e^{-t}\underset{\sim}{i} + (2e^{t} + te^{t})\ \underset{\sim}{j}$

4 A. $\left|\underset{\sim}{v}(0)\right| \cos\theta = \dfrac{\sqrt{2}}{4} \ldots$ ①

$\left|\underset{\sim}{v}(0)\right| \sin\theta = \dfrac{\sqrt{6}}{12} \ldots$ ②

② ÷ ①: $\tan\theta = \dfrac{\sqrt{6}}{12} \div \dfrac{\sqrt{2}}{4}$

$= \dfrac{4\sqrt{6}}{12\sqrt{2}}$

$= \dfrac{\sqrt{3}}{3}$

$= \dfrac{1}{\sqrt{3}}$

$\theta = \dfrac{\pi}{6}$

Substitute in ②:

$\left|\underset{\sim}{v}(0)\right| \cos\dfrac{\pi}{6} = \dfrac{\sqrt{2}}{4}$

$\left|\underset{\sim}{v}(0)\right| \times \dfrac{\sqrt{3}}{2} = \dfrac{\sqrt{2}}{4}$

$\left|\underset{\sim}{v}(0)\right| = \dfrac{2\sqrt{2}}{4\sqrt{3}}$

$= \dfrac{\sqrt{6}}{6}$

$= \dfrac{1}{\sqrt{6}}$

5 D. $\left|\underset{\sim}{v}(0)\right| = 30$ and $\theta = 60°$:

$y = 30t \sin 60° - \dfrac{1}{2}gt^2$

$= 15\sqrt{3}t - \dfrac{1}{2}gt^2 = 0$

$t\left(15\sqrt{3} - \dfrac{1}{2}gt\right) = 0$

$15\sqrt{3} = \dfrac{1}{2}gt$

$t = \dfrac{30\sqrt{3}}{g}$

Substitute in x:

$x = 30\left(\dfrac{30\sqrt{3}}{g}\right)\cos 60°$

$= \dfrac{450\sqrt{3}}{g}$

Part B

6 $\underset{\sim}{r}(t) = \left(2 - 4\sin\frac{t}{2}\right)\underset{\sim}{i} + \left(3 - 4\cos\frac{t}{2}\right)\underset{\sim}{j}$

$\underset{\sim}{v}(t) = -2\cos\frac{t}{2}\underset{\sim}{i} + 2\sin\frac{t}{2}\underset{\sim}{j}$

$\underset{\sim}{a}(t) = \sin\frac{t}{2}\underset{\sim}{i} + \cos\frac{t}{2}\underset{\sim}{j}$ ✓

$\underset{\sim}{a}(t).\underset{\sim}{v}(t) = \sin\frac{t}{2} \times \left(-2\cos\frac{t}{2}\right) + \cos\frac{t}{2} \times 2\sin\frac{t}{2}$

$= 0$ ✓

$\therefore$ **vectors are perpendicular.**

7 a Consider $\underset{\sim}{a}(t) = \ddot{x}(t)\underset{\sim}{i} + \ddot{y}(t)\underset{\sim}{j} = 0\underset{\sim}{i} - 10\underset{\sim}{j}$.

Consider $\tan\alpha = \frac{3}{4}$. By Pythagoras, $\sin\alpha = \frac{3}{5}$, $\cos\alpha = \frac{4}{5}$.

Now, when $t = 0$, $\left|\underset{\sim}{v}(0)\right| = 25$, $\sin\alpha = \frac{3}{5}$, $\cos\alpha = \frac{4}{5}$.

Using $\cos\alpha = \frac{4}{5}$, as $\frac{\dot{x}}{25} = \cos\alpha$, then $\dot{x} = 12$. ✓

Using $\sin\alpha = \frac{3}{5}$, as $\frac{\dot{y}}{25} = \sin\alpha$, then $\dot{y} = 15$.

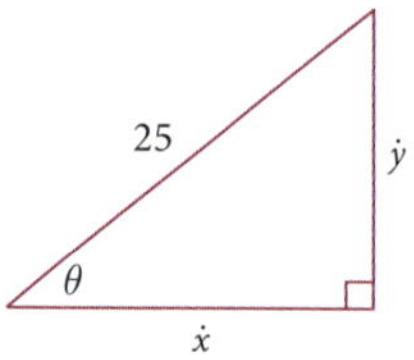

$\underset{\sim}{v}(t) = \int\ddot{x}(t)\,dt\underset{\sim}{i} + \int\ddot{y}(t)\,dt\underset{\sim}{j}$

Consider $\dot{x}(t) = \int 0\,dt$

$= c_1$

As $\dot{x}(0) = 12$,

$c_1 = 12$

$\therefore \dot{x}(t) = 12$

Consider $\dot{y}(t) = \int -10\,dt$

$= -10t + c_2$

As $\dot{y}(0) = 15$,

$-10(0) + c_2 = 15$

$\therefore c_2 = 15$

$\therefore \dot{y}(t) = 15 - 10t$

$\therefore \underset{\sim}{v}(t) = 12\underset{\sim}{i} + (15 - 10t)\underset{\sim}{j}$ ✓

Consider $x(t) = \int 12\,dt$

$= 12t + c_3$

As $x(0) = 0$,

$12(0) + c_3 = 0$

$\therefore x(t) = 12t$

Consider $y(t) = \int(15 - 10t)dt$

$= 15t - 5t^2 + c_4$

As $y(0) = 8$,

$15(0) - 5(0)^2 + c_4 = 8$

$\therefore c_4 = 8$

$\therefore y(t) = 15t - 5t^2 + 8$

$\therefore$ **$\underset{\sim}{r}(t) = 12t\underset{\sim}{i} + (15t - 5t^2 + 8)\underset{\sim}{j}$** ✓

b Let $\dot{y}(t) = 15 - 10t = 0$

$10t = 15$

$t = 1.5$ ✓

Substitute in $y(t)$:

$\therefore y(1.5) = 15(1.5) - 5(1.5)^2 + 8$

$= 19.25$

$\therefore$ **the maximum height is 19.25 m.** ✓

c As $x = 12t$, then $t = \frac{x}{12}$.

Substitute into y:

$y = 15\left(\frac{x}{12}\right) - 5\left(\frac{x}{12}\right)^2 + 8$

$\mathbf{y = \frac{5x}{4} - \frac{5x^2}{144} + 8}$ ✓

8 Consider $x(t) = \int 6e^{2t}\,dt$

$= 3e^{2t} + c_1$

As $x(0) = 4$,

$3e^{2(0)} + c_1 = 4$

$\therefore c_1 = 1$

$\therefore x(t) = 3e^{2t} + 1$ ✓

Consider $y(t) = \int 3e^{3t}\,dt$

$= e^t + c_2$

As $y(0) = -2$,

$e^{(0)} + c_2 = -2$

$\therefore c_2 = -3$

$\therefore y(t) = e^t - 3$

$\therefore$ **$\underset{\sim}{r}(t) = (3e^{2t} + 1)\underset{\sim}{i} + (e^t - 3)\underset{\sim}{j}$** ✓

9 a Substitute $x = 8$ and $t = 2$ in $x(t) = \left|\underset{\sim}{v}(0)\right|t\cos\alpha$ and let $\left|\underset{\sim}{v}(0)\right| = V$:

$8 = V(2)\cos\alpha$

$V = \frac{4}{\cos\alpha}$ ✓

Substitute $y = 6$ and $t = 2$ into $y(t) = \left|\underset{\sim}{v}(0)\right|t\sin\alpha - 5t^2$ and let $\left|\underset{\sim}{v}(0)\right| = V = \frac{4}{\cos\alpha}$:

$6 = \frac{4}{\cos\alpha}(2)\sin\alpha - 5(2)^2$

$6 = 8\tan\alpha - 20$

$8\tan\alpha = 26$

$\tan\alpha = \frac{13}{4}$ ✓

$\therefore \cos\alpha = \frac{4}{\sqrt{185}}$ (by Pythagoras)

Substitute into $V = \frac{4}{\cos\alpha}$:

$V = 4 \div \frac{4}{\sqrt{185}}$

$= \sqrt{185}$ ✓

$\therefore$ **$V = \sqrt{185}$ ms^{-1} and $\tan\alpha = \frac{13}{4}$**

b $\underset{\sim}{v}(t) = \left|\underset{\sim}{v}(0)\right|\cos\alpha\,\underset{\sim}{i} + \left(\left|\underset{\sim}{v}(0)\right|\sin\alpha - 10t\right)\underset{\sim}{j})$

Using $\left|\underset{\sim}{v}(0)\right| = \sqrt{185}$ ms^{-1}, $\sin\alpha = \frac{13}{\sqrt{185}}$, $\cos\alpha = \frac{4}{\sqrt{185}}$ and $t = 2$:

$\therefore \underset{\sim}{v}(t) = \sqrt{185} \times \frac{4}{\sqrt{185}}\underset{\sim}{i} + \left(\sqrt{185} \times \frac{13}{\sqrt{185}} - 10(2)\right)\underset{\sim}{j}$ ✓

$= \mathbf{4\underset{\sim}{i} - 7\underset{\sim}{j}}$ ✓

c Consider $y = \left|\underset{\sim}{v}(0)\right| t \sin\alpha - 5t^2$

$\therefore y = \sqrt{185}t\left(\frac{13}{\sqrt{185}}\right) - 5t^2 = 0$ ✓

$13t - 5t^2 = 0$

$t(13 - 5t) = 0$

$t = 0, 2.6$

$\therefore$ time of flight is 2.6 seconds. ✓

Substitute into x:

$x = \sqrt{185}(2.6)\left(\frac{4}{\sqrt{185}}\right)$

$= 10.4$

Difference $= 10.4 - 8$

$= 2.4$

$\therefore$ **the object lands 2.4 m beyond the wall.** ✓

10 $\underset{\sim}{r}_A(3) = (3(3) - 2)\underset{\sim}{i} + (2(3) + 1)\underset{\sim}{j}$

$= 7\underset{\sim}{i} + 7\underset{\sim}{j}$

$\underset{\sim}{r}_B(3) = 3^2\underset{\sim}{i} + (5 - 2(3))\underset{\sim}{j}$

$= 9\underset{\sim}{i} - \underset{\sim}{j}$

$\underset{\sim}{r}_A(2) - \underset{\sim}{r}_B(2) = (7\underset{\sim}{i} + 7\underset{\sim}{j}) - (9\underset{\sim}{i} - \underset{\sim}{j})$

$= -2\underset{\sim}{i} + 8\underset{\sim}{j}$ ✓

Distance $= \sqrt{(-2)^2 + 8^2}$

$= \sqrt{68}$

$\therefore$ **the particles are $\sqrt{68}$ units apart.** ✓

11 a As $\left|\underset{\sim}{v}(0)\right| = 20$ and $\alpha = 30°$, then $\underset{\sim}{v}(t) = 10\sqrt{3}\underset{\sim}{i} + (10 - 10t)\underset{\sim}{j}$.

As $x(t) = \int 10\sqrt{3}\, dt$, then $x(t) = 10\sqrt{3}t + c_1$. As $x(0) = 0$, then $c_1 = 0$ and $x(t) = 10\sqrt{3}t$. ✓

As $y(t) = \int(10 - 10t)\, dt$ then $y(t) = 10t - 5t^2 + c_2$. As $y(0) = 2$, then $c_2 = 2$ and $y(t) = 10t - 5t^2 + 2$.

$\therefore \underset{\sim}{r}(t) = 10\sqrt{3}t\underset{\sim}{i} + (10t - 5t^2 + 2)\underset{\sim}{j}$. ✓

b $\dot{y}(t) = 10 - 10t = 0$

$t = 1$

$y(1) = 10(1) - 5(1)^2 + 2$

$= 7$

$\therefore$ **the ball hits at a height of 7 m.** ✓

c Consider the position vector after the ball rebounds using the base of the wall as the new origin with, $\left|\underset{\sim}{v}(0)\right| = 8$ and $\alpha = 0°$:

$\underset{\sim}{r}(t) = 8t\cos 0°\underset{\sim}{i} + (8t\sin 0° - 5t^2 + 7)\underset{\sim}{j}$.

$= 8t\underset{\sim}{i} + (7 - 5t^2)\underset{\sim}{j}$. ✓

Let $y(t) = 0$:

$7 - 5t^2 = 0$

$5t^2 = 7$

$t = \sqrt{\frac{7}{5}}\ (t > 0)$

$\therefore$ **hits ground $\left(\sqrt{\frac{7}{5}} + 1\right)$ seconds after being thrown.** ✓

d First, substitute $t = 1$ into $x(t) = 10\sqrt{3}t$:

$x(1) = 10\sqrt{3}(1)$

$= 10\sqrt{3}$

$\therefore$ the ball travelled $10\sqrt{3}$ m to hit the wall.

Now, substitute $t = \sqrt{\frac{7}{5}}$ into $x(t) = 8t$:

$x\left(\sqrt{\frac{7}{5}}\right) = 8\left(\sqrt{\frac{7}{5}}\right)$

$\therefore$ the ball travelled $8\sqrt{\frac{7}{5}}$ m after hitting the wall. ✓

$\therefore$ distance from the wall the ball hits the ground

$= 10\sqrt{3} - 8\sqrt{\frac{7}{5}}$

$= 7.854\,780\,423\ldots$

$= 7.85$

$\therefore$ **the ball lands 7.85 m from the wall.** ✓

12

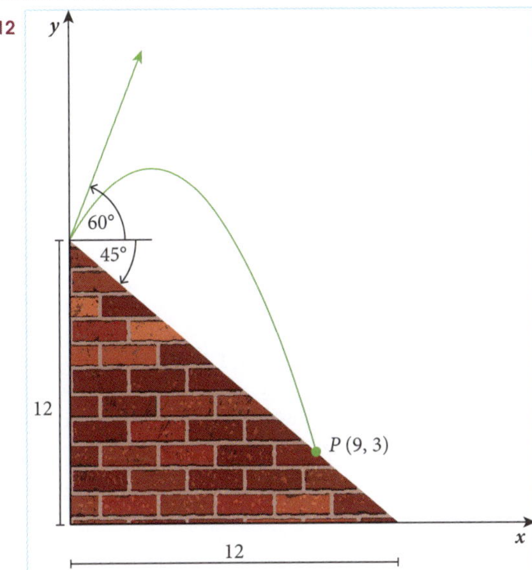

As $\frac{3}{4}$ of 12 = 9 the object lands at the point (9, 3). ✓

Substitute $x = 9$ into $x = \frac{Vt}{2}$:

$9 = \frac{Vt}{2}$

$t = \frac{18}{V}$ ✓

Substitute $y = 3$ and $t = \frac{18}{V}$ into $y = \frac{Vt\sqrt{3}}{2} - 5t^2 + 12$:

$3 = \frac{18}{V} \times \frac{V\sqrt{3}}{2} - 5\left(\frac{18}{V}\right)^2 + 12$

$3 = 9\sqrt{3} - \frac{1620}{V^2} + 12$

$\frac{1620}{V^2} = 9\sqrt{3} + 9$

$V^2 = \frac{1620}{9\sqrt{3} + 9}$

$V = 8.116\,931\,235\ldots$

$= 8.12$ (2 dec. pl.)

$\therefore$ **the initial speed was 8.12 ms^{-1}.** ✓

CHAPTER 12 TRIGONOMETRIC EQUATIONS

TOPIC: TRIGONOMETRIC FUNCTIONS

YEAR 12

1 Expressing in the form $r \sin(\theta \pm \alpha)$, $r \cos(\theta \pm \alpha)$

→ **In each of the following cases, $r = \sqrt{a^2 + b^2}$ and $\tan \alpha = \frac{b}{a}$ (α is acute),**

- $a \sin\theta + b\cos\theta = r\sin(\theta + \alpha)$
- $a \sin\theta - b\cos\theta = r\sin(\theta - \alpha)$
- $a \cos\theta + b\sin\theta = r\cos(\theta - \alpha)$
- $a \cos\theta - b\sin\theta = r\cos(\theta + \alpha)$

EXAMPLE 1

Express $\sqrt{3}\sin x - \cos x$ in the form $r\sin(x - \alpha)$.

$$r = \sqrt{(\sqrt{3})^2 + 1^2} = 2$$

$$\tan\alpha = \frac{1}{\sqrt{3}}$$

$a = \sqrt{3}$ and $b = 1$

$$\alpha = 30°$$

$$\therefore \sqrt{3}\sin x - \cos x = 2\sin(x - 30°).$$

EXAMPLE 2

Express $2\cos\theta + \sin\theta$ in the form $r\cos(\theta - \alpha)$, where α is in radians.

$$r = \sqrt{2^2 + 1^2} = \sqrt{5}$$

$$\tan\alpha = \frac{1}{2}$$

$a = 2$ and $b = 1$

$$\alpha = 0.463\,647\,609\ldots = 0.46 \text{ (2 dec. pl.)}$$

$$\therefore 2\cos\theta + \sin\theta = \sqrt{5}\cos(\theta - 0.46^c)$$

EXAMPLE 3

Sketch the graph of $y = \cos x - \sin x$, for $0 \le x \le 2\pi$.

Rewrite $\cos x - \sin x = r\cos(x + \alpha)$.

Now $r = \sqrt{1^2 + 1^2} = \sqrt{2}$ and $\tan\alpha = \frac{1}{1} \therefore \alpha = \frac{\pi}{4}$

$$\therefore y = \sqrt{2}\cos\left(x + \frac{\pi}{4}\right)$$

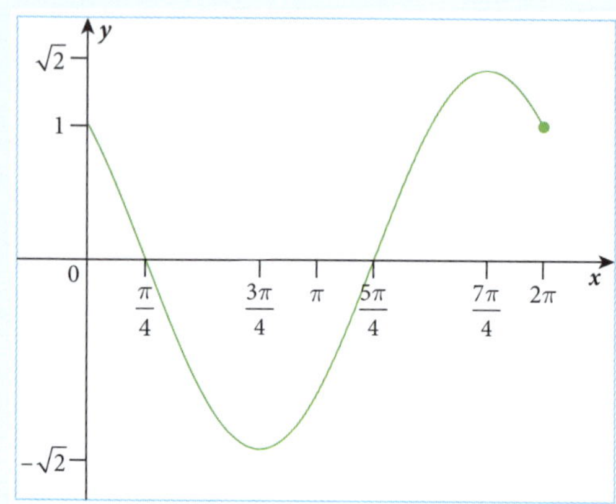

Remember dilations and translations.

EXAMPLE 4

Find the maximum value of $f(x) = 3\sin x + 2\cos x$.

Rewrite $3\sin x + 2\cos x$ in form $r\sin(x + \alpha)$.

$$r = \sqrt{3^2 + 2^2} = \sqrt{13}$$

The maximum value of $f(x) = \sin x$ is 1.

$\therefore 3\sin x + 2\cos x = \sqrt{13}\sin(x + \alpha)$

As $-1 \le \sin(x + \alpha) \le 1$,

then $-\sqrt{13} \le \sqrt{13}\sin(x + \alpha) \le \sqrt{13}$.

$\therefore$ the maximum value is $\sqrt{13}$.

→ PRACTICE

1 **Express:**

i $2\sin x + 2\sqrt{3}\cos x$ in the form $r\sin(x + \alpha)$, where α is in degrees

ii $3\sin x - 4\cos x$ in the form $r\sin(x - \alpha)$, where α is in degrees

2 **Express:**

i $\cos\theta + \sin\theta$ in the form $r\cos(\theta - \alpha)$, where α is in radians

ii $5\cos 2\theta - 3\sin 2\theta$ in the form $r\cos(2\theta + \alpha)$, where α is in radians

3 Sketch, for $0 \le x \le 2\pi$, the graph of:

i $y = \sin x + \cos x$

ii $y = 1 + \sqrt{3}\cos x + \sin x$

4 Find the:

i maximum value of $f(\theta) = \sqrt{3}\sin\theta + 2\cos\theta$

ii least value of $f(\theta) = 2 - 3\cos\theta + 4\sin\theta$

Answers ➲ p. 155

2 Equations of the form $a\cos x + b\sin x = c$

➔ **Equations in the form $a\sin x \pm b\cos x = c$** are rewritten in the form $r\sin(x \pm \alpha) = c$.

➔ **Equations in the form $a\cos x \pm b\sin x = c$** are rewritten in the form $r\cos(x \mp \alpha) = c$.

EXAMPLE 1

Solve, for $0° \le x \le 360°$, $\sqrt{3}\cos x + \sin x = 1$.

$$\sqrt{3}\cos x + \sin x = r\cos(x - \alpha)$$

$$r = \sqrt{(\sqrt{3})^2 + 1^2}$$

$$= 2$$

$$\tan\alpha = \frac{1}{\sqrt{3}}$$

α is acute.

$$\alpha = 30°$$

$$\therefore \sqrt{3}\cos x + \sin x = 2\cos(x - 30°)$$

$$\therefore 2\cos(x - 30°) = 1$$

$$\cos(x - 30°) = \frac{1}{2}$$

$$x - 30° = 60°, 300°$$

$$x = 90°, 330°$$

EXAMPLE 2

Solve, for $0 \le x \le 2\pi$, $\sin x + 2\cos x = 1$. Give answer to one decimal place.

$$\sin x + 2\cos x = r\sin(x + \alpha)$$

$$r = \sqrt{1^2 + 2^2}$$

$$= \sqrt{5}$$

$$\tan\alpha = 2$$

$$\alpha = 1.107\,148\,718\ldots$$

$$= 1.107^c \text{ (3 dec. pl.)}$$

When expressing in radians, quadrant solutions are $\pi - \theta$, etc.

$$\therefore \sqrt{5}\sin(x + 1.107^c) = 1$$

$$\sin(x + 1.107^c) = \frac{1}{\sqrt{5}}$$

$$x + 1.107^c = 0.464^c, 2.678^c, 6.747^c \text{ (3 dec. pl.)}$$

$$x = -0.6^c, 1.6^c, 5.6^c$$

$$x = 1.6^c, 5.6^c \text{ (as } 0 \le x \le 2\pi)$$

➔ PRACTICE

1 Solve, for $0° \le x \le 360°$:

i $\cos x - \sin x = -\sqrt{2}$

ii $6\sin x - 8\cos x - 5 = 0$

2 Solve, for $0 \le x \le 2\pi$:

i $\sqrt{3}\cos x - 3\sin x = \sqrt{3}$

ii $3\sin x + \sqrt{2}\cos x + 2 = 0$. Leave your answer correct to one decimal place.

Answers ➲ pp. 155–156

3 Equations involving sums, differences and double angles

➔ $\sin 2A = 2\sin A\cos A$

➔ $\cos 2A = \cos^2 A - \sin^2 A = 1 - 2\sin^2 A = 2\cos^2 A - 1$

➔ $\tan 2A = \dfrac{2\tan A}{1 - \tan^2 A}$

➔ $\sin A + \sin B = 2\sin\dfrac{A+B}{2}\cos\dfrac{A-B}{2}$

➔ $\sin A - \sin B = 2\cos\dfrac{A+B}{2}\sin\dfrac{A-B}{2}$

➔ $\cos A + \cos B = 2\cos\dfrac{A+B}{2}\cos\dfrac{A-B}{2}$

➔ $\cos A - \cos B = -2\sin\dfrac{A+B}{2}\sin\dfrac{A-B}{2}$

EXAMPLE 1

Solve $\sin 3x\cos x - \cos 3x\sin x = \frac{1}{2}$, for $0 \le x \le 2\pi$.

$$\sin(3x - x) = \frac{1}{2}$$

$$\sin 2x = \frac{1}{2}$$

Use compound angles results.

$$2x = \frac{\pi}{6}, \frac{5\pi}{6}, \frac{13\pi}{6}, \frac{17\pi}{6}$$

$$x = \frac{\pi}{12}, \frac{5\pi}{12}, \frac{13\pi}{12}, \frac{17\pi}{12}$$

EXAMPLE 2

Solve $\cos 2x \cos \frac{\pi}{6} + \sin 2x \sin \frac{\pi}{6} = \frac{1}{2}$, for $0 \le x \le \pi$.

$$\cos\left(2x - \frac{\pi}{6}\right) = \frac{1}{2}$$

$$2x - \frac{\pi}{6} = \frac{\pi}{3}, \frac{5\pi}{3}$$

$$2x = \frac{\pi}{2}, \frac{11\pi}{6}$$

$$x = \frac{\pi}{4}, \frac{11\pi}{12}$$

Domain is $0 \le x \le \pi$.

EXAMPLE 3

Solve, $\cos 2x + \cos x = 0$, for $0 \le x \le 2\pi$.

$$2\cos^2 x - 1 + \cos x = 0$$

$$2\cos^2 x + \cos x - 1 = 0$$

$$(2\cos x - 1)(\cos x + 1) = 0$$

$$\cos x = \frac{1}{2}, -1$$

$$x = \frac{\pi}{3}, \frac{5\pi}{3}, \pi$$

Use $\cos 2x = 2\cos^2 x - 1$.

EXAMPLE 4

Solve $\sin 3x - \sin x = 0$, for $0 \le x \le \pi$.

$$\sin 3x - \sin x = 2\cos\frac{3x + x}{2}\sin\frac{3x - x}{2}$$

$$= 2\cos 2x \sin x$$

$$\therefore 2\cos 2x \sin x = 0$$

$$\cos 2x = 0 \qquad \sin x = 0$$

$$2x = \frac{\pi}{2}, \frac{3\pi}{2} \qquad x = 0, \pi$$

$$x = \frac{\pi}{4}, \frac{3\pi}{4}$$

$$\therefore x = 0, \frac{\pi}{4}, \frac{3\pi}{4}, \pi$$

Use a trig sum to product result.

➔ PRACTICE

1 **Solve for $0 \le x \le \pi$:**

i $\cos 5x \cos 3x + \sin 5x \sin 3x = -\frac{\sqrt{3}}{2}$

ii $\frac{\tan 2x + \tan x}{1 - \tan 2x \tan x} = \sqrt{3}$

2 **Solve:**

i $\sin 2x \cos \frac{\pi}{4} + \cos 2x \sin \frac{\pi}{4} = \frac{\sqrt{3}}{2}$, for $0 \le x \le \pi$

ii $\cos \frac{x}{2} \cos \frac{\pi}{3} + \sin \frac{x}{2} \sin \frac{\pi}{3} = -\frac{1}{2}$, for $-\pi \le x \le \pi$

3 **Solve, for $0° \le \theta \le 360°$:**

i $\sin 2\theta + \sin \theta = 0$

ii $\cos 4A - \sin 2A = 0$

4 **Solve, for $0 \le x \le \pi$:**

i $\cos 5x + \cos 3x = 0$

ii $\sin 3x + \sin 2x + \sin x = 0$

Answers ➲ pp. 156–157

4 Using *t*-formulae to solve equations

➔ **For $t = \tan \frac{A}{2}$:**

- $\sin A = \frac{2t}{1+t^2}$
- $\cos A = \frac{1-t^2}{1+t^2}$
- $\tan A = \frac{2t}{1-t^2}$

➔ **When we use the *t*-formulae method** to solve equations we also need to check $\theta = 180°$ as a possible solution.

EXAMPLE 1

Using the *t*-formulae, solve $\sin\theta + \cos\theta = 1$, for $0° \le \theta \le 360°$. Give your answer to the nearest degree.

$$\frac{2t}{1+t^2} + \frac{1-t^2}{1+t^2} = 1$$

$$2t + 1 - t^2 = 1 + t^2$$

$$2t^2 - 2t = 0$$

$$2t(t - 1) = 0$$

$$t = 0, 1$$

$$\tan\frac{\theta}{2} = 0, 1$$

$$\frac{\theta}{2} = 0°, 45°, 180°$$

$$\theta = 0°, 90°, 360° \text{ (as } 0° \le \theta \le 360°)$$

Express $\sin\theta$ and $\cos\theta$ in terms of t.

Also, check possible solution of $\theta = 180°$:
$\sin 180° + \cos 180° \neq 1$. Not a solution.
$\therefore \theta = 0°, 90°, 360°$

EXAMPLE 2

Using the *t*-formulae, solve $3 \cos \theta - 4 \sin \theta + 2 = 0$ for $[0, 2\pi]$. Give your answer correct to three significant figures.

$$3\left(\frac{1-t^2}{1+t^2}\right) - 4\left(\frac{2t}{1+t^2}\right) + 2 = 0$$

$$3(1 - t^2) - 4(2t) + 2(1 + t^2) = 0$$

$$3 - 3t^2 - 8t + 2 + 2t^2 = 0$$

$$t^2 + 8t - 5 = 0$$

$$t = \frac{-8 \pm \sqrt{8^2 - 4(1)(-5)}}{2(1)}$$

$$= \frac{-8 \pm \sqrt{84}}{2}$$

$$\tan \frac{\theta}{2} = 0.5828\ldots, -8.5825\ldots$$

$$\frac{\theta}{2} = 0.527\,508\,977\ldots, 1.686\,788\,458\ldots$$

$\theta = 1.06^c, 3.37$ (3 sig. figs) (as $[0, 2\pi]$)

Also, check possible solution of
$\theta = \pi$: $3 \cos \pi - 4 \sin \pi + 2 = -3 + 2 \neq 0$. Not a solution.
$\therefore \theta = 1.06^c, 3.37^c$

Check possible solution of $\theta = \pi$.

→ PRACTICE

1 Using the *t*-formulae, solve these equations, for $0° \leq \theta \leq 360°$, leaving your answer to the nearest degree:
 i $\sin \theta - \cos \theta = 1$
 ii $\sqrt{3} \sin \theta + \cos \theta + 1 = 0$

2 Using the *t*-formulae, solve these equations, for $[0, 2\pi]$, giving your answer correct to three significant figures:
 i $12 \sin \theta - 5 \cos \theta - 3 = 0$
 ii $4 \cos \theta - 3 \sin \theta + 1 = 0$

Answers ⊃ p. 157

5 Expressing sin 3θ and cos 3θ in terms of θ

→ **sin 3x** $= 3 \sin x - 4 \sin^3 x$
→ **cos 3x** $= 4 \cos^3 x - 3 \cos x$

EXAMPLE 1

Show $\sin 3x = 3 \sin x - 4 \sin^3 x$.

$$\begin{aligned}
\text{LHS} &= \sin 3x \\
&= \sin (2x + x) \\
&= \sin 2x \cos x + \cos 2x \sin x \\
&= (2 \sin x \cos x) \cos x + (1 - 2 \sin^2 x) \sin x \\
&= 2 \sin x \cos^2 x + \sin x - 2 \sin^3 x \\
&= 2 \sin x(1 - \sin^2 x) + \sin x - 2 \sin^3 x \\
&= 2 \sin x - 2 \sin^3 x + \sin x - 2 \sin^3 x \\
&= 3 \sin x - 4 \sin^3 x \\
&= \text{RHS}
\end{aligned}$$

Use $3x = 2x + x$.

EXAMPLE 2

Solve $\cos 3x = \cos x$, for $[-\pi, \pi]$.

$$4 \cos^3 x - 3 \cos x = \cos x$$

$$4 \cos^3 x - 4 \cos x = 0$$

$$4 \cos x (\cos^2 x - 1) = 0$$

$$\cos x = 0, \cos x = \pm 1$$

$$x = -\frac{\pi}{2}, \frac{\pi}{2}, 0, \pi, -\pi$$

$$\therefore x = -\pi, -\frac{\pi}{2}, 0, \frac{\pi}{2}, \pi$$

Use the cos 3x result.

→ PRACTICE

1 i Show $\cos 3x = 4 \cos^3 x - 3 \cos x$.
 ii Express $\tan 3\theta$ in terms of $\tan \theta$.

2 i Solve $\sin 3x = \sin x$, for $[0°, 180°]$.
 ii Solve $\sin 4x - \sin 2x = 0$, for $[0, 2\pi]$.

Answers ⊃ p. 157

HSC EXAM-TYPE QUESTIONS

Total marks: 35

PART A

(Suggested time: 10 minutes)
(1 mark each)

1 Which of these are the values of r and α such that $\sqrt{3}\cos x - \sin x = r\cos(x + \alpha)$?

A $r = 2$ and $\alpha = \frac{\pi}{6}$ B $r = \sqrt{2}$ and $\alpha = \frac{\pi}{6}$

C $r = \sqrt{2}$ and $\alpha = \frac{\pi}{3}$ D $r = 2$ and $\alpha = \frac{\pi}{3}$

2 Which of these expressions is equivalent to $1 - \sin 2x \tan x$?

A $\sin^2 x - \cos^2 x$ B $2\cos x$

C $\sin x \cos x$ D $\cos 2x$

3 Find the number of solutions for the equation $\cos 3x + \cos x = 0$, given $[0, 2\pi]$.

A 2 B 4 C 6 D 8

4 What is the minimum value of $4 + 2\sin x - 3\cos x$?

A $-4 - \sqrt{13}$ B -1 C $4 - \sqrt{13}$ D 3

5 If $5\cos\theta + 3 = 0$ and $0 < \theta < \pi$, what is the value of $\tan\frac{\theta}{2}$?

A -2 B 2

C 3 or $\frac{1}{3}$ D -3 or $-\frac{1}{3}$

PART B

(Suggested time: 45 minutes)

Show all working.

6 A rectangle has a diagonal of length 6 cm.

Let α be the angle between the diagonal and the longer side of the rectangle.

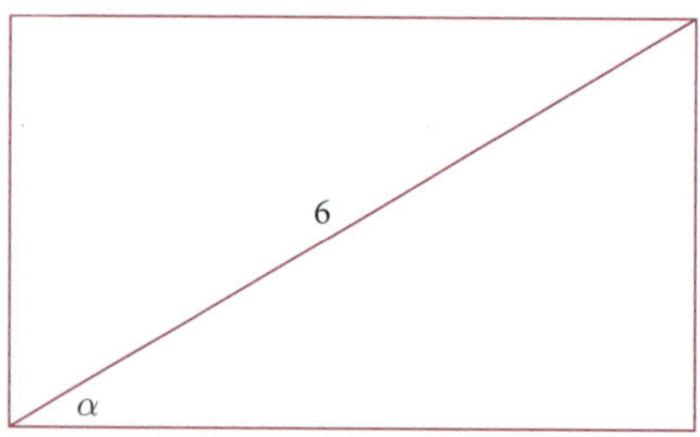

If the area (A) of the rectangle is 9 cm^2:

a show $A = 18\sin 2\alpha$ (2 marks)

b what is the size of α ? (1 mark)

7 a Express $3\sin 2t + 4\cos 2t$ in the form $R\sin(2t + \alpha)$ where $0° < \alpha < 90°$. (2 marks)

b Hence solve $3\sin 2t + 4\cos 2t - 5 = 0$, for $0° \leq t \leq 360°$. Leave your answer correct to the nearest minute. (2 marks)

8 Find the amplitude of the graph $y = 2\sin\theta - 2\sqrt{3}\cos\theta$. (2 marks)

9 Solve $2\cos 2x + 3 = 4\cos x$, where $[0, 2\pi]$. (3 marks)

10 Sketch the graph of $y = \cos\theta + \sqrt{3}\sin\theta$ for $0 \leq x \leq 2\pi$. (3 marks)

11 Solve $\cos 2A = \sin A$, for $0 \leq A \leq 2\pi$. (2 marks)

12 Solve, where $0 \leq x \leq \frac{\pi}{2}$, $\sin 6x - \sin 4x = 0$. (2 marks)

13 a Show that $\sin 3\theta = 3\sin\theta - 4\sin^3\theta$ for all values of θ. (2 marks)

b Hence, solve $\sin 3\theta + \sin 2\theta = \sin\theta$ for $0 \leq x \leq 2\pi$. (4 marks)

14 a Express $2\cos\theta + 2\sin\left(\theta + \frac{\pi}{6}\right)$ in the form $r\sin(\theta + \alpha)$, where $0 < \alpha < \frac{\pi}{2}$. (3 marks)

b Hence, or otherwise, solve

$2\cos\theta + 2\sin\left(\theta + \frac{\pi}{6}\right) = 3$, for $0 \leq x \leq 2\pi$. (2 marks)

ANSWERS

PRACTICE

1 Expressing in the form $r \sin(\theta \pm \alpha)$, $r \cos(\theta \pm \alpha)$ ➲ pp. 150–151

1 i $r = \sqrt{2^2 + (2\sqrt{3})^2}$
$= 4$

$\tan \alpha = \sqrt{3}$

$\alpha = 60°$

$\therefore \sqrt{3} \sin x + \cos x = 4 \sin(x + 60°)$.

ii $r = \sqrt{3^2 + 4^2}$
$= 5$

$\tan \alpha = \frac{4}{3}$

$\alpha = 53.130\,102\,35\ldots$
$= 53°$ (nearest degree)

$\therefore 3 \sin x - 4 \cos x = 5 \sin(x - 53°)$

2 i $r = \sqrt{1^2 + 1^2}$
$= \sqrt{2}$

$\tan \alpha = 1$

$\alpha = \frac{\pi}{4}$

$\therefore \cos\theta + \sin\theta = \cos\left(\theta - \frac{\pi}{4}\right)$

ii $r = \sqrt{5^2 + 3^2}$
$= \sqrt{34}$

$\tan \alpha = \frac{3}{5}$

$\alpha = 0.540\,4195$
$= 0.54$ (2 dec. pl.)

$\therefore 5 \cos 2\theta - 3 \sin 2\theta = \sqrt{34} \cos(2\theta + 0.54^c)$

3 i Rewrite $\sin x + \cos x = r \sin(x + \alpha)$.

Now $r = \sqrt{1^2 + 1^2} = \sqrt{2}$ and $\tan \alpha = \frac{1}{1} \therefore \alpha = \frac{\pi}{4}$.

$\therefore y = \sqrt{2} \sin\left(x + \frac{\pi}{4}\right)$

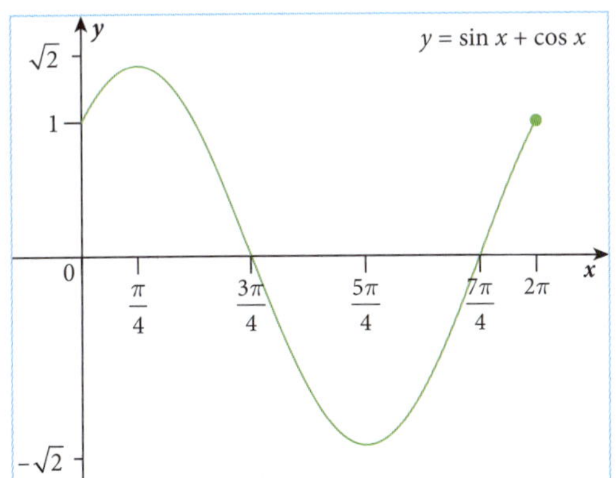

ii Rewrite $\sqrt{3} \cos x + \sin x = r \cos(x - \alpha)$.

Now $r = \sqrt{(\sqrt{3})^2 + 1^2} = 2$ and $\tan \alpha = \frac{1}{\sqrt{3}} \therefore \alpha = \frac{\pi}{6}$.

$\therefore y = 1 + 2 \cos\left(x - \frac{\pi}{6}\right)$

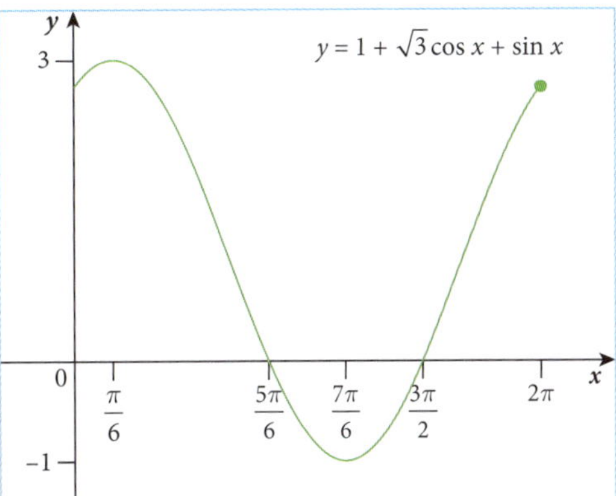

4 i Rewrite $\sqrt{3} \sin\theta + 2 \cos\theta = r \sin(\theta + \alpha)$.

Now $r = \sqrt{(\sqrt{3})^2 + 2^2} = \sqrt{7}$.

$\therefore f(\theta) = \sqrt{7} \sin(\theta + \alpha)$

As maximum value of sin x is 1, then maximum value of $\sqrt{7}$ sin $(\theta + \alpha)$ is $\sqrt{7}$.

ii $f(\theta) = 2 - (3 \cos\theta - 4 \sin\theta)$

Rewrite $3 \cos\theta - 4 \sin\theta = r \cos(\theta + \alpha)$.

Now $r = \sqrt{3^2 + 4^2} = 5$.

$\therefore f(\theta) = 2 - 5 \cos(\theta + \alpha)$

As the greatest value of cos x is 1, then the least value of $2 - 5 \cos(\theta + \alpha)$ is -3.

2 Equations of the form $a \cos x + b \sin x = c$ ➲ p. 151

1 i $\cos x - \sin x = r \cos(x + \alpha)$

$r = \sqrt{1^2 + 1^2}$
$= \sqrt{2}$

$\tan \alpha = 1$

$\alpha = 45°$

$\therefore \cos x - \sin x = \sqrt{2} \cos(x + 45°)$

$\therefore \sqrt{2} \cos(x + 45°) = -\sqrt{2}$

$\cos(x + 45°) = -1$

$x + 45° = 180°$

$\mathbf{x = 135°}$

ii $6 \sin x - 8 \cos x = 5$

$6 \sin x - 8 \cos x = r \sin(x - \alpha)$

$r = \sqrt{6^2 + 8^2}$
$= 10$

$\tan \alpha = \frac{8}{6}$

$\alpha = 53°8'$ (nearest min.)

$\therefore 6\sin x - 8\cos x = 10\sin(x - 53°8')$

$\therefore 10\sin(x - 53°8') = 5$

$\sin(x - 53°8') = 0.5$

$x - 53°8' = 30°, 150°, 390°$

$\mathbf{x = 83°8', 203°8'}$ **(as $0° \le x \le 360°$)**

2 i $\sqrt{3}\cos x - 3\sin x = r\cos(x + \alpha)$

$r = \sqrt{(\sqrt{3})^2 + 3^2}$

$= 2\sqrt{3}$

$\tan\alpha = \dfrac{3}{\sqrt{3}}$

$= \sqrt{3}$

$\alpha = \dfrac{\pi}{3}$

$\therefore \sqrt{3}\cos x - 3\sin x = 2\sqrt{3}\cos\left(x + \dfrac{\pi}{3}\right)$

$\therefore 2\sqrt{3}\cos\left(x + \dfrac{\pi}{3}\right) = \sqrt{3}$

$\cos\left(x + \dfrac{\pi}{3}\right) = \dfrac{1}{2}$

$x + \dfrac{\pi}{3} = \dfrac{\pi}{3}, \dfrac{5\pi}{3}, \dfrac{7\pi}{3}$

$\mathbf{x = 0, \dfrac{4\pi}{3}, 2\pi}$

ii $3\sin x + \sqrt{2}\cos x = r\sin(x - \alpha)$

$r = \sqrt{3^2 + (\sqrt{2})^2}$

$= \sqrt{11}$

$\tan\alpha = \dfrac{\sqrt{2}}{3}$

$\alpha = 0.440\,510\,663\ldots$

$= 0.441^c$ (3 dec. pl.)

$\therefore 3\sin x + \sqrt{2}\cos x + 2 = \sqrt{11}\sin(x + 0.441^c) + 2$

$\therefore \sqrt{11}\sin(x + 0.441^c) + 2 = 0$

$\sin(x + 0.441^c) = -\dfrac{2}{\sqrt{11}}$

$x + 0.441^c = 3.789^c, 5.636^c$ (3 dec. pl.)

$\mathbf{x = 3.3^c, 5.2^c}$

3 Equations involving sums, differences and double angles ➲ p. 152

1 i $\cos 5x\cos 3x + \sin 5x\sin 3x = -\dfrac{\sqrt{3}}{2}$

$\cos(5x - 3x) = -\dfrac{\sqrt{3}}{2}$

$\cos 2x = -\dfrac{\sqrt{3}}{2}$

$2x = \dfrac{5\pi}{6}, \dfrac{7\pi}{6}$

$\mathbf{x = \dfrac{5\pi}{12}, \dfrac{7\pi}{12}}$

ii $\dfrac{\tan 2x + \tan x}{1 - \tan 2x\tan x} = \sqrt{3}$

$\tan(2x + x) = \sqrt{3}$

$\tan 3x = \sqrt{3}$

$3x = \dfrac{\pi}{3}, \dfrac{4\pi}{3}, \dfrac{7\pi}{3}$

$\mathbf{x = \dfrac{\pi}{9}, \dfrac{4\pi}{9}, \dfrac{7\pi}{9}}$

2 i $\sin 2x\cos\dfrac{\pi}{4} + \cos 2x\sin\dfrac{\pi}{4} = \dfrac{\sqrt{3}}{2}$

$\sin\left(2x + \dfrac{\pi}{4}\right) = \dfrac{\sqrt{3}}{2}$

$2x + \dfrac{\pi}{4} = \dfrac{\pi}{3}, \dfrac{2\pi}{3}$

$2x = \dfrac{\pi}{12}, \dfrac{5\pi}{12}$

$\mathbf{x = \dfrac{\pi}{24}, \dfrac{5\pi}{24}}$

ii $\cos\dfrac{x}{2}\cos\dfrac{\pi}{3} + \sin\dfrac{x}{2}\sin\dfrac{\pi}{3} = -\dfrac{1}{2}$

$\cos\left(\dfrac{x}{2} - \dfrac{\pi}{3}\right) = -\dfrac{1}{2}$

$\dfrac{x}{2} - \dfrac{\pi}{3} = -\dfrac{4\pi}{3}, -\dfrac{2\pi}{3}, \dfrac{2\pi}{3}, \dfrac{4\pi}{3}$

$\dfrac{x}{2} = -\pi, -\dfrac{\pi}{3}, \pi, \dfrac{5\pi}{3}$

$\mathbf{x = -\dfrac{2\pi}{3}}$ **(as $-\pi \le x \le \pi$)**

3 i $\sin 2\theta + \sin\theta = 0$

$2\sin\theta\cos\theta + \sin\theta = 0$

$\sin\theta(2\cos\theta + 1) = 0$

$\sin\theta = 0$ $\quad\quad$ $\cos\theta = -\dfrac{1}{2}$

$\theta = 0°, 180°, 360°$ $\quad\quad$ $\theta = 120°, 240°$

$\mathbf{\theta = 0°, 120°, 180°, 240°, 360°}$

ii $\cos 4A - \sin 2A = 0$

$1 - 2\sin^2 2A - \sin 2A = 0$

$2\sin^2 2A + \sin 2A - 1 = 0$

$(2\sin 2A - 1)(\sin 2A + 1) = 0$

$\sin 2A = \dfrac{1}{2}$ $\quad\quad$ $\sin 2A = -1$

$2A = 30°, 150°, 390°, 510°$ $\quad\quad$ $2A = 270°, 630°$

$\therefore \mathbf{A = 15°, 75°, 195°, 255°, 135°, 315°}$

4 i $\cos 5x + \cos 3x = 2\cos\dfrac{5x + 3x}{2}\cos\dfrac{5x - 3x}{2}$

$= 2\cos 4x\cos x$

$\therefore 2\cos 4x\cos x = 0$

$\cos 4x = 0$ $\quad\quad$ $\cos x = 0$

$4x = \dfrac{\pi}{2}, \dfrac{3\pi}{2}, \dfrac{5\pi}{2}, \dfrac{7\pi}{2}$ $\quad\quad$ $x = \dfrac{\pi}{2}$

$x = \dfrac{\pi}{8}, \dfrac{3\pi}{8}, \dfrac{5\pi}{8}, \dfrac{7\pi}{8}$

$\therefore x = \mathbf{\dfrac{\pi}{8}, \dfrac{3\pi}{8}, \dfrac{\pi}{2}, \dfrac{5\pi}{8}, \dfrac{7\pi}{8}}$

ii $\sin 3x + \sin x + \sin 2x = 0$

$2\sin\frac{3x+x}{2}\cos\frac{3x-x}{2} + \sin 2x = 0$

$2\sin 2x\cos x + \sin 2x = 0$

$\sin 2x(2\cos x + 1) = 0$

$\sin 2x = 0 \qquad \cos x = -\frac{1}{2}$

$2x = 0, \pi, 2\pi \qquad x = \frac{2\pi}{3}$

$x = 0, \frac{\pi}{2}, \pi$

$\therefore \boldsymbol{x = 0, \frac{2\pi}{3}, \frac{\pi}{2}, \pi}$

4 Using *t*-formulae to solve equations ➲ p. 153

1 i $\sin\theta - \cos\theta = 1$

$\frac{2t}{1+t^2} - \frac{1-t^2}{1+t^2} = 1$

$2t - 1 + t^2 = 1 + t^2$

$2t = 2$

$t = 1$

$\tan\frac{\theta}{2} = 1$

$\frac{\theta}{2} = 45°$

$\theta = 90°$ (as $0° \le \theta \le 360°$)

Also, check possible solution of $\theta = 180°$:
$\sin 180° - \cos 180° = 0 - (-1) = 1$. Yes, $\theta = 180°$ is a solution.

$\therefore \boldsymbol{\theta = 90°, 180°}$.

ii $\sqrt{3}\sin\theta + \cos\theta + 1 = 0$

$\sqrt{3}.\frac{2t}{1+t^2} + \frac{1-t^2}{1+t^2} + 1 = 0$

$2\sqrt{3}t + 1 - t^2 + 1 + t^2 = 0$

$2\sqrt{3}t + 2 = 0$

$2\sqrt{3}t = -2$

$t = -\frac{1}{\sqrt{3}}$

$\tan\frac{\theta}{2} = -\frac{1}{\sqrt{3}}$

$\frac{\theta}{2} = 150°$

$\theta = 300°$ (as $0° \le \theta \le 360°$)

Also, check possible solution of $\theta = 180°$:
$\sqrt{3}\sin 180° + \cos 180° + 1 = 0$. Yes, a solution.

$\therefore \boldsymbol{\theta = 180°, 300°}$

2 i $12\sin\theta - 5\cos\theta - 3 = 0$

$12.\frac{2t}{1+t^2} - 5.\frac{1-t^2}{1+t^2} - 3 = 0$

$24t - 5 + 5t^2 - 3 - 3t^2 = 0$

$2t^2 + 24t - 8 = 0$

$t^2 + 12t - 4 = 0$

$t = \frac{-12 \pm \sqrt{12^2 - 4(1)(-4)}}{2(1)}$

$\tan\frac{\theta}{2} = 0.3245\ldots, -12.3245\ldots$

$\frac{\theta}{2} = 0.3138\ldots, 1.6517\ldots$

$\theta = 0.628^c, 3.30^c$ (3 sig. figs) (as $0 \le \theta \le 2\pi$)

Also, check possible solution of $\theta = \pi$: $12\sin\pi - 5\cos\pi - 3 \ne 0$.
No, not a solution.

$\therefore \boldsymbol{\theta = 0.628^c, 3.30^c}$

ii $4\cos\theta - 3\sin\theta + 1 = 0$

$4.\frac{1-t^2}{1+t^2} - 3.\frac{2t}{1+t^2} + 1 = 0$

$4 - 4t^2 - 6t + 1 + t^2 = 0$

$3t^2 + 6t - 5 = 0$

$t = \frac{-6 \pm \sqrt{6^2 - 4(3)(-5)}}{2(3)}$

$\tan\frac{\theta}{2} = 0.633\ldots, -2.633\ldots$

$\frac{\theta}{2} = 0.564\ldots, 1.933\ldots$

$\theta = 1.13^c, 3.87^c$ (3 sig. figs) (as $0 \le \theta \le 2\pi$)

Also, check possible solution of $\theta = \pi$: $4\cos\pi - 3\sin\pi + 1 \ne 0$.
No, not a solution.

$\therefore \boldsymbol{\theta = 1.13^c, 3.87^c}$

5 Expressing sin 3θ and cos 3θ in terms of θ ➲ p. 153

1 i LHS $= \cos 3x$

$= \cos(2x + x)$

$= \cos 2x\cos x - \sin 2x\sin x$

$= (2\cos^2 x - 1)\cos x - 2\sin x\cos x\sin x$

$= 2\cos^3 x - \cos x - 2\cos x\sin^2 x$

$= 2\cos^3 x - \cos x - 2\cos x(1 - \cos^2 x)$

$= 2\cos^3 x - \cos x - 2\cos x + 2\cos^3 x$

$= 4\cos^3 x - 3\cos x$

= RHS

ii $\tan 3\theta = \tan(2\theta + \theta)$

$= \frac{\tan 2\theta + \tan\theta}{1 - \tan 2\theta\tan\theta}$

$= \frac{\frac{2\tan\theta}{1-\tan^2\theta} + \tan\theta}{1 - \frac{2\tan\theta}{1-\tan^2\theta}\tan\theta}$

$= \frac{2\tan\theta + \tan\theta - \tan^3\theta}{1 - \tan^2\theta - 2\tan^2\theta}$ [multiplying top and bottom by $(1 - \tan^2\theta)$]

$= \boldsymbol{\frac{3\tan\theta - \tan^3\theta}{1 - 3\tan^2\theta}}$

2 i $\sin 3x = \sin x$

$3\sin x - 4\sin^3 x = \sin x$

$4\sin^3 x - 2\sin x = 0$

$2\sin x(2\sin^2 x - 1) = 0$

$\sin x = 0, \sin x = \pm\frac{1}{\sqrt{2}}$

x = 0°, 45°, 135° and 180°

ii $\sin 4x - \sin 2x = 0$

$2\sin 2x\cos 2x - \sin 2x = 0$

$\sin 2x(2\cos 2x - 1) = 0$

$\sin 2x = 0, 2\cos 2x = 1$

$\sin 2x = 0, \cos 2x = \frac{1}{2}$

$2x = 0, \pi, 2\pi, 3\pi, 4\pi, \frac{\pi}{3}, \frac{5\pi}{3}, \frac{7\pi}{3}, \frac{11\pi}{3}$

$\boldsymbol{x = 0, \frac{\pi}{2}, \pi, \frac{3\pi}{2}, 2\pi, \frac{\pi}{6}, \frac{5\pi}{6}, \frac{7\pi}{6}, \frac{11\pi}{6}}$

HSC EXAM-TYPE QUESTIONS

Part A

1 A. $r = \sqrt{(\sqrt{3})^2 + 1^2} = 2$

$\tan \alpha = \frac{1}{\sqrt{3}}$

$\alpha = \frac{\pi}{6}$

2 D. $1 - \sin 2x \tan x = 1 - 2 \sin x \cos x . \frac{\sin x}{\cos x}$

$= 1 - 2 \sin^2 x$

$= \cos 2x$

3 C. $\cos 3x + \cos x = 0$

$2 \cos 2x \cos x = 0$

$\cos 2x = 0, \cos x = 0$

$2x = \frac{\pi}{2}, \frac{3\pi}{2}, \frac{5\pi}{2}, \frac{7\pi}{2}, x = \frac{\pi}{2}, \frac{3\pi}{2}$

$x = \frac{\pi}{4}, \frac{3\pi}{4}, \frac{5\pi}{4}, \frac{7\pi}{4}, \frac{\pi}{2}, \frac{3\pi}{2}$

$\therefore$ 6 solutions

4 C. Consider $2 \sin x - 3 \cos x = r \sin(x + \alpha)$

$r = \sqrt{2^2 + 3^2} = \sqrt{13}$

As $-1 \le \sin(x + \alpha) \le 1$, then $-\sqrt{13} \le \sqrt{13} \sin(x + \alpha) \le \sqrt{13}$.

$\therefore$ the minimum value is $4 - \sqrt{13}$.

5 B. $5 \cos \theta + 3 = 0$

$\cos \theta = -\frac{3}{5}$

$\frac{1-t^2}{1+t^2} = -\frac{3}{5}$

$5 - 5t^2 = -3 - 3t^2$

$2t^2 = 8$

$t^2 = 4$

$t = \pm 2$

But as $\cos \theta < 0$ and $0 < \theta < \pi$, then θ is in second quadrant.

Hence, $\frac{\theta}{2}$ is acute and so $\tan \frac{\theta}{2} > 0$. $\therefore \tan \frac{\theta}{2} = 2$ ✓

Part B

6 a

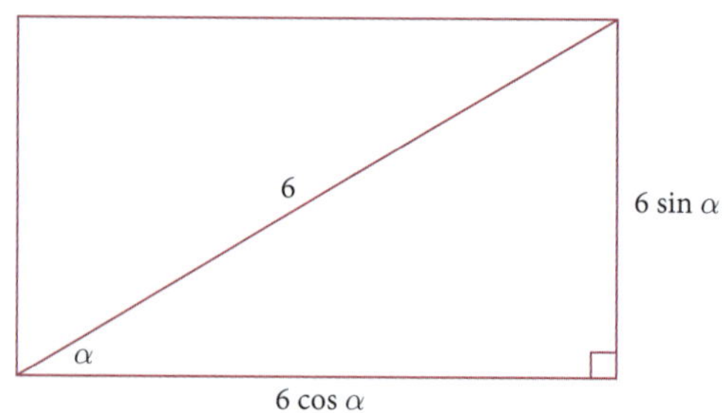

$A = 6 \cos \alpha \times 6 \sin \alpha$ ✓

$= 36 \sin \alpha \cos \alpha$

$= \mathbf{18 \sin 2\alpha}$ ✓

b $18 \sin 2\alpha = 9$

$\sin 2\alpha = \frac{1}{2}$

$2\alpha = \frac{\pi}{6}, \frac{5\pi}{6}$

$\alpha = \frac{\pi}{12}$ (as α is smallest <) ✓

7 a $3 \sin 2t + 4 \cos 2t = R \sin (2t + \alpha)$

$R = \sqrt{3^2 + 4^2}$

$= 5$ ✓

$\tan \alpha = \frac{4}{3}$

$\alpha = 53°8'$

$\therefore$ **3 sin 2*t* + 4 cos 2*t* = 5 sin (2*t* + 53°8')** ✓

b $5 \sin (2t + 53°8') - 5 = 0$

$5 \sin (2t + 53°8') = 5$

$\sin (2t + 53°8') = 1$ ✓

$2t + 53°8' = 90°, 450°$

$2t = 36°52', 396°52'$

***t* = 18°26', 198°26'** ✓

8 $r = \sqrt{2^2 + (2\sqrt{3})^2}$

$= 4$ ✓

$\therefore$ the equation of the graph is in the form $y = 4 \sin (\theta - \alpha)$.

$\therefore$ **the amplitude is 4.** ✓

9 $2 \cos 2x + 3 = 4 \cos x$

$2 \cos 2x + 3 - 4 \cos x = 0$

$2(2 \cos^2 x - 1) + 3 - 4 \cos x = 0$ ✓

$4 \cos^2 x - 2 + 3 - 4 \cos x = 0$

$4 \cos^2 x - 4 \cos x + 1 = 0$

$(2 \cos x - 1)^2 = 0$

$2 \cos x = 1$ ✓

$\cos x = \frac{1}{2}$

$\mathbf{x = \frac{\pi}{3}, \frac{5\pi}{3}}$ ✓

10 $\cos\theta + \sqrt{3}\sin\theta = r\cos(\theta - \alpha)$

$$r = \sqrt{1^2 + (\sqrt{3})^2}$$
$$= 2$$

$$\tan\alpha = \frac{\sqrt{3}}{1} = \sqrt{3}$$

$$\therefore \alpha = \frac{\pi}{3} \quad ✓$$

$$\therefore \mathbf{y = 2\cos\left(\theta - \frac{\pi}{3}\right)} \quad ✓$$

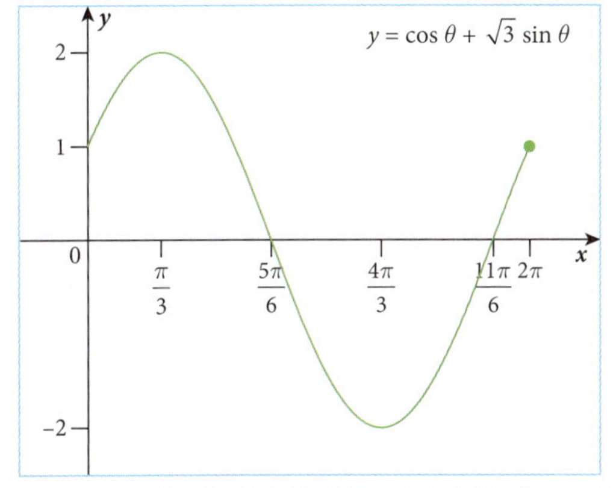

✓

11

$$1 - 2\sin^2 A = \sin A$$
$$2\sin^2 A + \sin A - 1 = 0 \quad ✓$$
$$(2\sin A - 1)(\sin A + 1) = 0$$
$$\sin A = \frac{1}{2}, -1$$
$$\mathbf{A = \frac{\pi}{6}, \frac{5\pi}{6}, \frac{3\pi}{2}} \quad ✓$$

12 $\sin 6x - \sin 4x = 2\cos\dfrac{6x+4x}{2}\sin\dfrac{6x-4x}{2}$

$$= 2\cos 5x \sin x$$

$$\therefore 2\cos 5x\sin x = 0 \quad ✓$$

$\cos 5x = 0$ $\sin x = 0$

$5x = \dfrac{\pi}{2}, \dfrac{3\pi}{2}, \dfrac{5\pi}{2}$ $x = 0$

$x = \dfrac{\pi}{10}, \dfrac{3\pi}{10}, \dfrac{\pi}{2}$

$$\therefore \mathbf{x = 0, \frac{\pi}{10}, \frac{3\pi}{10}, \frac{\pi}{2}} \quad ✓$$

13 a $\sin 3\theta = \sin(2\theta + \theta)$

$$= \sin 2\theta\cos\theta + \cos 2\theta\sin\theta \quad ✓$$
$$= 2\sin\theta\cos\theta\cos\theta + (1 - 2\sin^2\theta)\sin\theta$$
$$= 2\sin\theta\cos^2\theta + \sin\theta - 2\sin^3\theta$$
$$= 2\sin\theta(1 - \sin^2\theta) + \sin\theta - 2\sin^3\theta$$
$$= 2\sin\theta - 2\sin^3\theta + \sin\theta - 2\sin^3\theta$$
$$\mathbf{= 3\sin\theta - 4\sin^3\theta} \quad ✓$$

b

$$\sin 3\theta + \sin 2\theta = \sin\theta$$
$$3\sin\theta - 4\sin^3\theta + \sin 2\theta = \sin\theta \quad ✓$$
$$2\sin\theta - 4\sin^3\theta + 2\sin\theta\cos\theta = 0$$
$$2\sin\theta(1 - 2\sin^2\theta + \cos\theta) = 0$$
$$2\sin\theta(1 - 2[1 - \cos^2\theta] + \cos\theta) = 0 \quad ✓$$
$$2\sin\theta(1 - 2 + 2\cos^2\theta + \cos\theta) = 0$$
$$2\sin\theta(2\cos^2\theta + \cos\theta - 1) = 0$$
$$2\sin\theta(2\cos\theta - 1)(\cos\theta + 1) = 0 \quad ✓$$

$2\sin\theta = 0$ $2\cos\theta - 1 = 0$ $\cos\theta + 1 = 0$

$\sin\theta = 0$ $\cos\theta = \dfrac{1}{2}$ $\cos\theta = -1$

$\theta = 0, \pi, 2\pi$ $\theta = \dfrac{\pi}{3}, \dfrac{5\pi}{3}$ $\theta = \pi$

$$\therefore \theta = \mathbf{0, \pi, \frac{\pi}{3}, \frac{5\pi}{3}, 2\pi} \quad ✓$$

14 a $2\sin\left(\theta + \dfrac{\pi}{6}\right) + 2\cos\theta = 2\left[\sin\theta\cos\dfrac{\pi}{6} + \cos\theta\sin\dfrac{\pi}{6}\right] + 2\cos\theta$

$$= 2\left[\sin\theta.\frac{\sqrt{3}}{2} + \cos\theta.\frac{1}{2}\right] + 2\cos\theta$$
$$= \sqrt{3}\sin\theta + \cos\theta + 2\cos\theta$$
$$= \sqrt{3}\sin\theta + 3\cos\theta \quad ✓$$

$$\sqrt{3}\sin\theta + 3\cos\theta = r\sin(\theta + \alpha)$$
$$r = \sqrt{(\sqrt{3})^2 + 3^2}$$
$$= 2\sqrt{3} \quad ✓$$
$$\tan\alpha = \frac{3}{\sqrt{3}} = \sqrt{3}$$
$$\therefore \alpha = \frac{\pi}{3}$$
$$\therefore \mathbf{2\cos\theta + 2\sin\left(\theta + \frac{\pi}{6}\right) = 2\sqrt{3}\sin\left(\theta + \frac{\pi}{3}\right)} \quad ✓$$

b $2\sqrt{3}\sin\left(\theta + \dfrac{\pi}{3}\right) = 3$

$$\sin\left(\theta + \frac{\pi}{3}\right) = \frac{\sqrt{3}}{2} \quad ✓$$
$$\theta + \frac{\pi}{3} = \frac{\pi}{3}, \frac{2\pi}{3}, \frac{7\pi}{3}$$
$$\mathbf{\theta = 0, \frac{\pi}{3}, 2\pi} \quad ✓$$

TOPIC: CALCULUS

YEAR 12

CHAPTER 13 FURTHER CALCULUS SKILLS

1 Integration involving products of trigonometric functions

- $2\sin\frac{A+B}{2}\cos\frac{A-B}{2} = \sin A + \sin B$
- $2\cos\frac{A+B}{2}\sin\frac{A-B}{2} = \sin A - \sin B$
- $2\cos\frac{A+B}{2}\cos\frac{A-B}{2} = \cos A + \cos B$
- $-2\sin\frac{A+B}{2}\sin\frac{A-B}{2} = \cos A - \cos B$

EXAMPLE 1

Find $\int 2\sin 3x\cos x\, dx$.

$= \int (\sin 4x + \sin 2x)\, dx$

First express as a sum.

$= -\frac{1}{4}\cos 4x - \frac{1}{2}\cos 2x + c$

EXAMPLE 2

Evaluate $\int_0^{\frac{\pi}{3}} -2\sin\frac{5x}{2}\sin\frac{x}{2}\, dx$.

$= \int_0^{\frac{\pi}{3}} (\cos 3x - \cos 2x)\, dx$

$= \left[\frac{1}{3}\sin 3x - \frac{1}{2}\sin 2x\right]_0^{\frac{\pi}{3}}$

$= \left[\frac{1}{3}\sin 3\left(\frac{\pi}{3}\right) - \frac{1}{2}\sin 2\left(\frac{\pi}{3}\right) - 0\right]$

Show your line of substituting into the integrand.

$= \left[\frac{1}{3}(0) - \frac{1}{2}\left(\frac{\sqrt{3}}{2}\right) - 0\right]$

$= -\frac{\sqrt{3}}{4}$

→ PRACTICE

1 Find:

i $\int 2\cos 4x\cos x\, dx$

ii $\int \cos 3x\sin 2x\, dx$

2 Evaluate:

i $\int_0^{\frac{\pi}{6}} \cos\frac{3x}{2}\cos\frac{x}{2}\, dx$

ii $\int_0^{\frac{\pi}{2}} 4\sin x\cos\frac{x}{3}\, dx$

Answers ➲ p. 168

2 Integration using substitution

- $\int x^n\, dx = \frac{1}{n+1}x^{n+1} + c.\ (n \neq -1)$
- $\int f'(x)[f(x)]^n\, dx = \frac{1}{n+1}[f(x)]^{n+1} + c.\ (n \neq -1)$
- **When the integrand is more complex** an appropriate substitution may be used to make the integration possible.
- **If** $\frac{du}{dx} = g(x)$, then $dx = \frac{du}{g(x)}$.
- $\sin^2 x + \cos^2 x = 1$, $\sec^2 x = 1 + \tan^2 x$, $\text{cosec}^2 x = 1 + \cot^2 x$.

EXAMPLE 1

Find $\int x^3\sqrt{x^4+3}\, dx$ using the substitution $u = x^4 + 3$.

As $u = x^4 + 3$, $\therefore \frac{du}{dx} = 4x^3 \therefore dx = \frac{du}{4x^3}$

$\int x^3\sqrt{x^4+3}\, dx = \int x^3\sqrt{u}\,\frac{du}{4x^3}$

$= \frac{1}{4}\int u^{\frac{1}{2}}\, du$

Remember to express your answer in terms of x.

$= \frac{1}{4}\left[\frac{2u^{\frac{3}{2}}}{3}\right] + c$

$= \frac{\sqrt{(x^4+3)^3}}{6} + c$

EXAMPLE 2

Use the substitution $u = \tan x$, to find $\int \tan^2 x \sec^2 x \, dx$.

As $u = \tan x \therefore \frac{du}{dx} = \sec^2 x \therefore dx = \frac{du}{\sec^2 x}$

$$\int \tan^2 x \sec^2 x \, dx = \int u^2 \sec^2 x \frac{du}{\sec^2 x}$$
$$= \int u^2 \, du$$
$$= \frac{u^3}{3} + c$$
$$= \frac{\tan^3 x}{3} + c$$

$\frac{d}{dx}(\tan x) = \sec^2 x$

EXAMPLE 3

Find $\int_3^8 \frac{x}{\sqrt{x+1}} dx$ using the substitution $u = x + 1$.

As $u = x + 1 \therefore \frac{du}{dx} = 1 \therefore dx = du$. Also, $x = u - 1$.

When $x = 8$, $u = 9$ and when $x = 3$, $u = 4$.

$$\int_3^8 \frac{x}{\sqrt{x+1}} dx = \int_{u=4}^{u=9} \frac{u-1}{\sqrt{u}} du$$
$$= \int_4^9 (u^{\frac{1}{2}} - u^{-\frac{1}{2}}) du$$
$$= \left[\frac{2}{3}u^{\frac{3}{2}} - 2u^{\frac{1}{2}}\right]_4^9$$
$$= \left[\frac{2}{3}(9)^{\frac{3}{2}} - 2(9)^{\frac{1}{2}}\right] - \left[\frac{2}{3}(4)^{\frac{3}{2}} - 2(4)^{\frac{1}{2}}\right]$$
$$= 18 - 6 - \left(\frac{16}{3} - 4\right)$$
$$= 10\frac{2}{3}$$

Substitute to find the new limits of the integral.

EXAMPLE 4

Find $\int \frac{x}{\sqrt{1-x^2}} dx$ using the substitution $x = \sin u$.

As $x = \sin u \therefore \frac{dx}{du} = \cos u \therefore dx = \cos u \, du$.

$$\int \frac{x}{\sqrt{1-x^2}} dx = \int \frac{\sin u}{\sqrt{1-\sin^2 u}} . \cos u \, du$$
$$= \int \frac{\sin u}{\cos u} . \cos u \, du$$
$$= \int \sin u \, du$$
$$= -\cos u + c$$
$$= -\sqrt{1-x^2} + c$$

x, 1, u, $\sqrt{1-x^2}$

Form a triangle to find $\cos u$.

→ PRACTICE

1 **Find:**

i $\int 6x\sqrt{9-x^2} \, dx$ **using the substitution $u = 9 - x^2$**

ii $\int x\sqrt{x-2} \, dx$ **using the substitution $u = x - 2$**

2 **Use the substitution:**

i $u = \cos x$, **to find** $\int \frac{\sin x}{\cos^4 x} dx$

ii $u = \ln x$, **to find** $\int \frac{1}{x \ln x} dx$

3 **Find:**

i $\int_{-1}^{0} x^2\sqrt{1-2x^3} \, dx$ **using the substitution $u = 1 - 2x^3$**

ii $\int_0^2 \frac{8x}{\sqrt{4-x^2}} dx$ **using the substitution $u = 4 - x^2$**

4 **Find:**

i $\int \frac{x}{\sqrt{4-x^2}} dx$ **using the substitution $x = 2 \sin u$**

ii $\int \frac{dx}{x^2\sqrt{1-x^2}}$ **using the substitution $x = \cos \theta$**

Answers ⊃ pp. 168–169

3 Integration of $\sin^2 nx$ and $\cos^2 nx$

→ **$\cos 2nx = 1 - 2\sin^2 nx$**, and hence **$\sin^2 nx = \frac{1}{2}[1 - \cos 2nx]$**.

→ **$\cos 2nx = 2\cos^2 nx - 1$**, and hence **$\cos^2 nx = \frac{1}{2}[1 + \cos 2nx]$**.

EXAMPLE 1

Find $\int_{\frac{\pi}{6}}^{\frac{\pi}{3}} \sin^2 x \, dx$.

$$\int_{\frac{\pi}{6}}^{\frac{\pi}{3}} \sin^2 x \, dx = \int_{\frac{\pi}{6}}^{\frac{\pi}{3}} \frac{1}{2}[1 - \cos 2x] \, dx$$
$$= \frac{1}{2}\left[x - \frac{1}{2}\sin 2x\right]_{\frac{\pi}{6}}^{\frac{\pi}{3}}$$
$$= \frac{1}{2}\left[\frac{\pi}{3} - \frac{1}{2}\sin 2\left(\frac{\pi}{3}\right) - \left(\frac{\pi}{6} - \frac{1}{2}\sin 2\left(\frac{\pi}{6}\right)\right)\right]$$
$$= \frac{1}{2}\left[\frac{\pi}{3} - \frac{\sqrt{3}}{4} - \frac{\pi}{6} + \frac{\sqrt{3}}{4}\right]$$
$$= \frac{1}{2}\left[\frac{\pi}{6}\right]$$
$$= \frac{\pi}{12}$$

Learn $\sin^2 x = \frac{1}{2}(1 - \cos 2x)$

EXAMPLE 2

Find the primitive of $\cos^2 2\theta$.

$$\int \cos^2 2\theta \, d\theta = \int \frac{1}{2}[1 + \cos 4\theta] \, d\theta$$

$$= \frac{1}{2}\left[\theta + \frac{1}{4}\sin 4\theta\right] + c$$

Use $\cos^2 x = \frac{1}{2}[1 + \cos 2x]$ but replace x with 2θ.

EXAMPLE 3

Use the substitution $x = \sin\theta$ to find $\int \frac{x^2}{\sqrt{1-x^2}} \, dx$.

$x = \sin\theta$ then $\frac{dx}{d\theta} = \cos\theta$ and hence $dx = \cos\theta \, d\theta$.

Also, $\sin\theta = x$, then $\theta = \sin^{-1}x$. Also, by Pythagoras, $\cos\theta = \sqrt{1-x^2}$.

$$\int \frac{x^2}{\sqrt{1-x^2}} \, dx = \int \frac{\sin^2\theta}{\sqrt{1-\sin^2\theta}}.\cos\theta \, d\theta$$

$$= \int \frac{\sin^2\theta}{\cos\theta}.\cos\theta \, d\theta$$

$$= \int \sin^2\theta \, d\theta$$

If $x = \sin\theta$, then $\theta = \sin^{-1} x$.

$$= \frac{1}{2}\int (1 - \cos 2\theta) \, d\theta$$

$$= \frac{1}{2}\left[\theta - \frac{1}{2}\sin 2\theta\right] + c$$

$$= \frac{1}{2}[\theta - \sin\theta\cos\theta] + c \quad \text{(using } \sin 2\theta = 2\sin\theta\cos\theta\text{)}$$

$$= \frac{1}{2}\left[\sin^{-1}x - x.\sqrt{1-x^2}\right] + c$$

$$= \frac{1}{2}\sin^{-1}x - \frac{x\sqrt{1-x^2}}{2} + c$$

EXAMPLE 4

Find the exact area under the curve $y = \sin^2 \frac{x}{2}$ between $x = \frac{\pi}{4}$ and $x = \frac{\pi}{6}$.

$$\text{Area} = \int_{\frac{\pi}{6}}^{\frac{\pi}{4}} \sin^2 \frac{x}{2} \, dx$$

$$= \int_{\frac{\pi}{6}}^{\frac{\pi}{4}} \frac{1}{2}[1 - \cos x] \, dx$$

Substitute $n = \frac{1}{2}$ in $\sin^2 nx$ result.

$$= \frac{1}{2}[x - \sin x]_{\frac{\pi}{6}}^{\frac{\pi}{4}}$$

$$= \frac{1}{2}\left[\frac{\pi}{4} - \sin\frac{\pi}{4} - \left(\frac{\pi}{6} - \sin\frac{\pi}{6}\right)\right]$$

$$= \frac{1}{2}\left[\frac{\pi}{4} - \frac{1}{\sqrt{2}} - \frac{\pi}{6} + \frac{1}{2}\right]$$

$$= \frac{1}{2}\left[\frac{\pi}{12} - \frac{1}{\sqrt{2}} + \frac{1}{2}\right]$$

$\therefore$ the area is $\frac{1}{2}\left[\frac{\pi}{12} - \frac{1}{\sqrt{2}} + \frac{1}{2}\right]$ units^2.

→ PRACTICE

1 **Find:**

i $\int_{\frac{\pi}{4}}^{\frac{\pi}{2}} \cos^2 x \, dx$

ii $\int_{\frac{\pi}{6}}^{\frac{2\pi}{3}} 4\sin^2 x \, dx$

2 **Find the primitive of:**

i $\sin^2 \frac{\theta}{2}$

ii $\cos\alpha(\cos\alpha - 2)$

3 i **Use the substitution $x = 3\sin\theta$ to find $\int \sqrt{9-x^2} \, dx$.**

ii **Find $\int_0^{\frac{1}{2}} \sqrt{\frac{x}{1-x}} \, dx$ using the substitution $x = \sin^2 u$.**

4 **Find the exact area:**

i **under the curve $y = 1 + \cos^2 x$ between $x = \frac{\pi}{3}$ and $x = \frac{2\pi}{3}$**

ii **bounded by the curves $y = 2\sin^2 x$ and $y = \sin x$, between $x = 0$ and $x = \frac{\pi}{6}$.**

Answers ➲ pp. 169–170

4 Using the result $\frac{dy}{dx} = \frac{1}{\frac{dx}{dy}}$

→ $\frac{dy}{dx} = \frac{1}{\frac{dx}{dy}}$, or $\frac{dy}{dx} \times \frac{dx}{dy} = 1$

EXAMPLE 1

If $y = e^{2x}$, show that $\frac{dy}{dx} = \frac{1}{\frac{dx}{dy}}$.

Firstly, $\frac{dy}{dx} = 2e^{2x}$.

Also, $\ln y = \ln e^{2x}$

$$2x = \ln y$$

$$x = \frac{1}{2}\ln y$$

$$\frac{dx}{dy} = \frac{1}{2}.\frac{1}{y}$$

$$= \frac{1}{2}.\frac{1}{e^{2x}}$$

$$= \frac{1}{2e^{2x}}$$

$$\therefore \frac{dy}{dx} = \frac{1}{\frac{dx}{dy}}$$

$\frac{dy}{dx} \times \frac{dx}{dy} = 1$

EXAMPLE 2

If $x = y^2$, find $\frac{dy}{dx}$ in terms of y.

First, $\frac{dx}{dy} = 2y$

Identify any exclusions for y.

$$\therefore \frac{dy}{dx} = \frac{1}{2y}, y \neq 0$$

EXAMPLE 3

Use the result $\frac{dy}{dx} = \frac{1}{\frac{dx}{dy}}$ to find an expression for $\frac{dy}{dx}$ in terms of x, given $y = \sin^{-1} x$.

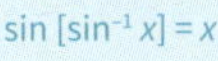

$$y = \sin^{-1} x$$

$$\sin y = x$$

$$\frac{dx}{dy} = \cos y$$

$$= \sqrt{1-x^2}$$

$$\frac{dy}{dx} = \frac{1}{\sqrt{1-x^2}}$$

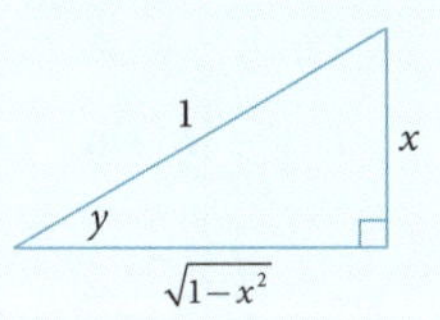

EXAMPLE 4

Find the equation of the tangent to the curve $x = 2y^2 - 3y - 1$ at the point where $y = 2$.

First, $x(2) = 2(2)^2 - 3(2) - 1$

$= 1 \therefore$ point is $(1, 2)$

Also, $\frac{dx}{dy} = 4y - 3$

$$\therefore \frac{dy}{dx} = \frac{1}{4y-3}, y \neq \frac{3}{4} \text{ (using } \frac{dy}{dx} = \frac{1}{\frac{dx}{dy}}\text{)}$$

$$\therefore \frac{dy}{dx}(2) = \frac{1}{4(2)-3}$$

$$= \frac{1}{5}$$

Using $(1, 2)$ and $m = \frac{1}{5}$:

Use $y - y_1 = m(x - x_1)$.

$$\therefore y - 2 = \frac{1}{5}(x - 1)$$

$$5y - 10 = x - 1$$

$$x - 5y + 9 = 0$$

→ PRACTICE

1 i If $y = e^{3x}$, show that $\frac{dy}{dx} = \frac{1}{\frac{dx}{dy}}$.

ii If $y = (3x - 1)^2$, show that $\frac{dy}{dx} \times \frac{dx}{dy} = 1$.

2 i If $x = 4y^3$, find $\frac{dy}{dx}$ in terms of y.

ii If $x = \sin\frac{y}{2}$ find the value of $\frac{dy}{dx}$ when $y = \frac{\pi}{3}$.

3 Use the result $\frac{dy}{dx} = \frac{1}{\frac{dx}{dy}}$ to find an expression for $\frac{dy}{dx}$, given:

i $y = \cos^{-1} x$

ii $y = \tan^{-1} 3x$

4 Find the equation of the:

i tangent to the curve $x = y^2 - 2y + 4$ at the point where $y = 3$

ii normal to the curve $x = \frac{4}{y-1}$ at the point $x = 2$

Answers ➲ pp. 170–171

5 Differentiation of inverse trigonometric functions

→ $\frac{d}{dx}\left(\sin^{-1}\frac{x}{a}\right) = \frac{1}{\sqrt{a^2-x^2}}$

→ $\frac{d}{dx}\left(\cos^{-1}\frac{x}{a}\right) = \frac{-1}{\sqrt{a^2-x^2}}$

→ $\frac{d}{dx}\left(\tan^{-1}\frac{x}{a}\right) = \frac{a}{a^2+x^2}$

→ $\frac{d}{dx}\left(\sin^{-1}\frac{f(x)}{a}\right) = \frac{1}{\sqrt{a^2-[f(x)]^2}}.f'(x)$

➔ $\frac{d}{dx}\left(\cos^{-1}\frac{f(x)}{a}\right) = \frac{-1}{\sqrt{a^2-[f(x)]^2}}.f'(x)$

➔ $\frac{d}{dx}\left(\tan^{-1}\frac{f(x)}{a}\right) = \frac{1}{a^2+[f(x)]^2}.f'(x)$

EXAMPLE 1

If $y = \sin^{-1} 4x$, find $\frac{dy}{dx}$.

$$\frac{dy}{dx} = \frac{1}{\sqrt{1-(4x)^2}}.4$$

$$= \frac{4}{\sqrt{1-16x^2}}$$

$\frac{d}{dx}[4x] = 4$

EXAMPLE 2

Find $\frac{d}{dx}[x^2 \tan^{-1} x]$.

$$\frac{d}{dx}[x^2 \tan^{-1} x] = 2x.\tan^{-1} x + x^2.\frac{1}{1+x^2}$$

$$= 2x \tan^{-1} x + \frac{x^2}{1+x^2}$$

Use product rule.

EXAMPLE 3

Find the second derivative of $y = \cos^{-1}\frac{2x}{3}$.

$$\frac{dy}{dx} = \frac{-1}{\sqrt{1-\left(\frac{2x}{3}\right)^2}}.\frac{2}{3}$$

$$= \frac{-1}{\sqrt{1-\frac{4x^2}{9}}}.\frac{2}{3}$$

$$= \frac{-2}{\sqrt{9-4x^2}}$$

$$= -2(9-4x^2)^{-\frac{1}{2}}$$

$$\frac{d^2y}{dx^2} = (9-4x^2)^{-\frac{3}{2}}.-8x$$

$$= \frac{-8x}{\sqrt{(9-4x^2)^3}}$$

$\frac{2x}{3} = \frac{2}{3}x$

EXAMPLE 4

Differentiate $\sin^{-1}(\cos x)$.

$$\frac{d}{dx}[\sin^{-1}(\cos x)] = \frac{1}{\sqrt{1-(\cos x)^2}}.-\sin x$$

$$= \frac{1}{\sqrt{1-\cos^2 x}}.-\sin x$$

$$= \frac{-\sin x}{\sqrt{\sin^2 x}}$$

$$= \frac{-\sin x}{|\sin x|}$$

$= -1$ when $\sin x > 0$ and 1 when $\sin x < 0$

$\sqrt{x^2} = |x|$

EXAMPLE 5

Find the gradient of the normal to the curve $y = 4 \tan^{-1} 2x$, at the point where $x = -1$.

$$\frac{dy}{dx} = 4 \times \frac{1}{1+(2x)^2}.2$$

$$= \frac{8}{1+4x^2}$$

$$\frac{dy}{dx}(-1) = \frac{8}{1+4(-1)^2}$$

$$= \frac{8}{5}$$

$\therefore$ the gradient of the normal is $-\frac{5}{8}$.

The gradient of the normal is the negative reciprocal of the gradient of the tangent.

➔ PRACTICE

1 Find $\frac{dy}{dx}$, if:

i $y = \tan^{-1} 3x$

ii $y = \sin^{-1}\frac{1}{x}$

2 Find:

i $\frac{d}{dx}[x^3 \sin^{-1} 2x]$

ii $\frac{d}{dx}\left[\frac{\cos^{-1} x}{x}\right]$

3 Find the second derivative of:

i $y = 4 \sin^{-1} 2x$

ii $y = \tan^{-1}\frac{x+1}{2}$

4 Differentiate:

i $\cos^{-1}(\sin x)$

ii $\sin\left(\cos^{-1}\frac{x}{2}\right)$

5 Find the:

i gradient of the normal to the curve $y = \sin^{-1}\sqrt{1-x^2}$ at the point where $x = 0$

ii equation of the tangent to the curve $y = \tan^{-1}\frac{x-1}{2}$, at the point where $x = 3$

Answers ➲ pp. 171–172

6 Integration resulting in inverse trigonometric functions

➔ $\int \frac{1}{\sqrt{a^2-x^2}}\,dx = \sin^{-1}\frac{x}{a} + c$

➔ $\int \frac{a}{a^2+x^2}\,dx = \tan^{-1}\frac{x}{a} + c$

EXAMPLE 1

Find $\int \frac{1}{16+x^2}\,dx$.

$$\int \frac{1}{16+x^2}\,dx = \frac{1}{4}\tan^{-1}\frac{x}{4} + c$$

$\int \frac{1}{a^2+x^2}\,dx = \frac{1}{a}\tan^{-1}\frac{x}{a} + c$

EXAMPLE 2

Evaluate $\int_{-\frac{1}{3}}^{\frac{1}{3}} \frac{1}{\sqrt{4-9x^2}}\,dx$.

$$\int_{-\frac{1}{3}}^{\frac{1}{3}} \frac{1}{\sqrt{4-9x^2}}\,dx = \int_{-\frac{1}{3}}^{\frac{1}{3}} \frac{1}{\sqrt{9\left(\frac{4}{9}-x^2\right)}}\,dx$$

$$= \frac{1}{3}\int_{-\frac{1}{3}}^{\frac{1}{3}} \frac{1}{\sqrt{\left(\frac{4}{9}-x^2\right)}}\,dx$$

$$= \frac{1}{3}\left[\sin^{-1}\frac{3x}{2}\right]_{-\frac{1}{3}}^{\frac{1}{3}}$$

Factorise to give expression involving x^2.

$$= \frac{1}{3}\left[\sin^{-1}\frac{3\left(\frac{1}{3}\right)}{2} - \sin^{-1}\frac{3\left(-\frac{1}{3}\right)}{2}\right]$$

$$= \frac{1}{3}\left[\sin^{-1}\frac{1}{2} - \sin^{-1}\left(\frac{-1}{2}\right)\right]$$

$$= \frac{1}{3}\left[\frac{\pi}{6} + \frac{\pi}{6}\right]$$

$$= \frac{\pi}{9}$$

EXAMPLE 3

Find $\int \frac{x-6}{x^2+4}\,dx$.

$$\int \frac{x-6}{x^2+4}\,dx = \int \frac{x}{x^2+4}\,dx - \int \frac{6}{x^2+4}\,dx$$

Split into 2 fractions.

$$= \frac{1}{2}\int \frac{2x}{x^2+4}\,dx - 3\int \frac{2}{4+x^2}\,dx$$

$$= \frac{1}{2}\ln(x^2+4) - 3\tan^{-1}\frac{x}{2} + c$$

EXAMPLE 4

Find the area bounded by the curve $y = \frac{2}{\sqrt{4-x^2}}$, the x-axis and the lines $x = \sqrt{3}$ and $x = -1$.

$$\text{Area} = \int_{-1}^{\sqrt{3}} \frac{2}{\sqrt{4-x^2}}\,dx$$

$$= 2\left[\sin^{-1}\frac{x}{2}\right]_{-1}^{\sqrt{3}}$$

$$= 2\left[\sin^{-1}\frac{\sqrt{3}}{2} - \sin^{-1}\frac{-1}{2}\right]$$

Always express answers in exact value where possible.

$$= 2\left[\frac{\pi}{3} - \left(-\frac{\pi}{6}\right)\right]$$

$$= \pi$$

$\therefore$ the area is π units^2.

→ PRACTICE

1 Find:

i $\displaystyle\int \frac{1}{\sqrt{25-x^2}}\,dx$

ii $\displaystyle\int \frac{1}{9+16x^2}\,dx$

2 Evaluate:

i $\displaystyle\int_{\sqrt{\frac{2}{3}}}^{\sqrt{6}} \frac{3}{4+2x^2}\,dx$

ii $\displaystyle\int_{\frac{1}{4}}^{\frac{1}{2}} \frac{1}{\sqrt{1-4x^2}}\,dx$

3 i Find $\displaystyle\int \frac{5x+1}{x^2+9}\,dx.$

ii Evaluate $\displaystyle\int_1^{\sqrt{3}} \frac{x+1}{x^2+3}\,dx.$

4 Find the area bounded by the curve:

i $y=\dfrac{\sqrt{3}}{3+x^2}$, the x-axis and the lines $x=-\sqrt{3}$ and $x=0$

ii $y=\dfrac{1}{\sqrt{4-25x^2}}$, the x-axis and the lines $x=\dfrac{2}{5}$ and $x=\dfrac{1}{5}$

Answers ➲ pp. 172–173

HSC EXAM-TYPE QUESTIONS

Total marks: 35

PART A

(Suggested time: 10 minutes)

(1 mark each)

1 If $y = \dfrac{2}{e^{3x}}$, which of these is an expression for $\dfrac{dx}{dy}$?

A $\dfrac{-e^{3x}}{6}$ B $\dfrac{-e^{3y}}{6}$ C $\dfrac{-6}{e^{3x}}$ D $\dfrac{-6}{e^{3y}}$

2 Which expression is equal to $\int \sin^2 \dfrac{\theta}{2}\, d\theta$?

A $\dfrac{1}{2}\left(\theta - \dfrac{1}{4}\sin\dfrac{\theta}{2}\right) + c$ B $\dfrac{1}{2}(\theta - 4\sin\theta) + c$

C $\dfrac{1}{2}\left(\theta - \dfrac{1}{2}\sin\dfrac{\theta}{2}\right) + c$ D $\dfrac{1}{2}(\theta - \sin\theta) + c$

3 Which integral is obtained when the substitution $u = 2 - x$ is applied to $\int_0^2 x\sqrt{2-x}\, dx$?

A $\int_0^2 u\sqrt{2-u}\, du$ B $\int_0^2 (2-u)\sqrt{u}\, du$

C $\int_0^2 (u-2)\sqrt{u}\, du$ D $\int_0^2 u\sqrt{u-2}\, du$

4 What is the derivative of $\cos^{-1} 2x$?

A $\dfrac{-2}{\sqrt{1-4x^2}}$ B $\dfrac{-2}{\sqrt{1-2x^2}}$

C $\dfrac{-2}{\sqrt{4-x^2}}$ D $\dfrac{-1}{\sqrt{1-4x^2}}$

5 What is the value of k such that $\int_0^k \dfrac{\sqrt{3}}{3+x^2}\, dx = \dfrac{\pi}{3}$?

A $\dfrac{1}{\sqrt{3}}$ B $\sqrt{3}$ C $2\sqrt{3}$ D 3

PART B

(Suggested time: 45 minutes)

Show all working.

6 If $y = (1 + x^2) \tan^{-1} x$, find $\dfrac{dy}{dx}$ when $x = -1$. (3 marks)

7 Given $0 < x < \pi$, differentiate $y = \sin^{-1} (\cos x)$ with respect to x. (2 marks)

8 Find $\int x^2\sqrt{x^3 - 1}\, dx$, by using the substitution $u = x^3 - 1$. (2 marks)

9 a Draw a sketch of the curve $y = \cos^2 2x$, using the domain $[0, \dfrac{\pi}{2}]$. (1 mark)

b Hence, find the area between the curve $y = \cos^2 2x$, the x-axis and the lines $x = 0$ and $x = \dfrac{\pi}{2}$. (2 marks)

10 a Use the result $\dfrac{dy}{dx} = \dfrac{1}{\frac{dx}{dy}}$ to find an expression for $\dfrac{dy}{dx}$ in terms of x, given $y = \sin^{-1} 2x$. (3 marks)

b Hence, or otherwise, show that the gradient of the tangent to $y = \sin^{-1} 2x$ at $x = \dfrac{1}{4}$ is $\dfrac{4\sqrt{3}}{3}$. (2 marks)

11 Evaluate $\int_0^{\frac{\pi}{6}} \cos 2x \cos x\, dx$. (3 marks)

12 Find the equation of the normal to the curve $y = \dfrac{1}{4}\tan^{-1}\dfrac{x}{2}$ at $x = 2$. (3 marks)

13 Find $\int_1^{\sqrt{3}} \dfrac{x}{\sqrt{4-x^2}}\, dx$, by using the substitution $x = 2 \sin u$. (3 marks)

14 If $f(x) = \tan^{-1} x + \tan^{-1} \dfrac{1}{x}$ for $x \neq 0$, show that $f(x) = \dfrac{\pi}{2}$, for all $x > 0$. (3 marks)

15 Find the area between the curve $y = \dfrac{1}{\sqrt{1-16x^2}}$, the x-axis, $x = 0$ and $x = \dfrac{1}{4}$. (3 marks)

ANSWERS

PRACTICE

1 Integration involving products of trigonometric functions ➲ p. 160

1 i $\int 2\cos 4x \cos x \, dx$

$= \int (\cos 5x + \cos 3x) \, dx$

$= \mathbf{\frac{1}{5}\sin 5x + \frac{1}{3}\sin 3x + c}$

ii $\int \cos 3x \sin 2x \, dx$

$= \frac{1}{2}\int 2\cos 3x \sin 2x \, dx$

$= \frac{1}{2}\int (\sin 5x - \sin x) \, dx$

$= \mathbf{-\frac{1}{10}\cos 5x - \frac{1}{2}\cos x + c}$

2 i $\int_0^{\frac{\pi}{6}} \cos\frac{3x}{2}\cos\frac{x}{2} \, dx$

$= \frac{1}{2}\int (\cos 2x + \cos x) dx$

$= \frac{1}{2}\left[\frac{1}{2}\sin 2x + \sin x\right]_0^{\frac{\pi}{6}}$

$= \frac{1}{2}\left[\frac{1}{2}\sin 2\left(\frac{\pi}{6}\right) + \sin\frac{\pi}{6} - 0\right]$

$= \frac{1}{2}\left[\frac{1}{2}\left(\frac{\sqrt{3}}{2}\right) + \frac{1}{2} - 0\right]$

$= \frac{1}{2}\left[\frac{\sqrt{3}}{4} + \frac{1}{2}\right]$

$= \mathbf{\frac{\sqrt{3}+2}{8}}$

ii $\int_0^{\frac{\pi}{2}} 4\sin x \cos\frac{x}{3} \, dx$

$= 2\int_0^{\frac{\pi}{2}} 2\sin x \cos\frac{x}{3} \, dx$

$= 2\int_0^{\frac{\pi}{2}} \left(\sin\frac{4x}{3} + \sin\frac{2x}{3}\right) dx$

$= 2\left[-\frac{3}{4}\cos\frac{4x}{3} - \frac{3}{2}\cos\frac{2x}{3}\right]_0^{\frac{\pi}{2}}$

$= -\frac{3}{2}\left[\cos\frac{4x}{3} + 2\cos\frac{2x}{3}\right]_0^{\frac{\pi}{2}}$

$= -\frac{3}{2}\left[\cos\frac{2\pi}{3} + 2\cos\frac{\pi}{3} - (\cos 0 + 2\cos 0)\right]$

$= -\frac{3}{2}\left[-\frac{1}{2} + 1 - (1+2)\right]$

$= -\frac{3}{2}\left[-\frac{5}{2}\right]$

$= \mathbf{\frac{15}{4}}$

2 Integration using substitution ➲ p. 161

1 i As $u = 9 - x^2 \therefore \frac{du}{dx} = -2x \therefore dx = \frac{du}{-2x}$

$\int 6x\sqrt{9-x^2} \, dx = \int 6x\sqrt{u} \frac{du}{-2x}$

$= -3\int u^{\frac{1}{2}} \, du$

$= -3\left[\frac{2u^{\frac{3}{2}}}{3}\right] + c$

$= \mathbf{-2\sqrt{(9-x^2)^3} + c}$

ii As $u = x - 2 \therefore \frac{du}{dx} = 1 \therefore dx = du$ Also, $x = u + 2$.

$\int x\sqrt{x-2} \, dx = \int (u+2)\sqrt{u} \, du$

$= \int (u^{\frac{3}{2}} + 2u^{\frac{1}{2}}) \, du$

$= \frac{2u^{\frac{5}{2}}}{5} + \frac{4u^{\frac{3}{2}}}{3} + c$

$= \mathbf{\frac{2\sqrt{(x-2)^5}}{5} + \frac{4\sqrt{(x-2)^3}}{3} + c}$

2 i As $u = \cos x \therefore \frac{du}{dx} = -\sin x \therefore dx = \frac{du}{-\sin x}$

$\int \frac{\sin x}{\cos^4 x} \, dx = \int \frac{\sin x}{u^4} \cdot \frac{du}{-\sin x}$

$= -\int u^{-4} \, du$

$= \frac{u^{-3}}{3} + c$

$= \mathbf{\frac{1}{3\cos^3 x} + c}$

ii As $u = \ln x \therefore \frac{du}{dx} = \frac{1}{x} \therefore dx = x \, du$

$\int \frac{1}{x\ln x} \, dx = \int \frac{1}{x.u} . x \, du$

$= \int \frac{1}{u} \, du$

$= \ln u + c$

$= \mathbf{\ln [\ln x] + c}$

3 i As $u = 1 - 2x^3 \therefore \frac{du}{dx} = -6x^2 \therefore dx = \frac{du}{-6x^2}$.

When $x = 0$, $u = 1$ and when $x = -1$, $u = 3$.

$\int_{-1}^{0} x^2\sqrt{1-2x^3} \, dx = \int_{u=3}^{u=1} x^2\sqrt{u} . \frac{du}{-6x^2}$

$= \frac{1}{6}\int_1^3 u^{\frac{1}{2}} \, du$

$= \frac{1}{6}\left[\frac{2}{3}u^{\frac{3}{2}}\right]_1^3$

$= \frac{1}{9}\left[3^{\frac{3}{2}} - 1\right]$

$= \mathbf{\frac{3\sqrt{3}-1}{9}}$

ii As $u = 4 - x^2 \therefore \dfrac{du}{dx} = -2x \therefore dx = \dfrac{du}{-2x}$.

When $x = 2$, $u = 0$ and when $x = 0$, $u = 4$.

$$\int_0^2 \frac{8x}{\sqrt{4-x^2}}\,dx = \int_{u=4}^{u=0} \frac{8x}{\sqrt{u}}.\frac{du}{-2x}$$
$$= 4\int_0^4 u^{-\frac{1}{2}}\,du$$
$$= 4\left[2u^{\frac{1}{2}}\right]_0^4$$
$$= 8(2 - 0)$$
$$= \mathbf{16}$$

4 i As $x = 2\sin u \therefore \dfrac{dx}{du} = 2\cos u \therefore dx = 2\cos u\,du$. Also, $\dfrac{x}{2} = \sin u$.

$$\int \frac{x}{\sqrt{4-x^2}}\,dx = \int \frac{2\sin u}{\sqrt{4-4\sin^2 u}}.2\cos u\,du$$
$$= \int \frac{2\sin u}{\sqrt{4(1-\sin^2 u)}}.2\cos u\,du$$
$$= \int \frac{2\sin u}{2\cos u}.2\cos u\,du$$
$$= 2\int \sin u\,du$$
$$= -2\cos u + c$$
$$= -2\left(\frac{\sqrt{4-x^2}}{2}\right) + c$$
$$= \mathbf{-\sqrt{4-x^2} + c}$$

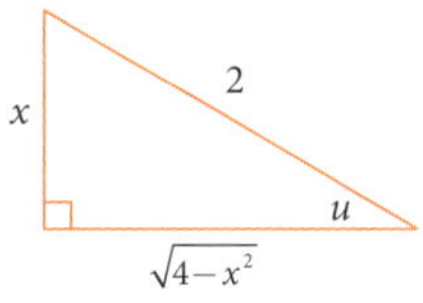

ii As $x = \cos\theta \therefore \dfrac{dx}{d\theta} = -\sin\theta \therefore dx = -\sin\theta\,d\theta$.

$$\int \frac{dx}{x^2\sqrt{1-x^2}} = \int \frac{1}{\cos^2\theta\sqrt{1-\cos^2\theta}}.-\sin\theta\,d\theta$$
$$= \int \frac{1}{\cos^2\theta.\sin\theta}.-\sin\theta\,d\theta$$
$$= -\int \sec^2\theta\,d\theta$$
$$= -\tan\theta + c$$
$$= -\frac{\mathbf{\sqrt{1-x^2}}}{\mathbf{x}} + \mathbf{c}$$

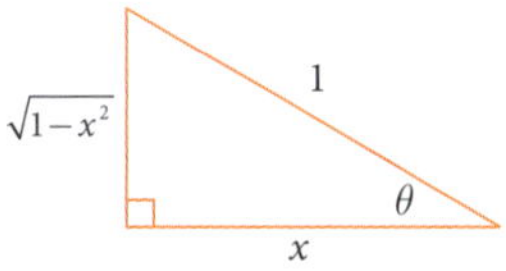

3 Integration of $\sin^2 nx$ and $\cos^2 nx$ ➲ p. 162

1 Find:

i $$\int_{\frac{\pi}{4}}^{\frac{\pi}{2}} \cos^2 x\,dx = \int_{\frac{\pi}{4}}^{\frac{\pi}{2}} \frac{1}{2}[1 + \cos 2x]\,dx$$
$$= \frac{1}{2}\left[x + \frac{1}{2}\sin 2x\right]_{\frac{\pi}{4}}^{\frac{\pi}{2}}$$
$$= \frac{1}{2}\left[\frac{\pi}{2} + \frac{1}{2}\sin 2\left(\frac{\pi}{2}\right) - \left(\frac{\pi}{4} + \frac{1}{2}\sin 2\left(\frac{\pi}{4}\right)\right)\right]$$
$$= \frac{1}{2}\left[\frac{\pi}{2} + 0 - \frac{\pi}{4} - \frac{1}{2}\right]$$
$$= \frac{1}{2}\left[\frac{\pi}{4} - \frac{1}{2}\right]$$
$$= \mathbf{\frac{\pi - 2}{8}}$$

ii $$\int_{\frac{\pi}{6}}^{\frac{2\pi}{3}} 4\sin^2 x\,dx = 4\int_{\frac{\pi}{6}}^{\frac{2\pi}{3}} \frac{1}{2}[1 - \cos 2x]\,dx$$
$$= 2\left[x - \frac{1}{2}\sin 2x\right]_{\frac{\pi}{6}}^{\frac{2\pi}{3}}$$
$$= 2\left[\frac{2\pi}{3} - \frac{1}{2}\sin 2\left(\frac{2\pi}{3}\right) - \left(\frac{\pi}{6} - \frac{1}{2}\sin 2\left(\frac{\pi}{6}\right)\right)\right]$$
$$= 2\left[\frac{2\pi}{3} + \frac{\sqrt{3}}{4} - \frac{\pi}{6} + \frac{\sqrt{3}}{4}\right]$$
$$= 2\left[\frac{\pi}{2} + \frac{\sqrt{3}}{2}\right]$$
$$= \mathbf{\pi + \sqrt{3}}$$

2 i $$\int \sin^2\frac{\theta}{2}\,d\theta = \int \frac{1}{2}[1 - \cos\theta]\,d\theta$$
$$= \mathbf{\frac{1}{2}(\theta - \sin\theta) + c}$$

ii $\cos\alpha\,(\cos\alpha - 2) = \cos^2\alpha - 2\cos\alpha$

$$\int (\cos^2\alpha - 2\cos\alpha)\,d\alpha = \int \left(\frac{1}{2}[1 + \cos 2\alpha] - 2\cos\alpha\right)d\alpha$$
$$= \frac{1}{2}\left[\alpha + \frac{1}{2}\sin 2\alpha\right] - 2\sin\alpha + c$$
$$= \mathbf{\frac{\alpha}{2} + \frac{1}{4}\sin 2\alpha - 2\sin\alpha + c}$$

3 i $x = 3\sin\theta$ then $\dfrac{dx}{d\theta} = 3\cos\theta$ and hence $dx = 3\cos\theta\,d\theta$.

Also, $\sin\theta = \dfrac{x}{3}$, then $\theta = \sin^{-1}\dfrac{x}{3}$. Also, by Pythagoras,

$\cos\theta = \dfrac{\sqrt{9-x^2}}{3}$.

$$\int \sqrt{9-x^2}\,dx = \int \sqrt{9-9\sin^2\theta}.\,3\cos\theta\,d\theta$$
$$= \int 3\sqrt{1-\sin^2\theta}.\,3\cos\theta\,d\theta$$
$$= 9\int \cos^2\theta\,d\theta$$
$$= \frac{9}{2}\int (1 + \cos 2\theta)\,d\theta$$
$$= \frac{9}{2}\left[\theta + \frac{1}{2}\sin 2\theta\right] + c$$
$$= \frac{9}{2}[\theta + \sin\theta\cos\theta] + c \quad \text{(using } \sin 2\theta = 2\sin\theta\cos\theta\text{)}$$
$$= \frac{9}{2}\left[\sin^{-1}\frac{x}{3} + \frac{x}{3}.\frac{\sqrt{9-x^2}}{3}\right] + c$$
$$= \mathbf{\frac{9}{2}\left[\sin^{-1}\frac{x}{3} + \frac{x\sqrt{9-x^2}}{9}\right] + c}$$

ii As $x = \sin^2 u \therefore \dfrac{dx}{du} = 2\sin u\cos u \therefore dx = 2\sin u\cos u\,du$.

If $\sin^2 u = \dfrac{1}{2}$, then $\sin u = \dfrac{1}{\sqrt{2}}$, and hence, $u = \dfrac{\pi}{4}$.

If $\sin^2 u = 0$, then $\sin u = 0$, and hence, $u = 0$.

$$\int_0^{\frac{1}{2}} \sqrt{\frac{x}{1-x}}\, dx = \int_0^{\frac{\pi}{4}} \sqrt{\frac{\sin^2 u}{1-\sin^2 u}}.2 \sin u \cos u\, du$$
$$= 2\int_0^{\frac{\pi}{4}} \frac{\sin u}{\cos u}. \sin u \cos u\, du$$
$$= 2\int_0^{\frac{\pi}{4}} \sin^2 u\, du$$
$$= 2\int_0^{\frac{\pi}{4}} \frac{1}{2}(1-\cos 2u)\, du$$
$$= \int_0^{\frac{\pi}{4}} (1-\cos 2u)\, du$$
$$= \left[u - \frac{1}{2}\sin 2u\right]_0^{\frac{\pi}{4}}$$
$$= \frac{\pi}{4} - \frac{1}{2}\sin 2\left(\frac{\pi}{4}\right) - 0$$
$$= \mathbf{\frac{\pi}{4} - \frac{1}{2}}$$

4 i Area $= \int_{\frac{\pi}{3}}^{\frac{2\pi}{3}} (1+\cos^2 x)\, dx$

$$= \int_{\frac{\pi}{3}}^{\frac{2\pi}{3}} \left(1 + \frac{1}{2}[1+\cos 2x]\right) dx$$
$$= \int_{\frac{\pi}{3}}^{\frac{2\pi}{3}} \left(\frac{3}{2} + \frac{1}{2}\cos 2x\right) dx$$
$$= \frac{1}{2}\left[3x + \frac{1}{2}\sin 2x\right]_{\frac{\pi}{3}}^{\frac{2\pi}{3}}$$
$$= \frac{1}{4}[6x + \sin 2x]_{\frac{\pi}{3}}^{\frac{2\pi}{3}}$$
$$= \frac{1}{4}\left[6\left(\frac{2\pi}{3}\right) + \sin 2\left(\frac{2\pi}{3}\right) - \left[6\left(\frac{\pi}{3}\right) + \sin 2\left(\frac{\pi}{3}\right)\right]\right]$$
$$= \frac{1}{4}\left[4\pi - \frac{\sqrt{3}}{2} - 2\pi - \frac{\sqrt{3}}{2}\right]$$
$$= \frac{1}{4}\left[2\pi - \sqrt{3}\right]$$
$$= \frac{2\pi - \sqrt{3}}{4}$$

$\therefore$ **the area is $\frac{2\pi - \sqrt{3}}{4}$ units².**

ii
$$2\sin^2 x = \sin x$$
$$2\sin^2 x - \sin x = 0$$
$$\sin x(2\sin x - 1) = 0$$
$$\sin x = 0, \frac{1}{2}$$
$$x = 0, \frac{\pi}{6}$$

Area $= \int_0^{\frac{\pi}{6}} (\sin x - 2\sin^2 x)\, dx$

$$= \int_0^{\frac{\pi}{6}} \left(\sin x - 2\times\frac{1}{2}(1-\cos 2x)\right) dx$$
$$= \int_0^{\frac{\pi}{6}} (\sin x - 1 + \cos 2x)\, dx$$
$$= \left[-\cos x - x + \frac{1}{2}\sin 2x\right]_0^{\frac{\pi}{6}}$$
$$= \left[-\cos\frac{\pi}{6} - \frac{\pi}{6} + \frac{1}{2}\sin 2\left(\frac{\pi}{6}\right)\right] - \left(-\cos 0 - 0 + 0\right)$$
$$= -\frac{\sqrt{3}}{2} - \frac{\pi}{6} + \frac{\sqrt{3}}{4} + 1$$
$$= 1 - \frac{\sqrt{3}}{4} - \frac{\pi}{6}$$
$$= \frac{12 - 3\sqrt{3} - 2\pi}{12}$$

$\therefore$ **the area is $\frac{12-3\sqrt{3}-2\pi}{12}$ units².**

4 Using the result $\frac{dy}{dx} = \frac{1}{\frac{dx}{dy}}$ ➲ p. 163

1 i $y = e^{3x}$

First, $\frac{dy}{dx} = 3e^{3x}$

Also, $\ln y = \ln e^{3x}$
$$3x = \ln y$$
$$x = \frac{1}{3}\ln y$$
$$\frac{dx}{dy} = \frac{1}{3}.\frac{1}{y}$$
$$= \frac{1}{3}.\frac{1}{e^{3x}}$$
$$= \frac{1}{3e^{3x}}$$
$$\therefore \mathbf{\frac{dy}{dx} = \frac{1}{\frac{dx}{dy}}}$$

ii
$$y = (3x-1)^2$$
$$\frac{dy}{dx} = 2(3x-1).3$$
$$= 6(3x-1)$$

Also, $\sqrt{y} = 3x - 1$
$$3x = 1 + \sqrt{y}$$
$$x = \frac{1}{3} + \frac{\sqrt{y}}{3}$$
$$\frac{dx}{dy} = \frac{1}{3}.\frac{1}{2}y^{-\frac{1}{2}}$$
$$= \frac{1}{6\sqrt{y}}$$
$$= \frac{1}{6(3x-1)}$$

As, $6(3x-1) \times \frac{1}{6(3x-1)} = 1$, then $\frac{dy}{dx} \times \frac{dx}{dy} = 1$.

2 i $x = 4y^3$

First, $\frac{dx}{dy} = 8y^2$

$\therefore \boldsymbol{\frac{dy}{dx} = \frac{1}{8y^2}, y \neq 0}$ **(using** $\boldsymbol{\frac{dy}{dx} = \frac{1}{\frac{dx}{dy}}}$**)**

ii $x = \sin\frac{y}{2}$

First, $\frac{dx}{dy} = \frac{1}{2}\cos\frac{y}{2}$.

$\therefore \frac{dy}{dx} = \frac{2}{\cos\frac{y}{2}}, y \neq n\pi$, for integer n (using $\frac{dy}{dx} = \frac{1}{\frac{dx}{dy}}$)

$$\therefore \frac{dy}{dx}\left(\frac{\pi}{3}\right) = \frac{2}{\cos\frac{\pi}{6}} = \frac{2}{\frac{\sqrt{3}}{2}} = \frac{4}{\sqrt{3}} = \boldsymbol{\frac{4\sqrt{3}}{3}}$$

3 i $y = \cos^{-1} x$

$\cos y = x$

$$\frac{dx}{dy} = -\sin y = -\sqrt{1-x^2}$$

$$\frac{dy}{dx} = \frac{-1}{\sqrt{1-x^2}}$$

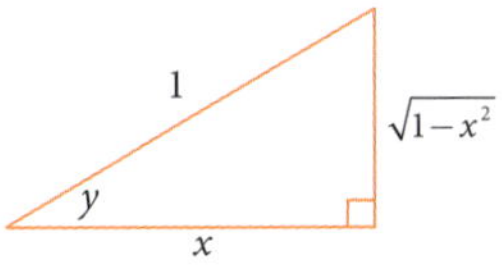

ii $y = \tan^{-1} 3x$

$\tan y = 3x$

$x = \frac{1}{3}\tan y$

$$\frac{dx}{dy} = \frac{1}{3}\sec^2 y = \frac{1}{3}(1 + \tan^2 y) = \frac{1}{3}(1 + 9x^2)$$

$$\boldsymbol{\frac{dy}{dx} = \frac{3}{1+9x^2}}$$

4 i First, $x(3) = (3)^2 - 2(3) + 4$

$= 7 \therefore$ point is $(7, 3)$

Also, $\frac{dx}{dy} = 2y - 2$

$\therefore \frac{dy}{dx} = \frac{1}{2y-2}, y \neq 1$ (using $\frac{dy}{dx} = \frac{1}{\frac{dx}{dy}}$)

$$\therefore \frac{dy}{dx}(3) = \frac{1}{2(3)-2} = \frac{1}{4}$$

Using $(7, 3)$ and $m = \frac{1}{4}$:

$\therefore y - 3 = \frac{1}{4}(x - 7)$

$4y - 12 = x - 7$

$\boldsymbol{x - 4y + 5 = 0}$

ii First, $2 = \frac{4}{y-1}$

$2y - 2 = 4$

$y = 3$

$\therefore$ the point is $(2, 3)$.

Also, as $x = 4(y-1)^{-1}$

$$\frac{dx}{dy} = -4(y-1)^{-2} = \frac{-4}{(y-1)^2}$$

$\therefore \frac{dy}{dx} = \frac{-(y-1)^2}{4}$ (using $\frac{dy}{dx} = \frac{1}{\frac{dx}{dy}}$)

$$\therefore \frac{dy}{dx}(3) = \frac{-(3-1)^2}{4} = -1$$

Using $(2, 3)$ and gradient of normal $m = 1$:

$\therefore y - 3 = 1(x - 2)$

$y - 3 = x - 2$

$\boldsymbol{x - y + 1 = 0}$

5 Differentiation of inverse trigonometric functions ⊃ p. 164–165

1 i $y = \tan^{-1} 3x$

$$\frac{dy}{dx} = \frac{1}{1+(3x)^2}.3 = \boldsymbol{\frac{3}{1+9x^2}}$$

ii $y = \sin^{-1}\frac{1}{x}$

$$\frac{dy}{dx} = \frac{1}{\sqrt{1-\left(\frac{1}{x}\right)^2}}.-x^{-2} = \frac{1}{\sqrt{1-\frac{1}{x^2}}}.\frac{-1}{x^2} = \frac{-1}{\sqrt{x^4-x^2}} = \boldsymbol{\frac{-1}{x\sqrt{x^2-1}}}$$

2 i $$\frac{d}{dx}\left[x^3\sin^{-1}2x\right] = 3x^2.\sin^{-1}2x + x^3.\frac{1}{\sqrt{1-(2x)^2}}.2 = \boldsymbol{3x^2\sin^{-1}2x + \frac{2x^3}{\sqrt{1-4x^2}}}$$

ii $$\frac{d}{dx}\left[\frac{\cos^{-1}x}{x}\right] = \frac{x.\frac{-1}{\sqrt{1-x^2}} - 1.\cos^{-1}x}{x^2} = \boldsymbol{\frac{-x-\sqrt{1-x^2}\cos^{-1}x}{x^2\sqrt{1-x^2}}}$$

3 i $\frac{dy}{dx} = 4.\frac{1}{\sqrt{1-(2x)^2}}.2$

$= \frac{8}{\sqrt{1-4x^2}}$

$= 8(1-4x^2)^{-\frac{1}{2}}$

$\frac{d^2y}{dx^2} = -4(1-4x^2)^{-\frac{3}{2}}.-8x$

$= \frac{\mathbf{32x}}{\sqrt{(\mathbf{1-4x^2})^3}}$

ii $\frac{dy}{dx} = \frac{1}{1+\left(\frac{x+1}{2}\right)^2}.\frac{1}{2}$

$= \frac{1}{1+\frac{(x+1)^2}{4}}.\frac{1}{2}$

$= \frac{1}{2+\frac{(x+1)^2}{2}}$

$= \frac{2}{4+(x+1)^2}$

$= \frac{2}{x^2+2x+5}$

$= 2(x^2 + 2x + 5)^{-1}$

$\frac{d^2y}{dx^2} = -2(x^2 + 2x + 5)^{-2}.(2x + 2)$

$= \frac{\mathbf{-4(x+1)}}{\mathbf{(x^2+2x+5)^2}}$

4 i $\frac{d}{dx}\left[\cos^{-1}(\sin x)\right] = \frac{-1}{\sqrt{1-(\sin x)^2}}.\cos x$

$= \frac{-1}{\sqrt{1-\sin^2 x}}.\cos x$

$= \frac{-\cos x}{\sqrt{\cos^2 x}}$

$= \frac{-\cos x}{|\cos^2 x|}$

= –1 when cos *x* > 0 and 1 when cos *x* < 0

ii $\frac{d}{dx}\left[\sin\left(\cos^{-1}\frac{x}{2}\right)\right] = \frac{-1}{\sqrt{1-\left(\frac{x}{2}\right)^2}}.\frac{1}{2}\cos\left(\cos^{-1}\frac{x}{2}\right)$

$= \frac{-1}{2\sqrt{1-\frac{x^2}{4}}}.\frac{x}{2}$

$= \frac{\mathbf{-x}}{\mathbf{2\sqrt{4-x^2}}}$

5 i $\frac{dy}{dx} = \frac{1}{\sqrt{1-(\sqrt{1-x^2})^2}}.\frac{1}{2}(1-x^2)^{-\frac{1}{2}}.-2x$

$= \frac{1}{\sqrt{1-1+x^2}}.\frac{-x}{\sqrt{1-x^2}}$

$= \frac{1}{x}.\frac{-x}{\sqrt{1-x^2}}$

$= \frac{-1}{\sqrt{1-x^2}}$

$\frac{dy}{dx}(0) = \frac{-1}{\sqrt{1-0^2}}$

$= -1$

∴ the gradient of the normal is 1.

ii $\frac{dy}{dx} = \frac{1}{1+\left(\frac{x-1}{2}\right)^2}.\frac{1}{2}$

$\frac{dy}{dx}(3) = \frac{1}{1+\left(\frac{3-1}{2}\right)^2}.\frac{1}{2}$

$= \frac{1}{4}$

Now, $y(3) = \tan^{-1}\frac{3-1}{2},$

$= \frac{\pi}{4}$

Using $\left(3, \frac{\pi}{4}\right)$ and $m = \frac{1}{4}$:

$y - \frac{\pi}{4} = \frac{1}{4}(x - 3)$

$4y - \pi = x - 3$

$\therefore \mathbf{x - 4y + \pi - 3 = 0}$

6 Integration resulting in inverse trigonometric functions ➲ p. 166

1 i $\int \frac{\mathbf{1}}{\sqrt{\mathbf{25-x^2}}}\,dx = \mathbf{\sin^{-1}\frac{x}{5} + c}$

ii $\int \frac{1}{9+16x^2}\,dx = \int \frac{1}{16\left(\frac{9}{16}+x^2\right)}\,dx$

$= \frac{1}{16}\int \frac{1}{\frac{9}{16}+x^2}\,dx$

$= \frac{1}{16} \times \frac{4}{3}\tan^{-1}\frac{4x}{3} + c$

$= \mathbf{\frac{1}{12}\tan^{-1}\frac{4x}{3} + c}$

2 i $\int_{\sqrt{\frac{2}{3}}}^{\sqrt{6}} \frac{3}{4+2x^2}\,dx = \frac{3}{2}\int_{\sqrt{\frac{2}{3}}}^{\sqrt{6}} \frac{1}{2+x^2}\,dx$

$= \frac{3}{2} \times \frac{1}{\sqrt{2}}\left[\tan^{-1}\frac{x}{\sqrt{2}}\right]_{\sqrt{\frac{2}{3}}}^{\sqrt{6}}$

$= \frac{3}{2\sqrt{2}}\left[\tan^{-1}\frac{\sqrt{6}}{\sqrt{2}} - \tan^{-1}\left(\sqrt{\frac{2}{3}} \div \sqrt{2}\right)\right]$

$= \frac{3\sqrt{2}}{4}\left[\tan^{-1}\sqrt{3} - \tan^{-1}\frac{1}{\sqrt{3}}\right]$

$= \frac{3\sqrt{2}}{4}\left[\frac{\pi}{3} - \frac{\pi}{6}\right]$

$= \mathbf{\frac{\pi\sqrt{2}}{8}}$

ii $\int_{\frac{1}{4}}^{\frac{1}{2}} \frac{1}{\sqrt{1-4x^2}}\,dx = \int_{\frac{1}{4}}^{\frac{1}{2}} \frac{1}{\sqrt{4\left(\frac{1}{4}-x^2\right)}}\,dx$

$= \frac{1}{2}\int_{\frac{1}{4}}^{\frac{1}{2}} \frac{1}{\sqrt{\left(\frac{1}{4}-x^2\right)}}\,dx$

$= \frac{1}{2}\left[\sin^{-1} 2x\right]_{\frac{1}{4}}^{\frac{1}{2}}$

$= \frac{1}{2}\left[\sin^{-1} 2\left(\frac{1}{2}\right) - \sin^{-1} 2\left(\frac{1}{4}\right)\right]$

$= \frac{1}{2}\left[\sin^{-1} 1 - \sin^{-1}\left(\frac{1}{2}\right)\right]$

$= \frac{1}{2}\left[\frac{\pi}{2} - \frac{\pi}{6}\right]$

$= \boldsymbol{\frac{\pi}{6}}$

3 i $\int \frac{5x+1}{x^2+9}\,dx$

$= \int \frac{5x}{x^2+9}\,dx + \int \frac{1}{9+x^2}\,dx$

$= \frac{5}{2}\int \frac{2x}{x^2+9}\,dx + \frac{1}{3}\int \frac{3}{9+x^2}\,dx$

$= \boldsymbol{\frac{5}{2}\ln(x^2+9) + \frac{1}{3}\tan^{-1}\frac{x}{3} + c}$

ii $\int_1^{\sqrt{3}} \frac{x+1}{x^2+3}\,dx = \int_1^{\sqrt{3}} \left(\frac{x}{x^2+3} + \frac{1}{x^2+3}\right)dx$

$= \left[\frac{1}{2}\ln(x^2+3) + \frac{1}{\sqrt{3}}\tan^{-1}\frac{x}{\sqrt{3}}\right]_1^{\sqrt{3}}$

$= \frac{1}{2}\ln[(\sqrt{3})^2+3] + \frac{1}{\sqrt{3}}\tan^{-1}\frac{\sqrt{3}}{\sqrt{3}}$

$- \left(\frac{1}{2}\ln[1^2+3] + \frac{1}{\sqrt{3}}\tan^{-1}\frac{1}{\sqrt{3}}\right)$

$= \frac{1}{2}\ln 6 + \frac{1}{\sqrt{3}} \times \frac{\pi}{4} - \frac{1}{2}\ln 4 - \frac{1}{\sqrt{3}} \times \frac{\pi}{6}$

$= \frac{1}{2}\ln 6 - \frac{1}{2}\ln 4 + \frac{\pi}{4\sqrt{3}} - \frac{\pi}{6\sqrt{3}}$

$= \frac{1}{2}\ln\frac{3}{2} + \frac{\pi}{12\sqrt{3}}$

$= \boldsymbol{\frac{1}{2}\ln\frac{3}{2} + \frac{\pi\sqrt{3}}{36}}$

4 i Area $= \sqrt{3}\left[\frac{1}{\sqrt{3}}\tan^{-1}\frac{x}{\sqrt{3}}\right]_{-\sqrt{3}}^{0}$

$= \left[\tan^{-1}\frac{x}{\sqrt{3}}\right]_{-\sqrt{3}}^{0}$

$= \left[\tan^{-1} 0 - \tan^{-1}\left(\frac{-\sqrt{3}}{\sqrt{3}}\right)\right]$

$= \left[0 - \left(\frac{-\pi}{4}\right)\right]$

$= \frac{\pi}{4}$

∴ the area is $\frac{\pi}{4}$ units².

ii Area $= \int_{\frac{1}{5}}^{\frac{2}{5}} \frac{1}{\sqrt{25\left(\frac{4}{25}-x^2\right)}}\,dx$

$= \frac{1}{5}\left[\sin^{-1}\frac{5x}{2}\right]_{\frac{1}{5}}^{\frac{2}{5}}$

$= \frac{1}{5}\left[\sin^{-1} 1 - \sin^{-1}\frac{1}{2}\right]$

$= \frac{1}{5}\left[\frac{\pi}{2} - \frac{\pi}{6}\right]$

$= \frac{\pi}{15}$

∴ the area is $\frac{\pi}{15}$ units².

HSC EXAM-TYPE QUESTIONS

Part A

1 A. $y = 2e^{-3x}$

$\frac{dy}{dx} = -6e^{-3x}$

$= \frac{-6}{e^{3x}}$

$\frac{dx}{dy} = \frac{-e^{3x}}{6}$

2 D. $\int \sin^2\frac{\theta}{2}\,d\theta = \int \frac{1}{2}[1-\cos\theta]\,d\theta$

$= \frac{1}{2}(\theta - \sin\theta) + c$

3 B. $u = 2 - x \;\therefore x = 2 - u$

$\frac{dx}{du} = -1$

$dx = -du$

Also, $x = 2$, $u = 0$ and $x = 0$, $u = 2$

$\int_0^2 x\sqrt{2-x}\,dx = \int_{u=2}^{u=0} (2-u)\sqrt{u} - du$

$= \int_0^2 \sqrt{u}(2-u)\,du$

4 A. $\frac{d}{dx}[\cos^{-1} 2x] = \frac{-1}{\sqrt{1-(2x)^2}}.2$

$= \frac{-2}{\sqrt{1-4x^2}}$

5 D. $\int_0^k \frac{\sqrt{3}}{3+x^2}\,dx = \frac{\pi}{3}$

$\sqrt{3}\int_0^k \frac{1}{3+x^2}\,dx = \frac{\pi}{3}$

$\sqrt{3}\left[\frac{1}{\sqrt{3}}\tan^{-1}\frac{x}{\sqrt{3}}\right]_0^k dx = \frac{\pi}{3}$

$\tan^{-1}\frac{k}{\sqrt{3}} - \tan^{-1} 0 = \frac{\pi}{3}$

$\tan^{-1}\frac{k}{\sqrt{3}} = \frac{\pi}{3}$

$\frac{k}{\sqrt{3}} = \tan\frac{\pi}{3}$

$\frac{k}{\sqrt{3}} = \sqrt{3}$

$k = 3$

Part B

6 $y = (1 + x^2)\tan^{-1} x$

$$\frac{dy}{dx} = 2x.\tan^{-1} x + (1 + x^2).\frac{1}{1+x^2} \quad ✓$$

$$= 2x\tan^{-1} x + 1 \quad ✓$$

$$\frac{dy}{dx}(-1) = 2(-1)\tan^{-1}(-1) + 1$$

$$= -2\left(-\frac{\pi}{4}\right) + 1$$

$$= \frac{\pi}{2} + 1$$

$$= \mathbf{\frac{\pi+2}{2}} \quad ✓$$

7 $y = \sin^{-1}(\cos x)$

$$\frac{dy}{dx} = \frac{1}{\sqrt{1-\cos^2 x}}.-\sin x \quad ✓$$

$$= \frac{1}{\sin x}.-\sin x$$

$$= \mathbf{-1} \quad ✓$$

8 $u = x^3 - 1$

$$\frac{du}{dx} = 3x^2$$

$$dx = \frac{du}{3x^2}$$

$$\int x^2\sqrt{x^3-1}\,dx = \int x^2\sqrt{u}\times\frac{du}{3x^2} \quad ✓$$

$$= \frac{1}{3}\int u^{\frac{1}{2}}\,du$$

$$= \frac{1}{3}\times\frac{2u^{\frac{3}{2}}}{3}$$

$$= \mathbf{\frac{2\sqrt{(x^3-1)^3}}{9} + c} \quad ✓$$

9 a

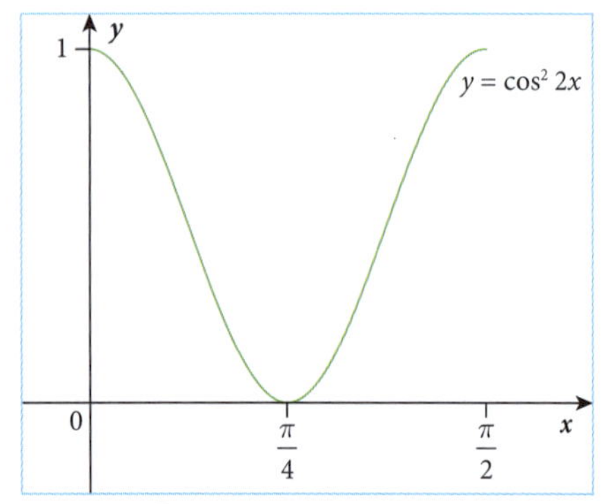

✓

b Area $= 2\int_0^{\frac{\pi}{4}} \cos^2 2x\,dx$

$$= 2\int_0^{\frac{\pi}{4}} \frac{1}{2}(1+\cos 4x)\,dx \quad ✓$$

$$= \left[x + \frac{1}{4}\sin 4x\right]_0^{\frac{\pi}{4}}$$

$$= \frac{\pi}{4} + \frac{1}{4}\sin\pi - \left(0 + \frac{1}{4}\sin 0\right)$$

$$= \mathbf{\frac{\pi}{4}}$$

∴ the area is $\frac{\pi}{4}$ unit². ✓

10 a $y = \sin^{-1} 2x$

$\sin y = 2x$ ✓

$$x = \frac{1}{2}\sin y$$

$$\frac{dx}{dy} = \frac{1}{2}\cos y \quad ✓$$

$$= \frac{1}{2}\times\frac{\sqrt{1-4x^2}}{1}$$

$$= \frac{\sqrt{1-4x^2}}{2}$$

$$\mathbf{\frac{dy}{dx} = \frac{2}{\sqrt{1-4x^2}}} \quad ✓$$

b $$\frac{dy}{dx}\left(\frac{1}{4}\right) = \frac{2}{\sqrt{1-4\left(\frac{1}{4}\right)^2}}$$

$$= \frac{2}{\sqrt{\frac{3}{4}}} \quad ✓$$

$$= 2 \div \frac{\sqrt{3}}{2}$$

$$= \frac{4}{\sqrt{3}}\times\frac{\sqrt{3}}{\sqrt{3}}$$

$$= \mathbf{\frac{4\sqrt{3}}{3}} \quad ✓$$

11 $$\int_0^{\frac{\pi}{6}} \cos 2x\cos x\,dx = \frac{1}{2}\int_0^{\frac{\pi}{6}} 2\cos 2x\cos x\,dx$$

$$= \frac{1}{2}\int_0^{\frac{\pi}{6}} (\cos 3x + \cos x)\,dx \quad ✓$$

$$= \frac{1}{2}\left[\frac{1}{3}\sin 3x + \sin x\right]_0^{\frac{\pi}{6}} \quad ✓$$

$$= \frac{1}{2}\left[\frac{1}{3}\sin\frac{\pi}{2} + \sin\frac{\pi}{6} - (\sin 0 + \sin 0)\right]$$

$$= \frac{1}{2}\left[\frac{1}{3} + \frac{1}{2}\right]$$

$$= \mathbf{\frac{5}{12}} \quad ✓$$

12 $$y(2) = \frac{1}{4}\tan^{-1}\frac{2}{2}$$

$$= \frac{1}{4}.\frac{\pi}{4}$$

$$= \frac{\pi}{16} \quad \therefore \text{point is} \left(2, \frac{\pi}{16}\right) \quad ✓$$

$$\frac{dy}{dx} = \frac{1}{4}.\frac{1}{1+\frac{x^2}{4}}.\frac{1}{2}$$

$$= \frac{1}{4+x^2}.\frac{1}{2}$$

$$= \frac{1}{2(4+x^2)} \quad ✓$$

$$\frac{dy}{dx}(2) = \frac{1}{2(4+2^2)}$$

$$= \frac{1}{16}$$

∴ gradient of normal = –16

Using $\left(2, \frac{\pi}{16}\right)$ and $m = -16$:

$$y - \frac{\pi}{16} = -16(x - 2)$$

$\mathbf{16x + y - 32 - \frac{\pi}{16} = 0}$ ✓

13 $x = 2\sin u$

$\frac{du}{dx} = 2\cos u$

$dx = 2\cos u\, du$ ✓

Also, as $x = \sqrt{3}$, then $\sin u = \frac{\sqrt{3}}{2} \therefore u = \frac{\pi}{3}$.

As $x = 1$, then $\sin u = \frac{1}{2} \therefore u = \frac{\pi}{6}$.

$$\int_1^{\sqrt{3}} \frac{x}{\sqrt{4-x^2}}\,dx = \int_{u=\frac{\pi}{6}}^{u=\frac{\pi}{3}} \frac{2\sin u}{\sqrt{4-4\sin^2 u}} \times 2\cos u\, du$$ ✓

$$= \int_{\frac{\pi}{6}}^{\frac{\pi}{3}} \frac{2\sin u}{2\sqrt{1-\sin^2 u}} \times 2\cos u\, du$$

$$= \int_{\frac{\pi}{6}}^{\frac{\pi}{3}} \frac{\sin u}{\cos u} \times 2\cos u\, du$$

$$= 2\int_{\frac{\pi}{6}}^{\frac{\pi}{3}} \sin u\, du$$

$$= -2\left[\cos u\right]_{\frac{\pi}{6}}^{\frac{\pi}{3}}$$

$$= -2\left(\cos\frac{\pi}{3} - \cos\frac{\pi}{6}\right)$$

$$= -2\left(\frac{1}{2} - \frac{\sqrt{3}}{2}\right)$$

$= \mathbf{\sqrt{3} - 1}$ ✓

14 $f(x) = \tan^{-1} x + \tan^{-1}\frac{1}{x}$

$$f'(x) = \frac{1}{1+x^2} + \frac{1}{1+\left(\frac{1}{x}\right)^2}\cdot -x^{-2}$$

$$= \frac{1}{1+x^2} + \frac{-1}{x^2\left(1+\frac{1}{x^2}\right)}$$

$$= \frac{1}{1+x^2} - \frac{1}{1+x^2}$$

$= 0$ ✓

Also, consider $f(1) = \tan^{-1} 1 + \tan^{-1}\frac{1}{1}$

$$= \frac{\pi}{4} + \frac{\pi}{4}$$

$= \frac{\pi}{2}$ ✓

$\therefore$ as $f'(x) = 0$, the sketch of the function is horizontal, and as $f(1) = \frac{\pi}{2}$ then $f(x) = \frac{\pi}{2}$, for all $x > 0$. ✓

15 Area $= \int_0^{\frac{1}{4}} \frac{1}{\sqrt{1-16x^2}}\,dx$

$$= \int_0^{\frac{1}{4}} \frac{1}{\sqrt{16\left(\frac{1}{16}-x^2\right)}}\,dx$$

$= \frac{1}{4}\int_0^{\frac{1}{4}} \frac{1}{\sqrt{\frac{1}{16}-x^2}}\,dx$ ✓

$= \frac{1}{4}\left[\sin^{-1} 4x\right]_0^{\frac{1}{4}}$ ✓

$$= \frac{1}{4}\left[\sin^{-1} 4\left(\frac{1}{4}\right) - \sin^{-1} 0\right]$$

$$= \frac{1}{4}\left[\frac{\pi}{2} - 0\right]$$

$$= \frac{\pi}{8}$$

$\therefore$ the area is $\frac{\pi}{8}$ units². ✓

CHAPTER 14 FURTHER AREA AND VOLUMES OF SOLIDS OF REVOLUTION

TOPIC: CALCULUS

YEAR 12

1 Area between two curves

➔ **Area** $= \int_a^b (f(x) - g(x))\,dx$, such that $f(x) > g(x)$.

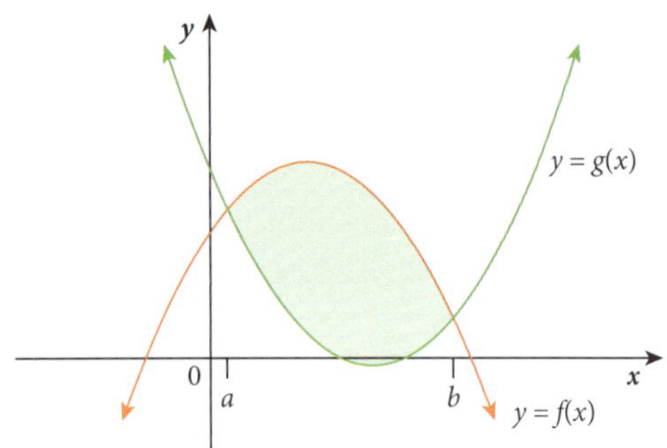

EXAMPLE 1

Find the area of the region bounded by the curves $y = \dfrac{4}{1+x^2}$ and the line $y = 1$.

$$\frac{4}{1+x^2} = 1$$
$$4 = 1 + x^2$$
$$x^2 = 3$$
$$x = \pm\sqrt{3}$$

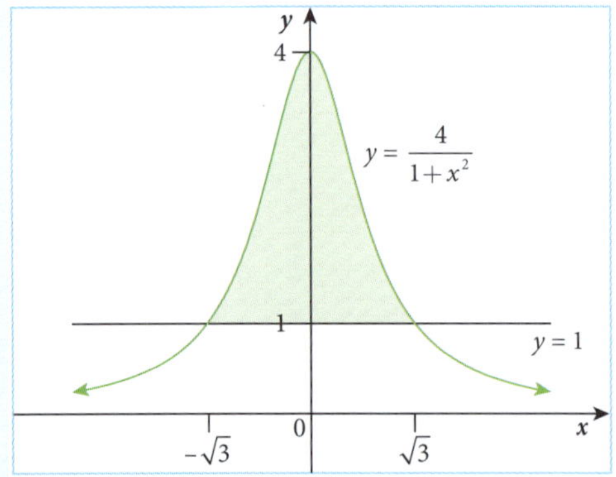

$$\text{Area} = \int_{-\sqrt{3}}^{\sqrt{3}} \left(\frac{4}{1+x^2} - 1\right) dx$$
$$= 2\int_0^{\sqrt{3}} \left(\frac{4}{1+x^2} - 1\right) dx$$

When symmetrical about y-axis use $2\int_0^b g(x)\,dx$.

$$= 2\left[4\tan^{-1} x - x\right]_0^{\sqrt{3}}$$
$$= 2\left[4\tan^{-1}\sqrt{3} - \sqrt{3} - (4\tan^{-1} 0 - 0)\right]$$
$$= 2\left(4 \times \frac{\pi}{3} - \sqrt{3} - 0\right)$$
$$= 2\left(\frac{4\pi}{3} - \sqrt{3}\right)$$

$\therefore$ the area is $2\left(\dfrac{4\pi}{3} - \sqrt{3}\right)$ units^2.

EXAMPLE 2

A company logo is made from the area between two functions. The functions are $f(x) = x^2 - \pi^2$ and $g(x) = 2\sin x$. Determine the exact area of the logo.

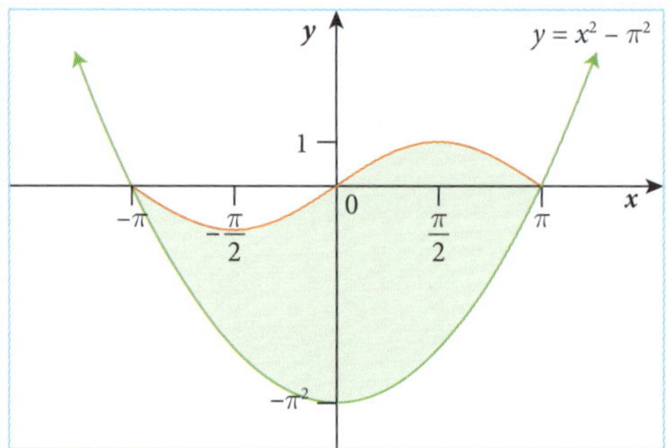

Draw a sketch.

$$\text{Area} = \int_{-\pi}^{\pi} (\sin x - (x^2 - \pi^2))\,dx$$
$$= \int_{-\pi}^{\pi} (\sin x - x^2 + \pi^2)\,dx$$
$$= \left[-\cos x - \frac{x^3}{3} + \pi^2 x\right]_{-\pi}^{\pi}$$
$$= -\cos\pi - \frac{\pi^3}{3} + \pi^2(\pi) - \left[-\cos(-\pi) - \frac{(-\pi)^3}{3} + \pi^2(-\pi)\right]$$
$$= 2\pi^3 - \frac{2\pi^3}{3}$$
$$= \frac{4\pi^3}{3}$$

$\therefore$ the area is $\dfrac{4\pi^3}{3}$ units^2.

Alternatively, area $= -2\int_0^{\pi} (x^2 - \pi^2)\, dx$

$$= \frac{4\pi^3}{3}$$

→ PRACTICE

1 Find the area of the region bounded by the:

i curve $y = \dfrac{1}{\sqrt{1-x^2}}$ and the line $y = 2$

ii $y = \dfrac{2}{1+x^2}$ and the line $y = x^2$

2 i A company's logo is to be designed using the area between the curves $y = x(x - \pi)$ and $y = \sin x$. Find the exact area of the logo.

ii The graphs of $y = \sin^2 x$ and $y = \cos^2 x$ in the domain $0 \le x \le \pi$ are drawn on the same number plane. The curves intersect at $x = a$ and $x = b$. Find the area between the curves over the domain $[a, b]$.

Answers ➲ pp. 188–189

2 Volumes of solids formed by rotating about the x-axis

→ **If $y = f(x)$ is rotated about the x-axis between $x = a$ and $x = b$**, the volume of the solid formed is given as $V = \pi \int_a^b y^2\, dx$.

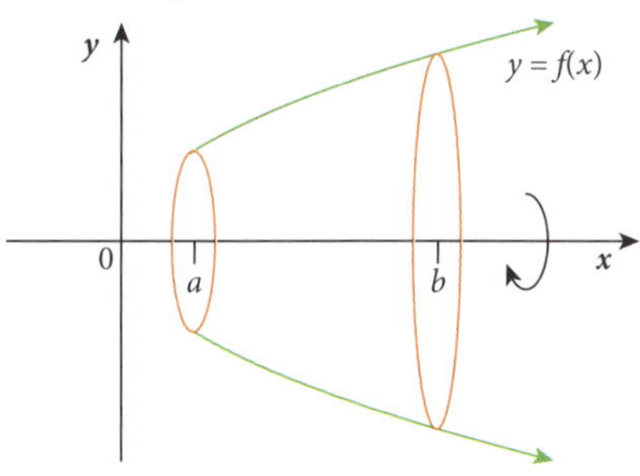

EXAMPLE 1

The diagram shows the region bounded by $y = \dfrac{1}{(x+1)^2}$, the x-axis, the y-axis, and the line $x = 2$.

The region is rotated about the x-axis to form a solid. Find the volume of the solid.

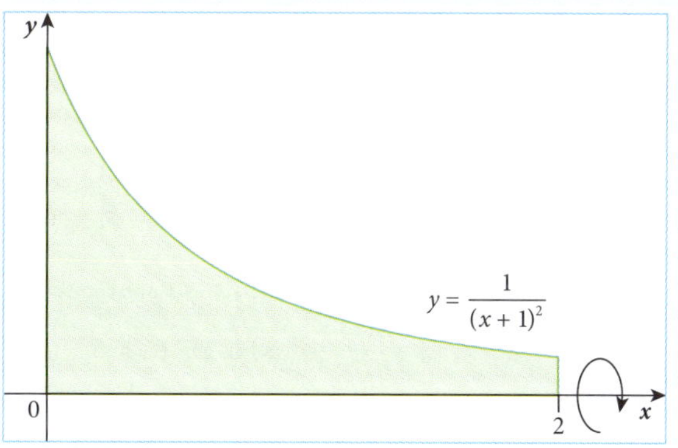

$$y^2 = \frac{1}{(x+1)^4} = (x+1)^{-4}$$

$$V = \pi \int_0^2 (x+1)^{-4}\, dx$$

$$= \pi \left[\left(\frac{x+1}{-3} \right)^{-3} \right]_0^2$$

$$= -\frac{\pi}{3} \left[\left(\frac{1}{(x+1)^3} \right)^{-3} \right]_0^2$$

$$= -\frac{\pi}{3} \left[\frac{1}{(2+1)^3} - \frac{1}{(0+1)^3} \right]_0^2$$

$$= -\frac{\pi}{3} \left[\frac{1}{27} - 1 \right]$$

$$= -\frac{\pi}{3} \left[-\frac{26}{27} \right]$$

Leave answer in terms of π.

$$= \frac{26\pi}{81}$$

$\therefore$ the volume is $\dfrac{26\pi}{81}$ units^3.

EXAMPLE 2

The diagram shows the region bounded by the curve $y = \sqrt{8 - 2x^2}$ and the x-axis. The region is rotated around the x-axis to form a solid. Find the exact volume of the solid formed.

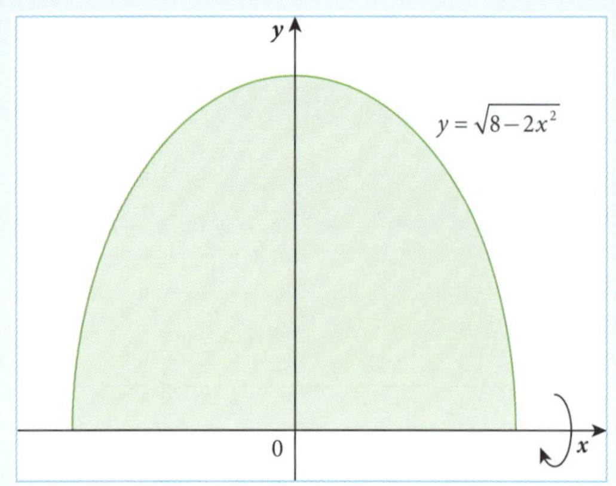

$$\sqrt{8-2x^2} = 0$$
$$8 - 2x^2 = 0$$
$$2x^2 = 8$$
$$x^2 = 4$$
$$x = \pm 2$$
$$y^2 = (\sqrt{8-2x^2})^2$$
$$= 8 - 2x^2$$
$$V = \pi \int_{-2}^{2} y^2\, dx$$
$$= \pi \int_{-2}^{2} (8-2x^2)\, dx$$
$$= 2\pi \int_{0}^{2} (8-2x^2)\, dx$$

When symmetrical about the y-axis use $2\int_0^b g(x)\, dx$.

$$= 2\pi \left[8x - \frac{2x^3}{3}\right]_0^2$$
$$= 2\pi \left[8(2) - \frac{2(2)^3}{3} - 0\right]$$
$$= 2\pi \left[16 - \frac{16}{3}\right]$$
$$= 2\pi \left[\frac{32}{3}\right]$$
$$= \frac{64\pi}{3}$$

$\therefore$ the volume is $\frac{64\pi}{3}$ units³.

EXAMPLE 3

The region bounded by the curve $y = 3 \sin 2x$ and the x-axis between $x = 0$ and $x = \frac{\pi}{2}$ is rotated around the x-axis to form a solid.

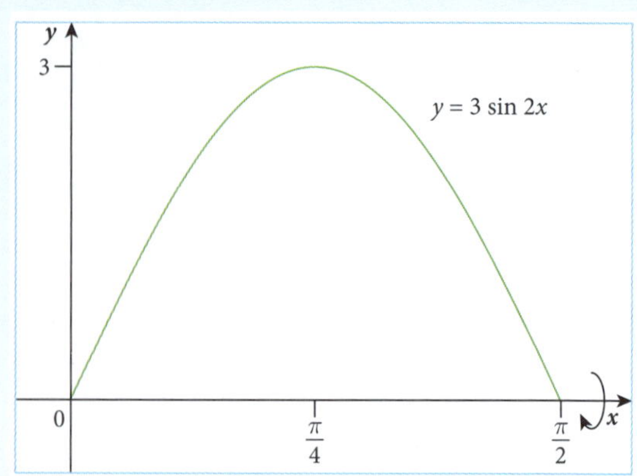

Find the exact volume of the solid formed.

$$y^2 = (3 \sin 2x)^2$$
$$= 9 \sin^2 2x$$
$$V = \pi \int_0^{\frac{\pi}{2}} y^2\, dx$$
$$= \pi \int_0^{\frac{\pi}{2}} 9 \sin^2 2x\, dx$$
$$= 9\pi \int_0^{\frac{\pi}{2}} \sin^2 2x\, dx$$
$$= 9\pi \int_0^{\frac{\pi}{2}} \frac{1}{2}(1 - \cos 4x)\, dx$$
$$= \frac{9\pi}{2}\left[x - \frac{1}{4}\sin 4x\right]_0^{\frac{\pi}{2}}$$
$$= \frac{9\pi}{2}\left[\frac{\pi}{2} - \frac{1}{4}\sin 4\left(\frac{\pi}{2}\right) - 0\right]$$
$$= \frac{9\pi}{2}\left[\frac{\pi}{2}\right]$$

Use $\sin^2 A = \frac{1}{2}(1 - \cos 2A)$.

$$= \frac{9\pi^2}{4}$$

$\therefore$ the volume is $\frac{9\pi^2}{4}$ units³.

→ PRACTICE

1 i **The diagram shows the region bounded by $y = x^2 + 2$, the x-axis, the y-axis, and the line $x = 4$. The region is rotated about the x-axis to form a solid.**

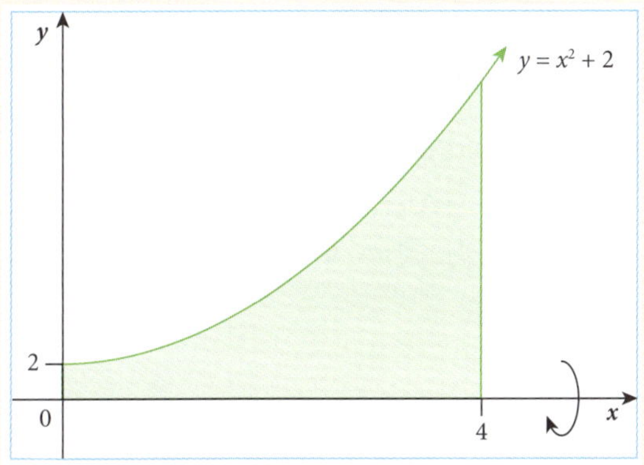

Find the volume of the solid, correct to two decimal places.

ii The diagram shows the region bounded by $y = \dfrac{1}{x-1}$, the x-axis and the lines $x = 2$ and $x = 4$. The region is rotated about the x-axis to form a solid.

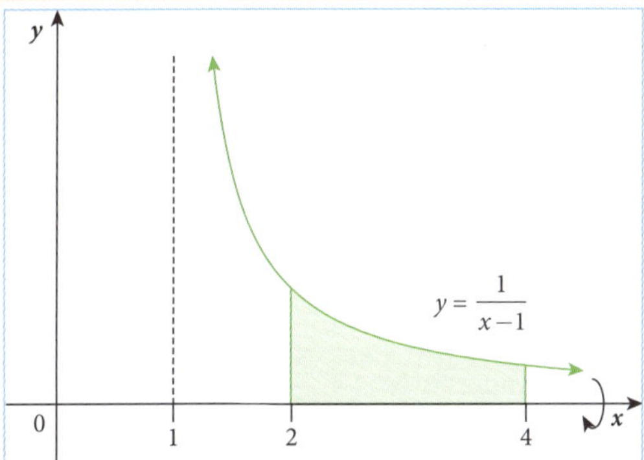

Find the exact volume of the solid.

2 i The diagram shows the region bounded by the curve $y = 2\sqrt{9-x^2}$ and the x-axis. The region is rotated around the x-axis to form a solid.

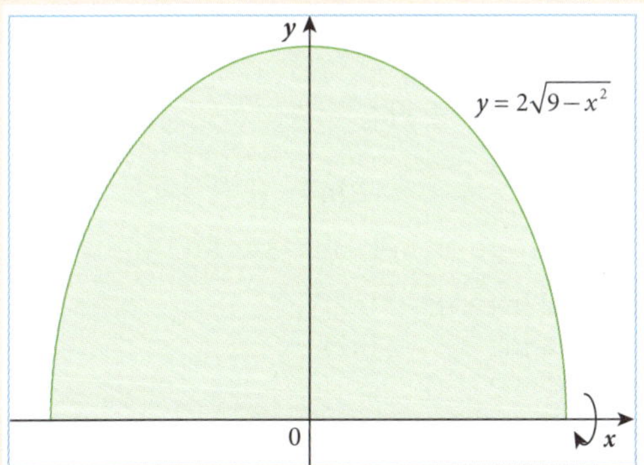

Find the exact volume of the solid formed.

ii The diagram shows the region bounded by the curve $y = x(x^2 - 4)$ and the x-axis. The region is rotated around the x-axis to form a solid.

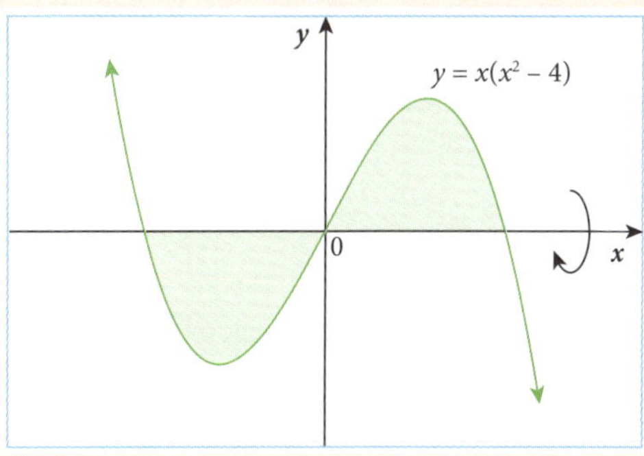

Find the volume of the solid formed, correct to three significant figures.

3 i The region bounded by the curve $y = 4\cos\dfrac{x}{2}$ and the x-axis between $x = 0$ and $x = \pi$ is rotated around the x-axis to form a solid.

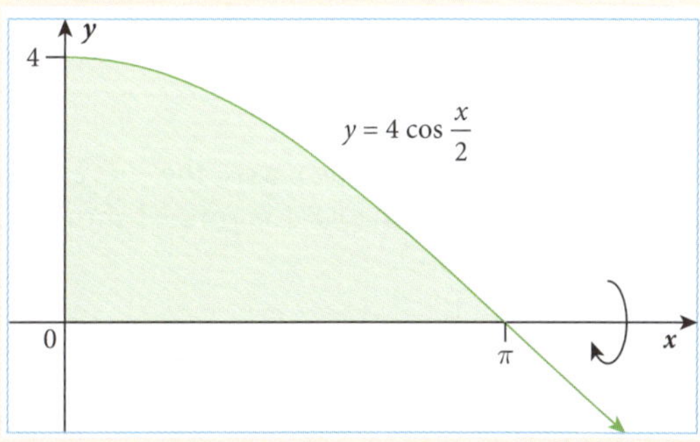

Find the exact volume of the solid formed.

ii The region bounded by the curve $y = \tan x$ and the x-axis between $x = \dfrac{\pi}{6}$ and $x = \dfrac{\pi}{3}$ is rotated around the x-axis to form a solid.

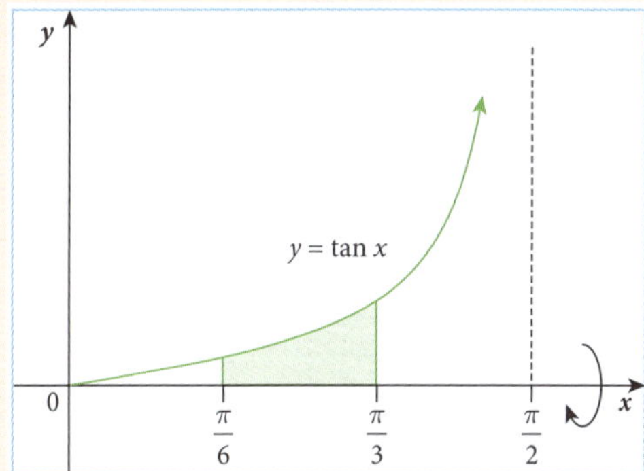

Find the exact volume of the solid formed.

Answers ➲ pp. 189–190

3 Volumes of solids formed by rotating about the y-axis

➔ **If $y = f(x)$ is rotated about the y-axis between $y = a$ and $y = b$,** the volume of the solid formed is given as $V = \pi\int_a^b x^2\,dy$.

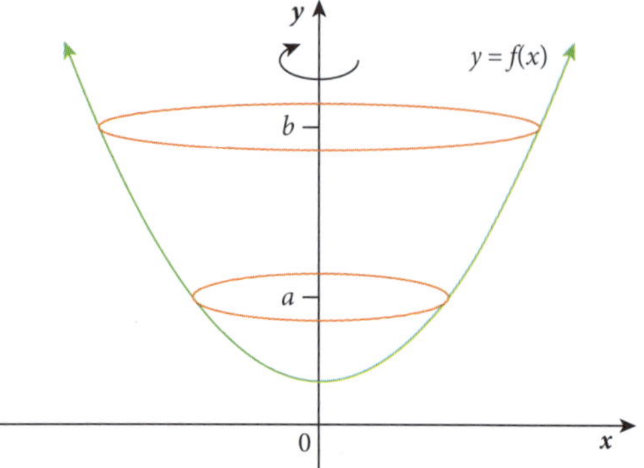

EXAMPLE 1

The region bounded by the y-axis, the curve $y = -1 + \sqrt{x}$ and the line $y = 1$ is rotated about the y-axis to form a solid.

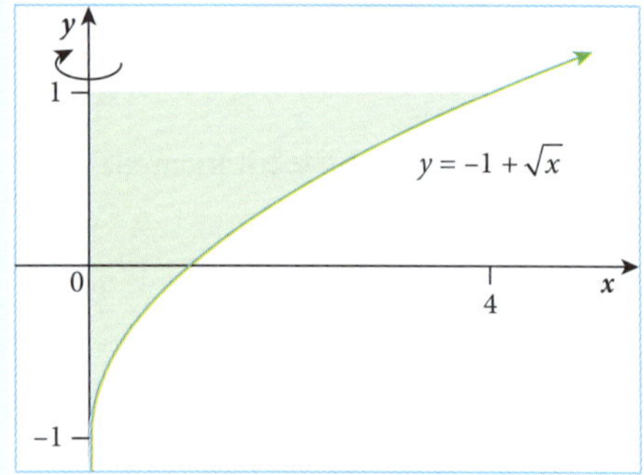

Find the volume of the solid.

$$y = -1 + \sqrt{x}$$
$$\sqrt{x} = y + 1$$
$$x^2 = (y+1)^4$$
$$V = \pi\int_{-1}^{1} x^2\,dy$$
$$= \pi\int_{-1}^{1}(y+1)^4\,dy$$
$$= \pi\left[\frac{(y+1)^5}{5}\right]_{-1}^{1}$$
$$= \frac{\pi}{5}\left[(1+1)^5 - (-1+1)^5\right]$$
$$= \frac{32\pi}{5}$$

$\therefore$ the volume is $\frac{32\pi}{5}$ units³.

$(\sqrt{a})^4 = a^2$

EXAMPLE 2

The region bounded by the y-axis and the curve $y = 2\ln(x+1)$ between $x = 1$ and $x = 3$ is rotated about the y-axis to form a solid.

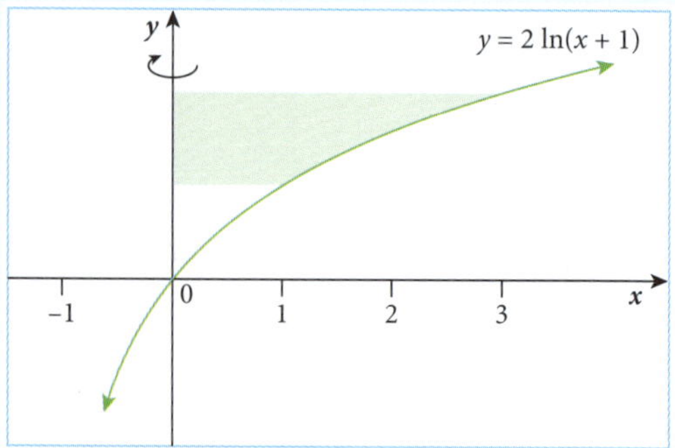

Find the exact volume of the solid.

$$y = 2\ln(x+1)$$
$$\ln(x+1) = \frac{y}{2}$$
$$x + 1 = e^{\frac{y}{2}}$$
$$x = e^{\frac{y}{2}} - 1$$
$$x^2 = (e^{\frac{y}{2}} - 1)^2$$
$$= e^y - 2e^{\frac{y}{2}} + 1$$

Substitute to find y-values.

Substitute values of x:

$y(1) = 2\ln(1+1)$
$= 2\ln 2$

$y(3) = 2\ln(3+1)$
$= 2\ln 4$

$$V = \pi\int_{2\ln 2}^{2\ln 4} x^2\,dy$$
$$= \pi\int_{2\ln 2}^{2\ln 4}(e^y - 2e^{\frac{y}{2}} + 1)\,dy$$
$$= \pi[e^y - 4e^{\frac{y}{2}} + y]_{2\ln 2}^{2\ln 4}$$
$$= \pi[e^{2\ln 4} - 4e^{\frac{2\ln 4}{2}} + 2\ln 4 - (e^{2\ln 2} - 4e^{\frac{2\ln 2}{2}} + 2\ln 2)]$$
$$= \pi[16 - 16 + 2\ln 4 - (4 - 8 + 2\ln 2)]$$
$$= \pi(2\ln 4 + 4 - 2\ln 2)$$
$$= \pi(4\ln 2 + 4 - 2\ln 2)$$
$$= \pi(2\ln 2 + 4)$$

$\therefore$ the volume is $\pi(2\ln 2 + 4)$ units³.

EXAMPLE 3

The region bounded by the y-axis and the curve $y = \frac{1}{2}\cos^{-1}\frac{x}{4}$ between $y = \frac{\pi}{4}$ and $y = \frac{\pi}{2}$ is rotated about the y-axis to form a solid.

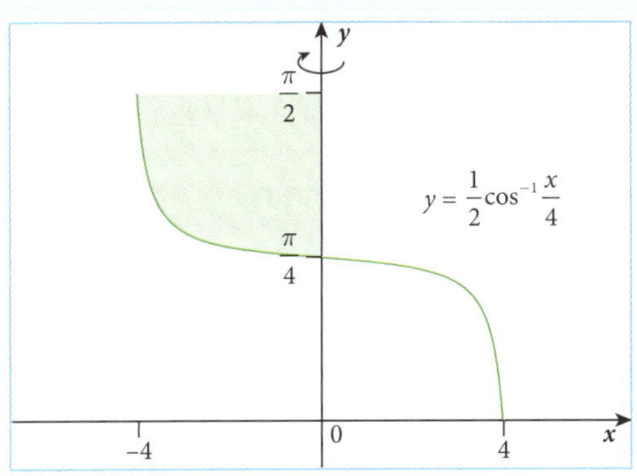

Find the volume of the solid.

$$y = \frac{1}{2}\cos^{-1}\frac{x}{4}$$

$$2y = \cos^{-1}\frac{x}{4}$$

cos (cos⁻¹ A) = A

$$\frac{x}{4} = \cos 2y$$

$$x = 4\cos 2y$$

$$x^2 = 16\cos^2 2y$$

$$V = \pi\int_{\frac{\pi}{4}}^{\frac{\pi}{2}} x^2\,dy$$

$$= \pi\int_{\frac{\pi}{4}}^{\frac{\pi}{2}} 16\cos^2 2y\,dy$$

$$= 16\pi\int_{\frac{\pi}{4}}^{\frac{\pi}{2}} \frac{1}{2}(1+\cos 4y)\,dy$$

$$= 8\pi\int_{\frac{\pi}{4}}^{\frac{\pi}{2}} (1+\cos 4y)\,dy$$

$$= 8\pi\left[y + \frac{1}{4}\sin 4y\right]_{\frac{\pi}{4}}^{\frac{\pi}{2}}$$

$$= 8\pi\left[\frac{\pi}{2} + \frac{1}{4}\sin 4\left(\frac{\pi}{2}\right) - \left(\frac{\pi}{4} + \frac{1}{4}\sin 4\left(\frac{\pi}{4}\right)\right)\right]$$

$$= 8\pi\left[\frac{\pi}{2} + 0 - \frac{\pi}{4} - 0\right]$$

$$= 2\pi^2$$

∴ the volume is $2\pi^2$ units³.

→ PRACTICE

1 i The region bounded by the x-axis, the y-axis and the curve $y = 2\sqrt{x-1}$ between $y = 2$ and $y = 4$ is rotated about the y-axis to form a solid.

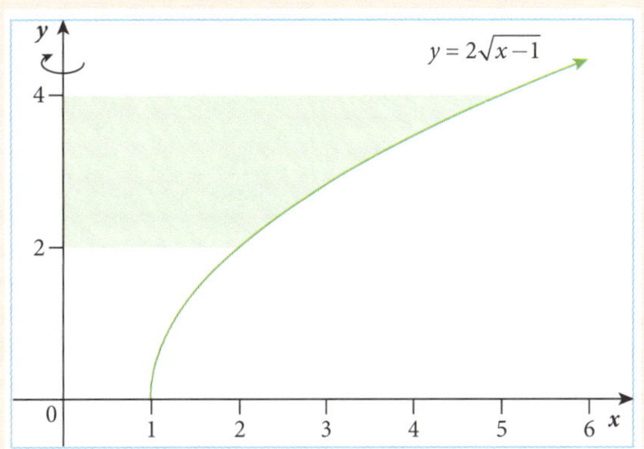

Find the volume of the solid, correct to one decimal place.

ii The region bounded by the x-axis, the y-axis and the parabola $y = (x - 3)^2$ is rotated about the y-axis to form a solid.

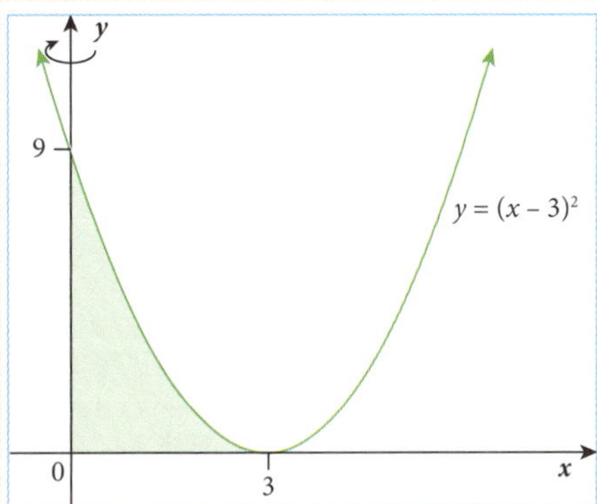

Find the exact volume of the solid.

2 i The region bounded by the y-axis and the curve $y = \log_e\left(\frac{x+2}{2}\right)$ between $x = 2$ and $x = 4$ is rotated about the y-axis to form a solid.

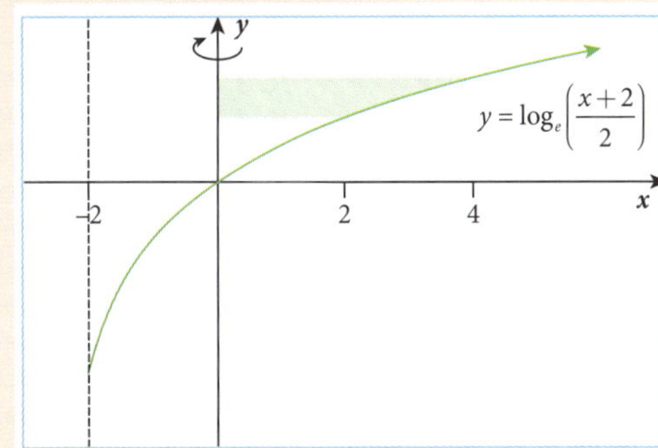

Find the volume of the solid, correct to two decimal places.

ii The region bounded by the y-axis and the curve $y = 2 + \log_e(x+3)$ between $x = 1$ and $x = 3$ is rotated about the y-axis to form a solid.

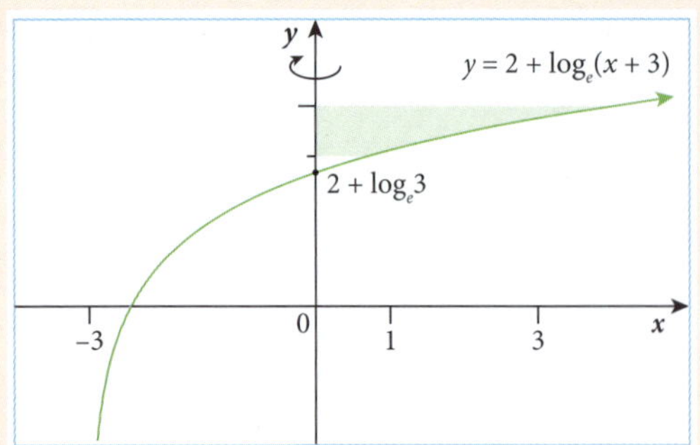

Find the volume of the solid, correct to two decimal places.

3 i The region bounded by the y-axis and the curve $y = 2\sin^{-1}\frac{x}{3}$ between $y = \frac{\pi}{3}$ and $y = \frac{2\pi}{3}$ is rotated about the y-axis to form a solid.

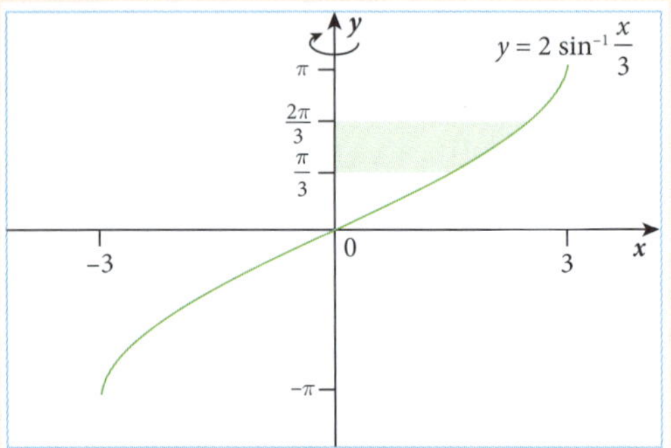

Find the volume of the solid.

ii The region bounded by the y-axis and the curve $y = \tan^{-1}\frac{x}{\sqrt{3}}$ between $x = 1$ and $x = 3$ is rotated about the y-axis to form a solid.

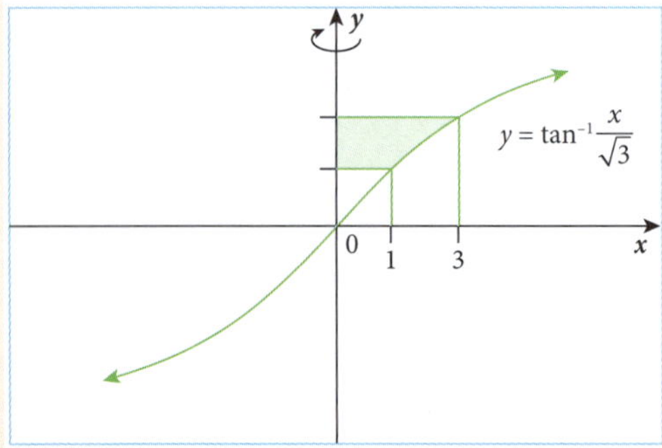

Show that the volume of the solid is $\pi\left[2\sqrt{3} - \frac{\pi}{2}\right]$ units³.

Answers ➲ pp. 190–191

4 Volumes of solids formed by rotating regions between two curves

➔ **If the area between $y = f(x)$ and $y = g(x)$ is rotated about the x-axis** between $x = a$ and $x = b$, the volume of the solid formed is given as $V = \pi\int_a^b ([f(x)]^2 - [g(x)]^2)\,dx$.

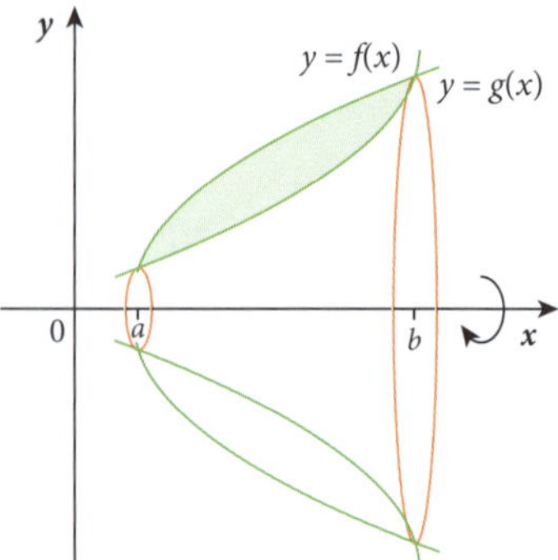

➔ **If the area between $y = f(x)$ and $y = g(x)$ is rotated about the y-axis** between $y = a$ and $y = b$, the volume of the solid formed is given as $V = \pi\int_a^b ([f(y)]^2 - [g(y)]^2)\,dy$.

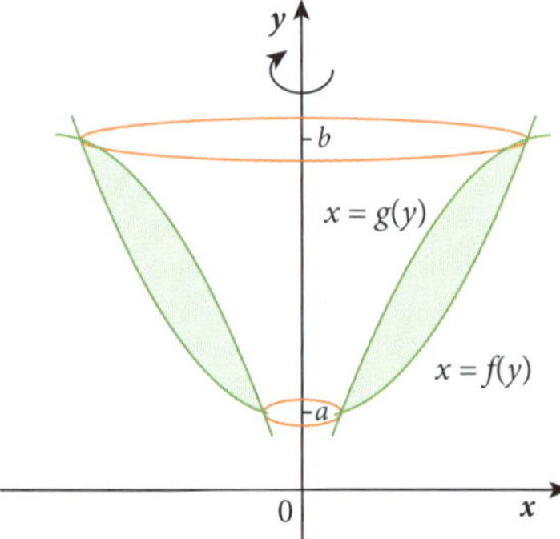

EXAMPLE 1

The area between the line $y = 2x + 1$ and the parabola $y = x^2 + 1$ is rotated about the x-axis.

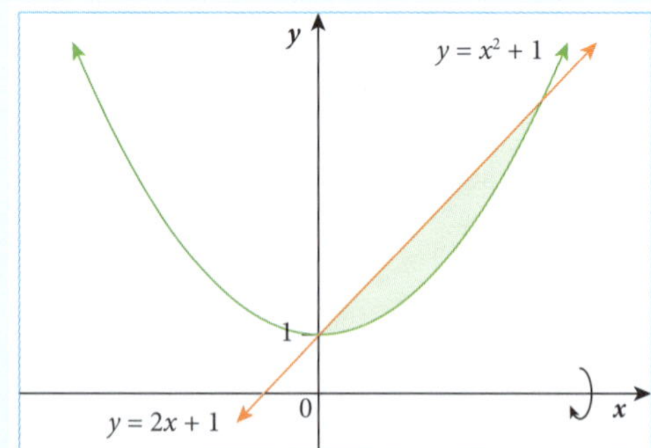

Find the volume of the solid formed, correct to two decimal places.

Solve simultaneously:

$$x^2 + 1 = 2x + 1$$
$$x^2 - 2x = 0$$
$$x(x - 2) = 0$$
$$x = 0, 2$$

Find the point of intersection of line and curve.

$$V = \pi\int_0^2 ((2x+1)^2 - (x^2+1)^2)\,dx$$
$$= \pi\int_0^2 (4x^2+4x+1-(x^4+2x^2+1))\,dx$$
$$= \pi\int_0^2 (-x^4+2x^2+4x)\,dx$$
$$= \pi\left[-\frac{x^5}{5}+\frac{2x^3}{3}+2x^2\right]_0^2$$
$$= \pi\left[-\frac{2^5}{5}+\frac{2(2)^3}{3}+2(2)^2-0\right]$$
$$= 21.781\,709\,06\ldots$$
$$= 21.78 \text{ (2 dec. pl.)}$$

$\therefore$ the volume is 21.78 units^3.

EXAMPLE 2

The area between the curves $y = x^2$ and $y = \sqrt{x}$ is rotated about the x-axis.

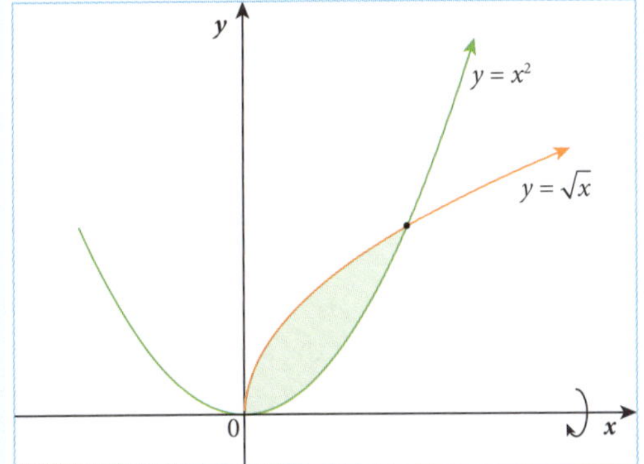

Find the exact volume of the solid formed.

Solve simultaneously:

$$x^2 = \sqrt{x}$$
$$x^4 = x$$
$$x^4 - x = 0$$
$$x(x^3 - 1) = 0$$
$$x = 0, 1$$

Solve simultaneous equations.

$$V = \pi\int_0^1 ((\sqrt{x})^2 - (x^2)^2)\,dx$$
$$= \pi\int_0^1 (x - x^4)\,dx$$
$$= \pi\left[\frac{x^2}{2}-\frac{x^5}{5}\right]_0^1$$
$$= \pi\left[\frac{1^2}{2}-\frac{1^5}{5}-0\right]$$
$$= \pi\left[\frac{1}{2}-\frac{1}{5}\right]$$
$$= \frac{3\pi}{10}$$

$\therefore$ the volume is $\frac{3\pi}{10}$ units^3.

EXAMPLE 3

The area between the curves $y = x^2 - 1$ and $y = 3x - 1$ is rotated about the y-axis.

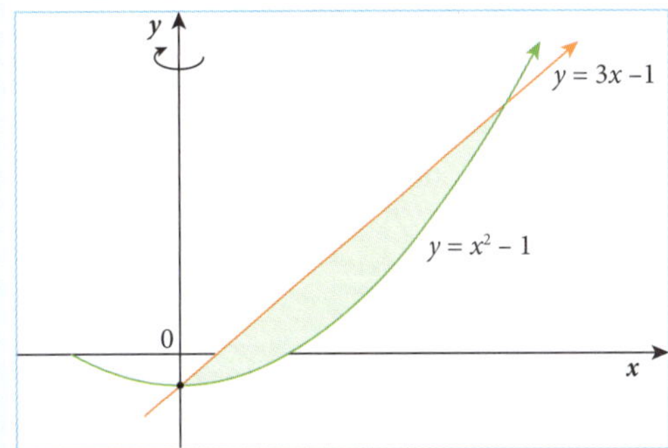

Find the exact volume of the solid formed.

Solve simultaneously:

$$x^2 - 1 = 3x - 1$$
$$x^2 - 3x = 0$$
$$x(x - 3) = 0$$
$$x = 0, 3$$

Find expressions for x^2.

Substitution into $y = 3x - 1$:

Hence, points of intersection at $(0, -1)$ and $(3, 8)$

As $y = x^2 - 1$, then $x^2 = y + 1$

As $y = 3x - 1$, then $3x = y + 1$

$$x = \frac{y+1}{3}$$
$$x^2 = \frac{(y+1)^2}{9}$$
$$V = \pi\int_{-1}^{8}\left(y+1-\frac{(y+1)^2}{9}\right)dy$$
$$= \pi\left[\frac{y^2}{2}+y-\frac{(y+1)^3}{27}\right]_{-1}^{8}$$
$$= \pi\left[\frac{8^2}{2}+8-\frac{(8+1)^3}{27}-\left(\frac{(-1)^2}{2}-1-\frac{(-1+1)^3}{27}\right)\right]$$
$$= \pi\left[32+8-27-\left(\frac{1}{2}-1-0\right)\right]$$
$$= \pi\left[13+\frac{1}{2}\right]$$
$$= \frac{27\pi}{2}$$

$\therefore$ the volume is $\frac{27\pi}{2}$ units^3.

EXAMPLE 4

The area between the curves $y = 1 - x^2$ and $y = \cos\frac{\pi x}{2}$ is rotated about the x-axis.

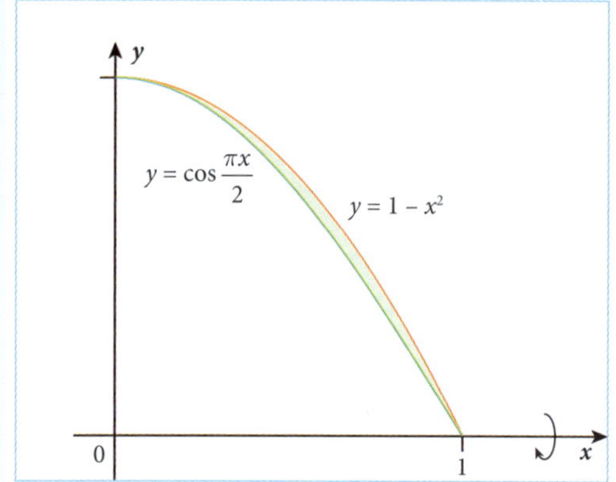

Find the exact volume of the solid formed.

$$V = \pi\int_0^1\left((1-x^2)^2 - \left(\cos\frac{\pi x}{2}\right)^2\right)dx$$

$$= \pi\int_0^1\left(1-2x^2+x^4-\cos^2\frac{\pi x}{2}\right)dx$$

$$= \pi\int_0^1\left(1-2x^2+x^4-\frac{1}{2}(1+\cos\pi x)\right)dx$$

$$= \pi\left[x-\frac{2x^3}{3}+\frac{x^5}{5}-\frac{1}{2}(x+\frac{1}{\pi}\sin\pi x)\right]_0^1$$

$$= \pi\left[1-\frac{2(1)^3}{3}+\frac{1^5}{5}-\frac{1}{2}(1+\frac{1}{\pi}\sin\pi(1))-(0-0)\right]$$

$$= \pi\left[1-\frac{2}{3}+\frac{1}{5}-\frac{1}{2}\right]$$

$$= \frac{\pi}{30}$$

$\therefore$ the volume is $\frac{\pi}{30}$ units^3.

$\int\cos\pi x\,dx = \frac{1}{\pi}\sin\pi x + c$

→ PRACTICE

1 i The area between the parabola $y = 16 - x^2$ and the line $y = x + 10$ is rotated about the x-axis.

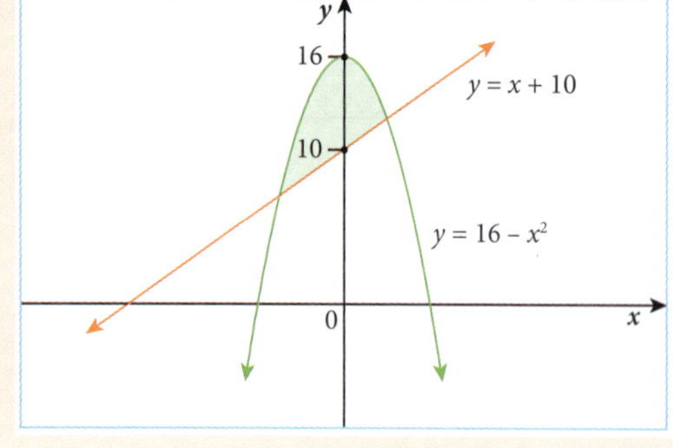

Show that the exact volume of the solid formed is 500π units^3.

ii The parabola $y = x^2 - 4x$ is graphed on the number plane. A straight line passes through the parabola at the origin and the point $(2, -4)$. The shaded region shown on the diagram is rotated around the x-axis.

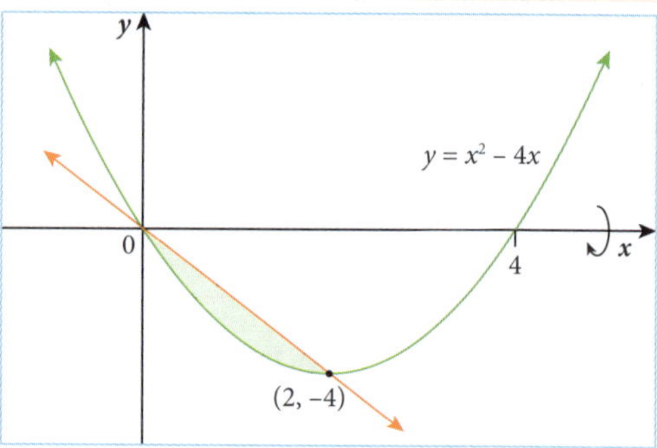

Find the exact volume of the solid formed.

2 i The area between the curves $y = x^3$ and $y = x^2$ is rotated about the x-axis.

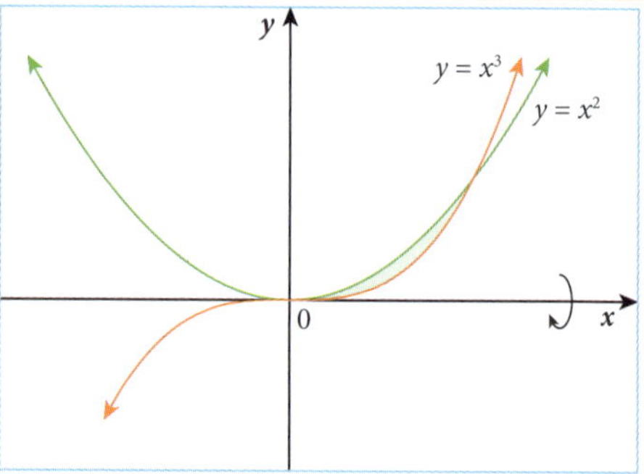

Find the exact volume of the solid formed.

ii The area between the curves $y = x^2$ and $y = 4x - x^2$ is rotated about the x-axis.

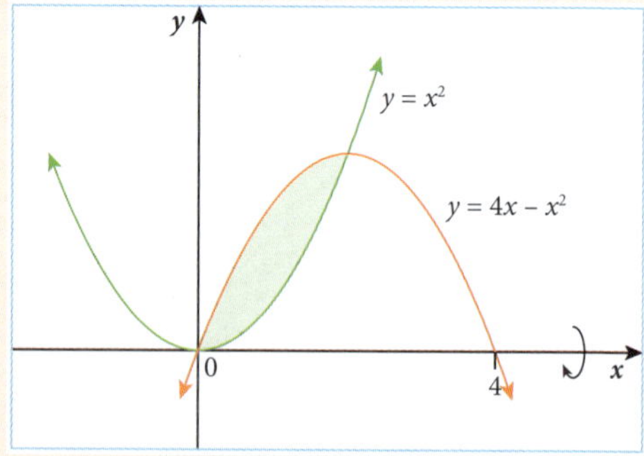

Find the volume of the solid formed, correct to two decimal places.

3 i The area between the curves $y = \frac{x^2}{2} - 3$ and $y = 4x - 3$ is rotated about the y-axis.

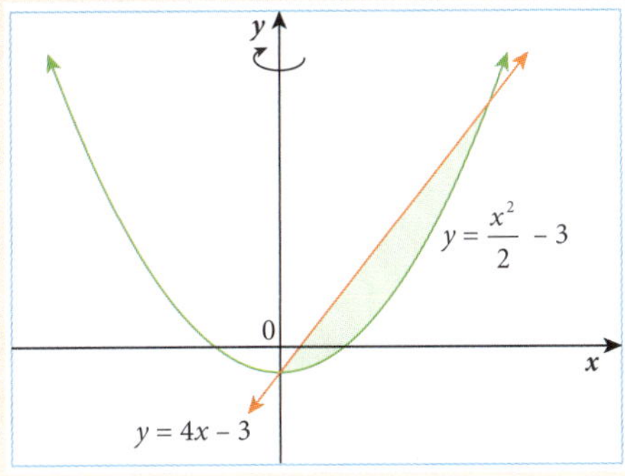

Find the volume of the solid formed. Leave your answer correct to three significant figures.

ii The area between the curves $y = x\sqrt{x}$ and $y = \sqrt{x}$ in the first quadrant is rotated about the y-axis.

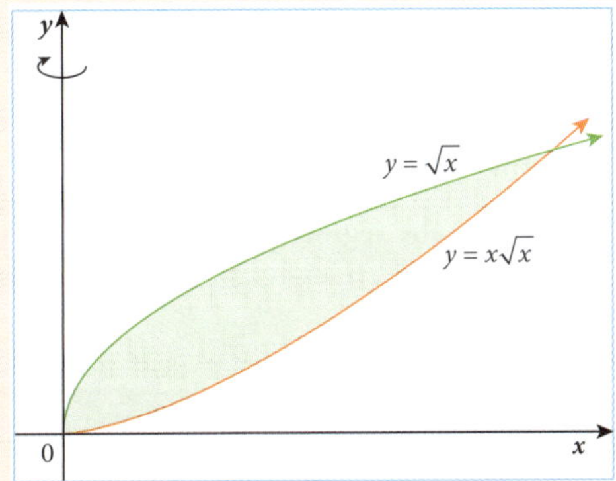

Find the exact volume of the solid formed.

4 i The area between the curves $y = \frac{2x}{3}$ and $y = \sin\frac{\pi x}{3}$ in the first quadrant is rotated about the x-axis.

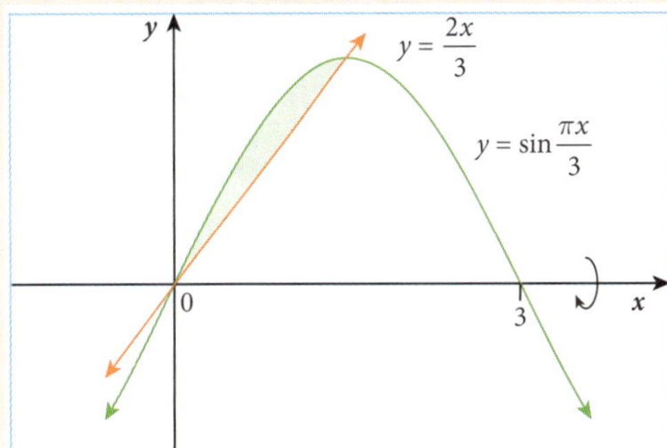

Find the exact volume of the solid formed.

ii The shaded area between the curves $y = \frac{1}{2}\sin\frac{x}{2}$ and $y = \sin x$ in the diagram below is rotated about the x-axis.

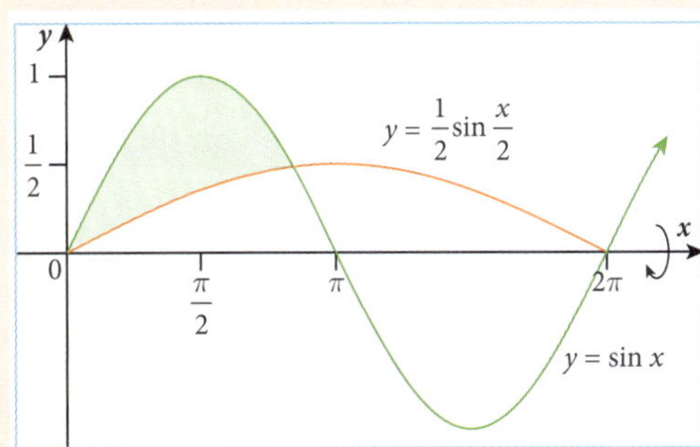

Find the volume of the solid formed. Leave your answer correct to two decimal places.

Answers ➲ pp. 191–193

HSC EXAM-TYPE QUESTIONS

Total marks: 35

PART A

(Suggested time: 10 minutes)

(1 mark each)

1 Which of these integrals can be used to find the area between the curves $y = x^2 - 4x + 4$ and $y = 10 - x^2$?

A $\int_{-1}^{3}(3+2x-x^2)\,dx$ B $\int_{-3}^{1}(6+4x-2x^2)\,dx$

C $\int_{-1}^{3}(6+4x-2x^2)\,dx$ D $\int_{-3}^{1}(3+2x-x)^2\,dx$

2 The region between $y = \sqrt{4-x^2}$ and the x-axis is rotated about the x-axis. What is the volume of the solid formed?

A $\frac{32\pi}{3}$ units^3 B $\frac{64\pi}{3}$ units^3

C $\frac{256\pi}{3}$ units^3 D $\frac{512\pi}{3}$ units^3

3 The region between the curve $y = e^x + 1$, the x-axis, the y-axis and $x = 1$ is to be rotated about the y-axis.

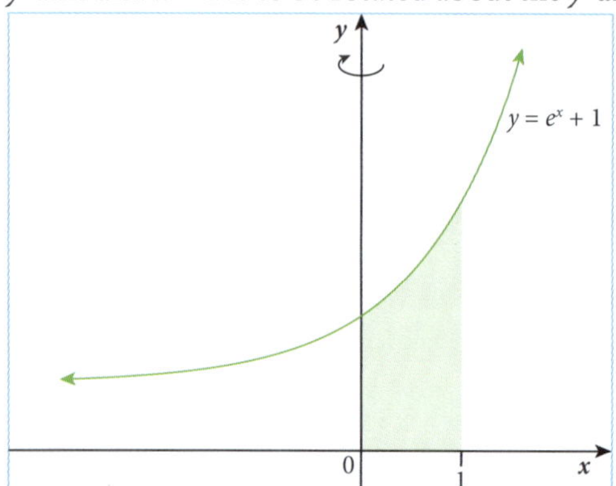

Which of these integrals represents the volume of the solid?

A $\pi\int_{2}^{e+1}(y-1)^2\,dy$ B $\pi\int_{2}^{e+1}\log_e(y-1)^2\,dy$

C $\pi\int_{2}^{e+1}[\log_e(y-1)]^2\,dy$ D $\pi\int_{2}^{3}2\log_e(y-1)\,dy$

4 Which of these is the area between $y = 2\cos 2x \sin x$ and the x-axis between $x = 0$ and $x = \frac{\pi}{6}$?

A $\left(\frac{\sqrt{3}}{2}-\frac{2}{3}\right)$ units^2 B $\left(\frac{\sqrt{3}}{2}-\frac{1}{2}\right)$ units^2

C $\left(\frac{\sqrt{3}}{2}-\frac{1}{3}\right)$ units^2 D $\left(\frac{\sqrt{3}}{2}-\frac{1}{6}\right)$ units^2

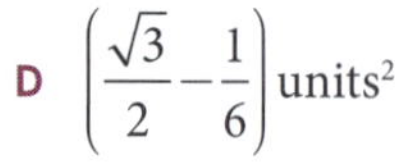

5 Which of these is the shaded area between $y = \log_e x$ and $y = 0$ represented on the diagram?

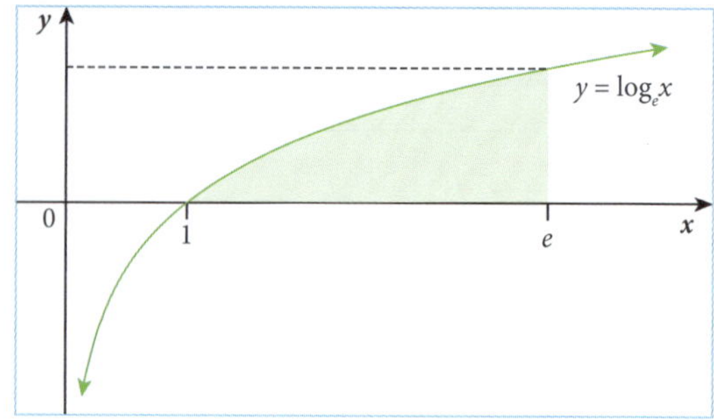

A $\frac{e-1}{2}$ units^2 B $(e - 1)$ units^2

C $(e + 1)$ units^2 D 1 unit^2

PART B

(Suggested time: 45 minutes)

Show all working.

6 Find the area of the region bounded by the curve $y = \log_e(x + 1)^2$, the y-axis between $y = \log_e 4$ and $y = 0$. (3 marks)

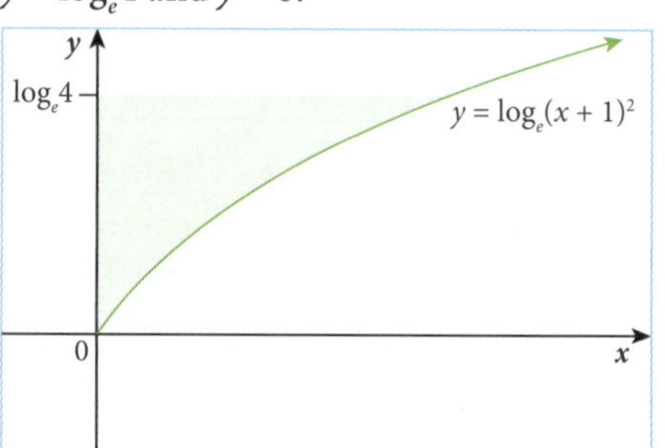

7 The diagram shows the graphs of $y = \sin^{-1} x$ and $y = \cos^{-1} x$.

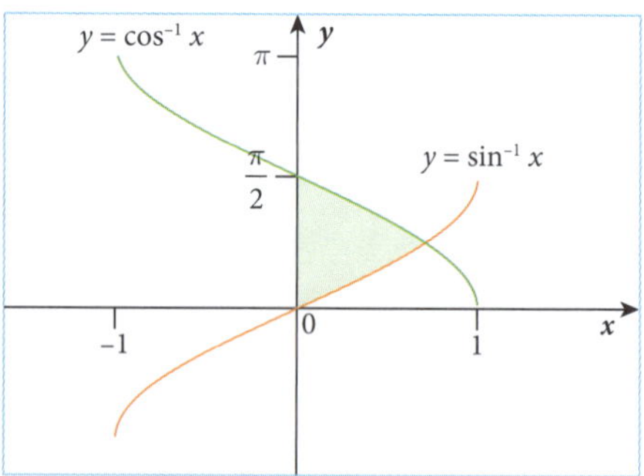

a Write down the point of intersection of $y = \sin^{-1} x$ and $y = \cos^{-1} x$. (1 mark)

b The area enclosed by the y-axis, $y = \sin^{-1} x$ and $y = \cos^{-1} x$ is rotated about the y-axis. Find the exact volume of the solid formed. (3 marks)

8 The area between $y = \frac{2\sqrt{3}}{3}$, $y = \sec\frac{x}{2}$ and the y-axis in the first quadrant is rotated around the x-axis. Find the volume of the solid formed, correct to two decimal places. (4 marks)

9 The graphs of the line $y = \frac{x}{3}$ and the curve $y = \frac{1}{\sqrt{4-x^2}}$ are drawn on a number plane.

a Show that the line and the graph do not intersect. (2 marks)

b Find the area between $y = \frac{x}{3}$, $y = \frac{1}{\sqrt{4-x^2}}$ and the lines $x = 0$ and $x = 1$. (2 marks)

10 The graph of $y = \sqrt{4-x^2}$ is graphed on the number plane below. The shaded region is rotated about the x-axis to form a solid.

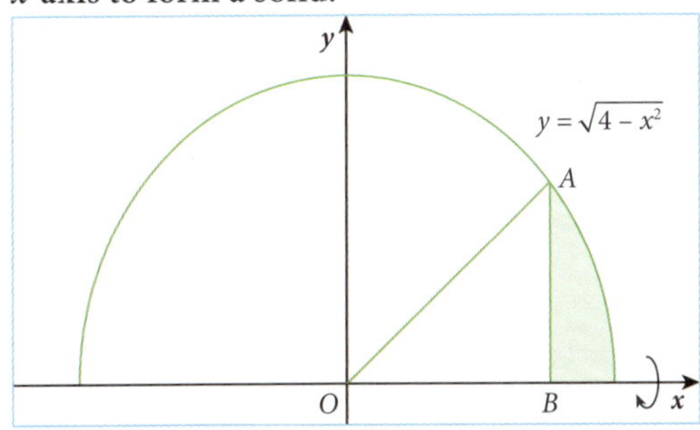

Given $\angle OAB = 60°$, find the exact volume of the solid. (3 marks)

11 The curves $y = 4\sin^2 x$ and $y = \sec^2 x$ are graphed on the same number plane for the domain $[0, \frac{\pi}{2}]$.

a Find the point of intersection $P(\alpha, \beta)$ of the two curves in the given domain. (2 marks)

b Find the area bounded by the two curves and the y-axis between $x = 0$ and $x = \alpha$. (3 marks)

12 A graphics company is designing a new logo based on the graphs of the functions $y = \frac{16}{4+x^2}$ and the line $y = 4 - 2x$.

a Without the use of calculus, sketch the graphs of the curve $y = \frac{16}{4+x^2}$ and the line $2x + y - 4 = 0$. (2 marks)

b Show that the area bounded by $y = \frac{16}{4+x^2}$ and $2x + y - 4 = 0$, between $x = 0$ and $x = 2$ is $2(\pi - 2)$ units^2. (2 marks)

13 The area bounded by the curve $y = \frac{10}{x}$ and the line $y = 7 - x$ is rotated around the x-axis. Find the exact area of the solid formed. (3 marks)

PRACTICE

1 Area between two curves ➲ p. 177

1 i $\frac{1}{\sqrt{1-x^2}} = 2$

$\frac{1}{1-x^2} = 4$

$1 = 4 - 4x^2$

$4x^2 = 3$

$x = \pm\frac{\sqrt{3}}{2}$

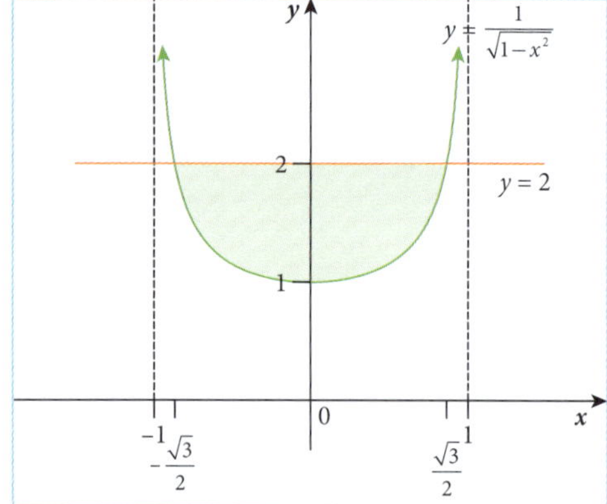

$$\text{Area} = 2\int_0^{\frac{\sqrt{3}}{2}} 2 - \frac{1}{\sqrt{1-x^2}}\,dx$$

$$= 2\left[2x - \sin^{-1}x\right]_0^{\frac{\sqrt{3}}{2}}$$

$$= 2\left[\frac{2\sqrt{3}}{2} - \sin^{-1}\frac{\sqrt{3}}{2} - (0 - \sin^{-1}0)\right]$$

$$= 2\left(\sqrt{3} - \frac{\pi}{3}\right)$$

$$= 2\sqrt{3} - \frac{2\pi}{3}$$

$\therefore$ the area is $\left(2\sqrt{3} - \frac{2\pi}{3}\right)$ units^2.

ii

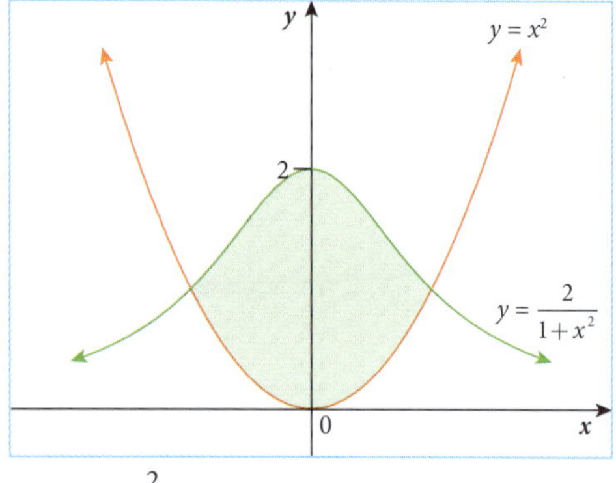

$\frac{2}{1+x^2} = x^2$

$2 = x^2 + x^4$

$x^4 + x^2 - 2 = 0$

$(x^2 + 2)(x^2 - 1) = 0$

$x^2 = -2, 1$

$x = \pm 1$ (no real solution for $x^2 = -2$)

$$\text{Area} = \int_{-1}^{1}\left(\frac{2}{1+x^2} - x^2\right)dx$$

$$= 2\int_0^1\left(\frac{2}{1+x^2} - x^2\right)dx$$

$$= 2\left[2\tan^{-1}x - \frac{x^3}{3}\right]_0^1$$

$$= 2\left[2\tan^{-1}1 - \frac{1^3}{3} - (2\tan^{-1}0 - 0)\right]$$

$$= 2\left(2 \times \frac{\pi}{4} - \frac{1}{3} - 0\right)$$

$$= \pi - \frac{2}{3}$$

$\therefore$ the area is $\left(\pi - \frac{2}{3}\right)$ units^2.

2 i

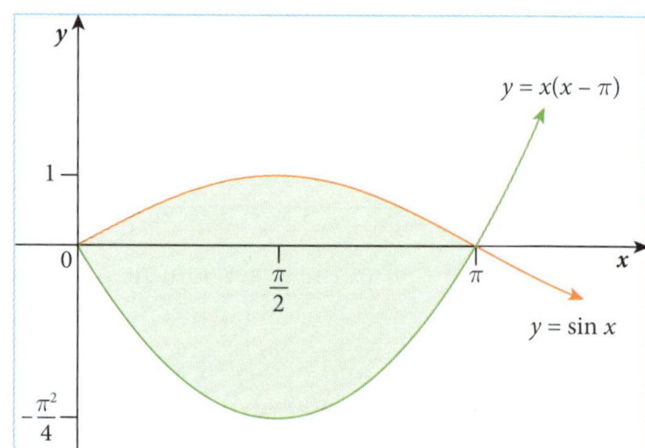

By inspection the points of intersection are (0, 0) and (π, 0).

$$\text{Area} = \int_0^{\pi}(\sin x - (x^2 - \pi x))\,dx$$

$$= \int_0^{\pi}(\sin x - x^2 + \pi x)\,dx$$

$$= \left[-\cos x - \frac{x^3}{3} + \frac{\pi x^2}{2}\right]_0^{\pi}$$

$$= \left[-\cos\pi - \frac{\pi^3}{3} + \frac{\pi(\pi^2)}{2} - (-\cos 0 - 0)\right]$$

$$= 1 - \frac{\pi^3}{3} + \frac{\pi^3}{2} + 1$$

$$= 2 + \frac{\pi^3}{6}$$

$\therefore$ the area is $\left(2 + \frac{\pi^3}{6}\right)$ units^2.

ii $\sin^2 x = \cos^2 x$

$\frac{\sin^2 x}{\cos^2 x} = \frac{\cos^2 x}{\cos^2 x}$

$\tan^2 x = 1$

$x = \frac{\pi}{4}, \frac{3\pi}{4}, \ldots$

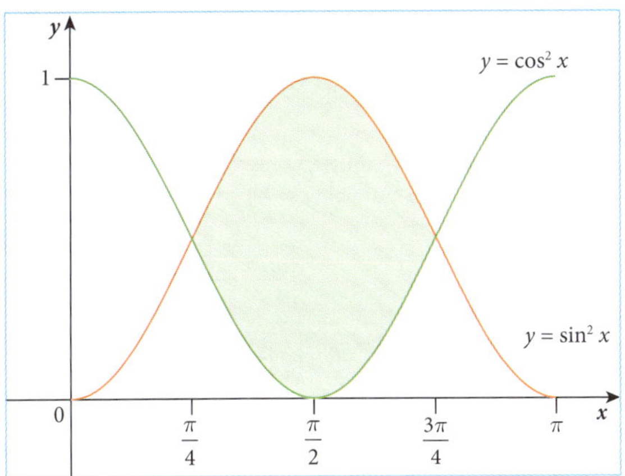

$$\text{Area} = \int_{\frac{\pi}{4}}^{\frac{3\pi}{4}} (\sin^2 x - \cos^2 x)\, dx$$

$$= 2\int_0^{\frac{\pi}{4}} (\cos^2 x - \sin^2 x)\, dx$$

$$= 2\int_0^{\frac{\pi}{4}} \cos 2x\, dx$$

$$= 2\left[\frac{1}{2}\sin 2x\right]_0^{\frac{\pi}{4}}$$

$$= \left[\sin 2x\right]_0^{\frac{\pi}{4}}$$

$$= \sin 2\left(\frac{\pi}{4}\right) - \sin 2(0)$$

$$= 1$$

∴ the area is 1 unit².

2 Volumes of solids formed by rotating about the *x*-axis ➲ pp. 178–179

1 i $y^2 = (x^2 + 2)^2$

$= x^4 + 4x^2 + 4$

$$V = \pi\int_0^4 (x^4 + 4x^2 + 4)\, dx$$

$$= \pi\left[\frac{x^5}{5} + \frac{4x^3}{3} + 4x\right]_0^4$$

$$= \pi\left[\frac{4^5}{5} + \frac{4(4)^3}{3} + 4(4) - 0\right]$$

$= 961.746\,231\ldots$

$= 961.75$ (2 dec. pl.)

∴ the volume is 961.75 units³.

ii $y^2 = \dfrac{1}{(x-1)^2}$

$= (x - 1)^{-2}$

$$V = \pi\int_2^4 (x-1)^{-2}\, dx$$

$$= \pi\left[\frac{(x-1)^{-1}}{-1}\right]_2^4$$

$$= \pi\left[\frac{-1}{x-1}\right]_2^4$$

$$= \pi\left[\frac{-1}{4-1} - \frac{-1}{2-1}\right]$$

$$= \pi\left[\frac{-1}{3} + 1\right]$$

$$= \frac{2\pi}{3}$$

∴ the volume is $\frac{2\pi}{3}$ units³.

2 i $2\sqrt{9-x^2} = 0$

$9 - x^2 = 0$

$x^2 = 9$

$x = \pm 3$

$y^2 = (2\sqrt{9-x^2})^2$

$= 4(9 - x^2)$

$$V = \pi\int_{-3}^{3} y^2\, dx$$

$$= 4\pi\int_{-3}^{3} (9 - x^2)\, dx$$

$$= 8\pi\int_0^3 (9 - x^2)\, dx$$

$$= 8\pi\left[9x - \frac{x^3}{3}\right]_0^3$$

$$= 8\pi\left[9(3) - \frac{3^3}{3} - 0\right]$$

$= 144\pi$

∴ the volume is 144π units³.

ii $x(x^2 - 4) = 0$

$x(x - 2)(x - 2) = 0$

$x = 0, \pm 2$

$y^2 = [x(x^2 - 4)]^2$

$= (x^3 - 4x)^2$

$= x^6 - 8x^4 + 16x^2$

$$V = \pi\int_{-2}^{0} y^2\, dx + \pi\int_0^2 y^2\, dx$$

$$= 2\pi\int_0^2 (x^6 - 8x^4 + 16x^2)\, dx$$

$$= 2\pi\left[\frac{x^7}{7} - \frac{8x^5}{5} + \frac{16x^3}{3}\right]_0^2$$

$$= 2\pi\left[\frac{2^7}{7} - \frac{8(2)^5}{5} + \frac{16(2)^3}{3} - 0\right]$$

$= 61.276\,016\,71\ldots$

$= 61.3$ (3 sig. figs)

∴ the volume is 61.3 units³.

3 i $y^2 = 16\cos^2\dfrac{x}{2}$

$$V = \pi\int_0^{\pi} y^2\, dx$$

$$= 16\pi\int_0^{\pi} \cos^2\frac{x}{2}\, dx$$

$$= 16\pi\int_0^{\pi} \frac{1}{2}(1 + \cos x)\, dx$$

$$= 8\pi\left[x + \sin x\right]_0^{\pi}$$

$$= 8\pi\left[\pi + \sin\pi - 0\right]$$

$= 8\pi^2$

∴ the volume is 8π² units³.

ii $y^2 = \tan^2 x$

$= \sec^2 x - 1$

$$V = \pi\int_{\frac{\pi}{6}}^{\frac{\pi}{3}} y^2\, dx$$

$$= \pi\int_{\frac{\pi}{6}}^{\frac{\pi}{3}} (\sec^2 x - 1)\, dx$$

$$= \pi[\tan x - x]_{\frac{\pi}{6}}^{\frac{\pi}{3}}$$

$$= \pi\left[\tan\frac{\pi}{3} - \frac{\pi}{3} - \left(\tan\frac{\pi}{6} - \frac{\pi}{6}\right)\right]$$

$$= \pi\left[\sqrt{3} - \frac{\pi}{3} - \frac{1}{\sqrt{3}} + \frac{\pi}{6}\right]$$

$$= \pi\left[\sqrt{3} - \frac{\pi}{6} - \frac{1}{\sqrt{3}} \times \frac{\sqrt{3}}{\sqrt{3}}\right]$$

$$= \pi\left[\frac{2\sqrt{3}}{3} - \frac{\pi}{6}\right]$$

$$= \pi\left[\frac{4\sqrt{3} - \pi}{6}\right]$$

$\therefore$ **the volume is $\pi\left[\frac{4\sqrt{3}-\pi}{6}\right]$ units^3.**

$$= \pi\left[81 - 108 + \frac{81}{2}\right]$$

$$= \frac{27\pi}{2}$$

$\therefore$ **the volume is $\frac{27\pi}{2}$ units^3.**

2 i $y = \log_e\left(\frac{x+2}{2}\right)$

$\frac{x+2}{2} = e^y$

$x + 2 = 2e^y$

$x = 2e^y - 2$

$x^2 = (2e^y - 2)^2$

$= 4e^{2y} - 8e^y + 4$

Substitute values of x:

$y(2) = \ln\left(\frac{2+2}{2}\right)$

$= \ln 2$

$y(4) = \ln\left(\frac{4+2}{2}\right)$

$= \ln 3$

$$V = \pi\int_{\ln 2}^{\ln 3} x^2\, dy$$

$$= \pi\int_{\ln 2}^{\ln 3} (4e^{2y} - 8e^y + 4)\, dy$$

$$= \pi\left[2e^{2y} - 8e^y + 4y\right]_{\ln 2}^{\ln 3}$$

$$= \pi\left[2e^{2\ln 3} - 8e^{\ln 3} + 4\ln 3 - (2e^{2\ln 2} - 8e^{\ln 2} + 4\ln 2)\right]$$

$$= \pi[18 - 24 + 4\ln 3 - 8 + 16 - 4\ln 2]$$

$$= \pi[2 + 4(\ln 3 - \ln 2)]$$

$= 11.378\,410\,13\ldots$

$= 11.38$ (2 dec. pl.)

$\therefore$ **the volume is 11.38 units^3.**

ii $y = 2 + \log_e(x+3)$

$y - 2 = \log_e(x+3)$

$x + 3 = e^{y-2}$

$x = e^{y-2} - 3$

$x^2 = (e^{y-2} - 3)^2$

$= e^{2y-4} - 6e^{y-2} + 9$

Substitute values of x:

$y(1) = 2 + \ln(1 + 3)$

$= 2 + \ln 4$

$y(3) = 2 + \ln(3 + 3)$

$= 2 + \ln 6$

$$V = \pi\int_{2+\ln 4}^{2+\ln 6} x^2\, dy$$

$$= \pi\int_{2+\ln 4}^{2+\ln 6} (e^{2y-4} - 6e^{y-2} + 9)\, dy$$

$$= \pi\left[\frac{1}{2}e^{2y-4} - 6e^{y-2} + 9y\right]_{2+\ln 4}^{2+\ln 6}$$

$$= \pi\left[\frac{1}{2}e^{2(2+\ln 6)-4} - 6e^{(2+\ln 6)-2} + 9(2+\ln 6) - \left(\frac{1}{2}e^{2(2+\ln 4)-4} - 6e^{(2+\ln 4)-2} + 9(2+\ln 4)\right)\right]$$

$$= \pi[18 - 36 + 18 + 9\ln 6 - 8 + 24 - 18 - 9\ln 4)]$$

$$= \pi[-2 + 9(\ln 6 - \ln 4)]$$

$= 5.181\,070\,537\ldots$

3 Volumes of solids formed by rotating about the y-axis ➲ pp. 181–182

1 i $y = 2\sqrt{x-1}$

$\frac{y}{2} = \sqrt{x-1}$

$\frac{y^2}{4} = x - 1$

$x = \frac{y^2}{4} + 1$

$= \frac{y^2+4}{4}$

$x^2 = \left(\frac{y^2+4}{4}\right)^2$

$= \frac{y^4 + 8y^2 + 16}{16}$

$$V = \pi\int_2^4 x^2\, dy$$

$$= \frac{\pi}{16}\int_2^4 (y^4 + 8y^2 + 16)\, dy$$

$$= \frac{\pi}{16}\left[\frac{y^5}{5} + \frac{8y^3}{3} + 16y\right]_2^4$$

$$= \frac{\pi}{16}\left[\frac{4^5}{5} + \frac{8(4)^3}{3} + 16(4) - \left(\frac{2^5}{5} + \frac{8(2)^3}{3} + 16(2)\right)\right]$$

$= 74.6$ (1 dec. pl.)

$\therefore$ **the volume is 74.6 units^3.**

ii y-intercept is 9.

$(x - 3)^2 = y$

$x - 3 = \pm\sqrt{y}$

$x = 3 \pm\sqrt{y}$

But as curve passes through (0, 9), then $x = 3 - \sqrt{y}$.

$$V = \pi\int_0^9 x^2\, dy$$

$$= \pi\int_0^9 (3 - \sqrt{y})^2\, dy$$

$$= \pi\int_0^9 (9 - 6\sqrt{y} + y)\, dy$$

$$= \pi\left[9y - 6.\frac{2y^{\frac{3}{2}}}{3} + \frac{y^2}{2}\right]_0^9$$

$$= \pi\left[9y - 4y^{\frac{3}{2}} + \frac{y^2}{2}\right]_0^9$$

$$= \pi\left[9(9) - 4(9)^{\frac{3}{2}} + \frac{9^2}{2} - 0\right]$$

$= 5.18$ (2 dec. pl.)

$\therefore$ **the volume is 5.18 units^3.**

3 i $y = 2\sin^{-1}\dfrac{x}{3}$

$\dfrac{y}{2} = \sin^{-1}\dfrac{x}{3}$

$\dfrac{x}{3} = \sin\dfrac{y}{2}$

$x = 3\sin\dfrac{y}{2}$

$x^2 = 9\sin^2\dfrac{y}{2}$

$$V = \pi\int_{\frac{\pi}{3}}^{\frac{2\pi}{3}} x^2\,dy$$

$$= \pi\int_{\frac{\pi}{3}}^{\frac{2\pi}{3}} 9\sin^2\frac{y}{2}\,dy$$

$$= 9\pi\int_{\frac{\pi}{3}}^{\frac{2\pi}{3}} \frac{1}{2}(1-\cos y)\,dy$$

$$= \frac{9\pi}{2}\int_{\frac{\pi}{3}}^{\frac{2\pi}{3}} (1-\cos y)\,dy$$

$$= \frac{9\pi}{2}\left[y-\sin y\right]_{\frac{\pi}{3}}^{\frac{2\pi}{3}}$$

$$= \frac{9\pi}{2}\left[\frac{2\pi}{3}-\sin\frac{2\pi}{3}-\left(\frac{\pi}{3}-\sin\frac{\pi}{3}\right)\right]$$

$$= \frac{9\pi}{2}\left[\frac{2\pi}{3}-\frac{\sqrt{3}}{2}-\frac{\pi}{3}+\frac{\sqrt{3}}{2}\right]$$

$$= \frac{9\pi}{2}\left[\frac{\pi}{3}\right]$$

$$= \frac{3\pi^2}{2}$$

$\therefore$ **the volume is $\dfrac{3\pi^2}{2}$ units^3.**

ii $y = \tan^{-1}\dfrac{x}{\sqrt{3}}$

$\dfrac{x}{\sqrt{3}} = \tan y$

$x = \sqrt{3}\tan y$

$x^2 = 3\tan^2 y$

Substitute values of x:

$y(1) = \tan^{-1}\dfrac{1}{\sqrt{3}}$

$= \dfrac{\pi}{6}$

$y(3) = \tan^{-1}\dfrac{3}{\sqrt{3}}$

$= \tan^{-1}\sqrt{3}$

$= \dfrac{\pi}{3}$

$$V = \pi\int_{\frac{\pi}{6}}^{\frac{\pi}{3}} x^2\,dy$$

$$= \pi\int_{\frac{\pi}{6}}^{\frac{\pi}{3}} 3\tan^2 y\,dy$$

$$= 3\pi\int_{\frac{\pi}{6}}^{\frac{\pi}{3}} (\sec^2 y-1)\,dy$$

$$= 3\pi\left[\tan y - y\right]_{\frac{\pi}{6}}^{\frac{\pi}{3}}$$

$$= 3\pi\left[\tan\frac{\pi}{3}-\frac{\pi}{3}-\left(\tan\frac{\pi}{6}-\frac{\pi}{6}\right)\right]$$

$$= 3\pi\left[\sqrt{3}-\frac{\pi}{3}-\frac{1}{\sqrt{3}}+\frac{\pi}{6}\right]$$

$$= 3\pi\left[\sqrt{3}-\frac{1}{\sqrt{3}}\times\frac{\sqrt{3}}{\sqrt{3}}-\frac{\pi}{6}\right]$$

$$= 3\pi\left[\sqrt{3}-\frac{\sqrt{3}}{3}-\frac{\pi}{6}\right]$$

$$= 3\pi\left[\frac{2\sqrt{3}}{3}-\frac{\pi}{6}\right]$$

$$= \pi\left[2\sqrt{3}-\frac{\pi}{2}\right]$$

$\therefore$ **the volume is $\pi\left[2\sqrt{3}-\dfrac{\pi}{2}\right]$ units^3.**

4 Volumes of solids formed by rotating regions between two curves ➲ pp. 184–185

1 i Solve simultaneously:

$16 - x^2 = x + 10$

$x^2 + x - 6 = 0$

$(x - 2)(x + 3) = 0$

$x = 2, -3$

$$V = \pi\int_{-3}^{2}((16-x^2)^2-(x+10)^2)\,dx$$

$$= \pi\int_{-3}^{2}(256-32x^2+x^4-(x^2+20x+100))\,dx$$

$$= \pi\int_{-3}^{2}(x^4-33x^2-20x+156)\,dx$$

$$= \pi\left[\frac{x^5}{5}-11x^3-10x^2+156x\right]_{-3}^{2}$$

$$= \pi\left[\frac{2^5}{5}-11(2)^3-10(2)^2+156(2)-\left(\frac{(-3)^5}{5}-11(-3)^3-10(-3)^2+156(-3)\right)\right]$$

$$= 500\pi$$

$\therefore$ **the volume is 500π units^3.**

ii As gradient of the line is $-\dfrac{4}{2} = -2$, the equation is $y = -2x$.

$$V = \pi\int_{0}^{2}((x^2-4x)^2-(-2x)^2)\,dx$$

$$= \pi\int_{0}^{2}(x^4-8x^3+16x^2-4x^2)\,dx$$

$$= \pi\int_{0}^{2}(x^4-8x^3+12x^2)\,dx$$

$$= \pi\left[\frac{x^5}{5}-2x^4+4x^3\right]_{0}^{2}$$

$$= \pi\left[\frac{2^5}{5}-2(2)^4+4(2)^3-0\right]$$

$$= \frac{32\pi}{5}$$

$\therefore$ **the volume is $\dfrac{32\pi}{5}$ units^3.**

2 i Solve simultaneously:

$x^3 = x^2$

$x^3 - x^2 = 0$

$x^2(x-1)=0$

$x=0,1$

$$V=\pi\int_0^1((x^2)^2-(x^3)^2)\,dx$$

$$=\pi\int_0^1(x^4-x^6)\,dx$$

$$=\pi\left[\frac{x^5}{5}-\frac{x^7}{7}\right]_0^1$$

$$=\pi\left[\frac{1^5}{5}-\frac{1^7}{7}-0\right]$$

$$=\pi\left[\frac{1}{5}-\frac{1}{7}\right]$$

$$=\frac{2\pi}{35}$$

$\therefore$ **the volume area is $\frac{2\pi}{35}$ units^3.**

ii Solve simultaneously:

$x^2=4x-x^2$

$2x^2-4x=0$

$2x(x-2)=0$

$x=0,2$

$$V=\pi\int_0^2((4x-x^2)^2-(x^2)^2)\,dx$$

$$=\pi\int_0^2(16x^2-8x^3+x^4-x^4)\,dx$$

$$=\pi\int_0^2(16x^2-8x^3)\,dx$$

$$=\pi\left[\frac{16x^3}{3}-2x^4\right]_0^2$$

$$=\pi\left[\frac{16(2)^3}{3}-2(2)^4-0\right]$$

$=33.510\,321\,64\ldots$

$=33.51$

$\therefore$ **the volume area is 33.51 units^3.**

3 i Solve simultaneously:

$\frac{x^2}{2}-3=4x-3$

$\frac{x^2}{2}-4x=0$

$x^2-8x=0$

$x(x-8)=0$

$x=0,8$

As $x=0$, then $y=-3$ and when $x=8$, $y=29$.

As $y=\frac{x^2}{2}-3$, then $x^2=2y+6$

As $y=4x-3$, then $4x=y+3$

$x=\frac{y+3}{4}$

$x^2=\frac{(y+3)^2}{16}$

$$V=\pi\int_{-3}^{29}\left(2y+6-\frac{(y+3)^2}{16}\right)dy$$

$$=\pi\left[y^2+6y-\frac{(y+3)^3}{48}\right]_{-3}^{29}$$

$$=\pi\left[29^2+6(29)-\frac{(29+3)^3}{48}-\left((-3)^2+6(-3)-\frac{(-3+3)^3}{48}\right)\right]$$

$$=\pi\left[841+174-\frac{32768}{48}-(9-18)\right]$$

$=1072.330\,292\ldots$

$=1070$

$\therefore$ **the volume is 1070 units^3.**

ii Solve simultaneously:

$x\sqrt{x}=\sqrt{x}$

$x\sqrt{x}-\sqrt{x}=0$

$\sqrt{x}(x-1)=0$

$x=0,1$

As $y=x\sqrt{x}$, then $y=x^{\frac{3}{2}}$

$y^{\frac{2}{3}}=x$

$x^2=y^{\frac{4}{3}}$

As $y=\sqrt{x}$, then $y^2=x$

$x^2=y^4$

$$V=\pi\int_0^1\left(y^{\frac{4}{3}}-y^4\right)dy$$

$$=\pi\left[\frac{3y^{\frac{7}{3}}}{7}-\frac{y^5}{5}\right]_0^1$$

$$=\pi\left[\frac{3(1)^{\frac{7}{3}}}{7}-\frac{1^5}{5}-0\right]$$

$$=\pi\left[\frac{3}{7}-\frac{1}{5}\right]$$

$$=\frac{8\pi}{15}$$

$\therefore$ **the volume is $\frac{8\pi}{15}$ units^3.**

4 i By inspection the point of intersection is $\left(\frac{3}{2},1\right)$.

$$V=\pi\int_0^{\frac{3}{2}}\left(\left(\sin\frac{\pi x}{3}\right)^2-\left(\frac{2x}{3}\right)^2\right)dx$$

$$=\pi\int_0^{\frac{3}{2}}\left(\sin^2\frac{\pi x}{3}-\frac{4x^2}{9}\right)dx$$

$$=\pi\int_0^{\frac{3}{2}}\left(\frac{1}{2}\left(1-\cos\frac{2\pi x}{3}\right)-\frac{4x^2}{9}\right)dx$$

$$=\pi\left[\frac{1}{2}\left(x-\frac{3}{2\pi}\sin\frac{2\pi x}{3}\right)-\frac{4x^3}{27}\right]_0^{\frac{3}{2}}$$

$$=\pi\left[\frac{1}{2}\left(\frac{3}{2}-\frac{3}{2\pi}\sin\left[\frac{2\pi}{3}\left(\frac{3}{2}\right)\right]\right)-\frac{4\left(\frac{3}{2}\right)^3}{27}-0\right]$$

$$=\pi\left[\frac{3}{4}-0-\frac{1}{2}-0\right]$$

$$=\frac{\pi}{4}$$

$\therefore$ **the volume is $\frac{\pi}{4}$ units^3.**

ii Solve simultaneously:

$$\frac{1}{2}\sin\frac{x}{2} = \sin x$$
$$\sin\frac{x}{2} = 2\sin x$$
$$\sin\frac{x}{2} = 2\sin 2\left(\frac{x}{2}\right)$$
$$\sin\frac{x}{2} = 2\left(2\sin\frac{x}{2}\cos\frac{x}{2}\right)$$
$$\sin\frac{x}{2} = 4\sin\frac{x}{2}\cos\frac{x}{2}$$
$$4\sin\frac{x}{2}\cos\frac{x}{2} - \sin\frac{x}{2} = 0$$
$$\sin\frac{x}{2}\left(4\cos\frac{x}{2} - 1\right) = 0$$
$$\sin\frac{x}{2} = 0,\quad 4\cos\frac{x}{2} - 1 = 0$$
$$\frac{x}{2} = 0 \qquad \cos\frac{x}{2} = \frac{1}{4}$$
$$x = 0 \qquad \frac{x}{2} = 1.318\,116\,072\ldots$$
$$x = 0 \qquad x = 2.636 \text{ (3 dec. pl.)}$$
$$V = \pi\int_0^{2.636}\left((\sin x)^2 - \left(\frac{1}{2}\sin\frac{x}{2}\right)^2\right)dx$$
$$= \pi\int_0^{2.636}\left(\sin^2 x - \frac{1}{4}\sin^2\frac{x}{2}\right)dx$$
$$= \pi\int_0^{2.636}\left(\frac{1}{2}(1-\cos 2x) - \frac{1}{8}(1-\cos x)\right)dx$$
$$= \frac{\pi}{8}\int_0^{2.636}(4 - 4\cos 2x - 1 + \cos x)\,dx$$
$$= \frac{\pi}{8}\int_0^{2.636}(3 - 4\cos 2x + \cos x)\,dx$$
$$= \frac{\pi}{8}\left[3x - 2\sin 2x + \sin x\right]_0^{2.636}$$
$$= \frac{\pi}{8}[3(2.636) - 2\sin 2(2.636) + \sin 2.636 - 0]$$
$$= 3.961\,253\,563\ldots$$
$$= 3.96 \text{ (correct to 1 dec. pl.)}$$

∴ **the volume is 3.96 units³.**

HSC EXAM-TYPE QUESTIONS

Part A

1 C. Let $x^2 - 4x + 4 = 10 - x^2$

$$2x^2 - 4x - 6 = 0$$
$$x^2 - 2x - 3 = 0$$
$$(x-3)(x+1) = 0$$
$$x = 3, -1$$
$$\text{Area} = \int_{-1}^{3}(10 - x^2 - (x^2 - 4x + 4))\,dx$$
$$= \int_{-1}^{3}(6 + 4x - 2x^2)\,dx$$

2 A. Semicircle rotates to form a sphere of radius 2 units.

$$\text{Volume} = \frac{4}{3}\pi r^3$$
$$= \frac{4}{3}\times\pi\times 2^3$$
$$= \frac{32\pi}{3}$$

3 C. $y = e^x + 1$

$e^x = y - 1$

$x = \log_e(y-1)$

$x^2 = [\log_e(y-1)]^2$

Also, $x = 0, y = 2; x = 1, y = e + 1$.

$$V = \pi\int_2^{e+1}[\log_e(y-1)]^2\,dy$$

4 A.
$$\int_0^{\frac{\pi}{6}} 2\cos 2x\sin x\,dx = \int_0^{\frac{\pi}{6}}(\sin 3x - \sin x)\,dx$$
$$= \left[-\frac{1}{3}\cos 3x + \cos x\right]_0^{\frac{\pi}{6}}$$
$$= -\frac{1}{3}\cos 3\left(\frac{\pi}{6}\right) + \cos\left(\frac{\pi}{6}\right) - \left(-\frac{1}{3}\cos 3(0) + \cos(0)\right)$$
$$= 0 + \frac{\sqrt{3}}{2} - \left(-\frac{1}{3} + 1\right)$$
$$= \frac{\sqrt{3}}{2} - \frac{2}{3}$$

5 D. As $x = e$, then $y = 1$.

$$\therefore \int_1^e \log_e x\,dx = e\times 1 - \int_0^1 e^y\,dy$$
$$= e - \left[e^y\right]_0^1$$
$$= e - (e - 1)$$
$$= 1$$

∴ area is 1 unit².

Part B

6
$$y = \log_e(x+1)^2$$
$$\therefore\ y = 2\log_e(x+1)$$
$$\frac{y}{2} = \log_e(x+1)$$
$$x + 1 = e^{\frac{y}{2}}$$
$$x = e^{\frac{y}{2}} - 1 \ ✓$$
$$\text{Area} = \int_0^{\log_e 4}(e^{\frac{y}{2}} - 1)\,dy$$
$$= \left[2e^{\frac{y}{2}} - y\right]_0^{\log_e 4} \ ✓$$
$$= \left[2e^{\frac{\log_e 4}{2}} - \log_e 4 - (2e^{\frac{0}{2}} - 0)\right]$$
$$= 2e^{\frac{\log_e 4}{2}} - \log_e 4 - 2$$
$$= 2e^{\frac{\log_e 4}{2}} - \log_e 4 - 2$$
$$= 4 - 2\log_e - 2$$
$$= 2 - 2\log_e 2$$
$$= 2(1 - \log_e 2)$$

∴ **area is 2(1 − log$_e$2) units².** ✓

7 a **By inspection,** $\left(\frac{1}{\sqrt{2}}, \frac{\pi}{4}\right)$. ✓

b $y = \sin^{-1} x$

$\sin y = x$

$x^2 = \sin^2 y$

Similarly, other expression is $x^2 = \cos^2 y$.

Considering symmetry:

$$\text{Volume} = 2\pi \int_0^{\frac{\pi}{4}} \sin^2 y \, dy \quad ✓$$

$$= 2\pi \int_0^{\frac{\pi}{4}} \frac{1}{2}(1 - \cos 2y)\, dy$$

$$= \pi \int_0^{\frac{\pi}{4}} (1 - \cos 2y)\, dy$$

$$= \pi \left[y - \frac{1}{2}\sin 2y \right]_0^{\frac{\pi}{4}} \quad ✓$$

$$= \pi \left(\frac{\pi}{4} - \frac{1}{2}\sin 2\left(\frac{\pi}{4}\right) - 0 \right)$$

$$= \pi \left(\frac{\pi}{4} - \frac{1}{2} \right)$$

$$= \pi \left(\frac{\pi - 2}{4} \right)$$

∴ **the volume is** $\pi\left(\frac{\pi - 2}{4}\right) u^3$. ✓

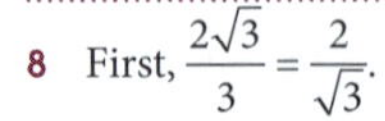

8 First, $\frac{2\sqrt{3}}{3} = \frac{2}{\sqrt{3}}$.

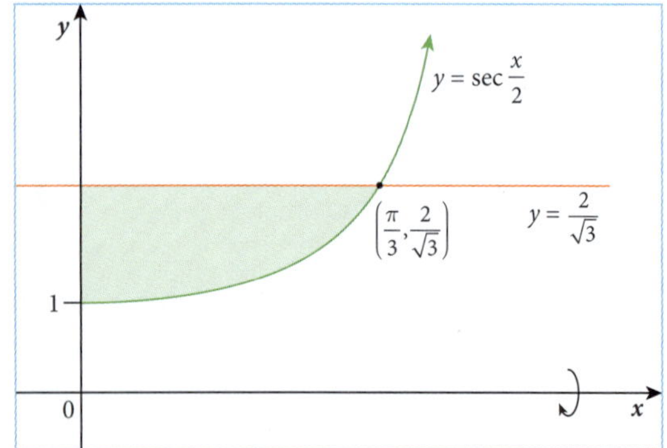

Let $\sec\frac{x}{2} = \frac{2}{\sqrt{3}}$ ✓

$\cos\frac{x}{2} = \frac{\sqrt{3}}{2}$

$\frac{x}{2} = \frac{\pi}{6}$

$x = \frac{\pi}{3}$

∴ point of intersection is $\left(\frac{\pi}{3}, \frac{2}{\sqrt{3}}\right)$. ✓

$$\text{Volume} = \pi \int_0^{\frac{\pi}{3}} \left(\left(\frac{2}{\sqrt{3}}\right)^2 - \left(\sec\frac{x}{2}\right)^2 \right) dx \quad ✓$$

$$= \pi \int_0^{\frac{\pi}{3}} \left(\frac{4}{3} - \sec^2\frac{x}{2} \right) dx$$

$$= \pi \left[\frac{4x}{3} - 2\tan\frac{x}{2} \right]_0^{\frac{\pi}{3}}$$

$$= \pi \left[\frac{4\left(\frac{\pi}{3}\right)}{3} - 2\tan\frac{\left(\frac{\pi}{3}\right)}{2} - 0 \right]$$

$$= \pi \left[\frac{4\pi}{9} - 2\tan\frac{\pi}{6} - 0 \right]$$

$= 0.758\,892\,116\ldots$

$= 0.76$ (2 dec. pl.)

∴ **the volume is 0.76 units³.** ✓

9 a $\frac{x}{3} = \frac{1}{\sqrt{4 - x^2}}$

$\frac{x^2}{9} = \frac{1}{4 - x^2}$

$4x^2 - x^4 = 9$ ✓

$x^4 - 4x^2 + 9 = 0$

Consider the discriminant:

$\Delta = (-4)^2 - 4(1)(9) < 0$

∴ **no solution and hence no point of intersection.** ✓

b $\text{Area} = \int_0^1 \left(\frac{1}{\sqrt{4 - x^2}} - \frac{x}{3} \right) dx$

$$= \left[\sin^{-1}\frac{x}{2} - \frac{x^2}{6} \right]_0^1 \quad ✓$$

$$= \sin^{-1}\frac{1}{2} - \frac{1^2}{6} - 0$$

$$= \frac{\pi}{6} - \frac{1}{6}$$

$$= \frac{\pi - 1}{6}$$

∴ **the area is** $\frac{\pi - 1}{6}$ **units².** ✓

10

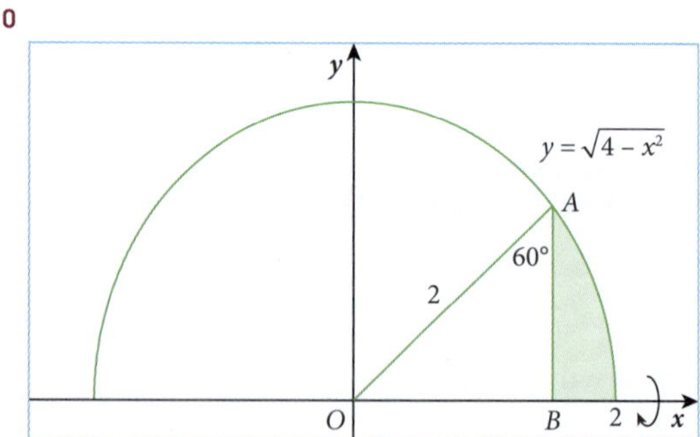

$\frac{OB}{2} = \sin 60°$

$OB = 2 \times \frac{\sqrt{3}}{2}$

$= \sqrt{3}$ ✓

$$V = \pi\int_{\sqrt{3}}^{2} y^2\,dx$$
$$= \pi\int_{\sqrt{3}}^{2}(4-x^2)\,dx \quad ✓$$
$$= \pi\left[4x-\frac{x^3}{3}\right]_{\sqrt{3}}^{2}$$
$$= \pi\left[4(2)-\frac{2^3}{3}-\left(4(\sqrt{3})-\frac{(\sqrt{3})^3}{3}\right)\right]$$
$$= \pi\left[8-\frac{8}{3}-4\sqrt{3}+\sqrt{3}\right]$$
$$= \pi\left[\frac{16}{3}-3\sqrt{3}\right]$$
$$= \pi\left[\frac{16-9\sqrt{3}}{3}\right]$$
$\therefore$ **the volume is** $\pi\left[\frac{16-9\sqrt{3}}{3}\right]$ **units**^{3}. ✓

11 a
$$4\sin^2 x = \sec^2 x$$
$$4\sin^2 x = \frac{1}{\cos^2 x}$$
$$4\sin^2 x\cos^2 x = 1$$
$$2\sin x\cos x = \pm 1$$
$$\sin 2x = \pm 1$$
$$2x = \frac{\pi}{2}, \frac{3\pi}{2}, \ldots$$
$$x = \frac{\pi}{4}\left(\text{as } 0 \le x \le \frac{\pi}{2}\right) \quad ✓$$
Substitute into $y = \sec^2 x$:
$$y = \sec^2\frac{\pi}{4}$$
$$= 2$$
$\therefore P\left(\frac{\pi}{4}, 2\right)$ ✓

b
$$\text{Area} = \int_0^{\frac{\pi}{4}}(\sec^2 x - 4\sin^2 x)\,dx$$
$$= \int_0^{\frac{\pi}{4}}\left(\sec^2 x - 4\times\frac{1}{2}(1-\cos 2x)\right)dx \quad ✓$$
$$= \left[\tan x - 2\left(x-\frac{1}{2}\sin 2x\right)\right]_0^{\frac{\pi}{4}} \quad ✓$$
$$= \tan\frac{\pi}{4} - 2\left(\frac{\pi}{4}-\frac{1}{2}\sin 2\left(\frac{\pi}{4}\right)\right) - 0$$
$$= 1 - 2\left(\frac{\pi}{4}-\frac{1}{2}\right)$$
$$= 2 - \frac{\pi}{2}$$
$$= \frac{4-\pi}{2}$$
$\therefore$ **the area is** $\frac{4-\pi}{2}$ **units**^{2}. ✓

12 a

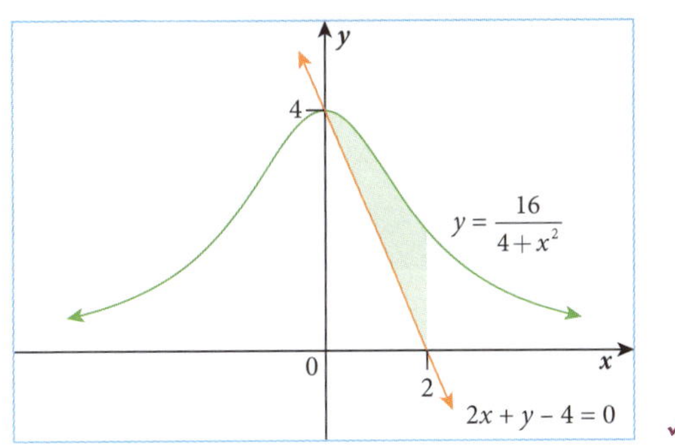

✓✓

b As $2x + y - 4 = 0$, $\therefore y = 4 - 2x$.
$$\text{Area} = \int_0^2\left(\frac{16}{4+x^2}-(4-2x)\right)dx$$
$$= \int_0^2\left(\frac{16}{4+x^2}-4+2x\right)dx$$
$$= \left[8\tan^{-1}\frac{x}{2}-4x+x^2\right]_0^2 \quad ✓$$
$$= 8\tan^{-1}1 - 4(2) + 2^2 - 0$$
$$= 2\pi - 8 + 4$$
$$= 2\pi - 4$$
$$= 2(\pi - 2)$$
$\therefore$ **the area is** $2(\pi - 2)$ **units**^{2}. ✓

13
$$\frac{10}{x} = 7 - x$$
$$10 = 7x - x^2$$
$$x^2 - 7x + 10 = 0$$
$$(x-5)(x-2) = 0$$
$$x = 5, 2 \quad ✓$$
$$\text{Volume} = \pi\int_2^5\left((7-x)^2-\frac{100}{x^2}\right)dx$$
$$= \pi\int_2^5((7-x)^2-100x^{-2})\,dx$$
$$= \pi\left[\frac{(7-x)^3}{-3}+100x^{-1}\right]_2^5 \quad ✓$$
$$= \pi\left[\frac{(7-5)^3}{-3}+\frac{100}{5}-\left(\frac{(7-2)^3}{-3}+\frac{100}{2}\right)\right]$$
$$= \pi\left[-\frac{8}{3}+\frac{100}{5}+\frac{125}{3}-\frac{100}{2}\right]$$
$$= 9\pi$$
$\therefore$ **the volume is** 9π **units**^{3}. ✓

CHAPTER 15 DIFFERENTIAL EQUATIONS

TOPIC: CALCULUS

YEAR 12

1 Verifying solutions of differential equations

- **A differential equation** is an equation involving derivatives, e.g. $\frac{dy}{dx} = 2x + 3$, $\ddot{y} + 2\dot{y} = 6$, etc.
- **Use differentiation** to verify a solution of a differential equation.

EXAMPLE 1

Verify that $y = 2 + e^{3x}$ is a solution to the differential equation $\frac{d^2y}{dx^2} - 2\frac{dy}{dx} - 3y + 6 = 0$.

$$\frac{dy}{dx} = 3e^{3x}$$

$$\frac{d^2y}{dx^2} = 9e^{3x}$$

Verify by substitution.

$$\begin{aligned}\frac{d^2y}{dx^2} - 2\frac{dy}{dx} - 3y + 6 &= 9e^{3x} - 2(3e^{3x}) - 3(2 + e^{3x}) + 6 \\ &= 9e^{3x} - 6e^{3x} - 6 - 3e^{3x} + 6 \\ &= 0\end{aligned}$$

$\therefore y = 2 + e^{3x}$ is a solution.

→ PRACTICE

1 i **Verify that $y = e^{x^3}$ is a solution to the differential equation $\frac{d^2y}{dx^2} - 3x^2\frac{dy}{dx} - 6xy = 0$.**

ii **Verify that $y = \tan^{-1} 2x$ is a solution to the differential equation $(1 + 4x^2)\frac{d^2y}{dx^2} + 8x\frac{dy}{dx} = 0$.**

Answers ⊃ p. 208

2 Direction (slope) field of a differential equation

- **A first-order differential equation** is an equation involving a first derivative, e.g. $\frac{dy}{dx} = 5x + 2$, $x^2 + 2y\frac{dy}{dx} = 6$, etc.
- **A direction field (or slope field)** is a graphical representation of the tangent lines to the solutions of a first-order differential equation.

EXAMPLE 1

A direction field representing a differential equation is shown below.

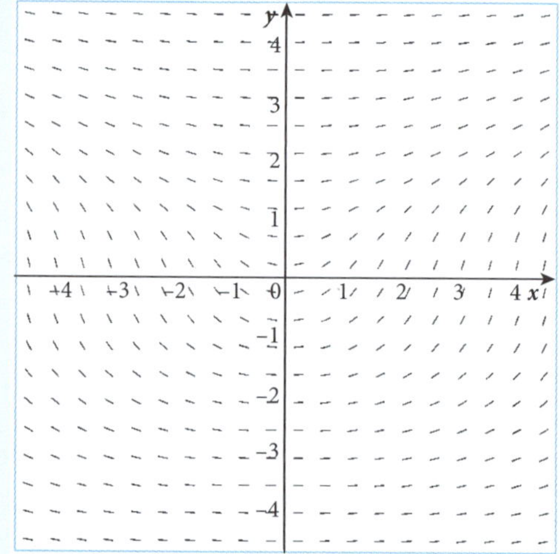

Which of these could be the differential equation?

A $\frac{dy}{dx} = \frac{x}{1+y^2}$

B $\frac{dy}{dx} = \frac{x}{2y}$

C $\frac{dy}{dx} = \frac{-y}{2x}$

D $\frac{dy}{dx} = \frac{-x}{1-y^2}$

Consider the slope for different x and y values.

Consider the slope at some points on the direction field:

x	y	gradient
0	2	horizontal
4	0	+steep
−4	−4	−gentle

By substitution, the correct equation is $\frac{dy}{dx} = \frac{x}{1+y^2}$.

$\therefore$ A

EXAMPLE 2

Which of these diagrams best represents the direction field of the differential equation $\frac{dy}{dx} = xy$?

A

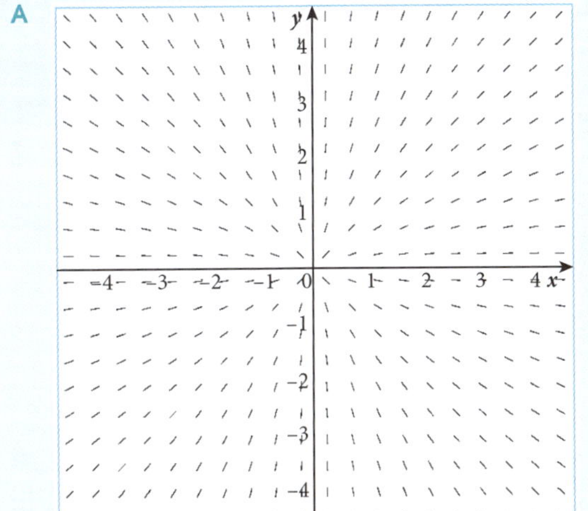

B

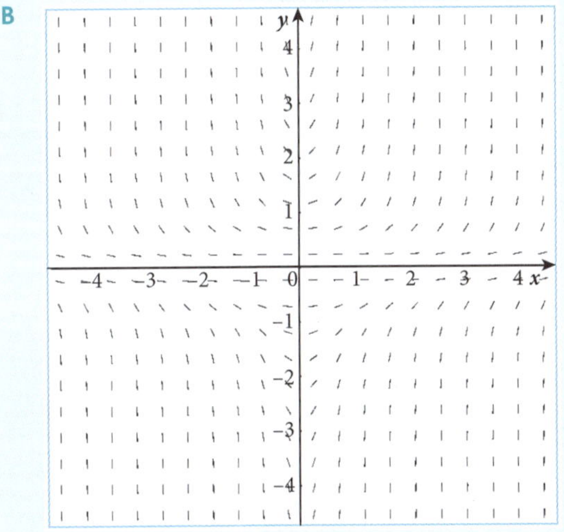

C

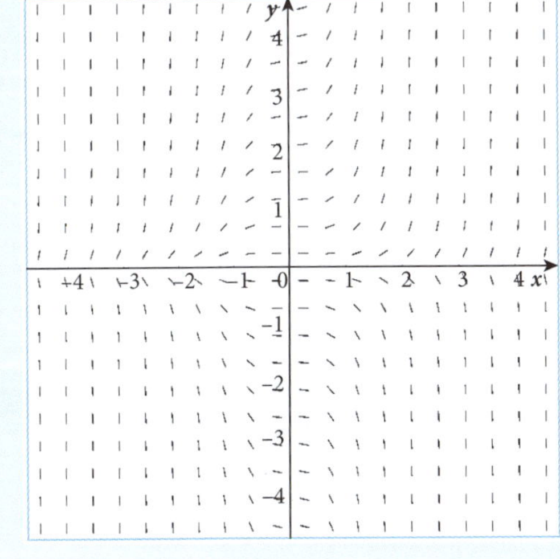

D

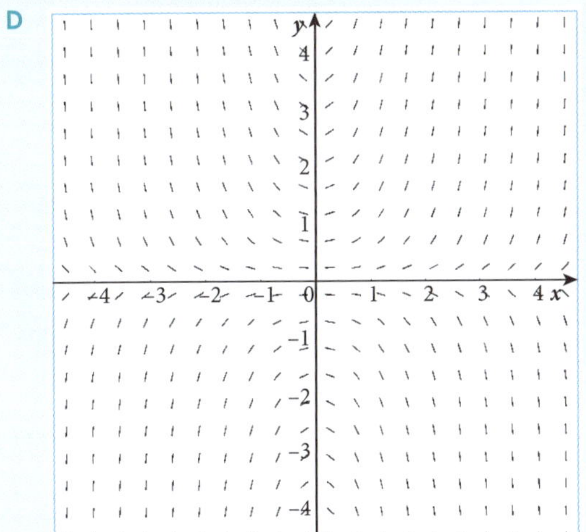

There must be zero gradients on the y-axis, positive gradients in the 1st and 3rd quadrants, and negative gradients in the 2nd and 4th quadrants.
$\therefore$ D

Consider $x = 0$ and $y = 0$ impact on gradient.

EXAMPLE 3

A slope field representing the differential equation $\frac{dy}{dx} = \frac{x}{1+y}$ is shown below.

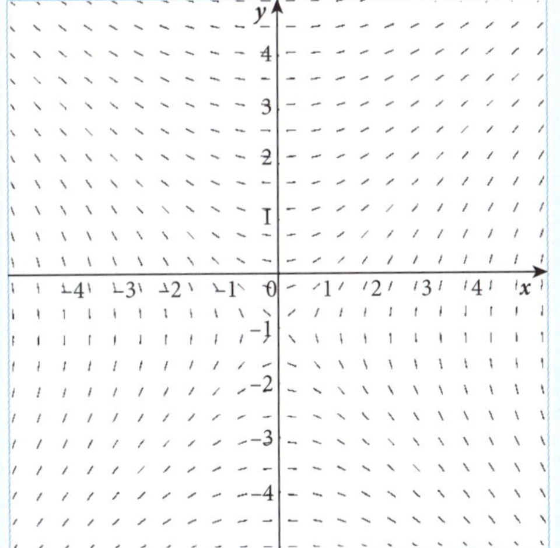

Sketch the solution curve of the differential equation corresponding to the condition $y(0) = 1$.

Start by plotting point.

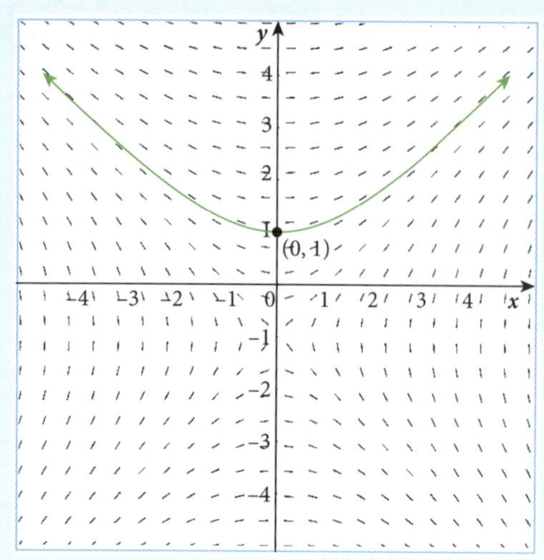

→ PRACTICE

1 i A direction field representing a differential equation is shown below.

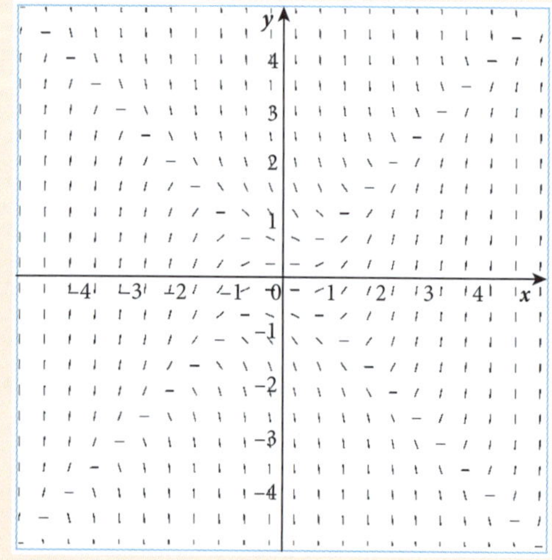

Which of these could be the differential equation?

A $\dfrac{dy}{dx} = \dfrac{1}{x^2 - y^2}$

B $\dfrac{dy}{dx} = \dfrac{1}{x^2 + y^2}$

C $\dfrac{dy}{dx} = x^2 - y^2$

D $\dfrac{dy}{dx} = x^2 + y^2$

ii A direction field representing a differential equation is shown below.

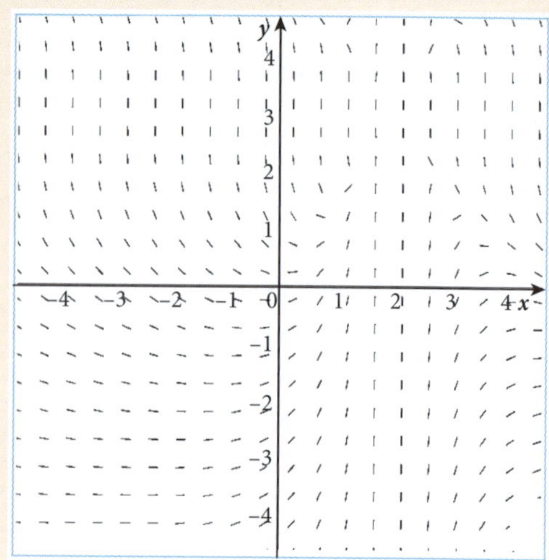

Which of these could be the differential equation?

A $\dfrac{dy}{dx} = \dfrac{(x+2)^2}{4} + \dfrac{(y+3)^2}{9}$

B $\dfrac{dy}{dx} = \dfrac{(x-2)^2}{4} + \dfrac{(y-3)^2}{9}$

C $\dfrac{dy}{dx} = \dfrac{4}{(x-2)^2} + \dfrac{9}{(y-3)^2}$

D $\dfrac{dy}{dx} = \dfrac{4}{(x-2)^2} - \dfrac{9}{(y-3)^2}$

2 i Which of these diagrams best represents the direction field of the differential equation $\dfrac{dy}{dx} = \dfrac{2x+y}{y-2x}$?

A

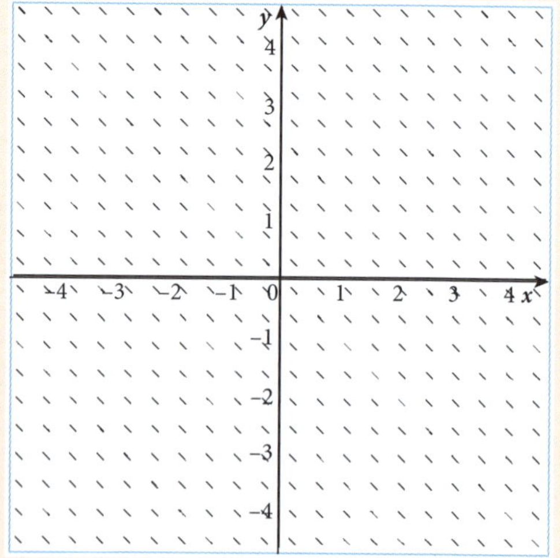

B

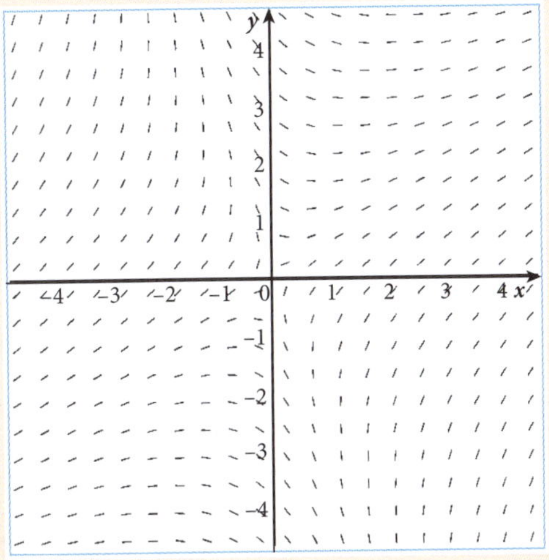

C

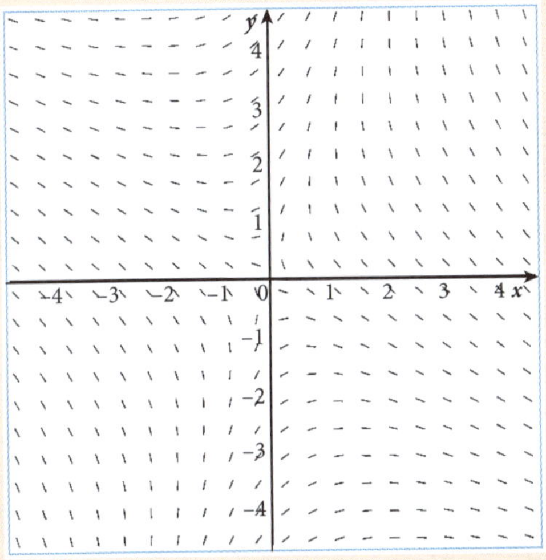

D

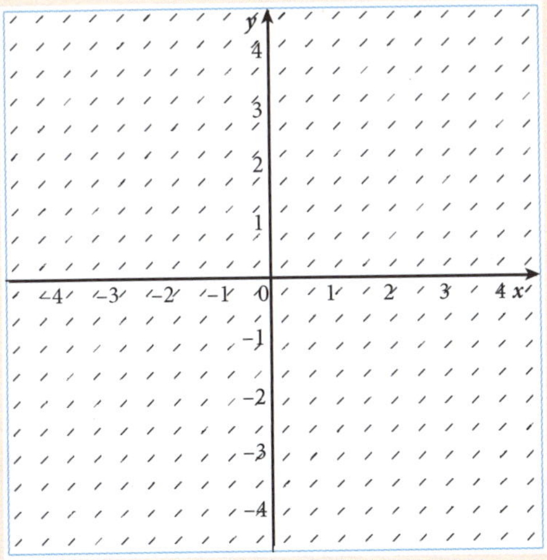

ii Which of these diagrams best represents the direction field of the differential equation $\frac{dy}{dx} = \frac{1}{x^2 - y^2}$?

A

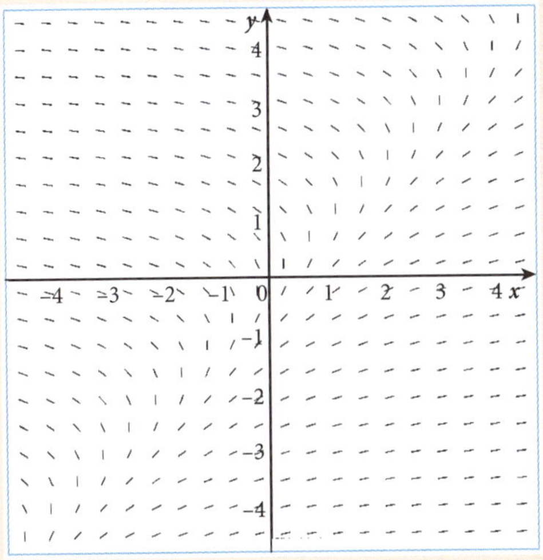

B

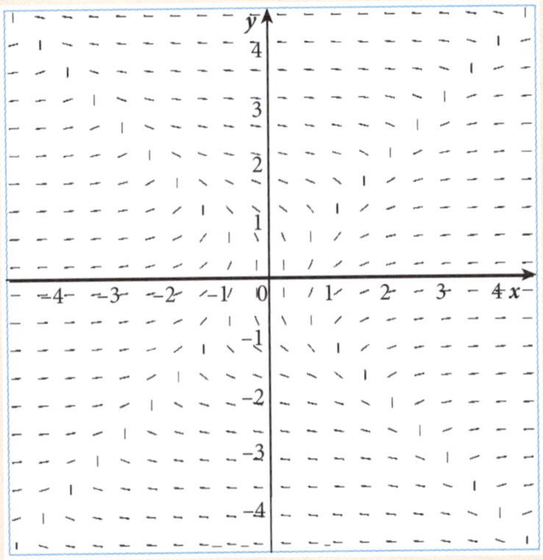

C

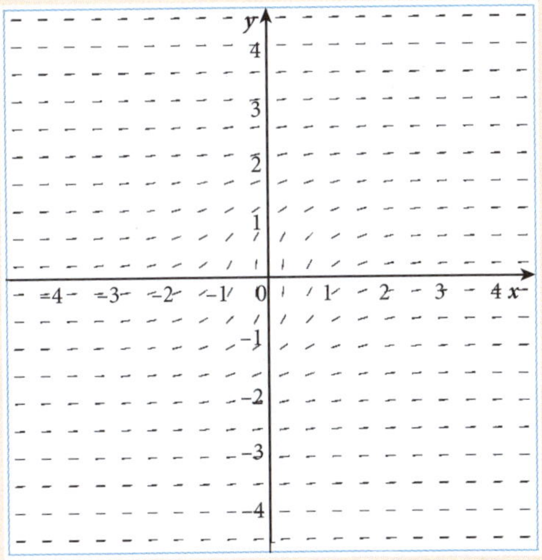

D

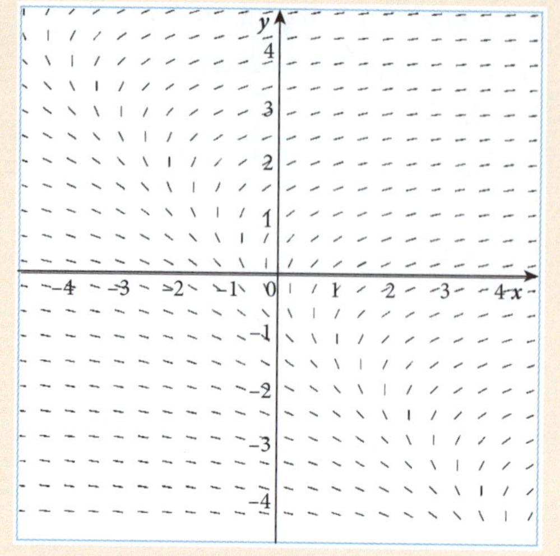

3 **i** A slope field representing the differential equation $\frac{dy}{dx} = \frac{x-1}{y+1}$ is shown below.

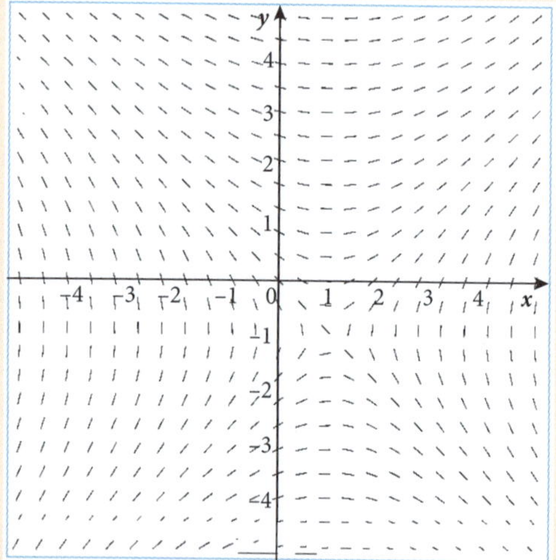

Sketch the solution curve of the differential equation corresponding to the condition $y(1) = 2$.

ii A slope field representing the differential equation $\frac{dy}{dx} = \frac{x+2}{y}$ is shown below.

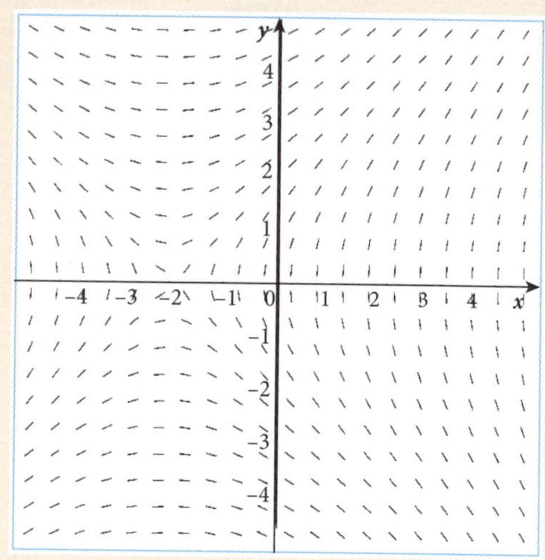

Sketch the solution curve of the differential equation corresponding to the condition $y(-2) = 1$.

Answers ➲ p. 208

3 Solving differential equations of the form $\frac{dy}{dx} = f(x)$

➔ If $\frac{dy}{dx} = f(x)$, then $\int dy = \int f(x)\,dx$. Hence, $y = \int f(x)\,dx$.

EXAMPLE 1

Find the general solution of the differential equation $2x\frac{dy}{dx} - 3x^2 = 5$.

$$2x\frac{dy}{dx} - 3x^2 = 5$$

$$2x\frac{dy}{dx} = 3x^2 + 5$$

$$\frac{dy}{dx} = \frac{3x^2}{2x} + \frac{5}{2x}$$

$$y = \int\left(\frac{3x}{2} + \frac{5}{2x}\right)dx$$

$$y = \frac{3x^2}{4} + \frac{5}{2}\log_e x + c$$

$\int dy = y$

EXAMPLE 2

Solve the differential equation $\frac{dy}{dx} + 4\sin 2x = 0$, with the condition $y\left(\frac{\pi}{3}\right) = 0$.

$$\frac{dy}{dx} + 4\sin 2x = 0$$

$$\frac{dy}{dx} = -4 \sin 2x$$

$$y = \int -4\sin 2x \, dx$$

$$= 2 \cos 2x + c$$

Substitute $y\left(\frac{\pi}{3}\right) = 0$:

$$0 = 2 \cos 2\left(\frac{\pi}{3}\right) + c$$

$$0 = 2 \cos \frac{2\pi}{3} + c$$

$y\left(\frac{\pi}{3}\right) = 0$ means $x = \frac{\pi}{3}$ and $y = 0$.

$$0 = 2\left(-\frac{1}{2}\right) + c$$

$$c = 1$$

$$\therefore y = 2 \cos 2x + 1$$

→ PRACTICE

1 i Find the general solution of the differential equation $x^2\frac{dy}{dx} + x = 3$.

ii Find the general solution of the differential equation $\sqrt{4-3x}\frac{dy}{dx} + 3x = 4$.

2 i Solve the differential equation $\frac{dy}{dx} - 4 \sin 3x \cos 3x = 0$, with the condition $y\left(\frac{\pi}{3}\right) = \frac{2}{3}$.

ii Solve the differential equation $(1 + \tan^2 x)\frac{dy}{dx} = 1$, with the condition $y\left(\frac{\pi}{4}\right) = \frac{1}{4}$.

Answers ➲ pp. 208–209

4 Solving differential equations of the form $\frac{dy}{dx} = g(y)$

→ If $\frac{dy}{dx} = g(y)$, then $\int dx = \int \frac{1}{g(y)} dy$.

Hence, $x = \int \frac{1}{g(y)} dy$.

EXAMPLE 1

Find the general solution of $2y + \frac{dy}{dx} = 0$.

$$2y + \frac{dy}{dx} = 0$$

$$\frac{dy}{dx} = -2y$$

$$\int \frac{1}{y} dy = \int -2 \, dx$$

$$\ln|y| = -2x + c$$

Write the constant c on RHS.

$$y = e^{-2x + c}$$

$y = Ae^{-2x}$, for constant A, where $A = e^c$.

EXAMPLE 2

Solve the differential equation $\frac{dy}{dx} = 1 - \frac{2y}{3}$, with the condition $x = 3$, $y = 1$.

$$\frac{dy}{dx} = \frac{3-2y}{3}$$

$$\int \frac{3}{3-2y} dy = \int dx$$

$$-\frac{3}{2}\ln|3 - 2y| = x + c$$

Substitute $y(3) = 1$:

Make y the subject.

$$-\frac{3}{2}\ln|3 - 2(1)| = 3 + c$$

$$0 = 3 + c$$

$$c = -3$$

$$-\frac{3}{2}\ln|3 - 2y| = x - 3$$

$$\ln|3 - 2y| = -\frac{2}{3}(x - 3)$$

$$\ln|3 - 2y| = \frac{6-2x}{3}$$

$$3 - 2y = e^{\frac{6-2x}{3}}$$

$$2y = 3 - e^{\frac{6-2x}{3}}$$

$$y = \frac{1}{2}(3 - e^{\frac{6-2x}{3}})$$

→ PRACTICE

1 i Find the general solution of $3y - \frac{dy}{dx} = 0$.

ii Show that the general solution of $\frac{dy}{dx} - 3\sqrt{y} = 0$ is given as $y = -\left(\frac{3x}{2}+c\right)^2$.

2 i Solve the differential equation $\frac{dy}{dx} - (2y - 1)^2 = 0$, with the condition $x = 2, y = 1$.

ii Solve the differential equation $\frac{dy}{dx} = y^2 + 4$, with the condition $y\left(\frac{\pi}{6}\right) = 2$.

Answers ➲ pp. 209–210

5 Solving differential equations of the form $\frac{dy}{dx} = f(x)\,g(y)$

➔ If $\frac{dy}{dx} = f(x)g(y)$, then $\int \frac{1}{g(y)}\,dy = \int f(x)dx$.

EXAMPLE 1

Solve the differential equation $\frac{dy}{dx} = \frac{-x}{1+y^2}$, with the condition $y(-1) = 1$.

$$\frac{dy}{dx} = \frac{-x}{1+y^2}$$

Separate the variables.

$$(1 + y^2)\,dy = -x\,dx$$

$$\int (1+y^2)\,dy = \int -x\,dx$$

$$y + \frac{y^3}{3} = -\frac{x^2}{2} + c$$

Substitute $y(-1) = 1$:

$$1 + \frac{1^3}{3} = -\frac{(-1)^2}{2} + c$$

$$1\frac{1}{3} = -\frac{1}{2} + c$$

$$c = \frac{11}{6}$$

$$\therefore y + \frac{y^3}{3} = -\frac{x^2}{2} + \frac{11}{6}$$

$$\therefore 6y + 2y^3 = -3x^2 + 11$$

$$\therefore 3x^2 + 2y^3 + 6y - 11 = 0$$

EXAMPLE 2

Given the differential equation $\frac{dy}{dx} = e^{-2y}\cos x$, with the condition $y(\pi) = 0$, show that $y = \frac{1}{2}\ln|2\sin x + 1|$.

$$\frac{dy}{dx} = e^{-2y}\cos x$$

$$e^{2y}\,dy = \cos x\,dx$$

$$\int e^{2y}\,dy = \int \cos x\,dx$$

2 x constant is still a constant.

$$\frac{1}{2}e^{2y} = \sin x + c$$

Substitute $y(\pi) = 0$:

$$\frac{1}{2}e^{2(0)} = \sin \pi + c$$

$$\frac{1}{2} = c$$

$$\frac{1}{2}e^{2y} = \sin x + \frac{1}{2}$$

$$e^{2y} = 2\sin x + 1$$

$$2y = \ln|2\sin x + 1|$$

$$y = \frac{1}{2}\ln|2\sin x + 1|$$

➔ PRACTICE

1 i For the differential equation $\frac{dy}{dx}\sin 2y - \cos x = 0$, with the condition $y\left(\frac{\pi}{2}\right) = 0$, show that $2\sin x + \cos 2y - 3 = 0$.

ii For the differential equation $\frac{dy}{dx} = (y + 5)(x - 2)$, with the condition $y(2) = -4$, show that $\frac{x^2}{2} - 2x + 2 = \ln|y + 5|$.

2 i For the differential equation $\sqrt{2-x^2}\frac{dy}{dx} = \frac{1}{2-y}$, with the condition $y(1) = 0$, show that $2y - \frac{y^2}{2} = \sin^{-1}\frac{x}{\sqrt{2}} - \frac{\pi}{4}$.

ii Solve the differential equation $\frac{dy}{dx} + y^2 = 2xy^2$, with the condition $y(2) = 1$.

Answers ➲ p. 210

6 Exponential growth and decay models

➔ This topic was introduced in Year 11. See page 65 in this study guide.

➔ $\frac{dP}{dt} = kP$

➔ $\frac{dP}{dt} = k(P - A)$

EXAMPLE 1

The population P of a city at time t years after a certain date increases at a rate proportional to the population at that time. Construct and solve a differential equation for P in terms of t.

$$\frac{dP}{dt} = kP$$

$$\int \frac{1}{P}\,dP = \int k\,dt$$

α is used to represent 'is proportional to'.

$$\ln|P| = kt + c$$
$$P = e^{kt+c}$$
$$P = Ae^{kt}, \text{ for constant } A, \text{ where } A = e^c.$$

EXAMPLE 2

According to Newton's Law of Cooling, the rate of change in the temperature of an object is proportional to the difference between its temperature and the temperature of the air (or surrounding matter). The air temperature is assumed to be constant.

A heated saucepan is placed in a room which has a temperature of 22 °C. The initial temperature of the saucepan is 70 °C and cools to 35 °C after 5 minutes. Let T be the temperature of the saucepan after t minutes.

Use this information to construct a differential equation for $\frac{dT}{dt}$ and solve it to find an equation for T in terms of t.

Using Newton's law of cooling,

$$\frac{dT}{dt} = -k(T-22)$$

$$\int \frac{1}{T-22}\,dT = \int -k\,dt$$

$$\ln(T-22) = -kt + c$$

Substitute $t(0) = 70$:

$$\ln(70-22) = -k(0) + c$$
$$c = \ln 48$$
$$\ln(T-22) = -kt + \ln 48$$
$$-\ln(T-22) = kt - \ln 48$$
$$\ln 48 - \ln(T-22) = kt$$
$$\ln\left(\frac{48}{T-22}\right) = kt$$
$$\frac{48}{T-22} = e^{kt}$$
$$T - 22 = 48e^{-kt}$$
$$T = 22 + 48e^{-kt}$$

Substitute $T = 35$, $t = 5$:

$$35 = 22 + 48e^{-k(5)}$$
$$48e^{-5k} = 13$$
$$e^{-5k} = \frac{13}{48}$$
$$-5k = \log_e \frac{13}{48}$$
$$k = -\frac{1}{5}\log_e \frac{13}{48}$$
$$= 0.261\,250\,33\ldots$$
$$= 0.2613 \text{ (4 dec. pl.)}$$
$$\therefore T = 22 + 48e^{-0.2613t}$$

Find the value of the constant of proportionality k.

→ PRACTICE

1 i The production output N, in units, in a factory increases at a rate proportional to the existing output over time t, in weeks. Construct and solve a differential equation for P in terms of t, assuming the initial output is 120 units.

ii The rate of decay of a mass m, in grams, of a radioactive substance over time t, in years, is proportional to the mass itself. Initially the mass of the substance is 60 g. Construct a differential equation and solve this equation to show $m = 60e^{-kt}$. Then, if the mass of substance decays to 45 g after 5 years, find the time taken to reduce the mass to 10 g.

2 i On a hot day a can of soft drink is taken from a coolroom which has a temperature of 3.2 °C. It is placed in a room with a constant temperature of 36 °C. After 20 minutes the temperature of the can is 15 °C. According to Newton's Law of Cooling, the rate of change in the temperature of an object is proportional to the difference between its temperature and the temperature of the air (or surrounding matter). The air temperature is assumed to be constant.

Let T be the temperature of the saucepan after t minutes.

Use this information to construct a differential equation for $\frac{dT}{dt}$ and solve it to find an equation for T in terms of t.

ii The rate of growth over t years of a town's population P is proportional to the excess of population over 12 000. In 2018 the population of the town was 14 400. Construct a differential equation and solve it to show $P = 12\,000 + 2400e^{kt}$. Then, if the town's population increases to 18 200 by 2030, according to this model, what will be the population in 2048?

Answers ➲ pp. 211–212

7 Mixing problems

➔ **These problems involve a differential equation** based on the concentration of a substance in a container.

➔ $\dfrac{dQ}{dt}$ **= rate of inflow – rate of outflow**

➔ **Solve the differential equation** to find the equation for Q in terms of t.

EXAMPLE 1

Water mixed with salt is called brine. Brine containing 3 g of salt per litre of water flows into a tank containing 300 litres of water which already contains 50 g of salt. The brine enters the tank at a rate of 15 L/min, completely mixes with the existing salty water and a mixture leaves the container at 15 L/min.

Let $Q(t)$ be the amount of salt, in grams, in the tank at any time t, in minutes.

a **Explain why $\dfrac{dQ}{dt} = 45 - \dfrac{Q}{20}$.**

b **Solve the differential equation to show $Q = 900 - 850\,e^{-0.05t}$.**

a $\dfrac{dQ}{dt}$ = rate of inflow – rate of outflow

Rate of inflow = 3 g/L × 15 L/min
= 45 g/min

Rate of outflow = Q g/300 L × 15 L/min
$= \dfrac{Q}{20}$ g/min

$\therefore \dfrac{dQ}{dt} = 45 - \dfrac{Q}{20}$

b $\dfrac{dQ}{dt} = 45 - \dfrac{Q}{20}$

$\dfrac{dQ}{dt}$ is rate of change of salt in the tank.

$= \dfrac{900 - Q}{20}$

$\dfrac{dt}{dQ} = \dfrac{20}{900 - Q}$

$dt = \dfrac{20}{900 - Q}\,dQ$

$\int dt = \int \dfrac{20}{900 - Q}\,dQ$

$t = 20\int \dfrac{1}{900 - Q}\,dQ$

$= -20\ln(900 - Q) + c$

Substitute $t = 0$, $Q = 50$:

$0 = -20\ln(900 - 50) + c$

$0 = -20\ln 850 + c$

$c = 20\ln 850$

$t = -20\ln(900 - Q) + 20\ln 850$

$= 20\ln\dfrac{850}{900 - Q}$

$0.05t = \ln\dfrac{850}{900 - Q}$

$\dfrac{850}{900 - Q} = e^{0.05t}$

$850\,e^{-0.05t} = 900 - Q$

$Q = 900 - 850\,e^{-0.05t}$

➔ PRACTICE

1 i **Suppose a tank contains 60 g of salt mixed in 1250 L of water. Brine with a concentration of 2 g of salt per litre is flowing into the tank at a rate of 5 L/minute. The tank is well stirred and brine flows out at a rate of 5 L/minute.**

a **Explain why $\dfrac{dM}{dt} = 10 - \dfrac{M}{250}$.**

b **Solve the differential equation to show $M = 2500 - 2440\,e^{-0.004t}$.**

ii **A metal vat contains 500 L of pure water. Ethanol is poured into the water at a rate of 10L/minute. Some rotors in the vat mix thoroughly the ethanol with the water. At the same time that the ethanol is being poured in, a drain is opened at the bottom of the vat to drain the mix of water and ethanol from the vat at the rate of 10L/minute.**

Let $E(t)$ be the amount of ethanol, in litres, in the vat at any time t, in minutes.

a **Explain why $\dfrac{dE}{dt} = 10 - \dfrac{E}{50}$.**

b **Solve the differential equation to show $Q = 100 - 100e^{-0.01t}$.**

c **Find the amount of ethanol present in the tank after 2 hours. Leave your answer to the nearest litre.**

Answers ➲ p. 212

8 The logistic growth model

→ **In this model the rate of change of the population *P*** is proportional to both the size of *P* and the difference between the maximum value of *P* and *P* at any given time *t*.

→ $\frac{dP}{dt} = kP(M - P)$

EXAMPLE 1

A remote location with no feral animals has been fenced and used to set up a breeding population of Tasmanian devils. Initially 24 devils were released in the area which is expected to support a population of 120 devils. After 3 years, the population has grown to 60.

Use the logistic model $\frac{dP}{dt} = kP(M - P)$ to show $P = \frac{30e^{120kt}}{1+0.25e^{120kt}}$ where $k = \frac{\log_e 4}{360}$.

[Hint: Use the result $\frac{1}{P(120-P)} = \frac{1}{120P} + \frac{1}{120(120-P)}$]

Maximum population $M = 120$

$$\frac{dP}{dt} = kP(120 - P)$$

$$\frac{1}{P(120-P)}\,dP = k\,dt$$

$$\int \frac{1}{P(120-P)}\,dP = \int k\,dt$$

$$\int \left(\frac{1}{120P} + \frac{1}{120(120-P)}\right) dP = kt + c$$

$$\frac{1}{120}\int \left(\frac{1}{P} + \frac{1}{(120-P)}\right) dP = kt + c$$

$$[\ln |P| - \ln|120 - P|] = 120(kt + c)$$

$$\frac{P}{120-P} = e^{120(kt+c)}$$

$$\frac{P}{120-P} = Ae^{120kt}, \text{ for constant } A, \text{ where } A = e^{120c}$$

Substitute $P = 24$, $t = 0$:

$$\frac{24}{120-24} = Ae^{120k(0)}$$

$$A = 0.25$$

Find an expression for *P* in terms of *t*.

$$\frac{P}{120-P} = 0.25e^{120kt}$$

$$P = (120 - P)0.25e^{120kt}$$

$$P = 30e^{120kt} - 0.25Pe^{120kt}$$

$$P + 0.25Pe^{120kt} = 30e^{120kt}$$

$$P(1 + 0.25e^{120kt}) = 30e^{120kt}$$

$$P = \frac{30e^{120kt}}{1+0.25e^{120kt}}$$

Substitute $P = 60$, $t = 3$:

$$60 = \frac{30e^{120(3)k}}{1+0.25e^{120(3)k}}$$

$$60(1 + 0.25e^{360k}) = 30e^{360k}$$

$$60 + 15e^{360k} = 30e^{360k}$$

$$15e^{360k} = 60$$

$$e^{360k} = 4$$

$$360k = \log_e 4$$

$$k = \frac{\log_e 4}{360}$$

→ PRACTICE

1 i **A leading academic has suggested that the maximum population that Australia can support is 60 million. The population in 1981 was 15 million and by 2018 the population had reached 25 million.**

a **Use the logistic model $\frac{dP}{dt} = kP(M - P)$ to show $P = \frac{60}{1+3e^{-60kt}}$ where $k = \frac{1}{2220}\log_e \frac{15}{7}$.**

[Use the result $\frac{1}{P(60-P)} = \frac{1}{60P} + \frac{1}{60(60-P)}$.]

b **Hence, find the population in 2050, to the nearest thousand.**

ii **The area A cm^2 is occupied by a bacterial colony. The colony has its growth modelled by the logistic equation $\frac{dA}{dt} = \frac{1}{20}A(40 - A)$, where $t \geq 0$ and t is measured in days. Initially, the area occupied by the bacteria colony is 0.2 cm^2.**

a **Show that $\frac{1}{A(40-A)} = \frac{1}{40}\left(\frac{1}{A} + \frac{1}{40-A}\right)$.**

b **Using the result from part a, solve the logistic equation and hence show $A = \frac{40}{1+199e^{-2t}}$.**

c **According to this model, what is the limiting area of the bacteria colony?**

Answers ⊃ p. 213

HSC EXAM-TYPE QUESTIONS

Total marks: 35

PART A

(Suggested time: 10 minutes)

(1 mark each)

1 Which of these diagrams best represents the direction field of the differential equation $\frac{dy}{dx} = \frac{x-y}{x+y}$?

A

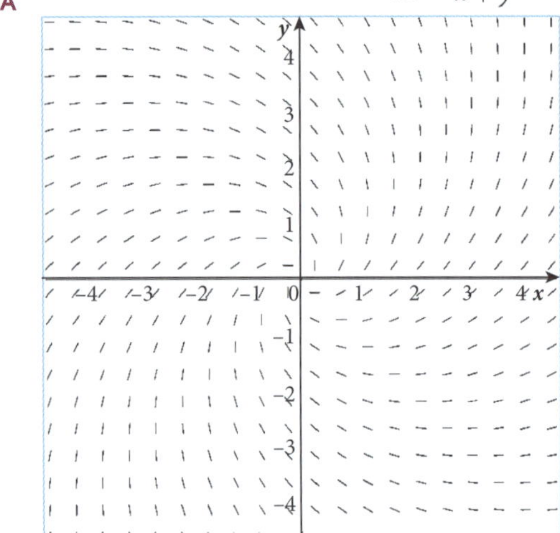

B

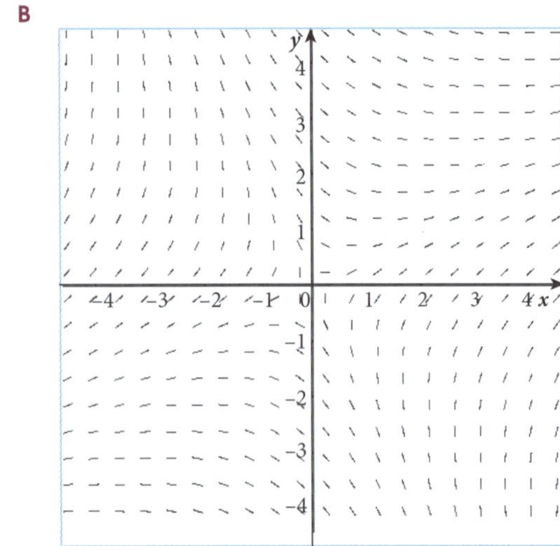

C

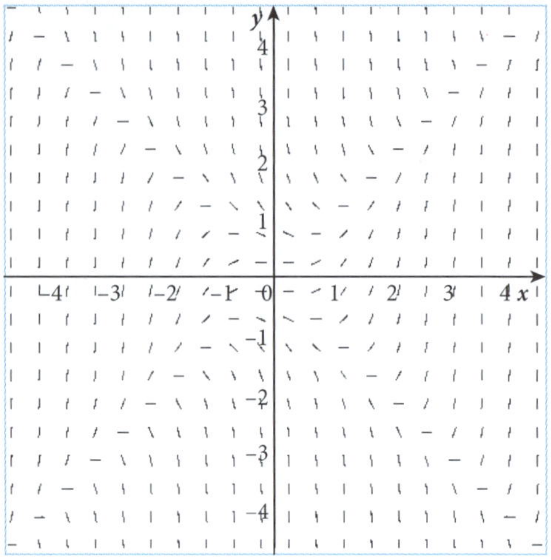

D

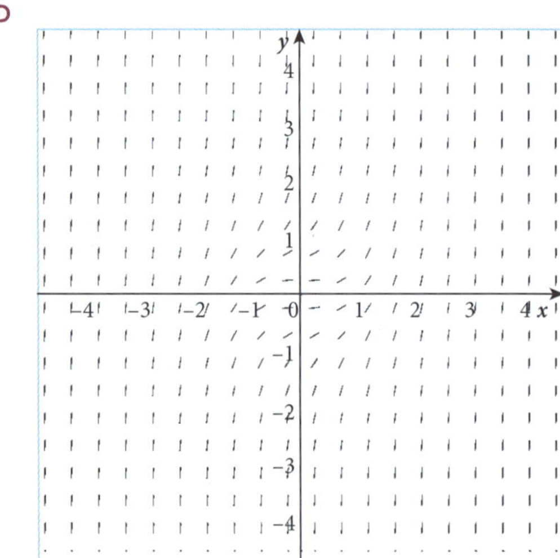

2 Which of these is the general solution of $\frac{dy}{dx} = 2xy + y$?

A $y = Ce^{x^2+1}$, where C is a constant

B $y = Ce^{x^2+x}$, where C is a constant

C $y = x^2 + x + C$, where C is a constant

D $y = e^{x^2+x} + C$, where C is a constant

3 What are the graphs of the solutions to $\frac{dy}{dx} = -\frac{x}{y}$?

A circles　　**B** parabolas

C hyperbolas　　**D** straight lines

4 Which of these is a solution of $\frac{d^2y}{dx^2} - 3\frac{dy}{dx} + 2y = 0$?

A $y = 1$　　**B** $y = x$

C $y = e^{-x}$　　**D** $y = e^{2x}$

5 What is the differential equation formed by $y = 2\cos x + 3\sin x + 4$?

A $\frac{d^2y}{dx^2} + y = 0$

B $\frac{d^2y}{dx^2} - y = 0$

C $\frac{d^2y}{dx^2} + y = 4$

D $\frac{d^2y}{dx^2} + y = -4$

PART B

(Suggested time: 45 minutes)

Show all working.

6 A slope field representing the differential equation $\frac{dy}{dx} = x^2 + y$ is shown below.

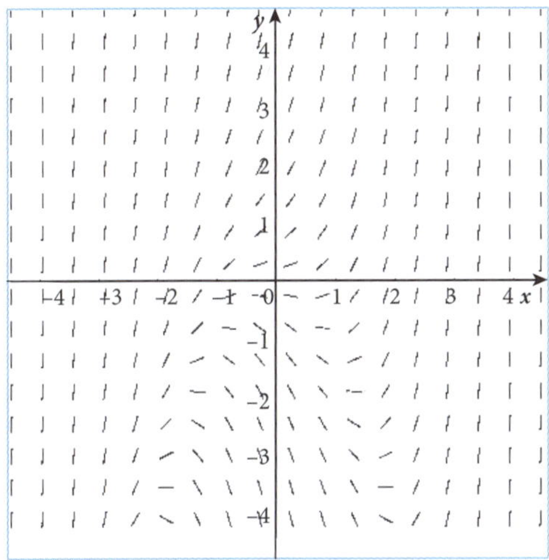

Sketch the solution curve of the differential equation corresponding to the condition $y(2) = -2$. (2 marks)

7 Consider the differential equation $\frac{dy}{dx} = \frac{y}{2x}$.

If $y = 1$ when $x = 4$, solve the differential equation, leaving your answer with y in terms of x. (2 marks)

8 Find the general solution of the differential equation $3x^2 \frac{dy}{dx} - x = 0$. (2 marks)

9 Solve the differential equation $(x^2 + 4)\frac{dy}{dx} - 2xy = 0$, with the condition $y(0) = 1$. (3 marks)

10 Find y given that $\frac{dy}{dx} = x + xy^2$ and $y(0) = 1$. (2 marks)

11 The number N of bacteria in a colony grows according to the differential equation $\frac{dN}{dt} = kN$, where k is a positive constant and t is the number of days. The number of bacteria present increases from 6000 to 8500 in 3 days.

a Solve the differential equation to show $N = 6000e^{kt}$. (2 marks)

b What will be the number of bacteria present in another 3 days? (2 marks)

12 A population of size P is decreasing according to the law $\frac{dP}{dt} = -0.003P$, where t is the time in days. The initial population is p_0.

a Solve the differential equation to show $P = p_0e^{-0.003t}$. (2 marks)

b Find to the nearest day how long it will take for the population to halve? (2 marks)

13 According to Newton's Law of Cooling, the rate of change in the temperature of an object is proportional to the difference between its temperature and the temperature of the air (or surrounding matter). The air temperature is assumed to be constant.

A cup of coffee is placed in a room which has a temperature of 18 °C. The initial temperature of the coffee is 75 °C and cools to 40 °C after 3 minutes. Let T be the temperature of the saucepan after t minutes.

a Use this information to construct a differential equation for $\frac{dT}{dt}$ and solve it to find an equation for T in terms of t. (3 marks)

b How long will it take for the temperature of the coffee to drop to 20 °C? Give your answer to the nearest 10 seconds. (2 marks)

14 Water mixed with salt is called brine. Brine containing 4 g of salt per litre of water flows into a tank containing 1000 L of water which already contains 200 g of salt. The brine enters the tank at a rate of 10 L/min, completely mixes with the existing salty water and a mixture leaves the container at 10 L/min.

Let $Q(t)$ be the amount of salt, in grams, in the tank at any time t, in minutes.

a Explain why $\frac{dQ}{dt} = 40 - \frac{Q}{100}$. (2 marks)

b Solve the differential equation to show $Q = 4000 - 3800\,e^{-0.01t}$. (3 marks)

c Find the amount of salt in the tank after 10 minutes. (1 mark)

PRACTICE

1 Verifying solutions of differential equations ⊃ p. 196

1 i
$$y = e^{x^3}$$
$$\frac{dy}{dx} = 3x^2e^{x^3}$$
$$\frac{d^2y}{dx^2} = 9x^4e^{x^3} + 6xe^{x^3}$$
$$\frac{d^2y}{dx^2} - 3x^2\frac{dy}{dx} - 6xy = 9x^4e^{x^3} + 6xe^{x^3} - 3x^2(3x^2e^{x^3}) - 6x(e^{x^3})$$
$$= 9x^4e^{x^3} + 6xe^{x^3} - 9x^4e^{x^3} - 6xe^{x^3}$$
$$= 0$$

$\therefore$ **$y = e^{x^3}$ is a solution.**

ii
$$y = \tan^{-1} 2x$$
$$\frac{dy}{dx} = \frac{2}{1+4x^2}$$
$$= 2(1 + 4x^2)^{-1}$$
$$\frac{d^2y}{dx^2} = -2(1 + 4x^2)^{-2}.8x$$
$$= \frac{-16x}{(1+4x^2)^2}$$
$$(1 + 4x^2)\frac{d^2y}{dx^2} + 8x\frac{dy}{dx} = (1 + 4x^2).\frac{-16x}{(1+4x^2)^2} + 8x.\frac{2}{1+4x^2}$$
$$= \frac{-16x}{1+4x^2} + \frac{16x}{1+4x^2}$$
$$= 0$$

$\therefore$ **$y = \tan^{-1} 2x$ is a solution.**

2 Direction (slope) field of a differential equation ⊃ pp. 198–200

1 i Consider the slope at some points on the direction field:

x	y	gradient
4	4	horizontal

By substitution and elimination of options, the correct equation is $\frac{dy}{dx} = x^2 - y^2$.

$\therefore$ **C**

ii Consider the slope at some points on the direction field:

x	y	gradient
2	3	vertical
–4	0	negative

By substitution and elimination of options, the correct equation is $\frac{dy}{dx} = \frac{4}{(x-2)^2} - \frac{9}{(y-3)^2}$.

$\therefore$ **D**

2 i There must be zero gradient at the point (–1, –2). This occurs in option C.

$\therefore$ C

ii At (1, 1), (–1, 1) etc, the gradient will be 'vertical'

as $\frac{1}{1^2-1^2} = \frac{1}{(-1)^2-1^2} = \frac{1}{0}$.

$\therefore$ **B**

3 i

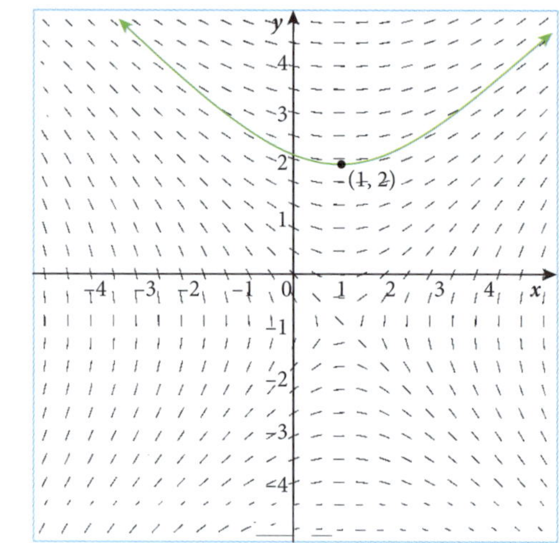

ii

3 Solving differential equations of the form $\frac{dy}{dx} = f(x)$ ⊃ p. 201

1 i
$$x^2\frac{dy}{dx} + x = 3$$
$$x^2\frac{dy}{dx} = 3 - x$$
$$\frac{dy}{dx} = \frac{3}{x^2} - \frac{1}{x}$$
$$y = \int\left(3x^{-2} - \frac{1}{x}\right)dx$$
$$y = -3x^{-1} - \ln|x| + c$$
$$\mathbf{y = -\frac{3}{x} - \ln|x| + c}$$

ii
$$\sqrt{4-3x}\frac{dy}{dx} + 3x = 4$$
$$\sqrt{4-3x}\frac{dy}{dx} = 4 - 3x$$

$$\frac{dy}{dx} = \sqrt{4-3x}$$
$$y = \int \sqrt{4-3x}\, dx$$
$$y = \int (4-3x)^{\frac{1}{2}}\, dx$$
$$y = \frac{2(4-3x)^{\frac{3}{2}}}{-9} + c$$
$$\mathbf{y = -\frac{2}{9}\sqrt{(4-3x)^3} + c}$$

2 i $\frac{dy}{dx} - 4\sin 3x\cos 3x = 0$

$$\frac{dy}{dx} = 4\sin 3x\cos 3x$$
$$\frac{dy}{dx} = 2\sin 6x$$
$$y = 2\int \sin 6x\, dx$$
$$y = 2 \times -\frac{1}{6}\cos 6x + c$$
$$y = -\frac{1}{3}\cos 6x + c$$

Substitute $y\left(\frac{\pi}{3}\right) = \frac{2}{3}$.

$$\frac{2}{3} = -\frac{1}{3}\cos 6\left(\frac{\pi}{3}\right) + c$$
$$\frac{2}{3} = -\frac{1}{3} + c$$
$$c = 1$$
$$\therefore \mathbf{y = 1 - \frac{1}{3}\cos 6x}$$

ii $(1 + \tan^2 x)\frac{dy}{dx} = 1$

$$\sec^2 x\frac{dy}{dx} = 1$$
$$\frac{dy}{dx} = \frac{1}{\sec^2 x}$$
$$\frac{dy}{dx} = \cos^2 x$$
$$y = \int \cos^2 x\, dx$$
$$y = \int \frac{1}{2}(1+\cos 2x)\, dx$$
$$y = \frac{1}{2}\left(x + \frac{1}{2}\sin 2x\right) + c$$
$$y = \frac{x}{2} + \frac{1}{4}\sin 2x + c$$

Substitute $y\left(\frac{\pi}{4}\right) = \frac{1}{4}$:

$$\frac{1}{4} = \frac{\pi}{8} + \frac{1}{4}\sin 2\left(\frac{\pi}{4}\right) + c$$
$$\frac{1}{4} = \frac{\pi}{8} + \frac{1}{4} + c$$
$$c = -\frac{\pi}{8}$$
$$\therefore \mathbf{y = \frac{x}{2} + \frac{1}{4}\sin 2x - \frac{\pi}{8}}$$

4 Solving differential equations of the form $\frac{dy}{dx} = g(y)$ ➲ pp. 201-202

1 i $3y - \frac{dy}{dx} = 0$

$$\frac{dy}{dx} = 3y$$
$$\int \frac{1}{y}\, dy = \int 3\, dx$$
$$\ln|y| = 3x + c$$
$$y = e^{3x+c}$$

$\mathbf{y = Ae^{3x}}$**, for constant $\mathbf{A}$, where $\mathbf{A = e^c}$.**

ii $\frac{dy}{dx} - 3\sqrt{y} = 0$

$$\frac{dy}{dx} = 3\sqrt{y}$$
$$\int \frac{1}{\sqrt{y}}\, dy = \int 3\, dx$$
$$\int y^{-\frac{1}{2}}\, dy = 3x + c$$
$$2\sqrt{y} = 3x + c$$
$$\sqrt{y} = \frac{3x+c}{2}$$
$$\mathbf{y = \left(\frac{3x+c}{2}\right)^2}$$

2 i $\frac{dy}{dx} - (2y-1)^2 = 0$

$$\frac{dy}{dx} = (2y-1)^2$$
$$\int (2y-1)^{-2}\, dy = \int dx$$
$$\frac{(2y-1)^{-1}}{-2} = x + c$$
$$\frac{-1}{2(2y-1)} = x + c$$

Substitute $y(2) = 1$:

$$\frac{-1}{2(2(1)-1)} = 2 + c$$
$$-\frac{1}{2} = 2 + c$$
$$c = -\frac{5}{2}$$
$$\frac{-1}{2(2y-1)} = x - \frac{5}{2}$$
$$\frac{-1}{2y-1} = 2x - 5$$
$$\frac{-1}{2x-5} = 2y - 1$$
$$2y = 1 + \frac{1}{5-2x}$$
$$2y = \frac{5-2x+1}{5-2x}$$
$$2y = \frac{6-2x}{5-2x}$$
$$\mathbf{y = \frac{3-x}{5-2x}, x \neq \frac{5}{2}}$$

ii $$\frac{dy}{dx} = y^2 + 4$$
$$\int \frac{1}{4+y^2}\,dy = \int dx$$
$$\frac{1}{2}\tan^{-1}\frac{y}{2} = x + c$$
Substitute $y\left(\frac{\pi}{6}\right) = 2$:
$$\frac{1}{2}\tan^{-1}\frac{2}{2} = \frac{\pi}{6} + c$$
$$\frac{1}{2}\times\frac{\pi}{4} = \frac{\pi}{6} + c$$
$$\frac{\pi}{8} = \frac{\pi}{6} + c$$
$$c = \frac{\pi}{8} - \frac{\pi}{6}$$
$$= -\frac{\pi}{24}$$
$$\therefore \frac{1}{2}\tan^{-1}\frac{y}{2} = x - \frac{\pi}{24}$$
$$x = \frac{1}{2}\tan^{-1}\frac{y}{2} + \frac{\pi}{24}$$
$$2x = \tan^{-1}\frac{y}{2} + \frac{\pi}{12}$$
$$\tan^{-1}\frac{y}{2} = 2x - \frac{\pi}{12}$$
$$\frac{y}{2} = \tan\left(2x - \frac{\pi}{12}\right)$$
$$\mathbf{y = 2\tan\left(2x - \frac{\pi}{12}\right)}$$

5 Solving differential equations of the form $\frac{dy}{dx} = f(x)\,g(y)$ ➲ p. 202

1 i $$\frac{dy}{dx}\sin 2y - \cos x = 0$$
$$\frac{dy}{dx}\sin 2y = \cos x$$
$$\sin 2y\,dy = \cos x\,dx$$
$$\int \sin 2y\,dy = \int \cos x\,dx$$
$$-\frac{1}{2}\cos 2y = \sin x + c$$
Substitute $y\left(\frac{\pi}{2}\right) = 0$:
$$-\frac{1}{2}\cos 2(0) = \sin\frac{\pi}{2} + c$$
$$-\frac{1}{2} = 1 + c$$
$$c = -\frac{1}{2} - 1$$
$$c = -\frac{3}{2}$$
$$-\frac{1}{2}\cos 2y = \sin x - \frac{3}{2}$$
$$-\cos 2y = 2\sin x - 3$$
$$\therefore \mathbf{2\sin x + \cos 2y - 3 = 0}$$

ii $$\frac{dy}{dx} = (y+5)(x-2)$$
$$\frac{1}{y+5}\,dy = (x-2)\,dx$$
$$\int \frac{1}{y+5}\,dy = \int (x-2)\,dx$$
$$\ln|y+5| = \frac{x^2}{2} - 2x + c$$
Substitute $y(2) = -4$:
$$\ln|-4+5| = \frac{2^2}{2} - 2(2) + c$$
$$0 = -2 + c$$
$$c = 2$$
$$\therefore \ln|y+5| = \frac{x^2}{2} - 2x + 2$$
$$\therefore \mathbf{\frac{x^2}{2} - 2x + 2 = \ln|y+5|}$$

2 i $$\sqrt{2-x^2}\frac{dy}{dx} = \frac{1}{2-y}$$
$$(2-y)\,dy = \frac{1}{\sqrt{2-x^2}}\,dx$$
$$\int (2-y)\,dy = \int \frac{1}{\sqrt{2-x^2}}\,dx$$
$$2y - \frac{y^2}{2} = \sin^{-1}\frac{x}{\sqrt{2}} + c$$
Substitute $y(1) = 0$:
$$2(0) - \frac{0^2}{2} = \sin^{-1}\frac{1}{\sqrt{2}} + c$$
$$0 = \frac{\pi}{4} + c$$
$$c = -\frac{\pi}{4}$$
$$\therefore \mathbf{2y - \frac{y^2}{2} = \sin^{-1}\frac{x}{\sqrt{2}} - \frac{\pi}{4}}$$

ii $$\frac{dy}{dx} + y^2 = 2xy^2$$
$$\frac{dy}{dx} = 2xy^2 - y^2$$
$$\frac{dy}{dx} = y^2(2x-1)$$
$$\frac{1}{y^2}\,dy = (2x-1)\,dx$$
$$\int y^{-2}\,dy = \int (2x-1)\,dx$$
$$\frac{y^{-1}}{-1} = \frac{x^2}{2} - x + c$$
$$-\frac{1}{y} = \frac{x^2}{2} - x + c$$
Substitute $y(2) = 1$:
$$-1 = \frac{2^2}{2} - 2 + c$$
$$-1 = c$$
$$-\frac{1}{y} = \frac{x^2}{2} - x - 1$$
$$-\frac{1}{y} = \frac{x^2 - 2x - 2}{2}$$
$$\mathbf{y = \frac{2}{2 + 2x - x^2}}$$

6 Exponential growth and decay models ➲ pp. 203–204

1 i As $\frac{dN}{dt} \alpha N$, then $\frac{dN}{dt} = kN$, for constant k.

$$\frac{dN}{dt} = kN$$

$$\frac{dt}{dN} = \frac{1}{kN}$$

$$dt = \frac{1}{kN} dN$$

$$t = \frac{1}{k}\int \frac{1}{N} dN$$

$$t = \frac{1}{k}\log_e N + c$$

$\log_e N = kt + c$

$N = e^{kt+c}$

$N = Ae^{kt}$, for constant A, where $A = e^c$.

Substitute $N = 120$, $t = 0$. Hence, $A = 120$.

$\mathbf{N = 120\, e^{kt}}$

ii As $\frac{dm}{dt} \alpha m$, then $\frac{dm}{dt} = -km$, for a constant k.

$$\frac{dm}{dt} = -km$$

$$\frac{dt}{dm} = \frac{-1}{km}$$

$$dt = \frac{-1}{km} dm$$

$$t = -\frac{1}{k}\int \frac{1}{m} dm$$

$$t = -\frac{1}{k}\log_e m + c$$

$\log_e m = -kt + c$

$m = e^{-kt+c}$

$\therefore m = Ae^{-kt}$, for constant A, where $A = e^c$.

Substitute $m = 60$, $t = 0$. Hence, $A = 60$.

$\therefore m = 60e^{-kt}$

Substitute $m = 45$, $t = 5$:

$45 = 60e^{-k(5)}$

$$e^{-5k} = \frac{45}{60}$$

$-5k = \log_e 0.75$

$$k = -\frac{1}{5}\log_e 0.75$$

$= 0.0575$ (4 dec. pl.)

Substitute $m = 10$, $k = 0.0575$:

$m = 60e^{-kt}$

$10 = 60e^{-kt}$

$$e^{-kt} = \frac{1}{6}$$

$$-kt = \log_e \frac{1}{6}$$

$$t = \frac{\log_e \frac{1}{6}}{-k}$$

$= 31.161\,034\,25\ldots$ (using $k = 0.0575$)

$= 31$ (nearest whole)

$\therefore$ **it will take 31 years.**

2 i $\frac{dT}{dt} = k(36 - T)$, for a constant k

$$\frac{dt}{dT} = \frac{1}{k(36-T)}$$

$$dt = \frac{1}{k(36-T)} dT$$

$$t = \int \frac{1}{k(36-T)} dT$$

$$= \frac{1}{k}\int \frac{1}{36-T} dT$$

$= -\frac{1}{k}\log_e(36 - T) + c$ (where $T < 36$)

Substitute $t(0) = 3.2$:

$$0 = -\frac{1}{k}\log_e(36 - 3.2) + c$$

$$c = \frac{1}{k}\log_e 32.8$$

$$t = -\frac{1}{k}\log_e(36 - T) + \frac{1}{k}\log_e 32.8$$

$$t = \frac{1}{k}\log_e \frac{32.8}{36-T}$$

$$kt = \log_e \frac{32.8}{36-T}$$

$$\frac{32.8}{36-T} = e^{kt}$$

$36 - T = 32.8e^{-kt}$

$T = 36 - 32.8e^{-kt}$

Substitute $T = 15$, $t = 20$:

$15 = 36 - 32.8e^{-k(20)}$

$32.8e^{-20k} = 21$

$$e^{-20k} = \frac{21}{32.8}$$

$$-20k = \log_e \frac{21}{32.8}$$

$$k = -\frac{1}{20}\log_e \frac{21}{32.8}$$

$= 0.022\,295\,303\ldots$

$= 0.0223$ (4 dec. pl.)

$\therefore \mathbf{T = 36 - 32.8e^{-0.0223t}}$

ii As $\frac{dP}{dt} \alpha (P - 12\,000)$

$\frac{dP}{dt} = k(P - 12\,000)$, for a constant k

$$\frac{dt}{dP} = \frac{1}{k(P-12\,000)}$$

$$dt = \frac{1}{k(P-12\,000)} dP$$

$$t = \int \frac{1}{k(P-12\,000)}\,dP$$
$$= \frac{1}{k}\int \frac{1}{P-12\,000}\,dP$$
$$= \frac{1}{k}\log_e(P - 12\,000) + c \text{ [where } P > 12\,000]$$
Substitute $t(0) = 14\,400$:
$$0 = \frac{1}{k}\log_e(14\,400 - 12\,000) + c$$
$$c = -\frac{1}{k}\log_e 2400$$
$$t = \frac{1}{k}\log_e(P - 12\,000) - \frac{1}{k}\log_e 2400$$
$$t = \frac{1}{k}\log_e \frac{P-12\,000}{2400}$$
$$kt = \log_e \frac{P-12\,000}{2400}$$
$$\frac{P-12\,000}{2400} = e^{kt}$$
$$P - 12\,000 = 2400e^{kt}$$
$$P = 12\,000 + 2400e^{kt}$$
Substitute $t = 12$, $P = 18\,200$:
$$18\,200 = 12\,000 + 2400e^{k(12)}$$
$$2400e^{12k} = 6200$$
$$e^{12k} = \frac{6200}{2400}$$
$$12k = \log_e \frac{31}{12}$$
$$k = \frac{\log_e \frac{31}{12}}{12}$$
$$= 0.079\,090\,046\ldots$$
$$= 0.0791 \text{ (4 dec. pl.)}$$
Substitute $t = 30$, $k = 0.0791$:
$$P = 12\,000 + 2400e^{kt}$$
$$= 12\,000 + 2400e^{k(30)}$$
$$= 37\,750.878\,35\ldots \text{ (using } k = 0.0791)$$
$$= 37\,751 \text{ (nearest whole)}$$
$\therefore$ **the population will be 37 751.**

7 Mixing problems ➲ p. 204

1 i a $\frac{dM}{dt}$ = rate of inflow − rate of outflow

Rate of inflow = 2 g/L × 5 L/min
= 10 g/min

Rate of outflow = M g/1250 L × 5 L/min
$= \frac{M}{250}$ g/min

$$\therefore \boldsymbol{\frac{dM}{dt} = 10 - \frac{M}{250}}$$

b $\frac{dM}{dt} = 10 - \frac{M}{250}$
$$= \frac{2500-M}{250}$$
$$\frac{dt}{dM} = \frac{250}{2500-M}$$
$$dt = \frac{250}{2500-M}\,dM$$
$$\int dt = \int \frac{250}{2500-M}\,dM$$
$$t = 250\int \frac{1}{2500-M}\,dM$$
$$= -250\ln(2500 - M) + c$$
Substitute $t = 0$, $M = 60$:
$$0 = -250\ln(2500 - 60) + c$$
$$0 = -250\ln 2440 + c$$
$$c = 250\ln 2440$$
$$t = -250\ln(2500 - M) + 250\ln 2440$$
$$= 250\ln \frac{2440}{2500-M}$$
$$0.004t = \ln \frac{2440}{2500-M}$$
$$\frac{2440}{2500-M} = e^{0.004t}$$
$$2440\,e^{-0.004t} = 2500 - M$$
$$\boldsymbol{M = 2500 - 2440\,e^{-0.004t}}$$

ii a $\frac{dE}{dt}$ = rate of inflow − rate of outflow

Rate of inflow = 10 L/min

Rate of outflow = E L/500 L × 10 L/min
$= \frac{E}{50}$ L/min

$$\therefore \boldsymbol{\frac{dE}{dt} = 10 - \frac{E}{50}}$$

b $\frac{dE}{dt} = 10 - \frac{E}{50}$
$$= \frac{500-E}{50}$$
$$\frac{dt}{dE} = \frac{500}{500-E}$$
$$dt = \frac{500}{500-E}\,dE$$
$$\int dt = \int \frac{50}{500-E}\,dE$$
$$t = 50\int \frac{1}{500-E}\,dE$$
$$= -50\ln(500 - E) + c$$
Substitute $t = 0$, $Q = 0$:
$$0 = -50\ln(500 - 0) + c$$
$$0 = -50\ln 500 + c$$
$$c = 50\ln 500$$
$$t = -50\ln(500 - E) + 50\ln 500$$
$$= 50\ln \frac{500}{500-E}$$
$$0.02t = \ln \frac{500}{500-E}$$
$$\frac{500}{500-E} = e^{0.02t}$$
$$500\,e^{-0.02t} = 500 - E$$
$$\boldsymbol{E = 500 - 500\,e^{-0.02t}}$$

c Substitute $t = 120$:
$$E = 500 - 500\,e^{-0.02(120)}$$
$$= 500 - 500\,e^{-0.02(120)}$$
$$= 454.641\,0234\ldots$$
$$= 455 \text{ (nearest whole)}$$
$\therefore$ **there will be 455 L of ethanol in the vat.**

8 The logistic growth model ➲ p. 205

1 i a As maximum population is 60 million, use $M = 60$.

$$\frac{dP}{dt} = kP(60 - P)$$

$$\frac{1}{P(60-P)}\,dP = k\,dt$$

$$\int \frac{1}{P(60-P)}\,dP = \int k\,dt$$

$$\int\left(\frac{1}{60P}+\frac{1}{60(60-P)}\right)dP = kt + c$$

$$\frac{1}{60}\int\left(\frac{1}{P}+\frac{1}{(60-P)}\right)dP = kt + c$$

$$\ln|P| - \ln|60 - P| = 60(kt + c)$$

$$\frac{P}{60-P} = e^{60(kt+c)}$$

$$\frac{P}{60-P} = Ae^{60kt}, \text{ for constant } A, \text{ where } A = e^{60c}$$

Substitute $P = 15$, $t = 0$:

$$\frac{15}{60-15} = Ae^{60k(0)}$$

$$A = \frac{1}{3}$$

$$\frac{P}{60-P} = \frac{1}{3}e^{60kt}$$

$$3P = e^{60kt}(60 - P)$$

$$3P = 60e^{60kt} - Pe^{60kt}$$

$$3P + Pe^{60kt} = 60e^{60kt}$$

$$P(3 + e^{60kt}) = 60e^{60kt}$$

$$P = \frac{60e^{60kt}}{3+e^{60kt}}$$

$$P = \frac{60}{3e^{-60kt}+1}$$

$$P = \frac{60}{1+3e^{-60kt}}$$

Substitute $P = 25$, $t = 37$:

$$25 = \frac{60}{1+3e^{-60k(37)}}$$

$$25(1 + 3e^{-2220k}) = 60$$

$$1 + 3e^{-2220k} = 2.4$$

$$3e^{-2220k} = 1.4$$

$$e^{-2220k} = \frac{1.4}{3}$$

$$-2220k = \log_e \frac{7}{15}$$

$$k = -\frac{1}{2220}\log_e \frac{7}{15}$$

$$= \mathbf{\frac{1}{2220}\log_e \frac{15}{7}}$$

b Substitute $t = 69$, $k = \dfrac{\log_e \frac{7}{15}}{-2220}$:

$$P = \frac{60}{3e^{-60kt}+1}$$

$$= 34.798\,575\,28\ldots$$

$$= 34.799 \text{ (3 dec. pl.)}$$

$\therefore$ **the population will be 34 799 000 in 2050.**

ii a $\text{RHS} = \dfrac{1}{40}\left(\dfrac{1}{A}+\dfrac{1}{40-A}\right)$

$$= \frac{1}{40}\left(\frac{40-A+A}{A(40-A)}\right)$$

$$= \frac{1}{40}\left(\frac{40}{A(40-A)}\right)$$

$$= \frac{1}{A(40-A)}$$

$$= \text{LHS}$$

b

$$\frac{dA}{dt} = \frac{1}{20}A(40 - A)$$

$$\int \frac{1}{A(40-A)}\,dA = \int \frac{1}{20}\,dt$$

$$\int \frac{1}{40}\left(\frac{1}{A}+\frac{1}{40-A}\right)dA = \frac{t}{20} + c$$

$$\frac{1}{40}(\log_e|A| - \log_e|40 - A|) = \frac{t}{20} + c$$

$$\log_e\left|\frac{A}{40-A}\right| = 2t + 20c$$

$$\frac{A}{40-A} = Ce^{2t} \quad (\text{where } C = e^{20c})$$

When $t = 0$, $A = 0.2$:

$$\frac{0.2}{40-0.2} = Ce^{2(0)}$$

$$C = \frac{1}{199}$$

$$\frac{A}{40 - A} = \frac{1}{199}e^{2t}$$

$$199A = (40 - A)e^{2t}$$

$$199A = 40e^{2t} - Ae^{2t}$$

$$Ae^{2t} + 199A = 40e^{2t}$$

$$A(e^{2t} + 199) = 40e^{2t}$$

$$A = \frac{40e^{2t}}{e^{2t}+199}$$

$$A = \frac{40}{1+199e^{-2t}} \quad \text{(by dividing top and bottom by } e^{2t}\text{)}$$

c As $t \to \infty$, $e^{-2t} \to 0$, then $A \to \dfrac{40}{1+199(0)} = 40$. The limiting area is 40 cm^2.

HSC EXAM-TYPE QUESTIONS

Part A

1 **B.** There must be zero gradient at points like (1, 1), (4, 4), etc and vertical at (−3, 3). This occurs in option B.

2 **B.**
$$\frac{dy}{dx} = 2xy + y$$
$$\frac{dy}{dx} = y(2x+1)$$
$$\int \frac{1}{y}\,dy = \int (2x+1)\,dx$$
$$\ln|y| = x^2 + x + c$$
$$y = e^{x^2+x+c}$$
$$= Ce^{x^2+x}, \text{ where } C = e^c$$

3 **A.**
$$\frac{dy}{dx} = -\frac{x}{y}$$
$$\int y\,dy = \int -x\,dx$$
$$\frac{y^2}{2} = -\frac{x^2}{2} + c$$
$$y^2 = -x^2 + C$$
$x^2 + y^2 = C$, which is a circle.

4 **D.** $y = e^{2x}, \frac{dy}{dx} = 2e^{2x}, \frac{d^2y}{dx^2} = 4e^{2x}$

$4e^{2x} - 3(2e^{2x}) + 2(e^{2x}) = 0$

5 **C.**
$$y = 2\cos x + 3\sin x + 4$$
$$\frac{dy}{dx} = -2\sin x + 3\cos x$$
$$\frac{d^2y}{dx^2} = -2\cos x - 3\sin x$$
$$\frac{d^2y}{dx^2} = -(2\cos x + 3\sin x)$$
$$= -(y-4)$$
$$= -y + 4$$
$$\frac{d^2y}{dx^2} + y = 4$$

Part B

6

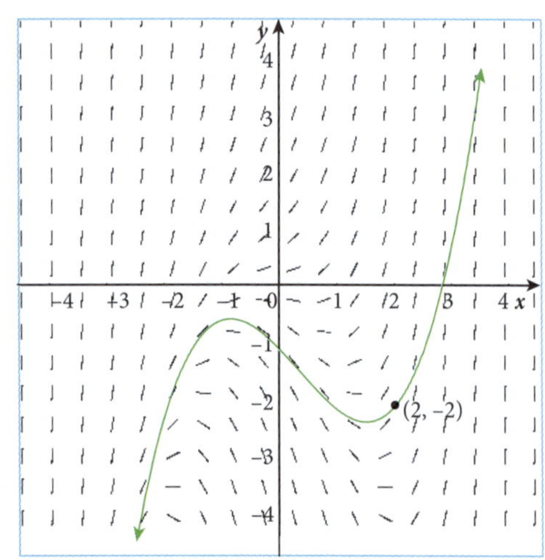

✓✓

7
$$\frac{dy}{dx} = \frac{y}{2x}$$
$$\int \frac{1}{y}dy = \int \frac{1}{2x}dx$$
$$\ln|y| = \frac{1}{2}\ln|x| + c \quad ✓$$
Substitute $y = 1, x = 4$:
$$0 = \frac{1}{2}\ln 4 + c$$
$$\therefore c = -\ln 2$$
$$\ln|y| = \ln\sqrt{x} - \ln 2$$
$$\ln|y| = \ln\frac{\sqrt{x}}{2}$$
$$\therefore \boldsymbol{y = \frac{\sqrt{x}}{2}} \quad ✓$$

8
$$3x^2\frac{dy}{dx} - x = 0$$
$$3x^2\frac{dy}{dx} = x$$
$$\frac{dy}{dx} = \frac{1}{3x} \quad ✓$$
$$y = \frac{1}{3}\int\frac{1}{x}dx$$
$$\boldsymbol{y = \frac{1}{3}\log_e x + c} \quad ✓$$

9
$$(x^2+4)\frac{dy}{dx} - 2xy = 0$$
$$(x^2+4)\frac{dy}{dx} = 2xy$$
$$\int\frac{1}{y}dy = \int\frac{2x}{x^2+4}dx \quad ✓$$
$$\ln|y| = \ln|x^2+4| + c$$
$$\ln\left|\frac{y}{x^2+4}\right| = c \quad ✓$$
Substitute $y(0) = 1$:
$$\ln\left|\frac{1}{0^2+4}\right| = c$$
$$c = \ln\frac{1}{4}$$
$$\ln|y| = \ln|x^2+4| + \ln\frac{1}{4}$$
$$\ln|y| = \ln|x^2+4| - \ln 4$$
$$\ln|y| = \ln\left|\frac{x^2+4}{4}\right|$$
$$\boldsymbol{y = \frac{x^2+4}{4}} \quad ✓$$

10 $\frac{dy}{dx} = x(1 + y^2)$

$\int \frac{dy}{1+y^2} = \int x\,dx$

$\tan^{-1} y = \frac{x^2}{2} + c$ ✓

Substitute $x = 0, y = 1$:

$\tan^{-1} 1 = \frac{0^2}{2} + c$

$\frac{\pi}{4} = 0 + c$

$c = \frac{\pi}{4}$

$\tan^{-1} y = \frac{x^2}{2} + \frac{\pi}{4}$

$\mathbf{y = \tan\left(\frac{x^2}{2} + \frac{\pi}{4}\right)}$ ✓

11 a $\frac{dN}{dt} = kN$

$\frac{dt}{dN} = \frac{1}{kN}$

$t = \frac{1}{k}\int \frac{1}{N}\,dN$

$kt = \ln N + c$, as $N > 0$

$\ln N = kt + c$

$N = e^{kt+c}$

$N = Ae^{kt}$, where $A = e^c$ ✓

Substitute $t(0) = 6000$:

$6000 = Ae^{k(0)}$

$A = 6000$

$\therefore \mathbf{N = 6000e^{kt}}$ ✓

b $N = 6000e^{kt}$

Substitute $t(3) = 8500$:

$8500 = 6000e^{k(3)}$

$e^{3k} = \frac{85}{60}$

$3k = \ln \frac{17}{12}$

$k = \frac{1}{3}\ln \frac{17}{12}$ ✓

Substitute $t = 6, k = \frac{1}{3}\ln\frac{17}{12}$:

$N = 6000e^{kt}$

$= 12\,041.666\,67\ldots$

$= 12\,042$ (nearest whole)

$\therefore$ **the number of bacteria is 12 042.** ✓

12 a $\frac{dP}{dt} = -0.003P$

$\int \frac{1}{P}\,dP = \int -0.003\,dt$

$\ln|P| = -0.003t + c$

$P = Ce^{-0.003t}$ where $C = e^c$ ✓

Substitute $t = 0, P = p_0$:

$p_0 = Ce^{-0.003(0)}$

$C = p_0$

$\therefore \mathbf{P = p_0 e^{-0.003t}}$ ✓

b Substitute $P = 1, P_0 = 2$:

$1 = 2e^{-0.003t}$

$e^{-0.003t} = 0.5$

$-0.003t = \ln 0.5$ ✓

$t = \frac{\ln 0.5}{-0.003}$

$= 231.049\,0602\ldots$

$= 231$ (nearest whole)

$\therefore$ **it will take 231 days.** ✓

13 a Using Newton's law of cooling,

$\frac{dT}{dt} = -k(T - 18)$

$\frac{dt}{dT} = \frac{-1}{k(T-18)}$

$t = \int \frac{-1}{k(T-18)}\,dT$

$= -\frac{1}{k}\int \frac{1}{T-18}\,dT$

$= -\frac{1}{k}\log_e|T - 18| + c$ ✓

Substitute $t = 0, T = 75$:

$0 = -\frac{1}{k}\log_e(75 - 18) + c$

$c = \frac{1}{k}\log_e 57$

$t = -\frac{1}{k}\log_e|T - 18| + \frac{1}{k}\log_e 57$

$t = \frac{1}{k}\log_e\left|\frac{57}{T-18}\right|$

$kt = \log_e\left|\frac{57}{T-18}\right|$

$\frac{57}{T-18} = e^{kt}$

$T - 18 = 57e^{-kt}$

$T = 18 + 57e^{-kt}$ ✓

Substitute $T = 40$, $t = 3$:

$$40 = 18 + 57e^{-k(3)}$$
$$57e^{-3k} = 22$$
$$e^{-3k} = \frac{22}{57}$$
$$-3k = \log_e \frac{22}{57}$$
$$k = -\frac{1}{3}\log_e \frac{22}{57}$$
$$= 0.317\,336\,271\ldots$$
$$= 0.3173 \text{ (4 dec. pl.)}$$

$\therefore$ $\mathbf{T = 18 + 57e^{-0.3173t}}$ ✓

b Substitute $T = 20$:

$$20 = 18 + 57e^{-0.3173t}$$
$$57e^{-0.3173t} = 2$$
$$e^{-0.3173t} = \frac{2}{57} \quad ✓$$
$$-0.3173t = \ln \frac{2}{57}$$
$$t = \frac{\ln \frac{2}{57}}{-0.3173}$$
$$= 10.557\,529\,43\ldots$$
$$= 10 \text{ min } 33.451\,765\,64\ldots \text{ s}$$
$$= 10 \text{ min } 30 \text{ s (nearest 10 seconds)}$$

$\therefore$ **it will take 10 min 30 s.** ✓

---

14 a $\frac{dQ}{dt}$ = rate of inflow – rate of outflow

Rate of inflow = 4 g/L × 10 L/min

= 40 g/min ✓

Rate of outflow = Q g/1000 L × 10 L/min

$= \frac{Q}{100}$ g/min

$\therefore \mathbf{\frac{dQ}{dt} = 40 - \frac{Q}{100}}$ ✓

b
$$\frac{dQ}{dt} = 40 - \frac{Q}{100}$$
$$= \frac{4000 - Q}{100}$$
$$\frac{dt}{dQ} = \frac{100}{4000 - Q} \quad ✓$$
$$\int dt = \int \frac{100}{4000 - Q}\,dQ$$
$$t = 100\int \frac{1}{4000 - Q}\,dQ$$
$$= -100 \ln|4000 - Q| + c$$

Substitute $t = 0$, $Q = 200$:

$$0 = -100 \ln(4000 - 200) + c$$
$$0 = -100 \ln 3800 + c$$
$$c = 100 \ln 3800 \quad ✓$$
$$t = -100 \ln|4000 - Q| + 100 \ln 3800$$
$$= 100 \ln\left|\frac{3800}{4000 - Q}\right|$$
$$0.01t = \ln\left|\frac{3800}{4000 - Q}\right|$$
$$\frac{3800}{4000 - Q} = e^{0.01t}$$
$$3800\,e^{-0.01t} = 4000 - Q$$

$\mathbf{Q = 4000 - 3800\,e^{-0.01t}}$ ✓

c Substitute $t = 10$:

$$Q(10) = 4000 - 3800\,e^{-0.01(10)}$$
$$= 561.617\,8116\ldots$$
$$= 562 \text{ (nearest whole)}$$

$\therefore$ **there is 562 g of salt present.** ✓

CHAPTER 16 THE BINOMIAL DISTRIBUTION

TOPIC: STATISTICAL ANALYSIS

YEAR 12

1 Bernoulli trials

- **Characteristics of Bernoulli trials** are:
  - **each trial has only two possible outcomes**: 'success' and 'failure'
  - **P(success) = p and P(failure) = $q = 1 - p$** and this is consistent for every trial
  - **trials are independent**.
- **X = number of successes in independent Bernoulli trials**. If there is only one trial, then the probability distribution table is detailed below:

x	0	1
$P(X = x)$	$1 - p$	p

- **Mean** = $E(X) = \mu = 0(1 - p) + 1(p)$
  $= p$
- **Variance** = $Var(X) = \sigma^2 = E[X^2] - [E(X)]^2$
  $= \{0^2(1 - p) + 1^2(p)\} - p^2$
  $= p(1 - p)$

EXAMPLE 1

A biased coin is tossed where the probability of tossing a tail is 0.6.

a Letting X be tossing a tail, complete a probability distribution table for tossing a tail.

x	0	1
$P(X = x)$		

b Find the mean and variance of tossing a tail.

a

x	0	1
$P(X = x)$	0.4	0.6

b Mean: $E(X) = 0.6$

Variance: $Var(X) = 0.6(1 - 0.4)$
$= 0.24$

$Var(X) = pq$

EXAMPLE 2

X is a discrete random variable that has a Bernoulli distribution. It is known that the variance for this distribution is 0.24. Find possible values of $E(X)$.

$$Var(X) = p(1 - p) = 0.24$$
$$p - p^2 = 0.24$$
$$p^2 - p + 0.24 = 0$$
$$(p - 0.6)(p - 0.4) = 0$$
$$p = 0.4, 0.6$$
$$\therefore E(X) = 0.4 \text{ or } 0.6$$

Solve a quadratic equation.

PRACTICE

1 i Jannah is playing for her soccer team. The chance that she scores a goal is 0.16. The ball has been passed to her and she shoots for goal. Let X be the random variable that defines Jannah getting a goal.

a Complete a probability distribution table for this discrete variable.

x	0	1
$P(X = x)$		

b Find $E(X)$ and $Var(X)$.

ii Michael is sitting a Biology multiple-choice test with questions that have four possible answer choices. Michael has no idea how to answer the final question and plans to make a guess. Let X be the random variable that Michael answers the question correctly.

a Construct a probability distribution table for X.

b Find $E(X)$ and Var(X).

2 i X is a discrete random variable that has a Bernoulli distribution. It is known that the variance for this distribution is 0.16. Find possible values of $E(X)$.

ii Adam is firing a pistol at a target. The chance of him hitting the target is less than 0.5. Letting Y be the random variable that Adam hits the target, it is known that the $\text{Var}(Y) = 0.2475$. What is the value of $E(Y)$?

Answers ➲ p. 226

2 Applications of the Bernoulli distribution

➔ **The geometric probability distribution** represents the number of failures before a success in a series of Bernoulli trials.

➔ **If there are x failures before a success**, then $P(X = x) = q^x p$.

x	0	1	2	3	4	5	...
$P(X = x)$	p	qp	q^2p	q^3p	q^4p	q^5p	...

EXAMPLE 1

A bag contains red, blue and green balls. Balls are randomly chosen from the bag, the colour noted and then returned to the bag. The probability of choosing a red ball is 0.15. Let X be the number of times a ball is drawn from the bag until a red ball is selected. Find, correct to four decimal places:

a $P(X = 4)$

b $P(3 \leq X \leq 6)$

a Let $p = 0.15$, $q = 0.85$.

$P(X = 4) = (0.85)^4(0.15)$

$= 0.078\,300\,937\ldots$

$= 0.0783$ (4 dec. pl.)

Find S_4 of geometric series.

b $P(X = 3) + P(X = 4) + P(X = 5) + P(X = 6)$

$= (0.85)^3(0.15) + (0.85)^4(0.15) + (0.85)^5(0.15) + (0.85)^6(0.15)$, which is a geometric series with $a = (0.85)^3(0.15)$, $r = 0.85$, $n = 4$.

$$= \frac{(0.85)^3(0.15)[1 - 0.85^4]}{1 - 0.85}$$

$= 0.293\,547\,911\ldots$

$= 0.2935$ (4 dec. pl.)

➔ PRACTICE

1 i Two people are playing a game which cannot start until one person throws a double using a pair of dice. Let X be the number of times the dice are rolled *before* the successful roll.

a Find the probability that the game will start on the tenth roll.

b Show that $P(5 \leq X \leq 8) = 0.201$ (correct to three decimal places).

ii The table shows the blood group frequency in Australia.

Blood Groups in Australia	
Type	**Percentage**
O positive	40
O negative	9
A positive	31
A negative	7
B positive	8
B negative	2
AB positive	2
AB negative	1

A blood technician randomly checks samples.

a What is the probability that she will need to check 45 samples before an A negative blood type is detected?

b What is the probability that she will need to check more than 100 samples before an AB positive blood type is detected?

Answers ➲ p. 226

3 Binomial distributions

➔ **$X \sim \text{Bin}(n, p)$ means random variable X is distributed binomially** in n trials and $P(\text{success}) = p$.

➔ **$P(X = r)$ is used to denote the probability** that there are r successes in n trials.

$P(X = r) = {}^nC_r p^r (1-p)^{n-r}$

EXAMPLE 1

An experiment consisted of tossing a biased coin four times and recording the number of heads obtained. This experiment was repeated 350 times and the results are shown in the table below.

Number of heads	0	1	2	3	4
Frequency	6	42	111	132	59

Based on these results, what is the probability that the coin shows heads when tossed, correct to two decimal places?

Consider $P(X = 4) = {}^4C_4 p^4 (1-p)^0$

$= p^4$

$\therefore p^4 \times 350 = 59$

$p^4 = 0.168\,571\,428\ldots$

$p = 0.640\,760\,883\ldots$

$= 0.64$ (2 dec. pl.)

Considering $P(X = 0)$ is also a good option.

EXAMPLE 2

A dice is rolled 30 times. Let X be the discrete random variable that describes the number of times a number greater than three is thrown. What is the probability of rolling a number greater than three on exactly 20 occasions? Give your answer correct to four decimal places.

$P(\text{greater than } 3) = 0.5$

Using $X \sim \text{Bin}(30, 0.5)$:

$P(X = 20) = {}^{30}C_{20}(0.5)^{20}(1-0.5)^{10}$

$= 0.027\,9816\ldots$

$= 0.0280$ (4 dec. pl.)

Bin (30, 0.5) means $n = 30$, $p = 0.5$.

EXAMPLE 3

A machine produces a computer component. It has been found that 96% of the components are found to be satisfactory, but the remainder are defective. From a sample of 15 components, what is the probability that at least three will be defective? Leave your answer correct to four decimal places.

Using $X \sim \text{Bin}(15, 0.04)$:

$P(X \geq 3) = 1 - [P(X = 0) + P(X = 1) + P(X = 2)]$

$= 1 - [{}^{15}C_0(0.04)^0(1-0.04)^{15} + {}^{15}C_1(0.04)^1(1-0.04)^{14} + {}^{15}C_2(0.04)^2(1-0.04)^{13}]$

$= 0.020\,291\,803\ldots$

$= 0.0203$ (4 dec. pl.)

$1 - 0.96 = 0.04$

EXAMPLE 4

Florrie buys a ticket each week in her staff raffle. There are 20 tickets sold each week and there is only one winning ticket drawn. For how many weeks will she need to buy raffle tickets so that she has a greater chance of winning exactly three raffles than of winning exactly two raffles?

Let X be the discrete random variable that describes the number of times Florrie wins the weekly raffle.

Using $X \sim \text{Bin}(n, 0.05)$:

$P(X = 3) > P(X = 2)$

${}^{n}C_3(0.05)^3(1-0.05)^{n-3} > {}^{n}C_2(0.05)^2(1-0.05)^{n-2}$

$$\frac{n!}{3!(n-3)!}0.05^3 0.95^{n-3} > \frac{n!}{2!(n-2)!}0.05^2 0.95^{n-2}$$

$$\frac{n!}{3!(n-3)!}0.05^3 0.95^{n-3} \times \frac{2!(n-2)!}{n!}\frac{1}{0.05^2}\frac{1}{0.95^{n-2}} > 1$$

$$\frac{0.05(n-2)}{3(0.95)} > 1$$

$n - 2 > 57$

$n > 59$

Solve an inequality.

$\therefore$ Florrie will need to buy tickets for 60 weeks.

→ PRACTICE

1 i An experiment consisted of tossing a biased coin four times and recording the number of tails (X) obtained. This experiment was repeated 200 times and the results are shown in the table below.

Number of tails (X)	0	1	2	3	4
Frequency	11	46	74	54	15

Based on these results, what is the probability that the coin shows tails when tossed, correct to two decimal places?

ii A biased coin was tossed five times and the number of heads (X) obtained was recorded. This experiment was repeated 360 times. Some of the results are recorded in the table below.

Number of heads (X)	0	1	2	3	4	5
Frequency				108	50	

Based on these results, find the probability (p) of tossing a head, correct to two decimal places, and hence find the expected number of occasions when there were two heads tossed. Leave your answer to the nearest whole.

2 i A dice is rolled 10 times. Let Y be the discrete random variable that describes the number of times a number less than five is thrown. What is the probability of rolling a number less than five 6 times? Give your answer correct to four decimal places.

ii A manufacturer makes light globes that have a probability of 0.04 of being defective. Quality control officers test random samples of eight light globes each hour and reject the light globes made in that hour if at least two light globes are defective. Find the probability that the light globes made in any hour will be rejected.

Give your answer correct to three significant figures.

3 i A biased coin falls heads 40% of the time. What is the probability that in eight tosses of the coin it lands as a tail at least three times, correct to four decimal places?

ii A goalkicker's record shows a success of five goals every eight attempts. In 12 shots what is the probability, to the nearest whole percentage, that he will be successful in at least 10 shots?

4 i As part of a marketing campaign a breakfast cereal maker places a plastic toy in every fourth box produced. A father wants to collect a toy for each of his four children. How many boxes should he buy so that he has a greater chance of collecting four toys than only three toys?

Let X be the discrete random variable that describes the number of times he collects a toy.

ii Leroy is practising free-throws with his basketball. The chance that he is successful with any attempt is 0.31. He gives himself one point for every successful shot. What is the least number of throws Leroy will need so that the probability of him scoring 10 points is higher than scoring eight points?

Let X be the discrete random variable that describes the number of points scored.

Answers ➲ pp. 226–227

4 Mean and variance of binomial distributions

➔ **Expected number** = Mean = $E(X) = \mu = np$

➔ **Variance** = $Var(X) = \sigma^2 = np(1-p)$

➔ **Standard deviation** = $\sigma = \sqrt{np(1-p)}$

EXAMPLE 1

The graph of a binomial distribution of 16 trials is shown below.

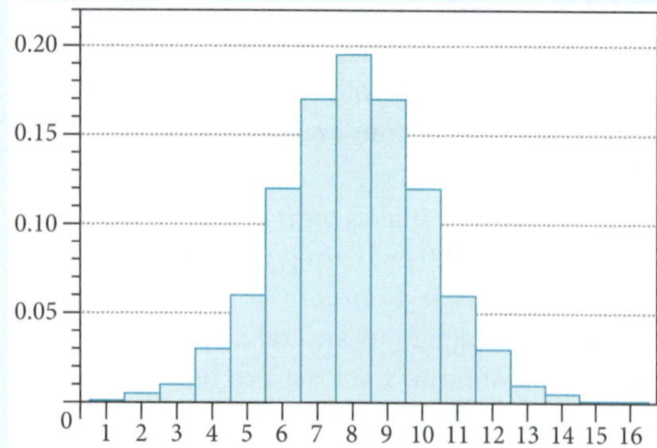

a Find the probability of success, p.

b Hence, find $Var(X)$.

a $E(X) = np = 8$, $n = 16$

$16p = 8$

$p = 0.5$

(Also, as the graph is symmetrical, then $p = 0.5$.)

$Var(X) = np(1-p) = npq$

b As $X \sim \text{Bin}(16, 0.5)$,

$Var(X) = np(1-p)$

$= 16 \times 0.5(1-0.5)$

$= 4$

EXAMPLE 2

A large metal fabrication company has a certain type of machine which has a probability of breaking down of 0.28. The company has six of the machines. Let X be the discrete random variable that describes the number of machines that have broken down.

a Find the expected number of machines that are broken down.

b Find the standard deviation, correct to four significant figures.

a $E(X) = np$

$= 6 \times 0.28$

$= 1.68$

$\sigma = \sqrt{Var(X)}$

b $\sigma = \sqrt{np(1-p)}$

$= \sqrt{6 \times 0.28(1-0.28)}$

$= 1.099\,818\,167\ldots$

$= 1.100$ (4 sig. figs)

EXAMPLE 3

A binomial random variable Y has a mean of 8.28 and a variance of 4.4712.

a Find the probability of success, p.

b Find the number of trials, n.

a $E(Y) = np = 8.28$

$Var(Y) = np(1-p) = 4.4712$

$8.28(1-p) = 4.4712$

$1 - p = 0.54$

$p = 0.46$

$Y \sim \text{Bin}(n, p)$

b Substitute into $np = 8.28$:

$n(0.46) = 8.28$

$n = 18$

→ PRACTICE

1 i The graph of a binomial distribution of 12 trials is shown below.

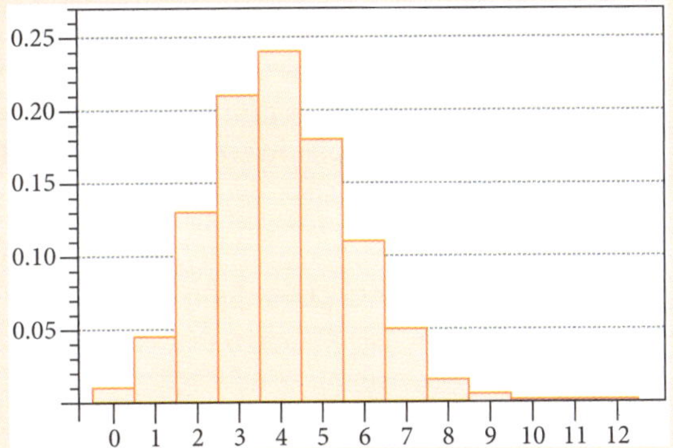

a Find the probability of success, p.

b Hence, find $Var(X)$.

ii A *portion* of the distribution of $X \sim \text{Bin}(n, 0.2)$ is shown below.

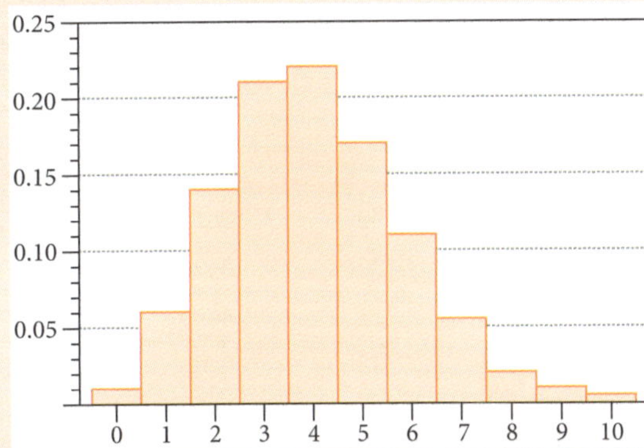

a What is the value of n?

b Find the standard deviation, correct to two decimal places.

2 i Bill Fumble is the fullback in the local rugby league team. Bill is not very good at catching balls that have been kicked high into the air by an opposition player. The probability of him catching a ball is 0.3. In a certain game the ball is kicked high into the air towards Bill on six occasions. Let X be the discrete random variable that describes the number of times Bill catches the ball.

a Find the expected number of times Bill will catch the ball.

b Find the standard deviation, correct to three decimal places.

ii The traffic light on the corner of Main Street and Beach Road has consistent timing for green, yellow and red lights at all times in the week. Every two minutes the lights are green for 75 seconds, yellow for three seconds and red for the remainder of the time. Carmen drives through the intersection once every day of the week, including weekends. Let X be the discrete random variable that describes the number of days she has a red light at the intersection.

a Find the expected number of times she will have a red light in the week.

b Find the standard deviation, correct to three significant figures.

3 i A binomial random variable Y has a mean of 9.1 and a variance of 3.185.

a Find the probability of success, p.

b Find the number of trials, n.

ii If $X \sim \text{Bin}(8, p)$ and $Var(X) = 1.5392$, find possible values of p.

Answers ⮌ p. 227

5 Normal approximation for the sample proportion

- **In a population**, the 'success' is recorded as p.
- **In a sample**, the 'success' proportion is recorded as $\hat{p}$, pronounced 'p hat'.
- As long as $np \geq 10$ and $n(1-p) \geq 10$ then $\hat{p}$ is said to be **normally distributed**.
- **Mean** $(\mu_{\hat{p}}) = \dfrac{np}{n} = p$ and a standard deviation $(\sigma_{\hat{p}}) = \dfrac{\sqrt{np(1-p)}}{n} = \sqrt{\dfrac{p(1-p)}{n}}$.
- **A table for a normal distribution** can be used to determine probabilities in the sample.

EXAMPLE 1

Given $n = 30$ and $p = 0.2$, find the standard deviation of $\hat{p}$, correct to three decimal places.

$$\sigma_{\hat{p}} = \frac{\sqrt{np(1-p)}}{n}$$
$$= \frac{\sqrt{30(0.2)(0.8)}}{30}$$
$$= 0.073\,029\,674\ldots$$
$$= 0.073 \text{ (3 dec. pl.)}$$

Checking, $np = 15 > 10$ and $n(1-p) = 24 > 10$

EXAMPLE 2

Approximately 11% of a country's population is left-handed. A survey of 300 adults was conducted.

a Show that the standard deviation of the distribution of sample proportions of the survey is 0.0181, correct to four decimal places.

b Estimate the probability that more than 15% of adults were right-handed.

Use the extract shown from a table giving values of $P(Z < z)$, where z has a normal distribution.

z	2.0	2.1	2.2
.00	.9772	.9821	.9861
.01	.9778	.9826	.9864
.02	.9783	.9830	.9868
.03	.9788	.9834	.9871
.04	.9793	.9838	.9875
.05	.9798	.9842	.9878
.06	.9803	.9846	.9881
.07	.9808	.9850	.9884
.08	.9812	.9854	.9887
.09	.9817	.9857	.9890

a As $n = 300$ and $p = 0.11$

$np = 300 \times 0.11$

$= 33$

$\mu_{\hat{p}} = p = 0.11$

$$\sigma_{\hat{p}} = \frac{\sqrt{np(1-p)}}{n}$$

$$= \frac{\sqrt{300(0.11)(1-0.11)}}{300}$$

$= 0.0181$ (4 dec. pl.)

$$\sigma_{\hat{p}} = \frac{\sqrt{\text{Var}(X)}}{n}$$

b $$z = \frac{x-\mu}{\sigma}$$

$$= \frac{0.15-0.11}{0.0181}$$

$= 2.21$ (2 dec. pl.)

Now, from the table, $P(z < 2.21) = 0.9864$.

$\therefore P(z > 2.21) = 1 - 0.9864$

$= 0.0136$

$\therefore$ the probability is 0.0136.

→ PRACTICE

1 i Given $n = 80$ and $p = 0.09$, find the standard deviation of $\hat{p}$, correct to three decimal places.

ii In Australia the proportion of people who are twins is 2.8%. Let $\hat{p}$ be the random variable that represents the sample proportion of people who are twins for samples of size n. Find the smallest integer value of n such that the standard deviation of $\hat{p}$ is less than or equal to 0.01.

2 i In Australia, 17% of the adult population smoke every day. A survey of 150 shoppers at a shopping centre was conducted.

a Show that the standard deviation of the distribution of sample proportions of such surveys is 0.0307, correct to four decimal places.

b Estimate the probability that more than 20% of those surveyed smoke every day.

Use the extract shown from a table giving values of $P(Z < z)$, where z has a normal distribution.

z	0.9	1.0	1.1
.00	.8159	.8413	.8643
.01	.8186	.8438	.8665
.02	.8212	.8461	.8686
.03	.8238	.8485	.8708
.04	.8264	.8508	.8729
.05	.8289	.8531	.8749
.06	.8315	.8554	.8770
.07	.8340	.8577	.8790
.08	.8365	.8599	.8810
.09	.8389	.8621	.8830

ii In the 2018 census, 29% of Australians nominated that they were born overseas. A survey of 250 people was conducted.

a Show that the mean and standard deviation of the distribution of sample proportions of such surveys are 0.29 and 0.0287 (correct to four decimal places) respectively.

b Estimate the probability that between 26% and 33% of those surveyed were born overseas.

Use the extract shown from a table giving values of $P(Z < z)$, where z has a normal distribution.

z	1.0	1.1	1.2	1.3
.00	.8413	.8643	.8849	.9032
.01	.8438	.8665	.8869	.9049
.02	.8461	.8686	.8888	.9066
.03	.8485	.8708	.8907	.9082
.04	.8508	.8729	.8925	.9099
.05	.8531	.8749	.8944	.9115
.06	.8554	.8770	.8962	.9131
.07	.8577	.8790	.8980	.9147
.08	.8599	.8810	.8997	.9162
.09	.8621	.8830	.9015	.9177

Answers ➲ p. 228

HSC EXAM-TYPE QUESTIONS

Total marks: 60

PART A

(Suggested time: 10 minutes)
(1 mark each)

1 The random variable X has the following probability distribution, where $0 < p < \frac{1}{2}$.

x	−1	0	1
$P(X = x)$	p	p	$1 - 2p$

Which of these is the variance of X?

A $5p$ **B** $2p - 3p^2$ **C** $2p - 9p^2$ **D** $5p - 9p^2$

2 Which of these could be the graph for $X \sim \text{Bin}(20, 0.4)$?

A

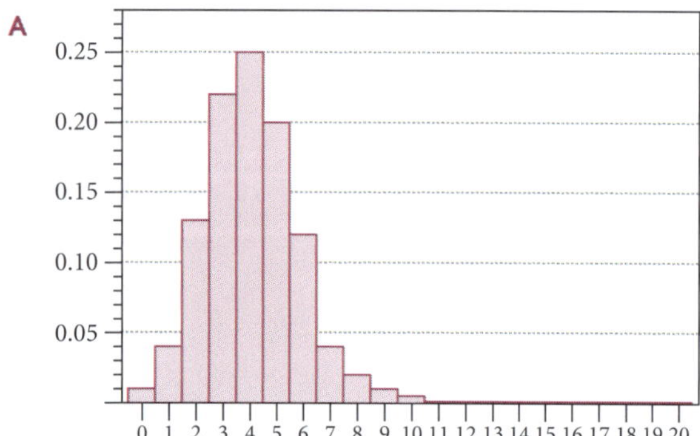

B

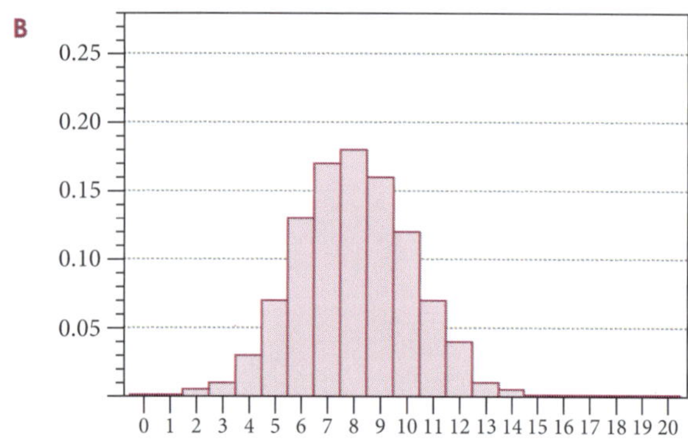

C

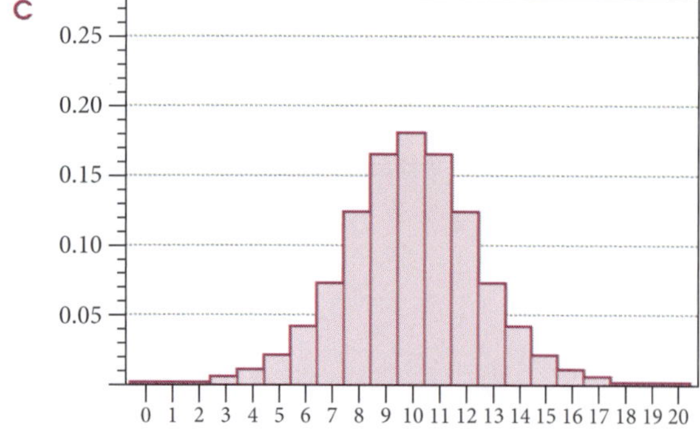

D

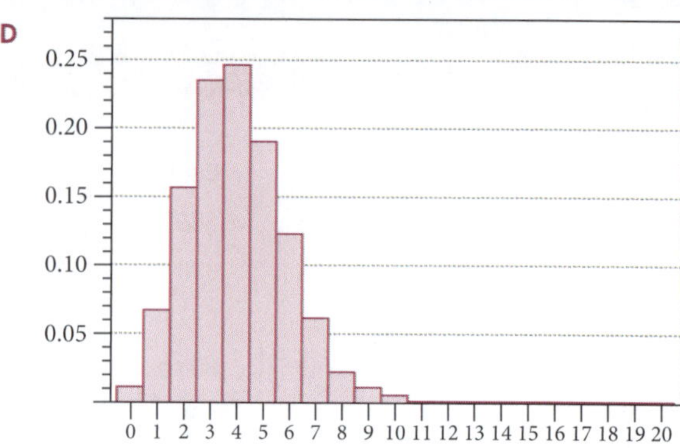

3 It is known that 4% of the components produced on an assembly line are defective. Components are randomly sampled from the line and tested. What is the probability that the 25th component tested is the first that is found to be defective?

A 1.5% **B** 1.6% **C** 15% **D** 16%

4 The probability of a child exposed to a certain disease catching that disease is 4%. Which of these is closest to the probability that the eightieth child exposed to the disease is the third to catch it?

A 0.0032 **B** 0.0085 **C** 0.2217 **D** 0.2268

5 Consider the following discrete probability distribution for the random variable X.

x	−1	0	1	3	5
$P(X = x)$	k	$2k$	k	$4k$	$2k$

What is the mean of the distribution?

A 1.8 **B** 2.2 **C** 2.6 **D** 3.2

PART B

(Suggested time: 45 minutes)

Show all working.

6 Given $X \sim \text{Bin}(4, p)$ and $P(X = 1) + P(X = 2) = 0.2$, show that $10p^4 - 30p^2 + 20p - 1 = 0$. (2 marks)

7 It is known that 35% of university students live in the family home. A random sample of 120 students is to be taken.

a Explain why this is a satisfactory sample size. (1 mark)

b Find the mean and standard deviation for the distribution of sample proportions of such surveys. (2 marks)

8 The binomial variable, X, has $E(X) = 3$ and $\text{Var}(X) = \frac{9}{4}$. Find $P(X \geq 2)$, correct to four decimal places. (3 marks)

9 The blood type O+ in a country's population is estimated at 31%. A survey of 1200 residents of the country was conducted.

a Show that the standard deviation of the distribution of sample proportions of such surveys is 0.0134 (correct to four decimal places). (1 mark)

b Estimate the probability that between 360 and 390 of those surveyed had the blood type O+. Use the extract shown from a table giving values of $P(Z < z)$, where z has a normal distribution. (3 marks)

z	0.7	0.8	0.9	1.0	1.1
.00	.7580	.7881	.8159	.8413	.8643
.01	.7611	.7910	.8186	.8438	.8665
.02	.7642	.7939	.8212	.8461	.8686
.03	.7673	.7967	.8238	.8485	.8708
.04	.7704	.7995	.8264	.8508	.8729
.05	.7734	.8023	.8289	.8531	.8749
.06	.7764	.8051	.8315	.8554	.8770
.07	.7794	.8078	.8340	.8577	.8790
.08	.7823	.8106	.8365	.8599	.8810
.09	.7852	.8133	.8389	.8621	.8830

10 The probability that a tomato seed germinates is 0.75. Each week a gardener plants 12 seeds. Let X be the number of seeds that successfully germinate.

a Explain why the average number of seeds the gardener could expect to germinate each week is 9. (1 mark)

b Find the probability that the number of seeds germinated in a week was more than the average. (2 marks)

11 Let Y be a discrete random variable with binomial distribution $Y \sim \text{Bin}(n, p)$. The mean is twice the standard deviation of this distribution. Given that $0 < p < 1$, what is the smallest number of trials n such that $p \leq 0.05$? (2 marks)

12 A large hardware store has 25 forklifts which are used for moving pallets around the shopfloor. The store manager has found that the probability of a forklift breaking down is 0.08. Let X be the discrete random variable that describes the number of forklifts that have broken down at any one time.

a Find the expected number of forklifts that are broken down. (1 mark)

b What is the probability that exactly 20% of the machines are broken down, correct to four decimal places? (1 mark)

c The manager decides to service 80% of the existing forklifts and remove the remainder from the store. This has meant the probability of a forklift breakdown has been reduced to 0.05. What is the probability that at any particular time there are at least two breakdowns? (2 marks)

13 A farmer supplies potatoes to a chip factory. Each day when a truckload arrives at the factory a specially designed machine randomly chooses five potatoes to check their mass. If two or more of these potatoes are underweight the whole truckload is rejected. It is known that 2.5% of all potatoes produced by the farmer are underweight. Let X be the discrete random variable that describes the number of potatoes that are underweight.

a Find the probability that a certain load of potatoes will be rejected. Give your answer correct to four decimal places. (2 marks)

b Rather than detecting five potatoes, the manager of the factory reprograms the machine to randomly choose n potatoes from each load to check. Find the smallest integer value of n such that the probability of at least one potato being underweight is greater than 0.4. (3 marks)

14 a If $X \sim \text{Bin}(n, p)$ and using the results for $P(X = x)$ and $P(X = x + 1)$, prove the recurrence relation $P(X = x + 1) = \dfrac{p(n-x)}{(x+1)(1-p)}.P(X = x)$. (2 marks)

b If $X \sim \text{Bin}(10, p)$ and $\dfrac{P(X=4)}{P(X=3)} = 0.75$, hence or otherwise, find the value of p. (2 marks)

PRACTICE

1 Bernoulli trials ⊃ pp. 217–218

1 i a

x	0	1
$P(X = x)$	**0.84**	**0.16**

b $E(X) = 0.16$

$Var(X) = 0.16(1 - 0.16)$

$= \mathbf{0.1344}$

ii a

x	0	1
$P(X = x)$	**0.75**	**0.25**

b $E(X) = 0.25$

$Var(X) = 0.25(1 - 0.25)$

$= \mathbf{0.1875}$

2 i $Var(X) = p(1 - p) = 0.16$

$p - p^2 = 0.16$

$p^2 - p + 0.16 = 0$

$(p - 0.8)(p - 0.2) = 0$

$p = 0.8, 0.2$

$\therefore$ **$E(X) = 0.8$ or 0.2**

ii $Var(Y) = p(1 - p) = 0.2475$

$p - p^2 = 0.2475$

$p^2 - p + 0.2475 = 0$

$$p = \frac{1 \pm \sqrt{(-1)^2 - 4(1)(0.2475)}}{2}$$

$p = 0.55, 0.45$

As $p < 0.5$, then $p = 0.45$.

$\therefore$ **$E(Y) = 0.45$**

2 Applications of the Bernoulli distribution ⊃ p. 218

1 i a As there are 9 failures, then need to find $P(X = 9)$.

Let $p = \frac{6}{36} = \frac{1}{6}, q = \frac{5}{6}$.

$P(X = 9) = \left(\frac{5}{6}\right)^9\left(\frac{1}{6}\right)$

$= 0.032\,301\,116\ldots$

$= \mathbf{0.0323}$ **(4 dec. pl.)**

b $P(X = 5) + P(X = 6) + P(X = 7) + P(X = 8)$

$= \left(\frac{5}{6}\right)^5\left(\frac{1}{6}\right) + \left(\frac{5}{6}\right)^6\left(\frac{1}{6}\right) + \left(\frac{5}{6}\right)^7\left(\frac{1}{6}\right) + \left(\frac{5}{6}\right)^8\left(\frac{1}{6}\right)$, which is geometric series with

$a = \left(\frac{5}{6}\right)^5\left(\frac{1}{6}\right), r = \frac{5}{6}, n = 4.$

$$= \frac{\left(\frac{5}{6}\right)^5\left(\frac{1}{6}\right)\left[1 - \left(\frac{5}{6}\right)^4\right]}{1 - \frac{5}{6}}$$

$= 0.208\,070\,872\ldots$

$= \mathbf{0.208}$ **(3 dec. pl.)**

ii a Let X be the number of samples checked before an A negative sample is detected.

As there are 44 failures, then need to find $P(X = 44)$.

Let $p = 0.07, q = 0.93$.

$P(X = 44) = (0.93)^{44}(0.07)$

$= 0.002\,873\,08\ldots$

$= \mathbf{0.0029}$ **(4 dec. pl.)**

b Let X be the number of samples checked before an AB positive sample is detected.

$P(\text{more than 100 checked}) = 1 - P(\text{100 or less checked})$

$= 1 - [P(X = 0) + P(X = 1) + \ldots + P(X = 99)]$

Let $p = 0.02, q = 0.98$.

$= 1 - [(0.98)^0(0.02) + (0.98)^1(0.02) + \ldots + (0.98)^{99}(0.02)]$

A geometric series with $a = 0.02, r = 0.98, n = 100$.

$$= 1 - \frac{0.02[1 - 0.98^{100}]}{1 - 0.98}$$

$= 0.132\,619\,55\ldots$

$= \mathbf{0.1326}$ **(4 dec. pl.)**

3 Binomial distributions ⊃ pp. 219–220

1 i Let $p = P(\text{tail})$.

Consider $P(X = 4) = {}^4C_4 p^4(1 - p)^0$

$= p^4$

$\therefore p^4 \times 200 = 15$

$p^4 = 0.075$

$p = 0.523\,317\,569\ldots$

$= \mathbf{0.52}$ **(2 dec. pl.)**

ii Let $p = P(\text{head})$.

$P(X = 3) = {}^5C_3 p^3(1 - p)^2 = 108 \div 360$

$10p^3(1 - p)^2 = 0.3 \ldots$ ①

$P(X = 4) = {}^5C_4 p^4(1 - p) = 50 \div 360$

$5p^4(1 - p) = 0.1389$ (4 dec. pl.) … ②

① ÷ ②: $\frac{2(1 - p)}{p} = 2.1598$ (4 dec. pl.)

$2 - 2p = 2.1598p$

$4.1598p = 2$

$p = 0.48$ (2 dec. pl.)

$\therefore P(X = 2) = {}^5C_2(0.48)^2(0.52)^3$

$= 0.323\,960\,832\ldots$

$= 0.3240$ (4 dec. pl.)

$E(X = 2) = 0.3240 \times 360$

$= 116.64$

$= \mathbf{117}$ **(nearest whole)**

2 i $P(\text{less than 5}) = \frac{2}{3}$.

Using $Y \sim \text{Bin}\left(10, \frac{2}{3}\right)$:

$P(Y = 6) = {}^{10}C_6\left(\frac{2}{3}\right)^6\left(1 - \frac{2}{3}\right)^4$

$= 0.227\,607\,58\ldots$

$= \mathbf{0.2276}$ **(4 dec. pl.)**

ii Let X be the discrete random variable that describes the number of times at least two light globes are defective.
Using $X \sim \text{Bin}(8, 0.04)$:

$P(X \geq 2) = 1 - [P(X = 0) + P(X = 1)]$

$= 1 - [{}^8C_0(0.04)^0(1-0.04)^8 + {}^8C_1(0.04)^1(1-0.04)^7]$

$= 0.038\,147\,228\ldots$

= 0.0381 (3 sig. figs)

3 i Using $X \sim \text{Bin}(8, 0.6)$:

$P(X \geq 3) = 1 - [P(X = 0) + P(X = 1) + P(X = 2)]$

$= 1 - [{}^8C_0(0.6)^0(1-0.6)^8 + {}^8C_1(0.6)^1(1-0.6)^7 + {}^8C_2(0.6)^2(1-0.6)^6]$

$= 0.950\,192\,64\ldots$

$= 0.9502$ (4 dec. pl.)

ii 5 out of 8 $= \frac{5}{8} = 0.625$

Using $X \sim \text{Bin}(12, 0.625)$:

$P(X \geq 10) = P(X = 10) + P(X = 11) + P(X = 12)$

$= {}^{12}C_{10}(0.625)^{10}(1-0.625)^2 + {}^{12}C_{11}(0.625)^{11}(1-0.625)^1 + {}^{12}C_{12}(0.625)^{12}(1-0.625)^0]$

$= 0.113\,544\,729\ldots$

= 11% (nearest whole)

4 i Using $X \sim \text{Bin}(n, 0.25)$:

$P(X = 4) > P(X = 3)$

$^nC_4(0.25)^4(1-0.25)^{n-4} > {}^nC_3(0.25)^3(1-0.25)^{n-3}$

$\frac{n!}{4!(n-4)!}0.25^4 0.75^{n-4} > \frac{n!}{3!(n-3)!}0.25^3 0.75^{n-3}$

$\frac{n!}{4!(n-4)!}0.25^4 0.75^{n-4} \times \frac{3!(n-3)!}{n!}\frac{1}{0.25^3}\frac{1}{0.75^{n-3}} > 1$

$\frac{0.25(n-3)}{4(0.75)} > 1$

$n - 3 > 12$

$n > 15$

∴ the father will need to buy at least 16 packets.

ii Using $X \sim \text{Bin}(n, 0.31)$:

$P(X = 10) > P(X = 8)$

$^nC_{10}(0.31)^{10}(1-0.31)^{n-10} > {}^nC_8(0.31)^8(1-0.31)^{n-8}$

$\frac{n!}{10!(n-10)!}0.31^{10}0.69^{n-10} > \frac{n!}{8!(n-8)!}0.31^8 0.69^{n-8}$

$\frac{n!}{10!(n-10)!}0.31^{10}0.69^{n-10} \times \frac{8!(n-8)!}{n!}\frac{1}{0.31^8}\frac{1}{0.69^{n-8}} > 1$

$\frac{0.31^2(n-8)(n-9)}{90(0.69)^2} > 1$

$(n - 8)(n - 9) > 446$ (nearest whole)

$n^2 - 17n + 72 - 446 > 0$

$n^2 - 17n - 374 > 0$

Consider $n^2 - 17n - 374 = 0$

$n = \frac{17 \pm \sqrt{(-17)^2 - 4(1)(-374)}}{2}$

$= 29.62$ (2 dec. pl.) (Also, $n > 0$)

∴ Leroy will need at least 30 throws.

4 Mean and variance of binomial distributions ➲ p. 221

1 i a $E(X) = np = 4$ and $n = 12$.

$p = 4 \div 12$

$= \mathbf{\frac{1}{3}}$

b As $X \sim \text{Bin}\left(12, \frac{1}{3}\right)$:

$\text{Var}(X) = np(1 - p)$

$= 12 \times \frac{1}{3}\left(1 - \frac{1}{3}\right)$

$= \mathbf{\frac{8}{3}}$

ii a $E(X) = 4$, and $p = 0.2$.

$n \times 0.2 = 4$

$\mathbf{n = 20}$

b $\sigma = \sqrt{np(1-p)}$

$= \sqrt{20 \times 0.2(1-0.2)}$

$= 1.788\,854\,382\ldots$

= 1.79 (2 dec. pl.)

2 i a $E(X) = np$

$= 6 \times 0.3$

= 1.8

b $\sigma = \sqrt{np(1-p)}$

$= \sqrt{6 \times 0.3(1-0.3)}$

$= 1.122\,497\,216\ldots$

= 1.122 (3 dec. pl.)

ii a Number of seconds red light = 120 − (75 + 3)

= 42

$P(\text{red}) = \frac{42}{120}$

$= 0.35$

$E(X) = np$

$= 7 \times 0.35$

= 2.45

b $\sigma = \sqrt{np(1-p)}$

$= \sqrt{7 \times 0.35(1-0.35)}$

$= 1.261\,942\,946\ldots$

= 1.26 (3 sig. figs)

3 i a $E(X) = np = 9.1$

$Var(X) = np(1 - p) = 3.185$

$9.1(1 - p) = 3.185$

$1 - p = 0.35$

$\mathbf{p = 0.65}$

b Substitute into $np = 9.1$:

$n(0.65) = 9.1$

$\mathbf{n = 14}$

ii $Var(X) = 8p(1 - p) = 1.5392$

$p(1 - p) = 0.1924$

$p - p^2 = 0.1924$

$p^2 - p + 0.1924 = 0$

$p = \frac{1 \pm \sqrt{(-1^2) - 4(1)(0.1924)}}{2}$

$= \frac{1 \pm 0.48}{2}$

= 0.26, 0.74

5 Normal approximation for the sample proportion ➲ pp. 222–223

1 i $\sigma_{\hat{p}} = \sqrt{\frac{p(1-p)}{n}}$

$= \sqrt{\frac{0.09(1-0.09)}{80}}$

$= 0.031\,996\,094\ldots$

$= \mathbf{0.032}$ **(3 dec. pl.)**

ii $p = 0.028$

$\sigma_{\hat{p}} = \sqrt{\frac{p(1-p)}{n}}$

$= \sqrt{\frac{0.028(1-0.028)}{n}} \le \frac{1}{100}$

$\frac{0.028(1-0.028)}{n} \le \frac{1}{10\,000}$

$\frac{0.027\,216}{n} \le \frac{1}{10\,000}$

$n \ge 272.16$

$\therefore$ ***n* is at least 273.**

2 i a As $n = 150$ and $p = 0.17$

$np = 150 \times 0.17$

$= 25.5$

$\mu_{\hat{p}} = p = 0.17$

$\sigma_{\hat{p}} = \frac{\sqrt{np(1-p)}}{n}$

$= \frac{\sqrt{150(0.17)(1-0.17)}}{150}$

$= \mathbf{0.0307}$ **(4 dec. pl.)**

b $z = \frac{x-\mu}{\sigma}$

$= \frac{0.2-0.17}{0.0307}$

$= 0.98$ (2 dec. pl.)

Now, from the table, $P(z < 0.98) = 0.8365$

$P(z > 0.98) = 1 - 0.8365$

$= 0.1635$

$\therefore$ **the probability is 0.1635.**

ii a As $n = 250$ and $p = 0.29$

$np = 250 \times 0.29$

$= 72.5$

$\mu_{\hat{p}} = \frac{np}{n} = p = 0.29$

$\sigma_{\hat{p}} = \frac{\sqrt{np(1-p)}}{n}$

$= \frac{\sqrt{250(0.29)(1-0.29)}}{250}$

$= \mathbf{0.0287}$ **(4 dec. pl.)**

b Using $x = 0.26$, $\mu = 0.29$, $\sigma = 0.0287$:

$z = \frac{x-\mu}{\sigma}$

$= \frac{0.26-0.29}{0.0287}$

$= -1.05$ (2 dec. pl.)

Now, from the table, $P(z < 1.05) = 0.8531$. This means $P(z < -1.05) = 1 - 0.8531 = 0.1469$.

Using $x = 0.33$, $\mu = 0.29$, $\sigma = 0.0287$:

$z = \frac{x-\mu}{\sigma}$

$= \frac{0.33-0.29}{0.0287}$

$= 1.39$ (2 dec. pl.)

Now, from the table, $P(z < 1.39) = 0.9177$

Hence, $\mathbf{P(-1.05 < z < 1.39) = 0.9177 - 0.1469 = 0.7708}$

HSC EXAM-TYPE QUESTIONS

Part A

1 **D.** $E(X) = -1(p) + 0(p) + 1(1 - 2p)$

$= -p + 1 - 2p$

$= 1 - 3p$

$E[X^2] = -1^2(p) + 0^2(p) + 1^2(1 - 2p)$

$= p + 1 - 2p$

$= 1 - p$

$Var(X) = E[X^2] - [E(X)]^2$

$= 1 - p - (1 - 3p)^2$

$= 1 - p - 1 + 6p - 9p^2$

$= 5p - 9p^2$

2 **B.** As $n = 20$ and $p = 0.4$, then $E(X) = 20 \times 0.4 = 8$, which is shown in the graph B.

3 **A.** $P(X = 25)$

Let $p = 0.04$, $q = 0.96$

$P(X = 25) = (0.96)^{24}(0.04)$

$= 0.015\,016\,529\ldots$

$= 1.5\%$ (1 dec. pl.)

4 **B.** Of the first 79 children, 2 have the disease. Then the 80th child gets the disease. Using $X \sim \text{Bin}(80, 0.04)$:

$P(X = 3) = {}^{79}C_2(0.04)^2(0.96)^{77}$

$= 0.212\,665\,296\ldots$

$= 0.2127$ (4 dec. pl.)

$P(\text{80th child is the 3rd infected}) = 0.2127 \times 0.04$

$= 0.008\,506\,611\ldots$

$= 0.0085$ (4 dec. pl.)

5 **B.** Sum of probabilitites = $k + 2k + k + 4k + 2k = 1$

$\therefore 10k = 1$

$k = \frac{1}{10}$

$E(X) = -1(k) + 0(2k) + 1(k) + 3(4k) + 5(2k)$

$= 22k$

Substitute $k = \frac{1}{10}$:

$\therefore E(X) = 2.2$

Part B

6 $\quad {}^4C_1 p(1-p)^3 + {}^4C_2 p^2(1-p)^2 = 0.2$ ✓

$4p(1 - 3p + 3p^2 - p^3) + 6p^2(1 - 2p + p^2) = 0.2$

$4p - 12p^2 + 12p^3 - 4p^4 + 6p^2 - 12p^3 + 6p^4 = \frac{1}{5}$

$2p^4 - 6p^2 + 4p = \frac{1}{5}$

$\mathbf{10p^4 - 30p^2 + 20p - 1 = 0}$ ✓

7 **a** **As $p = 0.35$ and $n = 120$ then $np = 42$ and $n(1 - p) = 78$ which are greater than 10 and so the sample is a satisfactory size.** ✓

b $\mu_{\hat{p}} = \frac{np}{n}$

$= \frac{120 \times 0.35}{120}$

$= 0.35$ ✓

$\sigma_{\hat{p}} = \frac{\sqrt{np(1-p)}}{n}$

$= \frac{\sqrt{120(0.35)(1-0.35)}}{120}$ ✓

$= 0.043\,541\,168\ldots$

$= 0.044$ (3 dec. pl.) ✓

8 $E(X) = np = 3$

$\text{Var}(X) = np(1 - p)$

$= 3(1 - p) = \frac{9}{4}$

$1 - p = \frac{9}{4} \div 3$

$= \frac{3}{4}$

$\therefore p = \frac{1}{4}$ ✓

$\therefore \frac{1}{4}n = 3$

$n = 12$ ✓

$P(X \geq 2) = 1 - P(X < 2)$

$= 1 - \left[{}^{12}C_0\left(\frac{1}{4}\right)^0\left(1-\frac{1}{4}\right)^{12} + {}^{12}C_1\left(\frac{1}{4}\right)^1\left(1-\frac{1}{4}\right)^{11}\right]$

$= 0.841\,618\,239\ldots$

$= \mathbf{0.8416}$ **(4 dec. pl.)** ✓

9 **a** As $n = 1200$ and $p = 0.31$

$\mu_{\hat{p}} = 0.31$

$\sigma_{\hat{p}} = \frac{\sqrt{np(1-p)}}{n}$

$= \frac{\sqrt{1200(0.31)(1-0.31)}}{1200}$

$= \mathbf{0.0134}$ **(4 dec. pl.)** ✓

b Consider $x = \frac{360}{1200} = 0.3$.

Using $x = 0.3$, $\mu = 0.31$, $\sigma = 0.0134$:

$z = \frac{x - \mu}{\sigma}$

$= \frac{0.3 - 0.31}{0.0134}$

$= -0.75$ (2 dec. pl.) ✓

Now, from the table, $P(z < 0.75) = 0.7734$. This means $P(z < -1.05) = 1 - 0.7734 = 0.2266$.

Consider $x = \frac{390}{1200} = 0.325$.

Using $x = 0.325$, $\mu = 0.31$, $\sigma = 0.0134$:

$z = \frac{0.325 - 0.31}{0.0134}$

$= 1.12$ (2 dec. pl.) ✓

Now, from the table, $P(z < 1.12) = 0.8686$.

Hence, $P(-0.75 < z < 1.12) = 0.8686 - 0.2266 = 0.642$ ✓

10 **a** $X \sim \text{Bin}(12, 0.75)$

$E(X) = np$

$= 12 \times 0.75$

$= \mathbf{9}$ ✓

b $P(X > 9) = P(X = 10) + P(X = 11) + P(X = 12)$ ✓

$= {}^{12}C_{10}(0.75)^{10}(1-0.75)^2 + {}^{12}C_{11}(0.75)^{11}(1-0.75)^1 + {}^{12}C_{12}(0.75)^{12}(1-0.75)^0$

$= 0.390\,675\,008\ldots$

$= \mathbf{0.3907}$ **(4 dec. pl.)** ✓

11 $\quad \bar{x} = 2\sigma$

$np = 2\sqrt{np(1-p)}$

$n^2p^2 = 4np(1 - p)$

$np = 4 - 4p$ (as $np \neq 0$)

$p(n + 4) = 4$

$p = \frac{4}{n+4} \leq 0.05$ ✓

$\frac{4}{n+4} \leq \frac{1}{20}$

$80 \leq n + 4$ (as $n > 0$)

$n \geq 76$

$\therefore$ **76 trials** ✓

12 a $E(X) = np$

$= 25 \times 0.08$

$= \mathbf{2}$ ✓

b 20% of 25 = 5

Using $X \sim \text{Bin}(25, 0.08)$:

$P(X = 5) = {}^{25}C_5(0.08)^5(1-0.08)^{20}$

$= 0.032\,850\,826\ldots$

$= \mathbf{0.0329}$ **(4 dec. pl.)** ✓

c 80% of 25 = 20

Using $X \sim \text{Bin}(20, 0.05)$:

$P(X \geq 2) = 1 - [P(X = 0) + P(X = 1)]$ ✓

$= 1 - [{}^{20}C_0(0.05)^0(1-0.05)^{20} + {}^{20}C_1(0.05)^1(1-0.05)^{19}]$

$= 0.264\,160\,475\ldots$

$= \mathbf{0.2642}$ **(4 dec. pl.)** ✓

13 a $n = 5, p = 0.025$

$P(X \geq 2) = 1 - [P(X = 0) + P(X = 1)]$ ✓

$= 1 - [{}^5C_0(0.025)^0(0.975)^5 + {}^5C_1(0.025)^1(0.975)^4]$

$= 0.005\,943\,32\ldots$

$= \mathbf{0.0059}$ **(4 dec. pl.)** ✓

b $1 - P(X = 0) = 1 - {}^nC_0(0.025)^0(0.975)^n > 0.4$ ✓

$1 - 0.975^n > 0.4$

$0.975^n < 0.6$

$n \ln 0.975 < \ln 0.6$

$n > \dfrac{\ln 0.6}{\ln 0.975}$ (dividing by negative) ✓

$n > 20.176\,5344\ldots$

$\therefore$ **the machine needs to check 21 potatoes.** ✓

14 a Use $P(X = x) = {}^nC_x p^x(1-p)^{n-x}$ … ①

and $P(X = x + 1) = {}^nC_{x+1}p^{x+1}(1-p)^{n-x-1}$ … ②

② ÷ ①: $\dfrac{P(X=x+1)}{P(X=x)} = \dfrac{{}^nC_{x+1}p^{x+1}(1-p)^{n-x-1}}{{}^nC_x p^x(1-p)^{n-x}}$ ✓

$= \dfrac{n!}{(x+1)!(n-x-1)!}.p^{x+1}.(1-p)^{n-x-1} \div \left[\dfrac{n!}{x!(n-x)!}.p^x.(1-p)^{n-x}\right]$

$= \dfrac{n!}{(x+1)!(n-x-1)!}.p^{x+1}.(1-p)^{n-x-1} \times \dfrac{x!(n-x)!}{n!}.\dfrac{1}{p^x}.\dfrac{1}{(1-p)^{n-x}}$

$= \dfrac{p(n-x)}{(x+1)(1-p)}$

$\therefore \mathbf{P(X = x + 1) = \dfrac{p(n-x)}{(x+1)(1-p)}.P(X = x)}$ ✓

b Substitute $n = 10$, $x = 3$ and $\dfrac{P(X=4)}{P(X=3)} = 0.75$, and use

$\dfrac{P(X=x+1)}{P(X=x)} = \dfrac{p(n-x)}{(x+1)(1-p)}$

$\dfrac{P(X=4)}{P(X=3)} = \dfrac{p(10-3)}{(3+1)(1-p)} = 0.75$ ✓

$\dfrac{7p}{4(1-p)} = \dfrac{3}{4}$

$7p = 3 - 3p$

$10p = 3$

$\mathbf{p = 0.3}$ ✓

SAMPLE HSC EXAMINATION 1

Try to complete these papers as if they are the real thing. These are the instructions you need to follow in the HSC Exam:

General instructions

- Reading time: 10 minutes
- Working time: 2 hours
- Write using black pen.
- NESA approved calculators may be used.
- A reference sheet is provided at the back of this book.
- For questions in Section II, show relevant mathematical reasoning and/or calculations.

Total marks: 70

Section I: 10 marks

Section II: 60 marks

Section I: 10 marks

Allow about 15 minutes for this section.

1 If $f(x) = \dfrac{1}{\sin^{-1} x}$, which of these is $f'(x)$?

A $f'(x) = \dfrac{-1}{\sin^{-1} x}$

B $f'(x) = \dfrac{-1}{\sqrt{1-x^2}\sin^{-1} x}$

C $f'(x) = \dfrac{-1}{\sqrt{1-x^2}\cos^{-1} x}$

D $f'(x) = \dfrac{-1}{\sqrt{1-x^2}(\sin^{-1} x)^2}$

2 In how many ways can eight people from a group of 15 people be chosen and then arranged in a circle?

A $\dfrac{15!}{7\times 8!}$ B $\dfrac{15!}{8\times 7!}$ C $\dfrac{15!}{7!}$ D $\dfrac{15!}{8!}$

3 Using the substitution $u = \sqrt{x-1}$, $\displaystyle\int \frac{dx}{(x+1)\sqrt{x-1}}$ can be expressed as:

A $\displaystyle\int \frac{du}{u\sqrt{2+u}}$

B $\displaystyle 2\int \frac{du}{2+u^2}$

C $\displaystyle\int \frac{du}{u(2+u^2)}$

D $\displaystyle\int \frac{du}{u(2+u)}$

4 A direction field representing a differential equation is shown below.

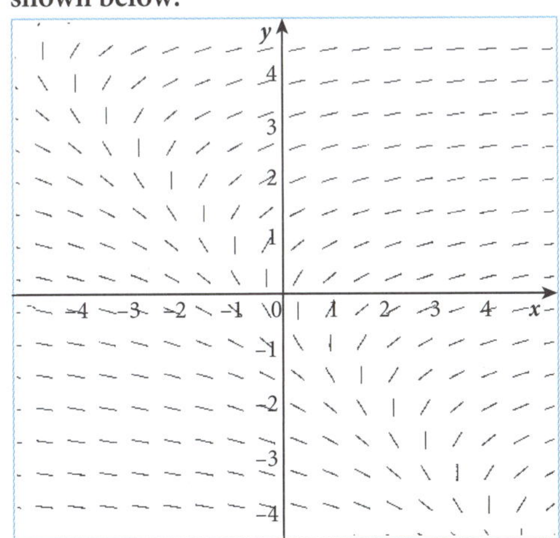

Which of these could be the differential equation?

A $\dfrac{dy}{dx} = \dfrac{1}{x+y}$

B $\dfrac{dy}{dx} = \dfrac{1}{x-y}$

C $\dfrac{dy}{dx} = \dfrac{-1}{xy}$

D $\dfrac{dy}{dx} = -\dfrac{x}{y}$

5 When the radius of a sphere is 5 cm, the volume is increasing at the rate of 750π cm³/minute. What is the rate of increase of the radius?

A 7.5 cm/min B 15 cm/min

C 75 cm/min D 150 cm/min

6 An experiment consisted of tossing a biased coin six times and recording the number of heads obtained. This experiment was repeated 400 times and the results are shown in the table below.

Number of heads	0	1	2	3	4	5	6
Frequency	3	22	71	120	114	58	12

Based on these results, what is the probability that the coin shows heads when tossed, correct to two decimal places?

A 0.54 B 0.56 C 0.58 D 0.59

7 The velocity, in cms^{-1}, of a body is given by $v = \sqrt{2t-3}$, where t is time in seconds.

What is the acceleration after 6 seconds?

A $\dfrac{1}{3}\,\text{cms}^{-2}$ B $3\,\text{cms}^{-2}$

C $4\dfrac{1}{2}\,\text{cms}^{-2}$ D $19\dfrac{1}{2}\,\text{cms}^{-2}$

8 It is known that 5% of the components produced on an assembly line are defective. Components are randomly sampled from the line and tested. Which of these is closest to the probability that the 20th component tested is the first that is found to be defective?

A 1% B 1.8% C 1.9% D 2.1%

9 $\dfrac{{}^{n-1}C_{k-1}}{{}^{n}C_k} =$

A $\dfrac{n}{k}$ **B** $\dfrac{n-1}{k-1}$ **C** $\dfrac{k}{n}$ **D** $\dfrac{k-1}{n-1}$

10 In a large population the probability that a person is left-handed is 0.1. Let $\hat{p}$ be the random variable that represents the sample proportion of left-handed people for sample of size n drawn from the population. Find the smallest value of n such that the standard deviation of $\hat{p}$ is less than or equal to 0.05.

A $n = 34$ **B** $n = 36$ **C** $n = 38$ **D** $n = 39$

Section II: 60 marks

Allow about 1 hour and 45 minutes for this section.

Question 11 (15 marks)

a The polynomial $P(x) = x^3 - 6x^2 + kx + 54$ has roots α, β and γ.

i Write down the values of $\alpha + \beta + \gamma$ and $\alpha\beta\gamma$. (1 mark)

ii It is known that two of the roots are equal in magnitude but opposite in sign. Find the third root and hence find the value of k. (2 marks)

b i Express cos 3A in terms of cos A. (2 marks)

ii Hence, find the exact value of $\cos\left(3\tan^{-1}\dfrac{2}{\sqrt{5}}\right)$. (3 marks)

c Two vectors are given as $\underset{\sim}{a} = 6\underset{\sim}{i} - m\underset{\sim}{j}$ and $\underset{\sim}{b} = 4\underset{\sim}{i} + n\underset{\sim}{j}$, where m and n are positive integers. If $\underset{\sim}{a}$ and $\underset{\sim}{b}$ are perpendicular and $|\underset{\sim}{a}| = 13$, find the values of m and n. (2 marks)

d Find the area between the curves $y = \cos^2\dfrac{x}{2}$ and $y = \cos x$, in the domain $0 \le x \le 2\pi$. (3 marks)

e Find the domain of the function $y = 3\cos^{-1}\dfrac{1}{x}$. (2 marks)

Question 12 (15 marks)

a By expressing $\sqrt{2}\sin\left(\theta+\dfrac{\pi}{4}\right)+\sqrt{2}\cos\left(\theta-\dfrac{\pi}{4}\right)$ in the form $R\sin(\theta + \alpha)$, where $R > 0$ and $0 < \alpha < \dfrac{\pi}{2}$, solve $\sqrt{2}\sin\left(\theta+\dfrac{\pi}{4}\right)+\sqrt{2}\cos\left(\theta-\dfrac{\pi}{4}\right)+\sqrt{6} = 0$, for $0 < \alpha < 2\pi$. (5 marks)

b The position vector of a particle at time $t \ge 0$ is given by $\underset{\sim}{r}(t) = (t-1)\underset{\sim}{i} + (t^2 - 3t + 4)\underset{\sim}{j}$, where distance is measured in metres and time in seconds.

i Show that the cartesian equation of the path followed by the particle is $y = x^2 - x + 2$. (1 mark)

ii Find the speed of the particle when $t = 2$. (2 marks)

c The region bounded by the curve $y = \dfrac{x\sqrt{x}}{4-x^2}$, the x-axis and the lines $x = 0$ and $x = 1$ is rotated around the x-axis. Use the substitution $u = 4 - x^2$ to find the exact volume of the solid obtained. (4 marks)

d Find the exact value of $\displaystyle\int_{\frac{\pi}{12}}^{\frac{\pi}{6}} \sin 4x \sin 2x \, dx$. (3 marks)

Question 13 (15 marks)

a A container of water is heated to 100 °C and then placed in a room with a constant temperature of 16 °C. After 10 minutes the temperature of the water is 64 °C. Use Newton's law of cooling $\dfrac{dT}{dt} = -k(T - 16)$, where T °C is the temperature of the water at time t minutes after the water is placed in the room.

i Show that $T = 16 + 84e^{-kt}$ is a solution to the differential equation $\dfrac{dT}{dt} = -k(T - 16)$. (1 mark)

ii Show that $k = \dfrac{1}{10}\ln\dfrac{7}{4}$. (2 marks)

iii Find the length of time until the temperature drops to 20 °C. (2 marks)

b i If $P(x) = x^3 - 3x^2 + 2x - 120$, explain why $(x - 6)$ is the only linear factor of $P(x)$. (2 marks)

ii Hence, solve for positive n, ${}^{n}C_3 = 20$. (2 marks)

c Prove $\dfrac{2+\cos x-\cos 2x}{3\sin x-\sin 2x} = \dfrac{\cos x+1}{\sin x}$. (2 marks)

d In the following diagram, point P divides the line segment AB in the ratio 1:2 and M is the midpoint of line segment OP. $\overrightarrow{OA} = \underset{\sim}{a}$ and $\overrightarrow{OB} = \underset{\sim}{b}$.

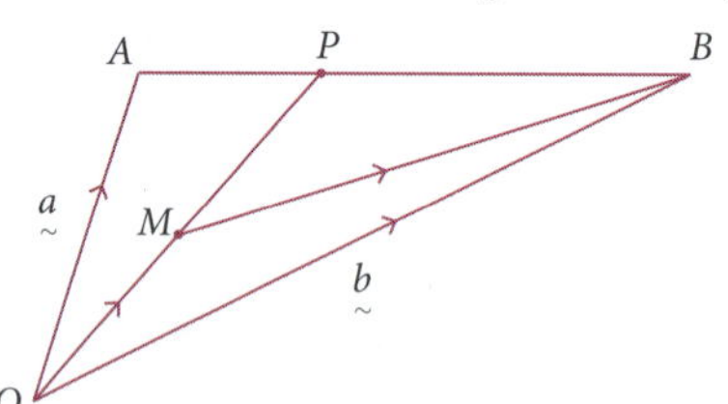

Show $\overrightarrow{MB} = \dfrac{1}{6}(5\underset{\sim}{b} - 2\underset{\sim}{a})$. (2 marks)

e During a military exercise 40 missiles are fired at targets. The probability that a missile hits its target is 85%. Let X be the discrete random variable that describes the number of missiles that have successfully hit their target. What is the probability that at least 95% of the missiles fired hit their targets, correct to four decimal places? (2 marks)

Question 14 (15 marks)

a Use mathematical induction to prove that,

for $n \geq 1$, $1^3 + 2^3 + 3^3 + \ldots + n^3 = \dfrac{n^2}{4}(n+1)^2$. (3 marks)

b Solve the differential equation $\sqrt{4-x^2}\dfrac{dy}{dx} = \dfrac{1}{y+1}$, given that $y(\sqrt{3}) = 0$. Express y in terms of x. (4 marks)

c William and Kate are in the school playground. William is sitting on a bench and tosses a ball at a velocity of 8 ms^{-1} from a height of 1.2 m at an angle of 60° to the horizontal towards Kate who is sitting 6 m away. Use the position vector

$\underset{\sim}{r}(t) = |\underset{\sim}{v}(0)|t\cos\theta\,\underset{\sim}{i} + (|\underset{\sim}{v}(0)|t\sin\theta - 5t^2 + 1.2)\underset{\sim}{j}$.

i What is the maximum height of the ball? (2 marks)

ii At what height does Kate catch the ball? (2 marks)

Assume Kate does not catch the ball.

iii How far past Kate will the ball hit the ground? Give your answer to the nearest centimetre. (2 marks)

iv Find the speed and the angle at which the ball hits the ground. (2 marks)

SAMPLE HSC EXAMINATION 2

Section I: 10 marks

Allow about 15 minutes for this section.

1 X is a discrete random variable that has a Bernoulli distribution. It is known that the variance for this distribution is 0.21. Find possible values of $E(X)$.

A 0.2 or 0.8 **B** 0.3 or 0.7
C 0.07 or 0.14 **D** 0.41 or 0.2

2 The polynomial $2x^3 + 8x^2 - 5x - 6$ has zeros α, β and γ. What is the value of $\alpha\beta\gamma(\alpha + \beta + \gamma)$?

A -24 **B** -12 **C** 12 **D** 24

3 A machine produces a certain engine component. It has been found that 97% of the components are satisfactory, but the remainder are defective. From a sample of 20 components, which of these is closest to the probability that at least two will be defective?

A 9% **B** 10% **C** 11% **D** 12%

4 If $\underset{\sim}{u} = -3\underset{\sim}{i} + \underset{\sim}{j}$ and $\underset{\sim}{v} = 2\underset{\sim}{i} + 4\underset{\sim}{j}$, find the vector projection of $\underset{\sim}{v}$ in the direction of $\underset{\sim}{u}$.

A $-\frac{1}{5}(-3\underset{\sim}{i} + \underset{\sim}{j})$ **B** $-\frac{1}{10}(-3\underset{\sim}{i} + \underset{\sim}{j})$

C $-\frac{1}{5}(2\underset{\sim}{i} + 4\underset{\sim}{j})$ **D** $-\frac{1}{10}(2\underset{\sim}{i} + 4\underset{\sim}{j})$

5 Which of these diagrams best represents the direction field of the differential equation $\frac{dy}{dx} = \frac{x-1}{y-1}$?

A

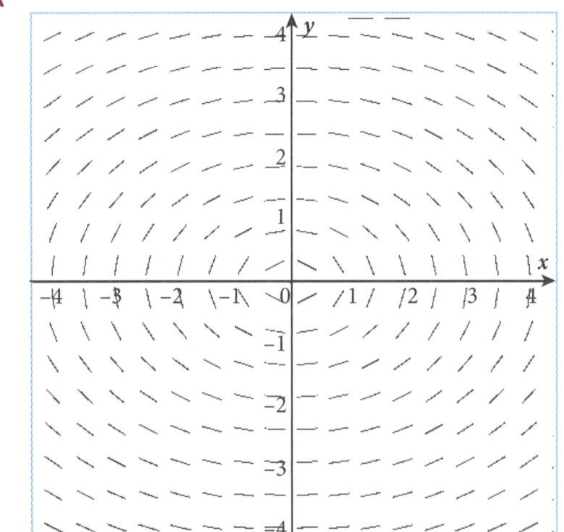

B

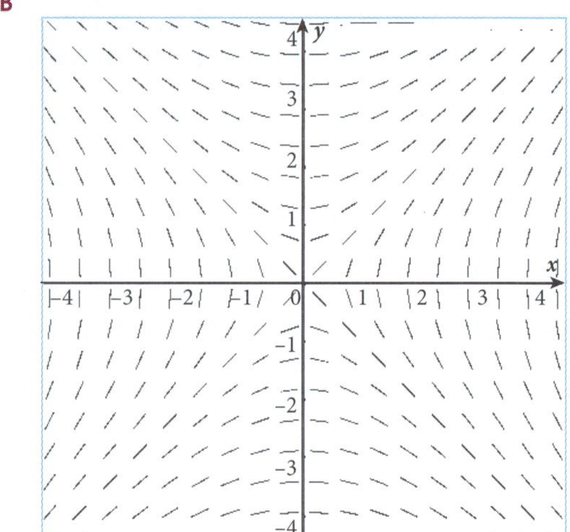

C

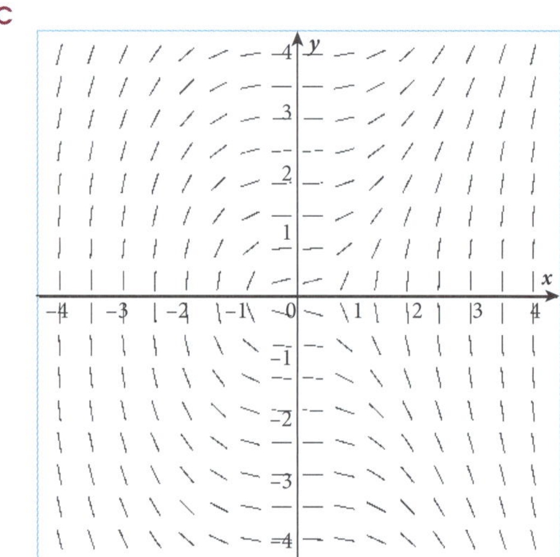

D

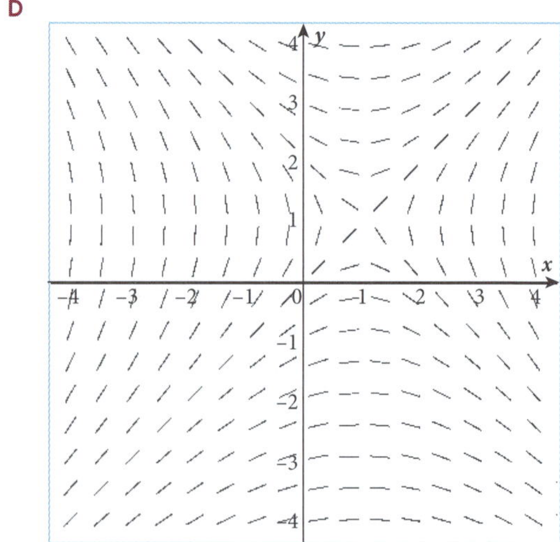

6 What is the remainder when $2x^3 - 4x^2 - 6x + 4$ is divided by $x - 2$?

A $-x^3 + 2x^2 + 3x - 2$ B -16
C -8 D $x^3 - 2x^2 - 3x + 2$

7 Two vectors are given by $\underset{\sim}{a} = 3\underset{\sim}{i} - m\underset{\sim}{j}$ and $\underset{\sim}{b} = -2\underset{\sim}{i} + n\underset{\sim}{j}$ where $m, n > 0$.
If $|\underset{\sim}{a}| = 5$, and $\underset{\sim}{a}$ is perpendicular to $\underset{\sim}{b}$, what is the value of n?

A -6 B $-\frac{3}{2}$ C $-\frac{2}{3}$ D $\frac{2}{3}$

8 Given $f(x) = 1 - x^2$, which of these is the graph of $\frac{1}{\sqrt{f(x)}}$?

A
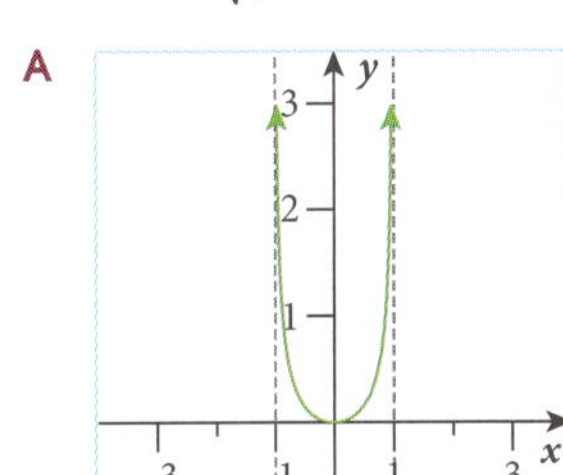
B
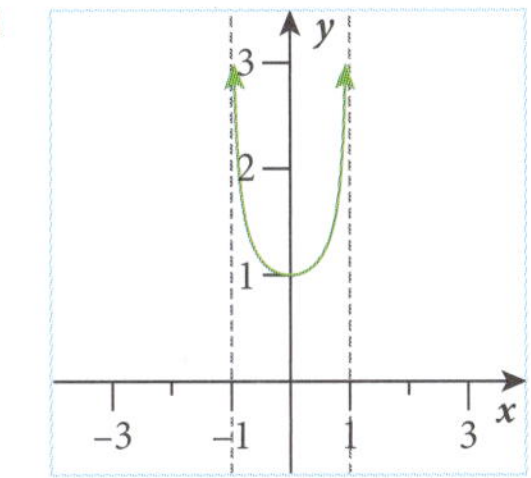
C
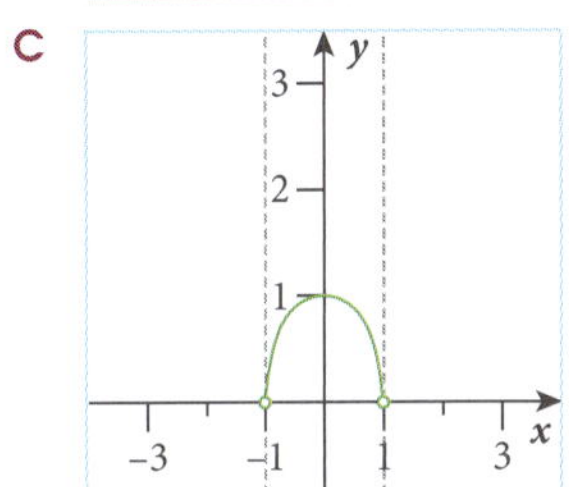
D
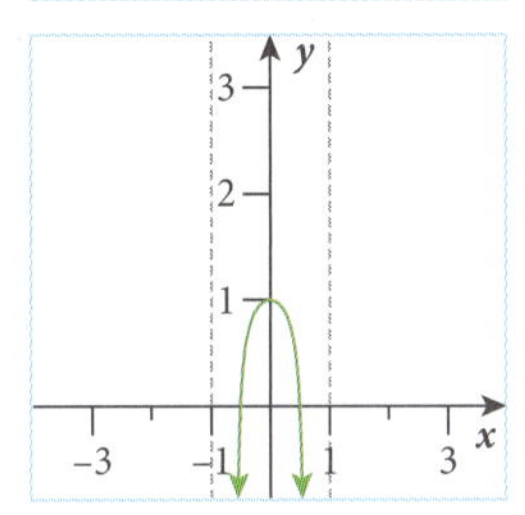

9 Consider the function $f(x) = \frac{x}{x-2}$, for $x > 2$. The function $f(x)$ has an inverse $f^{-1}(x)$.
Which of these is the point of intersection of $y = f(x)$ and $y = f^{-1}(x)$?

A $(0, 0)$ B $(3, 3)$ C $(3, 5)$ D $(4, 2)$

10 How many solutions are there to the trigonometric equation $\cos 3x + \cos x = 0$, over the domain $0 \leq x \leq \pi$?

A 1   B 2   C 3   D 4

Section II: 60 marks

Allow about 1 hour and 45 minutes for this section.

Question 11 (15 marks)

a i Sketch the functions $y = x^2$ and $y = \frac{1}{x}$ on the same number plane, showing any point(s) of intersection. (2 marks)

ii Without the use of calculus, use the same number plane in part a to sketch $y = \frac{x^3+1}{x}$. Label any intercepts on the sketch. (2 marks)

b Given $\underset{\sim}{a} = 4\underset{\sim}{i} + (2p + q)\underset{\sim}{j}$, $\underset{\sim}{b} = (p - 3q)\underset{\sim}{i} - 4\underset{\sim}{j}$ and $\underset{\sim}{a} + \underset{\sim}{b} = 6\underset{\sim}{i} + 7\underset{\sim}{j}$, write $-2\underset{\sim}{a} + 3\underset{\sim}{b}$ in component form. (3 marks)

c Prove by mathematical induction that $n(n^2 - 3n + 5)$ is divisible by 3 for all positive integers n. (3 marks)

d Solve $8 \sin \theta + 15 \cos \theta + 15 = 0$, where $0° \leq \theta \leq 360°$. (3 marks)

e A curve is defined parametrically by the equations $x = 2\left(t - \frac{1}{t}\right)$ and $y = 2\left(t^2 + 1 + \frac{1}{t^2}\right)$.
Find the cartesian equation of the curve. (2 marks)

Question 12 (15 marks)

a A multi-storey golf driving range allows players to hit from an elevated position. Carrie hits a golf ball at an initial velocity of 20 ms^{-1} from a height of 8 m above the ground at an angle of 60°.

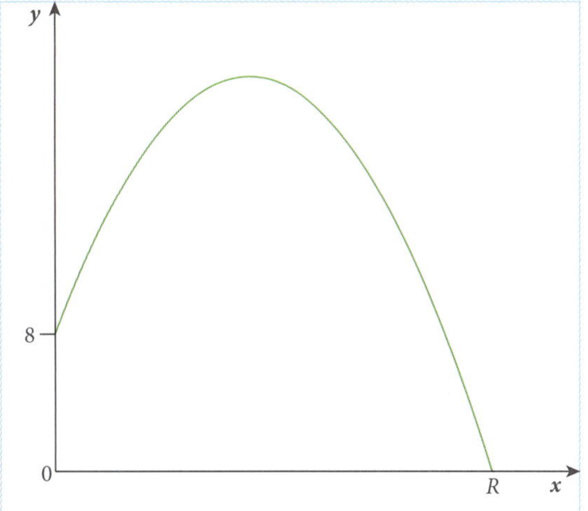

The position vector has equation
$\underset{\sim}{r}(t) = |\underset{\sim}{v}(0)| t \cos \theta \, \underset{\sim}{i} + (|\underset{\sim}{v}(0)| t \sin \theta - 5t^2 + 8)\underset{\sim}{j}$, where θ is the angle of projection and $|\underset{\sim}{v}(0)|$ is the initial velocity.

Let O be the origin, directly beneath the point of projection.

i Find the maximum height above the ground that the ball reaches. (2 marks)

The ball lands on the ground at R.

ii Find the range OR, correct to one decimal place. (2 marks)

iii Find the speed of the ball when it lands at R, to the nearest ms^{-1}. (2 marks)

b If $\tan x \tan y = 3$ and $\cos(x + y) = -\dfrac{2}{5}$, find the value of $\cos(x - y)$. (3 marks)

c On a number plane draw the graph of the function $y = 4\tan^{-1}(1 - x)$, stating the domain and range of the function. (3 marks)

d The diagram shows trapezium $PQRS$ with $PQ \parallel SR$. M and N are the midpoints of diagonals SQ and RP respectively. If $\vec{SP} = \underset{\sim}{a}$, $\vec{PQ} = \underset{\sim}{b}$, $\vec{QR} = \underset{\sim}{c}$, show that $MN \parallel SR$ by vector method. (3 marks)

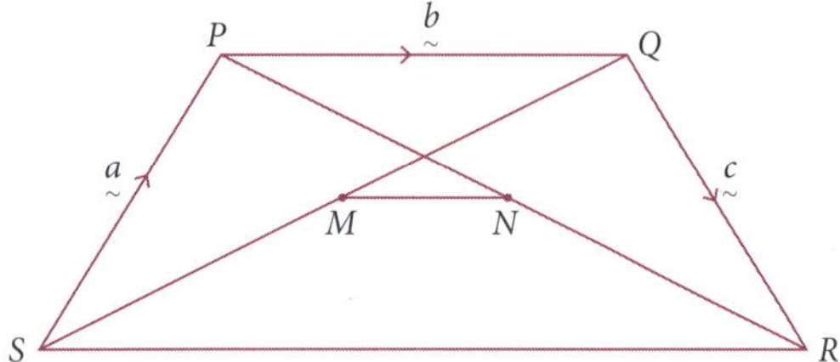

Question 13 (15 marks)

a Show that $\dfrac{\tan 2A}{1+\sec 2A} = \tan A$. (2 marks)

b The population in Tasmania is known to be 50% female. What is the probability that a random sample of 100 people will contain more than 55% females? (2 marks)

c The diagram shows a vessel in the shape of a hollow cone of vertical angle 60°. Water is poured into the empty vessel at the rate of 5 mL/s.

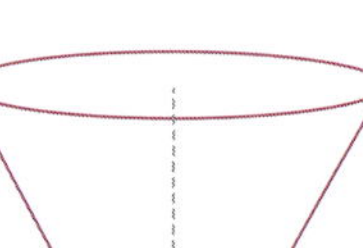

When the depth of water is 10 cm:

i show that the rate at which the water is rising is $\dfrac{3}{20\pi}$ cm/s. (2 marks)

ii find the rate at which the area of the surface of the vessel that is wet is increasing. (2 marks)

(Curved surface area of cone $= \pi r \ell$)

d Using the substitution $x = \tan\theta$,

show $\displaystyle\int_{-1}^{1} \frac{dx}{(1+x^2)^2} = \frac{\pi+2}{4}$. (3 marks)

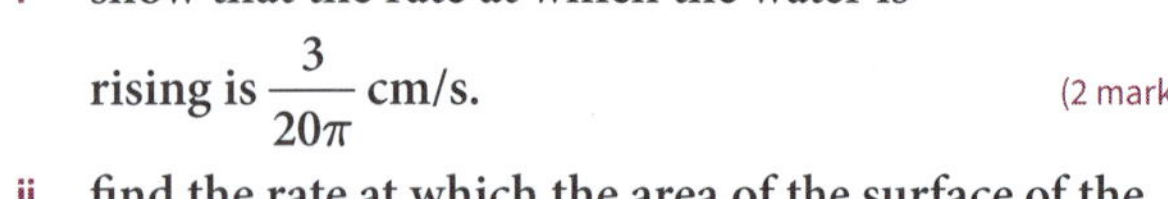

e Solve the differential equation $\sqrt{16-x^2}\,\dfrac{dy}{dx} = 4 - y$, with the condition $y(2) = 3$, and find the exact value of y when $x = 0$. (4 marks)

Question 14 (15 marks)

a $T(\alpha, \beta)$ is the point of intersection of the graphs $y = \tan x$ and $y = A\sin x$ (where $A > 1$), and $0 < \alpha < \dfrac{\pi}{2}$.

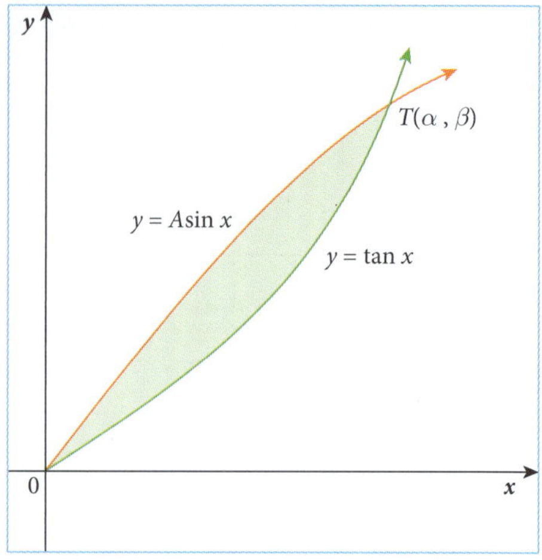

i Prove $\cos\alpha = \dfrac{1}{A}$. (2 marks)

ii Show that the shaded area is $(A - 1 - \ln A)$ units². (3 marks)

b The functions $y = \sin x$ and $y = 3\cos\dfrac{x}{2}$ intersect at the point $P(a, b)$, where $0 \le a \le 2\pi$. A solid is formed by rotating the area bounded by the curves $y = \sin x$ and $y = 3\cos\dfrac{x}{2}$ between $x = 0$ and $x = a$.

Find the exact volume of the solid. (4 marks)

c The number N of flying foxes in a colony after t months can be modelled by $N = \dfrac{3000}{1+2e^{-t}}$.

i What is the initial flying fox population? (1 mark)

ii Show that $\dfrac{dN}{dt} = \dfrac{6000e^t}{(e^t+2)^2}$. (2 marks)

iii Find the maximum growth rate of the flying fox population. (3 marks)

ANSWERS

SAMPLE HSC EXAMINATION 1

Section I

1 D. $f(x) = (\sin^{-1})^{-1}$

$$f'(x) = -1(\sin^{-1})^{-2}.\frac{1}{\sqrt{1-x^2}}$$

$$= \frac{-1}{\sqrt{1-x^2}(\sin^{-1}x)^2}$$

2 D. 15 choose 8: ${}^{15}C_8 = \frac{15!}{8! \times 7!}$

Arranged in circle: $\frac{15!}{8! \times 7!} \times 7!$

$$= \frac{15!}{8!}$$

3 B. $u = \sqrt{x-1}$

$u^2 = x - 1$

$x = u^2 + 1$

$\frac{dx}{du} = 2u$

$dx = 2u\, du$

Also, $x + 1 = u^2 + 2$.

$$\int \frac{dx}{(x+1)\sqrt{x-1}} = \int \frac{2u}{(u^2+2).u} du$$

$$= 2\int \frac{du}{2+u^2}$$

4 A. Consider the slope at some points on the direction field:

x	y	gradient
2	-2	vertical
4	4	nearly horizontal

By substitution, the correct equation is $\frac{dy}{dx} = \frac{1}{x+y}$.

5 A. $V = \frac{4}{3}\pi r^3$, hence $\frac{dV}{dr} = 4\pi r^2$.

When $r = 5$, then $\frac{dV}{dr} = 100\pi$.

$$\frac{dV}{dt} = \frac{dV}{dr} \times \frac{dr}{dt}$$

$$750\pi = 100\pi \times \frac{dr}{dt}$$

$$\therefore \frac{dr}{dt} = 7.5$$

6 B. Consider $P(X = 6) = {}^6C_6 p^6(1-p)^0$

$= p^6$

$\therefore p^6 \times 400 = 12$

$p^6 = 0.03$

$p = 0.557\,425\,556\ldots$

$= 0.56$ (2 dec. pl.)

7 A. $v = (2t-3)^{\frac{1}{2}}$

$$a = \frac{1}{2}(2t-3)^{-\frac{1}{2}}.2$$

$$= \frac{1}{\sqrt{2t-3}}$$

$$a(6) = \frac{1}{\sqrt{2(6)-3}}$$

$$= \frac{1}{3}$$

8 C. Let $p = 0.05$, $q = 0.95$

$P(X = 20) = (0.95)^{19}(0.05)$

$= 0.018\,867\,68\ldots$

$= 1.9\%$ (1 dec. pl.)

9 C. $$\frac{{}^{n-1}C_{k-1}}{{}^nC_k} = \frac{(n-1)!}{(k-1)!(n-k)!} \div \frac{n!}{k!(n-k)!}$$

$$= \frac{(n-1)!}{(k-1)!(n-k)!} \times \frac{k!(n-k)!}{n!}$$

$$= \frac{k}{n}$$

10 B. Use standard deviation = $\sqrt{\frac{p(1-p)}{n}}$

$$\therefore \sqrt{\frac{0.1(1-0.1)}{n}} \leq 0.05$$

$$\frac{0.1(1-0.1)}{n} \leq 0.0025$$

$0.0025n \geq 0.09$ (as $n > 0$)

$$n \geq \frac{0.09}{0.0025}$$

$n \geq 36$

$\therefore$ the smallest value is 36.

Section II

11 a i $\boldsymbol{\alpha + \beta + \gamma = 6}$ **and** $\boldsymbol{\alpha\beta\gamma = -54}$ ✓

ii Let roots be α, $-\alpha$ and β.

$\therefore \alpha - \alpha + \beta = 6$

$\therefore \beta = 6$

$\therefore$ **third root is $x = 6$.** ✓

$\therefore P(6) = 6^3 - 6(6)^2 + k(6) + 54 = 0$

$216 - 216 + 6k + 54 = 0$

$k = -9$

$\therefore$ **$k = -9$** ✓

b i $\cos 3A = \cos(2A + A)$

$= \cos 2A \cos A - \sin 2A \sin A$

$= (2\cos^2 A - 1)\cos A - 2\sin A \cos A \sin A$ ✓

$= 2\cos^3 A - \cos A - 2\cos A \sin^2 A$

$= 2\cos^3 A - \cos A - 2\cos A(1 - \cos^2 A)$

$= 2\cos^3 A - \cos A - 2\cos A + 2\cos^3 A$

$= 4\cos^3 A - 3\cos A$ ✓

ii For $\cos\left(3\tan^{-1}\frac{2}{\sqrt{5}}\right)$, let $A = \tan^{-1}\frac{2}{\sqrt{5}}$

$\therefore \tan A = \frac{2}{\sqrt{5}}$

$\therefore \cos A = \frac{\sqrt{5}}{3}$ (using Pythagoras' theorem) ✓

Now, $\cos\left(3\tan^{-1}\frac{2}{\sqrt{5}}\right) = \cos 3A$ ✓

$= 4\cos^3 A - 3\cos A$

$= 4\left(\frac{\sqrt{5}}{3}\right)^3 - 3\left(\frac{\sqrt{5}}{3}\right)$

$= \frac{20\sqrt{5}}{27} - \sqrt{5}$

$= -\frac{\mathbf{7\sqrt{5}}}{\mathbf{27}}$ ✓

c $|\underset{\sim}{a}| = \sqrt{6^2+(-m)^2} = 10$

$36 + m^2 = 100$

$m^2 = 64$

$m = 8 \ (m > 0)$ ✓

$\therefore \underset{\sim}{u} = 6\underset{\sim}{i} - 8\underset{\sim}{j}$

$\underset{\sim}{u}.\underset{\sim}{v} = 6 \times 4 + (-8) \times n = 0$

$24 - 8n = 0$

$-8n = -24$

$n = 3$ ✓

$\therefore$ **$m = 8$ and $n = 3$**

d Let $\cos^2\frac{x}{2} = \cos x$

$\cos^2\frac{x}{2} = \cos 2\left(\frac{x}{2}\right)$

$\cos^2\frac{x}{2} = 2\cos^2\frac{x}{2} - 1$

$\cos^2\frac{x}{2} = 1$

$\cos\frac{x}{2} = \pm 1$

$\frac{x}{2} = 0, \pi$

$x = 0, 2\pi$ ✓

$\text{Area} = \int_0^{2\pi}\left(\cos^2\frac{x}{2} - \cos x\right)dx$

$= \int_0^{2\pi}\left(\frac{1}{2}(1+\cos x) - \cos x\right)dx$

$= \frac{1}{2}\int_0^{2\pi}(1-\cos x)\,dx$ ✓

$= \frac{1}{2}[x - \sin x]_0^{2\pi}$

$= \frac{1}{2}[2\pi - \sin 2\pi - 0]$

$= \pi$

$\therefore$ **the area is π units^2.** ✓

e As domain for $y = \cos^{-1} x$ is $-1 \leq x \leq 1$, then the domain of $y = 3\cos^{-1}\frac{1}{x}$ is $-1 \leq \frac{1}{x} \leq 1$.

$\therefore$ Multiplying by x^2:

$-x^2 \leq x \leq x^2$

Consider $x \leq x^2$:

$\therefore x^2 - x \geq 0$

$x(x - 1) \geq 0$

By inspection, $x \leq 0$ and $x \geq 1$. But $x \neq 0$, hence $x < 0$ and $x \geq 1$. ✓

Also, consider $-x^2 \leq x$:

$\therefore x^2 + x \geq 0$

$x(x + 1) \geq 0$

By inspection, $x \leq -1$ and $x \geq 0$. But $x \neq 0$, hence $x \leq -1$ and $x > 0$.

Now, checking on the number line:

✓ ✗ ✗ ✓

−1 0 1

$\therefore \mathbf{x \leq -1, x \geq 1}$ ✓

12 a $\sqrt{2}\sin\left(\theta+\frac{\pi}{4}\right) + \sqrt{2}\cos\left(\theta-\frac{\pi}{4}\right)$

$= \sqrt{2}\left[\sin\theta\cos\frac{\pi}{4} + \cos\theta\sin\frac{\pi}{4}\right] + \sqrt{2}\left[\cos\theta\cos\frac{\pi}{4} + \sin\theta\sin\frac{\pi}{4}\right]$

$= \sqrt{2}\left[\frac{1}{\sqrt{2}}\sin\theta + \frac{1}{\sqrt{2}}\cos\theta\right] + \sqrt{2}\left[\frac{1}{\sqrt{2}}\cos\theta + \frac{1}{\sqrt{2}}\sin\theta\right]$

$= \sin\theta + \cos\theta + \cos\theta + \sin\theta$

$= 2\sin\theta + 2\cos\theta$ ✓

$R = \sqrt{2^2+2^2}$

$= \sqrt{8}$

$= 2\sqrt{2}$ ✓

$\tan\alpha = \frac{2}{2}$

$= 1$

$\alpha = \frac{\pi}{4}$ ✓

$\therefore \sqrt{2}\sin\left(\theta+\frac{\pi}{4}\right) + \sqrt{2}\cos\left(\theta-\frac{\pi}{4}\right) = 2\sqrt{2}\sin\left(\theta+\frac{\pi}{4}\right)$

$2\sqrt{2}\sin\left(\theta+\frac{\pi}{4}\right) + \sqrt{6} = 0$ ✓

$2\sqrt{2}\sin\left(\theta+\frac{\pi}{4}\right) = -\sqrt{6}$

$2\sqrt{2}\sin\left(\theta+\frac{\pi}{4}\right) = -\frac{\sqrt{6}}{2\sqrt{2}}$

$\sin\left(\theta+\frac{\pi}{4}\right) = -\frac{\sqrt{3}}{2}$

$\theta + \frac{\pi}{4} = \frac{4\pi}{3}, \frac{5\pi}{3}$

$\boldsymbol{\theta = \frac{13\pi}{12}, \frac{17\pi}{12}}$ ✓

b i $x = t - 1$

$\therefore t = x + 1$

Substitute into y:

$y = (x + 1)^2 - 3(x + 1) + 4$

$y = x^2 + 2x + 1 - 3x - 3 + 4$

$\mathbf{y = x^2 - x + 2}$ ✓

ii $\underset{\sim}{v}(t) = \underset{\sim}{i} + (2t - 3)\underset{\sim}{j}$

$\underset{\sim}{v}(2) = \underset{\sim}{i} + (2(2) - 3)\underset{\sim}{j}$ ✓

$\underset{\sim}{v}(2) = \underset{\sim}{i} + \underset{\sim}{j}$

$\text{Speed} = |\underset{\sim}{v}(2)| = \sqrt{1^2 + 1^2}$
$= \sqrt{2}$

$\therefore$ **speed of particle is $\sqrt{2}$ ms^{-1}.** ✓

c $y = \dfrac{x\sqrt{x}}{4 - x^2} \therefore y^2 = \dfrac{x^3}{(4 - x^2)^2}$ ✓

As $u = 4 - x^2$, then $\dfrac{du}{dx} = -2x$, hence $dx = \dfrac{du}{-2x}$. Also, $x^2 = 4 - 4$

Also, $x = 1$ gives $u = 3$ and $x = 0$ gives $u = 4$.

$$\text{Vol} = \pi \int_0^1 \frac{x^3}{(4 - x^2)^2}\, dx \quad ✓$$

$$= \pi \int_{u=4}^{u=3} \frac{x^3}{u^2} \cdot \frac{du}{-2x}$$

$$= \frac{\pi}{2} \int_3^4 \frac{x^2}{u^2}\, du$$

$$= \frac{\pi}{2} \int_3^4 \frac{4 - u}{u^2}\, du$$

$$= \frac{\pi}{2} \int_3^4 \left(4u^{-2} - \frac{1}{u}\right) du$$

$$= \frac{\pi}{2} \left[-\frac{4}{u} - \ln|u|\right]_3^4$$

$$= \frac{\pi}{2} \left[-\frac{4}{4} - \ln 4 - \left(-\frac{4}{3} - \ln 3\right)\right]$$

$$= \frac{\pi}{2} \left[-1 - \ln 4 + \frac{4}{3} + \ln 3\right]$$

$$= \frac{\pi}{2} \left[\ln\frac{3}{4} + \frac{1}{3}\right]$$

$\therefore$ **volume is $\dfrac{\pi}{2}\left[\ln\dfrac{3}{4} + \dfrac{1}{3}\right]$ units^3.** ✓

d $\displaystyle\int_{\frac{\pi}{12}}^{\frac{\pi}{6}} \sin 4x \sin 2x\, dx$

$$= \int_{\frac{\pi}{12}}^{\frac{\pi}{6}} \frac{1}{2}(\cos 2x - \cos 6x)\, dx \quad ✓$$

$$= \frac{1}{2}\left[\frac{1}{2}\sin 2x - \frac{1}{6}\sin 6x\right]_{\frac{\pi}{12}}^{\frac{\pi}{6}} \quad ✓$$

$$= \frac{1}{12}\left[3\sin 2x - \sin 6x\right]_{\frac{\pi}{12}}^{\frac{\pi}{6}}$$

$$= \frac{1}{12}\left[3\sin 2\left(\frac{\pi}{6}\right) - \sin 6\left(\frac{\pi}{6}\right) - \left[3\sin 2\left(\frac{\pi}{12}\right) - \sin 6\left(\frac{\pi}{12}\right)\right]\right]$$

$$= \frac{1}{12}\left[\frac{3\sqrt{3}}{2} - 0 - \left(\frac{3}{2} - 1\right)\right]$$

$$= \frac{1}{12}\left[\frac{3\sqrt{3}}{2} - \frac{1}{2}\right]$$

$$= \mathbf{\frac{3\sqrt{3} - 1}{24}} \quad ✓$$

13 a i $T = 16 + 84e^{-kt}$

First, $T - 16 = 84e^{-kt}$

Now, $\dfrac{dT}{dt} = -k.84e^{-kt}$

$= \mathbf{-k(T - 16)}$ ✓

ii Substitute $T = 64$, $t = 10$:

$64 = 16 + 84e^{-k(10)}$

$48 = 84e^{-10k}$

$e^{-10k} = \dfrac{48}{84}$ ✓

$-10k = \ln\dfrac{4}{7}$

$k = -\dfrac{\ln\frac{4}{7}}{10}$

$= \mathbf{\dfrac{1}{10}\ln\dfrac{7}{4}}$ ✓

iii Substitute $T = 20$, $k = \dfrac{1}{10}\ln\dfrac{7}{4}$:

$20 = 16 + 84e^{-kt}$

$4 = 84e^{-kt}$

$e^{-kt} = \dfrac{1}{21}$ ✓

$-kt = \ln\dfrac{1}{21}$

$t = \dfrac{\ln\frac{1}{21}}{-k}$

$= 54.403\,798\,17\ldots$

$= 54$ (nearest whole)

$\therefore$ **it will take 54 minutes.** ✓

b i $P(x) = x^3 - 3x^2 + 2x - 120$

$$\begin{array}{r} x^2 + 3x + 20 \\ x - 6\overline{)x^3 - 3x^2 + 2x - 120} \\ \underline{x^3 - 6x^2} \qquad\qquad \\ 3x^2 + 2x \qquad \\ \underline{3x^2 - 18x} \qquad \\ 20x - 120 \\ \underline{20x - 120} \\ 0 \end{array}$$

$P(x) = (x - 6)(x^2 + 3x + 20)$ ✓

For $x^2 + 3x + 20$, $\Delta = 3^2 - 4(1)(20) < 0$.

Hence, no real roots, so cannot factorise $x^2 + 3x + 20$ into linear factors.

$\therefore$ **$(x - 6)$ is the only linear factor of $P(x)$.** ✓

ii ${}^nC_3 = 20$

$\dfrac{n!}{3!(n-3)!} = 20$

$\dfrac{n(n-1)(n-2)}{6} = 20$

$n(n - 1)(n - 2) = 120$ ✓

$(n^2 - n)(n - 2) - 120 = 0$

$n^3 - 3n^2 + 2n - 120 = 0$

$(n - 6)(n^2 + 3n + 20) = 0$ (from part i)

$\therefore \mathbf{n = 6}$ ✓

c $\text{LHS} = \dfrac{2+\cos x - \cos 2x}{3\sin x - \sin 2x}$

$= \dfrac{2+\cos x - (2\cos^2 x - 1)}{3\sin x - 2\sin x \cos x}$

$= \dfrac{-2\cos^2 x + \cos x + 3}{3\sin x - 2\sin x \cos x}$ ✓

$= \dfrac{-(2\cos^2 x - \cos x - 3)}{3\sin x - 2\sin x \cos x}$

$= \dfrac{-(2\cos x - 3)(\cos x + 1)}{\sin x(3 - 2\cos x)}$

$= \dfrac{(3 - 2\cos x)(\cos x + 1)}{\sin x(3 - 2\cos x)}$

$= \dfrac{\cos x + 1}{\sin x}$

= **RHS** ✓

d $\overrightarrow{AB} = \underset{\sim}{b} - \underset{\sim}{a}$

$\overrightarrow{AP} = \frac{1}{3}(\underset{\sim}{b} - \underset{\sim}{a})$

$\overrightarrow{OP} = \underset{\sim}{a} + \frac{1}{3}(\underset{\sim}{b} - \underset{\sim}{a})$ ✓

$= \frac{1}{3}(2\underset{\sim}{a} + \underset{\sim}{b})$

$\overrightarrow{OM} = \frac{1}{6}(2\underset{\sim}{a} + \underset{\sim}{b})$

$\overrightarrow{MB} = \underset{\sim}{b} - \frac{1}{6}(2\underset{\sim}{a} + \underset{\sim}{b})$

$= \mathbf{\frac{1}{6}(5\underset{\sim}{b} - 2\underset{\sim}{a})}$ ✓

e 95% of 40 = 38

Using $X \sim \text{Bin}(40, 0.85)$: ✓

$P(X \geq 38) = {}^{40}C_{38}(0.85)^{38}(1-0.85)^2 + {}^{40}C_{39}(0.85)^{39}(1-0.85) + {}^{40}C_{40}(0.85)^{40}$

= 0.048 598 665...

= **0.0486 (4 dec. pl.)** ✓

14 a If $n = 1$: LHS = 1^3 = 1   RHS = $\frac{1^2}{4}(1+1)^2$ = 1

∴ true for $n = 1$. ✓

Assume true for $n = k$:

i.e. $S_k = 1^3 + 2^3 + 3^3 + \ldots + k^3 = \frac{k^2}{4}(k+1)^2$

Then if $n = k + 1$:

i.e. need to prove $S_k + T_{k+1} = S_{k+1}$

i.e. $1^3 + 2^3 + 3^3 + \ldots + k^3 + (k+1)^3 = \dfrac{(k+1)^2}{4}(k+2)^2$ ✓

$\text{LHS} = \frac{k^2}{4}(k+1)^2 + (k+1)^3$

$= \dfrac{(k+1)^2}{4}(k^2 + 4(k+1))$

$= \dfrac{(k+1)^2}{4}(k^2 + 4k + 4)$

$= \dfrac{(k+1)^2}{4}(k+2)^2$

= RHS

∴ hence the claim is true for $n = k + 1$

Since shown true for $n = 1$, is true for $n = 2, 3, 4, \ldots$ and so true for all integers $n \geq 1$. ✓

b $\sqrt{4-x^2}\dfrac{dy}{dx} = \dfrac{1}{y+1}$

$\int (y+1)\, dy = \int \dfrac{1}{\sqrt{4-x^2}}\, dx$

$\dfrac{y^2}{2} + y = \sin^{-1}\dfrac{x}{2} + c$ ✓

Using $y(\sqrt{3}) = 0$:

$0 = \sin^{-1}\dfrac{\sqrt{3}}{2} + c$

$c = -\dfrac{\pi}{3}$ ✓

$\dfrac{y^2}{2} + y = \sin^{-1}\dfrac{x}{2} - \dfrac{\pi}{3}$

$y^2 + 2y = 2\sin^{-1}\dfrac{x}{2} - \dfrac{2\pi}{3}$

$y^2 + 2y + 1 = 2\sin^{-1}\dfrac{x}{2} - \dfrac{2\pi}{3} + 1$ ✓

$(y+1)^2 = 2\sin^{-1}\dfrac{x}{2} - \dfrac{2\pi}{3} + 1$

$y + 1 = \sqrt{2\sin^{-1}\dfrac{x}{2} - \dfrac{2\pi}{3} + 1}$, as $y(\sqrt{3}) = 0$.

$\mathbf{y = -1 + \sqrt{2\sin^{-1}\dfrac{x}{2} - \dfrac{2\pi}{3} + 1}}$ ✓

c i Use $|\underset{\sim}{v}(0)| = 8$ and $\theta = 60°$.

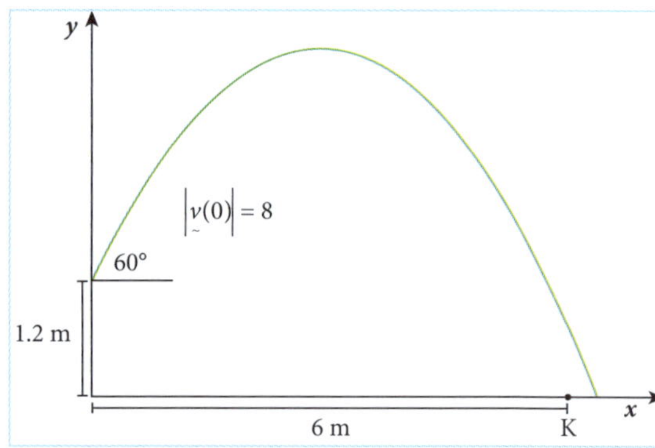

$\underset{\sim}{r}(t) = 8t\cos 60°\, \underset{\sim}{i} + (8t\sin 60° - 5t^2 + 1.2)\underset{\sim}{j}$

$= 4t\underset{\sim}{i} + (4\sqrt{3}t - 5t^2 + 1.2)\underset{\sim}{j}$

Consider $y = 4\sqrt{3}t - 5t^2 + 1.2$

$\dfrac{dy}{dt} = 4\sqrt{3} - 10t = 0$

$t = \dfrac{4\sqrt{3}}{10}$

$= \dfrac{2\sqrt{3}}{5}$ ✓

Substitute into y:

$y\left(\dfrac{2\sqrt{3}}{5}\right) = 4\sqrt{3}\left(\dfrac{2\sqrt{3}}{5}\right) - 5\left(\dfrac{2\sqrt{3}}{5}\right)^2 + 1.2$

$= 3.6$

∴ the ball reaches a maximum height of 3.6 m. ✓

ii Let $x = 6$:

$6 = 4t$

$t = 1.5$

Substitute $t = 1.5$: ✓

As $y(1.5) = 4\sqrt{3}(1.5) - 5(1.5)^2 + 1.2$

$= 0.342\,304\,845\ldots$

$= 0.34$ (2 dec. pl.)

$\therefore$ **Kate catches the ball at a height of 0.34 m.** ✓

iii Let $y = 0$:

$4\sqrt{3}t - 5t^2 + 1.2 = 0$

$5t^2 - 4\sqrt{3}t - 1.2 = 0$

$$t = \frac{4\sqrt{3} \pm \sqrt{(-4\sqrt{3})^2 - 4(5)(-1.2)}}{2(5)}$$

$= 1.541\,348\,46\ldots$ $(t > 0)$

$= 1.5413$ (4 dec. pl.) ✓

Substitute $t = 1.5413$ into x:

$x = 4(1.5413)$

$= 6.1652$

$= 6.17$ (2 dec. pl.)

Difference $= 6.17 - 6$

$= 0.17$

$\therefore$ **the ball hits the ground 17 cm past Kate.** ✓

iv $\underset{\sim}{r}(t) = 4t\underset{\sim}{i} + (4\sqrt{3}t - 5t^2 + 1.2)\underset{\sim}{j}$

$\underset{\sim}{v}(t) = 4\underset{\sim}{i} + (4\sqrt{3} - 10t)\underset{\sim}{j}$

$|\underset{\sim}{v}(1.5413)| = \sqrt{4^2 + (4\sqrt{3} - 10(1.5413))^2}$

$= 9.380\,393\,181\ldots$

$= 9.38$ (2 dec. pl.) ✓

$$\tan\theta = \left|\frac{4\sqrt{3} - 10(1.5413)}{4}\right|$$

$= 64°46'$ ✓

The ball hits the ground at a speed of 9.38 ms^{-1} at an acute angle of 64°46' to the horizontal.

SAMPLE HSC EXAMINATION 2

Section I

1 B. $Var(X) = p(1 - p) = 0.21$

$p - p^2 = 0.21$

$p^2 - p + 0.21 = 0$

$(p - 0.7)(p - 0.3) = 0$

$p = 0.7, 0.3$

$\therefore E(X) = 0.7$ or 0.3

2 B. $\alpha + \beta + \gamma = -\frac{b}{a} = -4$

$\alpha\beta\gamma = -\frac{d}{a} = 3$

$\alpha\beta\gamma(\alpha + \beta + \gamma) = (-4) \times 3$

$= -12$

3 D. Using $X \sim \text{Bin}(15, 0.04)$:

$P(X \geq 2) = 1 - [P(X = 0) + P(X = 1)]$

$= 1 - [{}^{20}C_0(0.03)^0(1 - 0.03)^{20} + {}^{20}C_1(0.03)^1(1 - 0.03)^{19}]$

$= 0.119\,838\,022\ldots$

$= 0.12$ (2 dec. pl.)

$= 12\%$

4 A. $\underset{\sim}{u}.\underset{\sim}{v} = -3 \times 2 + 1 \times 4$

$= -2$

$|\underset{\sim}{u}|^2 = (-3)^2 + 1^2$

$= 10$

$\therefore \underset{\sim}{v}_u = -\frac{2}{10}(-3\underset{\sim}{i} + \underset{\sim}{j})$

$= -\frac{1}{5}(-3\underset{\sim}{i} + \underset{\sim}{j})$

5 D. Consider some points: gradient at (0, 0) is 1; gradient at (1, 4) is 0

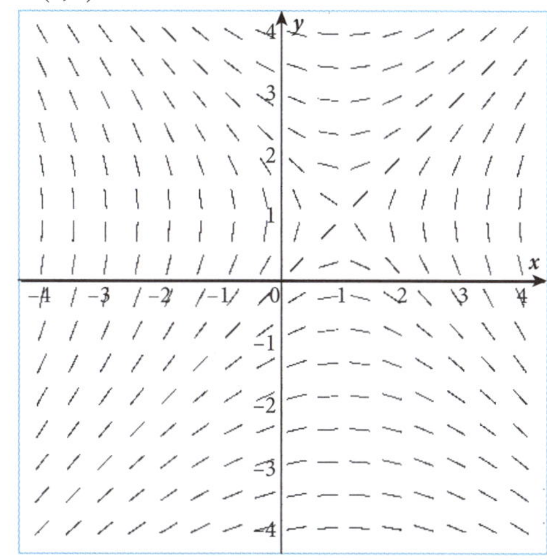

6 C. Let $P(x) = 2x^3 - 4x^2 - 6x + 4$

$P(2) = 2(2)^3 - 4(2)^2 - 6(2) + 4$

$= 16 - 16 - 12 + 4$

$= -8$

7 B. As $|\underset{\sim}{a}| = 5$, then $\sqrt{3^2 + (-m)^2} = 5$. $\therefore m = 4$.

Using $\underset{\sim}{a} = 3\underset{\sim}{i} - 4\underset{\sim}{j}$ and $\underset{\sim}{b} = -2\underset{\sim}{i} + n\underset{\sim}{j}$:

$\underset{\sim}{a} \cdot \underset{\sim}{b} = 3 \times (-2) + (-4) \times n = 0$

$-6 - 4n = 0$

$-4n = 6$

$n = -\frac{3}{2}$

8 B. $y = \frac{1}{\sqrt{1 - x^2}}$ has a domain of $-1 < x < 1$ and range $y \geq 1$.

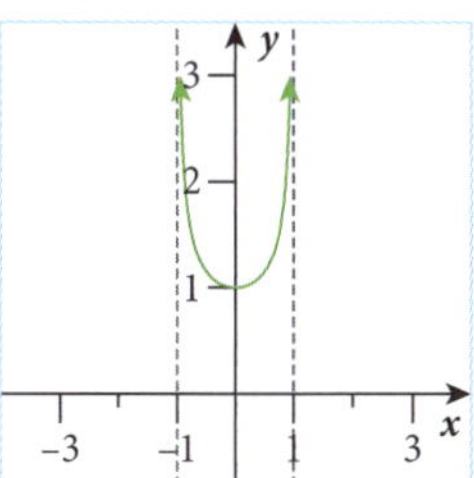

9 B. Solve $f(x) = x$:

$\frac{x}{x - 2} = x$

$x^2 - 2x = x$

$x^2 - 3x = 0$

$x(x - 3) = 0$

$x = 3$ (as $x > 2$)

$\therefore$ point of intersection is (3, 3).

10 C. $\cos 3x + \cos x = 0$

$2\cos 2x \cos x = 0$

$\cos 2x = 0 \quad \cos x = 0$

$2x = \frac{\pi}{2}, \frac{3\pi}{2} \qquad x = \frac{\pi}{2}$

$x = \frac{\pi}{4}, \frac{3\pi}{4}$

$x = \frac{\pi}{4}, \frac{\pi}{2}, \frac{3\pi}{4}$

$\therefore$ There are three solutions.

Section II

11 a i and ii

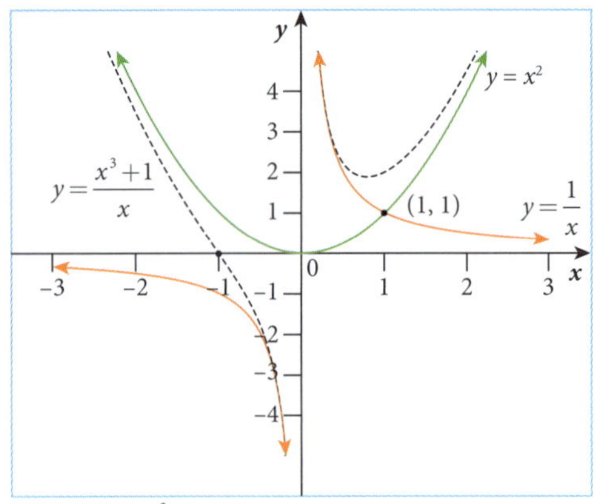

✓✓✓✓

As $x^2 + \frac{1}{x} = \frac{x^3+1}{x}$, then sketch by adding ordinates.

b $\underset{\sim}{a} + \underset{\sim}{b} = (4 + p - 3q)\underset{\sim}{i} + (2p + q - 4)\underset{\sim}{j} = 6\underset{\sim}{i} + 7\underset{\sim}{j}$

$\therefore 4 + p - 3q = 6$

$p - 3q = 2 \ldots$ ①

Also, $2p + q - 4 = 7$

$2p + q = 11 \ldots$ ②

$2 \times$ ①: $2p - 6q = 4 \ldots$ ③

② − ③: $7q = 7$

$q = 1$

Substitute into ②:

$2p + 1 = 11$

$p = 5$ ✓

$\therefore \underset{\sim}{a} = 4\underset{\sim}{i} + (2(5) + 1)\underset{\sim}{j}$

$= 4\underset{\sim}{i} + 11\underset{\sim}{j}$

Also, $\underset{\sim}{b} = (p - 3q)\underset{\sim}{i} - 4\underset{\sim}{j}$

$= (5 - 3(1))\underset{\sim}{i} - 4\underset{\sim}{j}$

$= 2\underset{\sim}{i} - 4\underset{\sim}{j}$ ✓

$-2\underset{\sim}{a} + 3\underset{\sim}{b} = -2(4\underset{\sim}{i} + 11\underset{\sim}{j}) + 3(2\underset{\sim}{i} - 4\underset{\sim}{j})$

$= -8\underset{\sim}{i} - 22\underset{\sim}{j} + 6\underset{\sim}{i} - 12\underset{\sim}{j}$

$= \mathbf{-2\underset{\sim}{i} - 34\underset{\sim}{j}}$ ✓

c If $n = 1$: $1(1^2 - 3(1) + 5) = 3$, which is divisible by 3.

$\therefore$ true for $n = 1$. ✓

Assume true for $n = k$:

$k(k^2 - 3k + 5) = 3M$, where M is an integer.

Then if $n = k + 1$:

$(k + 1)[(k + 1)^2 - 3(k + 1) + 5] = (k + 1)(k^2 + 2k + 1 - 3k - 3 + 5)$ ✓

$= (k + 1)(k^2 - k + 3)$

$= k^3 - k^2 + 3k + k^2 - k + 3$

$= k^3 + 2k + 3$

$= k(k^2 - 3k + 5) + 3k^2 - 3k + 3$

$= 3M + 3(k^2 - k + 1)$

$= 3(M + k^2 - k + 1)$, which is divisible by 3.

$\therefore$ hence the claim is true for $n = k + 1$

Since shown true for $n = 1$, is true for $n = 2, 3, 4, \ldots$ and so true for all integers $n \geq 1$. ✓

d $8\sin\theta + 15\cos\theta + 15 = 0$

$8\sin\theta + 15\cos\theta = -15$

Rewriting $8\sin\theta + 15\cos\theta$ in the form $R\sin(\theta + a)$:

$R = \sqrt{8^2 + 15^2}$

$= 17$

$\tan\alpha = \frac{15}{8}$

$\alpha = 61°\,56'$ (nearest minute) ✓

$17\sin(\theta + 61°\,56') = -15$ ✓

$\sin(\theta + 61°\,56') = -\frac{15}{17}$

$\theta + 61°\,56' = 241°\,56', 298°\,4'$

$\boldsymbol{\theta = 180°, 236°}$ **(nearest degree)** ✓

e $x = 2\left(t - \frac{1}{t}\right)$

$t - \frac{1}{t} = \frac{x}{2}$

$\left(t - \frac{1}{t}\right)^2 = \frac{x^2}{4}$ ✓

$t^2 - 2 + \frac{1}{t^2} = \frac{x^2}{4}$

$t^2 - 2 + \frac{1}{t^2} + 3 = \frac{x^2}{4} + 3$

$t^2 + 1 + \frac{1}{t^2} = \frac{x^2}{4} + 3$

$2\left(t^2 + 1 + \frac{1}{t^2}\right) = 2\left(\frac{x^2}{4} + 3\right)$

$\boldsymbol{y = \frac{x^2}{2} + 6}$ ✓

12 a i $\underset{\sim}{r}(t) = |\underset{\sim}{v}(0)|t\cos\theta\,\underset{\sim}{i} + (|\underset{\sim}{v}(0)|t\sin\theta - 5t^2 + 8)\underset{\sim}{j}$

Substitute $|\underset{\sim}{v}(0)| = 20$, $\theta = 60°$:

$\underset{\sim}{r}(t) = 20t\cos 60°\,\underset{\sim}{i} + (20t\sin 60° - 5t^2 + 8)\underset{\sim}{j}$

$= 10t\,\underset{\sim}{i} + (10t\sqrt{3} - 5t^2 + 8)\underset{\sim}{j}$

Consider $y = 10t\sqrt{3} - 5t^2 + 8$.

$\dot{y} = 10\sqrt{3} - 10t = 0$

$t = \sqrt{3}$ ✓

Substitute $t = \sqrt{3}$:

$y(\sqrt{3}) = 10(\sqrt{3})\sqrt{3} - 5(\sqrt{3})^2 + 8$
$= 30 - 15 + 8$
$= 23$

$\therefore$ **the maximum height is 23 m.** ✓

ii Let $y = 10t\sqrt{3} - 5t^2 + 8 = 0$

$5t^2 - 10t\sqrt{3} - 8 = 0$

$$t = \frac{10\sqrt{3} \pm \sqrt{(-10\sqrt{3})^2 - 4(5)(-8)}}{2(5)}$$

$$= \frac{10\sqrt{3} \pm \sqrt{460}}{10}$$

$= 3.876\,811\,867\ldots\ (t > 0)$

$= 3.877$ (3 dec. pl.) ✓

Substitute $t = 3.877$ in x:

$x = 10t$
$= 10(3.877)$
$= 38.8$ (1 dec. pl.)

$\therefore$ **the range is 38.8 m.** ✓

iii $\underset{\sim}{v}(t) = 10\underset{\sim}{i} + (10\sqrt{3} - 10t)\underset{\sim}{j}$

Substitute $t = 3.877$:

$\underset{\sim}{v}(3.877) = 10\underset{\sim}{i} + (10\sqrt{3} - 38.77)\underset{\sim}{j}$

$\left|\underset{\sim}{v}(3.877)\right| = \sqrt{10^2 + (10\sqrt{3} - 38.77)^2}$
$= 23.666\,024\,25\ldots$
$= 24$ (nearest whole)

$\therefore$ **the speed of the ball is 24 ms^{-1}.** ✓

b $\tan x \tan y = 3$

$\therefore \dfrac{\sin x \sin y}{\cos x \cos y} = 3$

$\sin x \sin y = 3 \cos x \cos y$ ✓

$\cos(x + y) = \cos x \cos y - \sin x \sin y = -\dfrac{2}{5}$

$\cos x \cos y - 3 \cos x \cos y = -\dfrac{2}{5}$

$-2 \cos x \cos y = -\dfrac{2}{5}$

$\cos x \cos y = \dfrac{1}{5}$ ✓

$\cos(x - y) = \cos x \cos y + \sin x \sin y$

$= \dfrac{1}{5} + 3\left(\dfrac{1}{5}\right)$

$= \dfrac{\mathbf{4}}{\mathbf{5}}$ ✓

c

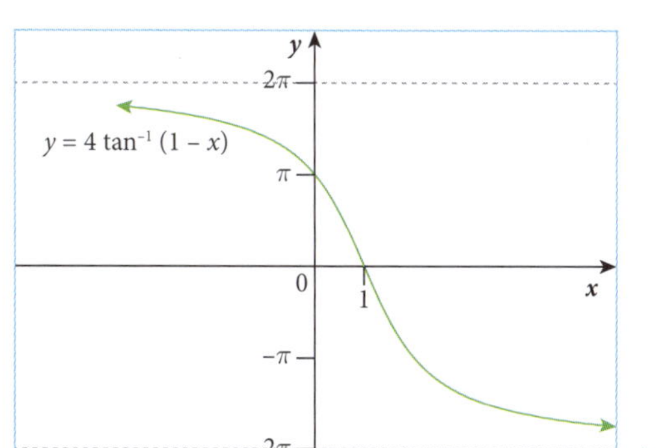

Domain is all real x, range is $-2\pi < y < 2\pi$. ✓

d

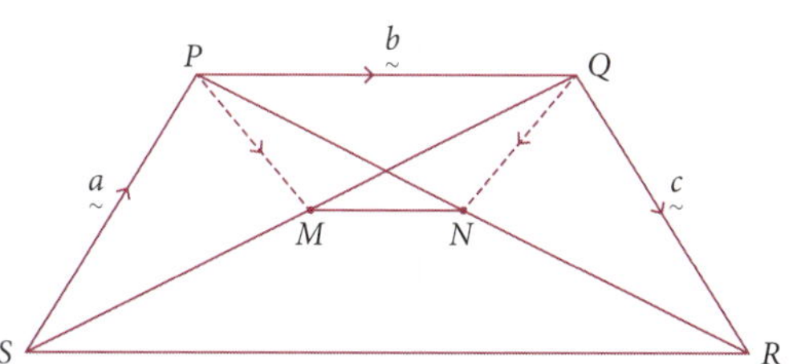

$\overrightarrow{SQ} = \underset{\sim}{a} + \underset{\sim}{b}$

$\overrightarrow{SM} = \dfrac{1}{2}(\underset{\sim}{a} + \underset{\sim}{b})$

$\overrightarrow{PM} = \dfrac{1}{2}(\underset{\sim}{a} + \underset{\sim}{b}) - \underset{\sim}{a}$

$= \dfrac{1}{2}(\underset{\sim}{b} - \underset{\sim}{a})$ ✓

Similarly, $\overrightarrow{QN} = \dfrac{1}{2}(\underset{\sim}{c} - \underset{\sim}{b})$

$\therefore \overrightarrow{MN} = -\overrightarrow{PM} + \underset{\sim}{b} + \dfrac{1}{2}(\underset{\sim}{c} - \underset{\sim}{b})$

$= \dfrac{1}{2}(\underset{\sim}{a} - \underset{\sim}{b}) + \underset{\sim}{b} + \dfrac{1}{2}(\underset{\sim}{c} - \underset{\sim}{b})$

$= \dfrac{1}{2}(\underset{\sim}{a} + \underset{\sim}{c})$

$= \dfrac{1}{2}(\underset{\sim}{a} + \underset{\sim}{b} + \underset{\sim}{c} - \underset{\sim}{b})$ ✓

But $\overrightarrow{SR} = \underset{\sim}{a} + \underset{\sim}{b} + \underset{\sim}{c} = \overrightarrow{SR} = k\underset{\sim}{b}$, for scalar k (as $PQ \parallel SR$)

$\therefore \overrightarrow{MN} = \dfrac{1}{2}(k\underset{\sim}{b} - \underset{\sim}{b})$

$= \dfrac{1}{2}(k - 1)\underset{\sim}{b}$

$\therefore \overrightarrow{MN}$ is a scalar multiple of PQ.

Hence, $MN \parallel SR$. ✓

13 a LHS $= \dfrac{\tan 2A}{1+\sec 2A}$

$= \dfrac{\frac{\sin 2A}{\cos 2A}}{1+\frac{1}{\cos 2A}}$ ✓

$= \dfrac{\sin 2A}{\cos 2A + 1}$

$= \dfrac{2\sin A\cos A}{2\cos^2 A - 1 + 1}$

$= \dfrac{\sin A}{\cos A}$

$= \tan A$

= **RHS** ✓

b Let X be the number of females in the sample. Consider, $X \sim \text{Bin}(100, 0.5)$:

$\sigma_{\hat{p}} = \sqrt{\dfrac{100(0.5)(1-0.5)}{100}}$

$= 0.05$

$= 5\%$ ✓

$\therefore 55\% = \bar{x} + \sigma$

$\therefore$ z-score = 1

$\therefore$ **$P(z > 1) = 16\%$** ✓

c i Let r = radius of water surface

$\therefore \dfrac{r}{h} = \tan 30°$

$r = \dfrac{h}{\sqrt{3}}$

$V = \dfrac{1}{3}\pi\left(\dfrac{h}{\sqrt{3}}\right)^2 h$

$V = \dfrac{\pi h^3}{9}$

$\dfrac{dV}{dh} = \dfrac{\pi h^2}{3}$ ✓

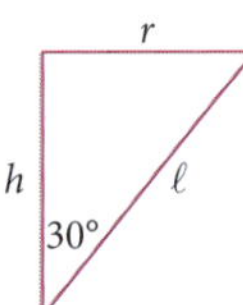

Substitute $h = 10$:

$\dfrac{dV}{dh} = \dfrac{100\pi}{3}$

$\dfrac{dV}{dt} = \dfrac{dV}{dh} \times \dfrac{dh}{dt}$

Substitute $\dfrac{dV}{dt} = 5$ (as 5 mL = 5 cm³) and $\dfrac{dV}{dh} = \dfrac{100\pi}{3}$:

$5 = \dfrac{100\pi}{3} \times \dfrac{dh}{dt}$

$\dfrac{dh}{dt} = \dfrac{3}{20\pi}$

$\therefore$ **the height is increasing by $\dfrac{3}{20\pi}$ cm/s.** ✓

ii Let ℓ = slant height

$\therefore \dfrac{h}{\ell} = \cos 30°$

$\ell = \dfrac{2h}{\sqrt{3}}$

$S = \pi\left(\dfrac{h}{\sqrt{3}}\right)\left(\dfrac{2h}{\sqrt{3}}\right)$

$= \dfrac{2\pi h^2}{3}$

$\dfrac{dS}{dh} = \dfrac{4\pi h}{3}$

Substitute $h = 10$:

$\dfrac{dS}{dh} = \dfrac{40\pi}{3}$ ✓

$\dfrac{dS}{dt} = \dfrac{dS}{dh} \times \dfrac{dh}{dt}$

Substitute $\dfrac{dS}{dh} = \dfrac{40\pi}{3}$ and $\dfrac{dh}{dt} = \dfrac{3}{20\pi}$:

$\dfrac{dS}{dt} = \dfrac{40\pi}{3} \times \dfrac{3}{20\pi}$

$\dfrac{dS}{dt} = 2$

$\therefore$ **the wet surface area is increasing by 2 cm²/s.** ✓

d $\dfrac{dx}{d\theta} = \sec^2\theta$

$dx = \sec^2\theta\, d\theta$

If $x = 1, \theta = \dfrac{\pi}{4}$ and $x = -1, \theta = -\dfrac{\pi}{4}$.

$\displaystyle\int_{-1}^{1} \frac{dx}{(1+x^2)^2} = \int_{-\frac{\pi}{4}}^{\frac{\pi}{4}} \frac{\sec^2\theta}{(1+\tan^2\theta)^2}\, d\theta$ ✓

$\displaystyle = \int_{-\frac{\pi}{4}}^{\frac{\pi}{4}} \frac{\sec^2\theta}{(\sec^2\theta)^2}\, d\theta$

$\displaystyle = \int_{-\frac{\pi}{4}}^{\frac{\pi}{4}} \frac{1}{\sec^2\theta}\, d\theta$

$\displaystyle = \int_{-\frac{\pi}{4}}^{\frac{\pi}{4}} \cos^2\theta\, d\theta$

$\displaystyle = \frac{1}{2}\int_{-\frac{\pi}{4}}^{\frac{\pi}{4}} (1+\cos 2\theta)\, d\theta$ ✓

$= \dfrac{1}{2}\left[x + \dfrac{1}{2}\sin 2\theta\right]_{-\frac{\pi}{4}}^{\frac{\pi}{4}}$

$= \dfrac{1}{2}\left[\dfrac{\pi}{4} + \dfrac{1}{2}\sin 2\left(\dfrac{\pi}{4}\right) - \left(\left(-\dfrac{\pi}{4}\right) + \dfrac{1}{2}\sin 2\left(-\dfrac{\pi}{4}\right)\right)\right]$

$= \dfrac{1}{2}\left[\dfrac{\pi}{4} + \dfrac{1}{2} + \dfrac{\pi}{4} + \dfrac{1}{2}\right]$

$= \dfrac{1}{2}\left[\dfrac{\pi}{2} + 1\right]$

$= \mathbf{\dfrac{\pi + 2}{4}}$ ✓

e $\sqrt{16 - x^2}\dfrac{dy}{dx} = 4 - y$

$\displaystyle\int \frac{1}{4-y}\, dy = \int \frac{1}{\sqrt{16-x^2}}\, dx$ ✓

$-\ln|4 - y| = \sin^{-1}\dfrac{x}{4} + c$

Substitute $x = 2, y = 3$:

$-\ln|4 - 3| = \sin^{-1}\dfrac{2}{4} + c$

$0 = \dfrac{\pi}{6} + c$

$c = -\dfrac{\pi}{6}$ ✓

$-\ln|4 - y| = \sin^{-1}\dfrac{x}{4} - \dfrac{\pi}{6}$

$\ln|4 - y| = \dfrac{\pi}{6} - \sin^{-1}\dfrac{x}{4}$

Substitute $x = 0$:

$\ln|4 - y| = \frac{\pi}{6} - \sin^{-1}\frac{0}{4}$ ✓

$\ln|4 - y| = \frac{\pi}{6}$

$4 - y = e^{\frac{\pi}{6}}$

$\mathbf{y = 4 - e^{\frac{\pi}{6}}}$ ✓

14 a i

$A \sin x = \tan x$

$A \sin x = \frac{\sin x}{\cos x}$

$A \sin x \cos x = \sin x$

$A \sin x \cos x - \sin x = 0$ ✓

$\sin x\,(A \cos x - 1) = 0$

$\sin x = 0, \cos x = \frac{1}{A}$

But $0 < \alpha < \frac{\pi}{2}$,

$\therefore \cos x = \frac{1}{A}$

As $T(\alpha, \beta)$, then $\cos \alpha = \frac{1}{A}$. ✓

ii Area $= \int_0^{\alpha}(A\sin x - \tan x)\,dx$ ✓

$= \int_0^{\alpha}\left(A\sin x - \frac{\sin x}{\cos x}\right)dx$

$= \left[-A\cos x + \ln\cos x\right]_0^{\alpha}$ ✓

$= -A \cos a + \ln \cos a - (-A \cos 0 + \ln \cos 0)$

$= -1 + \ln\frac{1}{A} - (-A + \ln 1)$ (as $\cos \alpha = \frac{1}{A}$)

$= -1 - \ln A + A$

$= A - \ln A - 1$

$\therefore$ the area is $(A - 1 - \ln A)$ units^2. ✓

b

$\sin x = 3 \cos \frac{x}{2}$

$2 \sin \frac{x}{2} \cos \frac{x}{2} = 3 \cos \frac{x}{2}$ ✓

$2 \sin \frac{x}{2} \cos \frac{x}{2} - 3 \cos \frac{x}{2} = 0$

$\cos \frac{x}{2}(2 \sin \frac{x}{2} - 3) = 0$

$\cos \frac{x}{2} = 0 \quad 2 \sin \frac{x}{2} - 3 = 0$

$\frac{x}{2} = \frac{\pi}{2} \quad 2 \sin \frac{x}{2} \neq 3$

$x = \pi \quad \therefore$ no solution

$\therefore P(\pi, 0)$ ✓

$V = \pi\int_0^{\pi}\left(3\cos\frac{x}{2}\right)^2 - (\sin x)^2\,dx$

$= \pi\int_0^{\pi}\left(9\cos^2\frac{x}{2} - \sin^2 x\right)dx$

$= \pi\int_0^{\pi}\left(\frac{9}{2}(1+\cos x) - \frac{1}{2}(1-\cos 2x)\right)dx$ ✓

$= \frac{\pi}{2}\left[9(x+\sin x) - \left(x - \frac{1}{2}\sin 2x\right)\right]_0^{\pi}$

$= \frac{\pi}{2}(9\pi - \pi)$

$= 4\pi^2$

$\therefore$ volume is $4\pi^2$ units^3. ✓

c i Substitute $t = 0$:

$N = \frac{3000}{1+2e^0}$

$= \mathbf{1000}$ ✓

ii $N = \frac{3000}{1+2e^{-t}}$

$= 3000(1 + 2e^{-t})^{-1}$

$\frac{dN}{dt} = -3000(1 + 2e^{-t})^{-2}.-2e^{-t}$

$= \frac{6000}{e^t(1+2e^{-t})^2}$ ✓

$= \frac{6000e^t}{e^{2t}(1+2e^{-t})^2}$

$= \frac{6000e^t}{e^{2t}(1+4e^{-t}+4e^{-2t})}$

$= \frac{6000e^t}{e^{2t}+4e^t+4}$

$= \mathbf{\frac{6000e^t}{(e^t+2)^2}}$ ✓

iii $\frac{dN}{dt} = \frac{6000e^t}{(e^t+2)^2}$

$\frac{d^2N}{dt^2} = \frac{(e^t+2)^2 6000e^t - 6000e^t.2(e^t+2).e^t}{(e^t+2)^4}$

$= \frac{(e^t+2)6000e^t - 6000e^t.2.e^t}{(e^t+2)^3}$

$= \frac{6000e^{2t} + 12000e^t - 12000e^{2t}}{(e^t+2)^3}$

$= \frac{12000e^t - 6000e^{2t}}{(e^t+2)^3}$ ✓

$= \frac{6000e^t(2-e^t)}{(e^t+2)^3} = 0$

$2 - e^t = 0$ (as $e^t \geq 0$)

$e^t = 2$

$t = \ln 2$ ✓

x	0.6	ln 2	0.7
$\frac{d^2N}{dt^2}$	> 0	0	< 0

$\therefore$ maximum.

$\frac{dN}{dt}(\ln 2) = \frac{6000e^{\ln 2}}{(e^{\ln 2}+2)^2}$

$= \frac{12000}{(2+2)^2}$

$= 750$

$\therefore$ maximum growth of 750 per month after ln 2 months. ✓

INDEX

A

$\alpha \cos x \pm b \sin x = c$ 151
$\alpha \sin x \pm b \cos x = c$ 151
acceleration 63
- vectors 135

addition
- vectors 111–12

angles
- differences 42
- double-angle formulae 43, 151
- sums 42
- vectors, between 121

area 246
- between two curves 176–7
- surface 246

'at rest' 63

B

Bernoulli trials 217
- application 218
- *X*, value of 217

binomial distribution 217–23, 247
- application 218
- mean 220
- probability, and 218
- standard deviation 220
- variance 220
- *X* 218

binomial expansion 85

C

calculus 160–6, 176–85, 196–205
- rates of change 63–70
- reference sheet 248

cartesian equations 12
chain rule 69
coefficients, determination of 85
column vector 109
combinations 84
combinatorics 81–6
- reference sheet 249

complex numbers 249
component form of a vector 109, 112
counting
- multiplication principle 81

cubics 30
curves
- area between two 176–7
- regions between two, rotating 182–4

D

decay, exponential 65–6, 67–8
- models, using 202–3

denominator
- squaring the 7

diagonals of a parallelogram 125
differential equations 196–205
- definition 196
- direction (slope) field 196–8
- $\frac{dy}{dx} = f(x)$ 200–1
- $\frac{dy}{dx} = f(x)g(y)$ 202
- $\frac{dy}{dx} = g(x)$ 201
- exponential growth and decay models, using 202–3
- first-order 196
- logistic growth model 205
- mixing problems 204
- verifying solutions 196

differentiation 196
- inverse trigonometric functions 163–4

direction
- change of 63
- displacement 63
- field 196–8
- vectors 120
- velocity 63

displacement 63
- direction 63
- magnitude 63
- vectors 109, 123

domain 53
- graph 53

dot product 121
$\frac{dy}{dx}$ 162–3
- differential equations, and solving 200–2

E

equations 150–3
- differences 151
- differential *see* differential equations
- double angles 151
- sums 151
- *t*-formulae 44, 152–3

exams (sample)
- year 11 93–5
- year 12 (HSC) 231–6

F

factor theorem 29
factorial notation 82
false 'proofs by induction' 101–2
financial mathematics 246
force of an object 123
- downwards 136

functions 4–13
- inverse *see* inverse functions
- polynomial, graphing 32–3
- reference sheet 246
- trigonometric *see* trigonometric functions

G

geometric probability distribution 218
geometric proofs 125
graphs
- domain 53
- graphical relationships 4–5
- inverse functions 10
- polynomial functions 32–3
- range 53

growth, exponential 65–6, 67–8
- models, using 202–3

I

inductive statement
- divisibility, relating to 101
- series, relating to 100

inequalities
- absolute values 9
- quadratic 6–7
- unknown in denominator, where 7–8
- values 6

initial statement
- divisibility, relating to 101
- series, relating to 100

'initially' 63
integration
- $\cos^2 nx$ 161–2
- inverse trigonometric functions 165
- products of trigonometric functions 160
- $\sin^2 nx$ 161–2
- substitution, using 160–1

inverse functions 10–11
- differentiation 163–4
- graphs 10
- integration, and 165
- properties 54–5
- trigonometric 53–6, 163–5

L

location of a particle
- position vectors 134

logistic growth model 205

M

magnitude
- direction 63
- vectors 120, 123
- velocity 63, 123

mathematical induction
- divisibility 101
- series 100

maximum height, finding 136
mean
  binomial distribution 220
  normal distribution 221
mechanics 249
multiplication
  vectors 109, 111–12
multiplication principle 81

N

n objects, arranging 82
n! 82
negatives
  vectors 109
non-zero vectors 120
normal distribution 221–2, 247
  table 221
notation, factorial 82

O

ordered pair 109
orthogonal 122
outcomes
  success or failure 217, 218

P

P(failure) = $q = 1 - p$ 217
P(success) = p 217
parallel 122
parallelogram
  geometric proofs relating to 125
parametric form 12
  functions 12
  relations 12
Pascal's triangle 85
permutations 83
perpendicular 122
pigeonhole principle 83
polynomials 28–34
  division of 28–9
  factor theorem 29
  functions, graphing 32–3
  general form 28
  multiplicity of roots 31–2
  products of roots 30–1
  remainder theorem 29
  roots 30–2
  sums of roots 30–1
population
  recording 'success' 221
position vectors 109
  function of time, as 134
probability 247
projectile motion 136–9
projection, vector 122
pronumerals 12
proof 100–2
  false 101–2
  geometric 125

Q

quadratic inequalities 6–7
quadratics 30
quadrilateral
  geometric proofs relating to 125
quartics 30

R

$r\cos(\theta \pm \alpha)$ 150
$r\sin(\theta \pm \alpha)$ 150
range 53
  finding 136
  graph 53
rates of change
  exponential decay 65–6, 67–8
  exponential growth 65–6, 67–8
  logistic growth model 205
  proportional 65, 67
  related 69
  time 63–4
reference sheet 246–9
related graphs 4
  development of 4
remainder theorem 29
roots
  multiplicity of 31–2
  products of 30–1
  sums of 30–1

S

scalar 109
  product 121
slope field 196–8
solids
  volumes of *see* volumes of solids
standard deviation
  binomial distribution 220
statements 100
statistical analysis 217–23
  reference sheet 247
subtraction
  vectors 111–12

T

tan (t)
  t-formulae 44, 152–3
time
  position vectors 134
  rates of change 63–4
  velocity and acceleration vectors 135
trigonometric functions 42–6
  differentiation 163–4
  equations 150–3
  integration 160–2
  inverse 53–6, 163–5
  products, integration involving 160
  properties 54–5
  reference sheet 247
trigonometric products
  differences, as 45–6
  sums, as 45

U

unit vectors 109

V

variance
  binomial distribution 220
vectors 109–13, 120–5, 134–40
  acceleration 135
  addition 111–12
  column 109
  component form 109, 112
  definition 109
  direction 120
  displacement 109, 123
  dot product 121
  force of an object 123
  length 109
  magnitude 120, 123
  multiplication 109, 111–12
  negatives 109
  non-zero 120
  ordered pair 109
  orthogonal 122
  parallel 122
  perpendicular 122
  position 109, 134, 135
  projection 122
  reference sheet 249
  scalar product 121
  subtraction 111–12
  unit 109
  velocity 123, 135
  writing 109
velocity 63
  displacement 63
  increasing at an increasing rate 63
  magnitude 63, 123
  vectors 123, 135
volume 246
volumes of solids
  area between two curves 176–7
  regions between two curves, rotating 182–4
  x-axis, rotating about 177–8
  y-axis, rotating about 180–1

NSW Education Standards Authority

2020 HIGHER SCHOOL CERTIFICATE EXAMINATION

Mathematics Advanced
Mathematics Extension 1
Mathematics Extension 2

REFERENCE SHEET

Measurement

Length

$$l = \frac{\theta}{360} \times 2\pi r$$

Area

$$A = \frac{\theta}{360} \times \pi r^2$$

$$A = \frac{h}{2}\left(a + b\right)$$

Surface area

$$A = 2\pi r^2 + 2\pi rh$$

$$A = 4\pi r^2$$

Volume

$$V = \frac{1}{3}Ah$$

$$V = \frac{4}{3}\pi r^3$$

Functions

$$x = \frac{-b \pm \sqrt{b^2 - 4ac}}{2a}$$

For $ax^3 + bx^2 + cx + d = 0$:

$$\alpha + \beta + \gamma = -\frac{b}{a}$$

$$\alpha\beta + \alpha\gamma + \beta\gamma = \frac{c}{a}$$

and $\alpha\beta\gamma = -\frac{d}{a}$

Relations

$$\left(x - h\right)^2 + \left(y - k\right)^2 = r^2$$

Financial Mathematics

$$A = P\left(1 + r\right)^n$$

Sequences and series

$$T_n = a + (n - 1)d$$

$$S_n = \frac{n}{2}\left[2a + (n - 1)d\right] = \frac{n}{2}(a + l)$$

$$T_n = ar^{n-1}$$

$$S_n = \frac{a\left(1 - r^n\right)}{1 - r} = \frac{a\left(r^n - 1\right)}{r - 1}, \; r \neq 1$$

$$S = \frac{a}{1 - r}, \; |r| < 1$$

Logarithmic and Exponential Functions

$$\log_a a^x = x = a^{\log_a x}$$

$$\log_a x = \frac{\log_b x}{\log_b a}$$

$$a^x = e^{x \ln a}$$

Trigonometric Functions

$$\sin A = \frac{\text{opp}}{\text{hyp}}, \quad \cos A = \frac{\text{adj}}{\text{hyp}}, \quad \tan A = \frac{\text{opp}}{\text{adj}}$$

$$A = \frac{1}{2}ab\sin C$$

$$\frac{a}{\sin A} = \frac{b}{\sin B} = \frac{c}{\sin C}$$

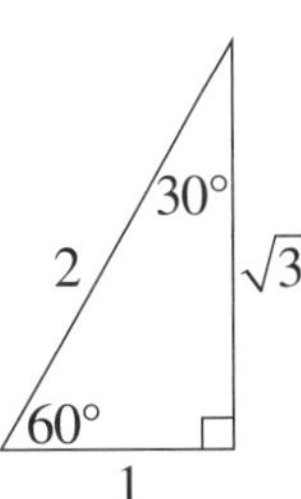

$$c^2 = a^2 + b^2 - 2ab\cos C$$

$$\cos C = \frac{a^2 + b^2 - c^2}{2ab}$$

$$l = r\theta$$

$$A = \frac{1}{2}r^2\theta$$

30°, 2, $\sqrt{3}$, 60°, 1

Trigonometric identities

$$\sec A = \frac{1}{\cos A}, \ \cos A \neq 0$$

$$\operatorname{cosec} A = \frac{1}{\sin A}, \ \sin A \neq 0$$

$$\cot A = \frac{\cos A}{\sin A}, \ \sin A \neq 0$$

$$\cos^2 x + \sin^2 x = 1$$

Compound angles

$$\sin(A + B) = \sin A\cos B + \cos A\sin B$$

$$\cos(A + B) = \cos A\cos B - \sin A\sin B$$

$$\tan(A + B) = \frac{\tan A + \tan B}{1 - \tan A\tan B}$$

If $t = \tan\frac{A}{2}$ then $\sin A = \frac{2t}{1 + t^2}$

$$\cos A = \frac{1 - t^2}{1 + t^2}$$

$$\tan A = \frac{2t}{1 - t^2}$$

$$\cos A\cos B = \frac{1}{2}\left[\cos(A - B) + \cos(A + B)\right]$$

$$\sin A\sin B = \frac{1}{2}\left[\cos(A - B) - \cos(A + B)\right]$$

$$\sin A\cos B = \frac{1}{2}\left[\sin(A + B) + \sin(A - B)\right]$$

$$\cos A\sin B = \frac{1}{2}\left[\sin(A + B) - \sin(A - B)\right]$$

$$\sin^2 nx = \frac{1}{2}(1 - \cos 2nx)$$

$$\cos^2 nx = \frac{1}{2}(1 + \cos 2nx)$$

Statistical Analysis

$$z = \frac{x - \mu}{\sigma}$$

An outlier is a score
less than $Q_1 - 1.5 \times IQR$
or
more than $Q_3 + 1.5 \times IQR$

Normal distribution

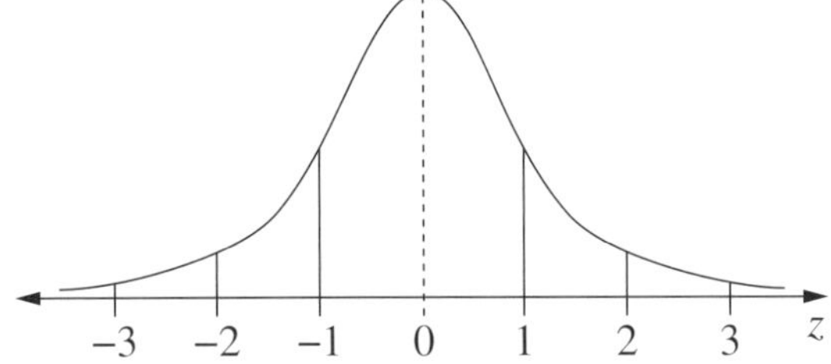

- approximately 68% of scores have z-scores between –1 and 1
- approximately 95% of scores have z-scores between –2 and 2
- approximately 99.7% of scores have z-scores between –3 and 3

$$E(X) = \mu$$

$$\text{Var}(X) = E\left[(X - \mu)^2\right] = E\left(X^2\right) - \mu^2$$

Probability

$$P(A \cap B) = P(A)P(B)$$

$$P(A \cup B) = P(A) + P(B) - P(A \cap B)$$

$$P(A|B) = \frac{P(A \cap B)}{P(B)}, \ P(B) \neq 0$$

Continuous random variables

$$P(X \leq r) = \int_a^r f(x)\,dx$$

$$P(a < X < b) = \int_a^b f(x)\,dx$$

Binomial distribution

$$P(X = r) = {}^nC_r p^r(1 - p)^{n-r}$$

$$X \sim \text{Bin}(n, p)$$

$$\Rightarrow \ P(X = x)$$

$$= \binom{n}{x}p^x(1 - p)^{n-x}, \ x = 0, 1, \ldots, n$$

$$E(X) = np$$

$$\text{Var}(X) = np(1 - p)$$

Differential Calculus

Function	Derivative
$y = f(x)^n$	$\frac{dy}{dx} = nf'(x)[f(x)]^{n-1}$
$y = uv$	$\frac{dy}{dx} = u\frac{dv}{dx} + v\frac{du}{dx}$
$y = g(u)$ where $u = f(x)$	$\frac{dy}{dx} = \frac{dy}{du} \times \frac{du}{dx}$
$y = \frac{u}{v}$	$\frac{dy}{dx} = \frac{v\frac{du}{dx} - u\frac{dv}{dx}}{v^2}$
$y = \sin f(x)$	$\frac{dy}{dx} = f'(x)\cos f(x)$
$y = \cos f(x)$	$\frac{dy}{dx} = -f'(x)\sin f(x)$
$y = \tan f(x)$	$\frac{dy}{dx} = f'(x)\sec^2 f(x)$
$y = e^{f(x)}$	$\frac{dy}{dx} = f'(x)e^{f(x)}$
$y = \ln f(x)$	$\frac{dy}{dx} = \frac{f'(x)}{f(x)}$
$y = a^{f(x)}$	$\frac{dy}{dx} = (\ln a)f'(x)a^{f(x)}$
$y = \log_a f(x)$	$\frac{dy}{dx} = \frac{f'(x)}{(\ln a)f(x)}$
$y = \sin^{-1} f(x)$	$\frac{dy}{dx} = \frac{f'(x)}{\sqrt{1-[f(x)]^2}}$
$y = \cos^{-1} f(x)$	$\frac{dy}{dx} = -\frac{f'(x)}{\sqrt{1-[f(x)]^2}}$
$y = \tan^{-1} f(x)$	$\frac{dy}{dx} = \frac{f'(x)}{1+[f(x)]^2}$

Integral Calculus

$$\int f'(x)[f(x)]^n dx = \frac{1}{n+1}[f(x)]^{n+1} + c$$

where $n \neq -1$

$$\int f'(x)\sin f(x)dx = -\cos f(x) + c$$

$$\int f'(x)\cos f(x)dx = \sin f(x) + c$$

$$\int f'(x)\sec^2 f(x)dx = \tan f(x) + c$$

$$\int f'(x)e^{f(x)}dx = e^{f(x)} + c$$

$$\int \frac{f'(x)}{f(x)}dx = \ln|f(x)| + c$$

$$\int f'(x)a^{f(x)}dx = \frac{a^{f(x)}}{\ln a} + c$$

$$\int \frac{f'(x)}{\sqrt{a^2-[f(x)]^2}}dx = \sin^{-1}\frac{f(x)}{a} + c$$

$$\int \frac{f'(x)}{a^2+[f(x)]^2}dx = \frac{1}{a}\tan^{-1}\frac{f(x)}{a} + c$$

$$\int u\frac{dv}{dx}dx = uv - \int v\frac{du}{dx}dx$$

$$\int_a^b f(x)dx$$

$$\approx \frac{b-a}{2n}\left\{f(a) + f(b) + 2\left[f(x_1) + \cdots + f(x_{n-1})\right]\right\}$$

where $a = x_0$ and $b = x_n$

Combinatorics

$$^nP_r = \frac{n!}{(n-r)!}$$

$$\binom{n}{r} = {}^nC_r = \frac{n!}{r!\,(n-r)!}$$

$$(x+a)^n = x^n + \binom{n}{1}x^{n-1}a + \cdots + \binom{n}{r}x^{n-r}a^r + \cdots + a^n$$

Vectors

$$\left|\underset{\sim}{u}\right| = \left|x\underset{\sim}{i} + y\underset{\sim}{j}\right| = \sqrt{x^2+y^2}$$

$$\underset{\sim}{u}\cdot\underset{\sim}{v} = \left|\underset{\sim}{u}\right|\left|\underset{\sim}{v}\right|\cos\theta = x_1x_2 + y_1y_2,$$

where $\underset{\sim}{u} = x_1\underset{\sim}{i} + y_1\underset{\sim}{j}$

and $\underset{\sim}{v} = x_2\underset{\sim}{i} + y_2\underset{\sim}{j}$

$$\underset{\sim}{r} = \underset{\sim}{a} + \lambda\underset{\sim}{b}$$

Complex Numbers

$$z = a + ib = r(\cos\theta + i\sin\theta) = re^{i\theta}$$

$$\left[r(\cos\theta + i\sin\theta)\right]^n = r^n(\cos n\theta + i\sin n\theta) = r^ne^{in\theta}$$

Mechanics

$$\frac{d^2x}{dt^2} = \frac{dv}{dt} = v\frac{dv}{dx} = \frac{d}{dx}\left(\frac{1}{2}v^2\right)$$

$$x = a\cos(nt+\alpha) + c$$

$$x = a\sin(nt+\alpha) + c$$

$$\ddot{x} = -n^2(x-c)$$

Reprinted 2022, 2024

ISBN 978 1 74125 683 3

Pascal Press
PO Box 250
Glebe NSW 2037
(02) 9198 1748
www.pascalpress.com.au

Publisher: Vivienne Joannou
Project editor: Rosemary Peers
Edited by Rosemary Peers
Answers checked by Peter Little
Cover and page design by Sonia Woo
Typeset by Julianne Billington (lj Design)
Printed by Vivar Printing/Green Giant Press

Students
All care has been taken in the preparation of this study guide but please check with your teacher or the NSW Education Standards Authority about the exact requirements of the course you are studying as these can change from year to year.

The validity and appropriateness of the internet addresses (URLs) in this book were checked at the time of publication. Due to the dynamic nature of the internet, the publisher cannot accept responsibility for the continued validity or content of these web addresses.

EXCEL YEAR 11 & 12 MATHS EXTENSION 1 PASSCARDS

YOUR PATH TO SUCCESS!

This set of cards is designed for you to use as the final step in your revision program. The author has carefully selected the most important facts of the course for you to focus on just before your exam or test. You can use your ***Excel*** Year 11 & 12 Passcards:

✓ **ON THE GO.** If you are by yourself, read over each card again and again until you completely master its content. If you're with a friend, revise as a team by turning the bullet points into questions and quizzing each other on key points. Your answers will be there on the cards.

✓ **AT HOME.** Read each card thoroughly and make sure you understand all the points. You should also know more detailed information on each topic—if you are not completely sure of a topic, revise it in your ***Excel*** Year 11 and 12 study guide.

✓ **ON YOUR DIGITAL DEVICE.** Download a **FREE** digital copy of these cards at **www.pascalpress.com.au/free**

CONTENTS (2)

35 Pigeonhole principle
36 Permutations
37 Combinations
38 The binomial expansion
39 Combinatorial proofs

PROOF (YEAR 12)

PROOF BY MATHEMATICAL INDUCTION
40 Mathematical induction relating to series
41 Mathematical induction relating to divisibility
42 False 'proofs by induction'

VECTORS (YEAR 12)

INTRODUCTION TO VECTORS
43 Vector terminology
44 Expressing vectors in different forms
45 Operations with vectors
46 Magnitude of vectors
47 Direction of vectors
48 Scalar (dot) product of vectors
49 Angle between two vectors
50 Parallel and perpendicular vectors
51 Vector projection
52 Displacement as a vector
53 Force as a vector
54 Velocity as a vector
55 Geometric proofs involving vectors
56 Position vector as a function of time
57 Velocity vector as a function of time
58 Acceleration vector as a function of time
59 Determining the position vector using the velocity vector
60 Projectile motion concepts
61 Finding the greatest height of a projectile
62 Finding the range of the projectile
63 The cartesian equation of the trajectory of the projectile
64 Angle and velocity of projectile on landing

TRIGONOMETRIC FUNCTIONS (YEAR 12)

TRIGONOMETRIC EQUATIONS
65 Expressing in the form $r\sin(\theta \pm \alpha)$, $r\cos(\theta \pm \alpha)$
66 Equations of the form $a\cos x + b\sin x = c$
67 Equations involving sums and differences
68 Equations involving double angles
69 Equations involving products as sums and differences
70 Using t-formulae to solve equations
71 Expressing $\sin 3\theta$ in terms of θ
72 Expressing $\cos 3\theta$ in terms of θ
73 Expressing $\tan 3\theta$ in terms of θ

CALCULUS (YEAR 12)

FURTHER CALCULUS SKILLS
74 Integration involving products of trigonometric functions
75 Integration using substitution

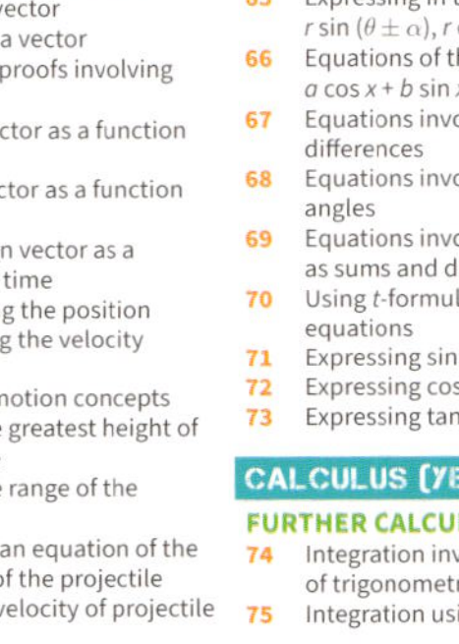

FURTHER WORK WITH FUNCTIONS

1 Graphing $y = f(x - c)$ and $y = \dfrac{1}{f(x)}$

➔ **Given $y = f(x)$, graph $y = f(x - c)$.**
- Translate the graph horizontally c units to the right.

➔ **Given $y = f(x)$, graph $y = \dfrac{1}{f(x)}$.**
- Where $y = f(x)$ cuts x-axis, then $y = \dfrac{1}{f(x)}$ is undefined.
- $y = f(x)$ and $y = \dfrac{1}{f(x)}$ intersect when $y = \pm 1$.

TOPIC: FUNCTIONS See ➲ *Excel* Year 11 & 12 Mathematics Extension 1 pp. 4–6

FURTHER WORK WITH FUNCTIONS

3 Graphing $y = |f(x)|$ and $y = f(|x|)$

➔ **Given $y = f(x)$, graph $y = |f(x)|$.**
- The section of the graph of $y = f(x)$ which is above the x-axis is identical for $y = |f(x)|$.
- The section of the graph of $y = f(x)$ which is below the x-axis is redrawn above the x-axis.

➔ **Given $y = f(x)$, graph $y = f(|x|)$.**
- The graph of $y = f(x)$ on the right of the y-axis is reflected about the y-axis.

TOPIC: FUNCTIONS See ➲ *Excel* Year 11 & 12 Mathematics Extension 1 pp. 4–6

FURTHER WORK WITH FUNCTIONS

5 Quadratic inequalities

➔ **When an inequality is solved** there is a range of values.

➔ **A quadratic inequality** is an inequality involving a quadratic expression or function.

EXAMPLE: Solve $2x^2 + x - 6 \leq 0$.

Solution:

$(2x - 3)(x + 2) \leq 0$

Let $(2x - 3)(x + 2) = 0$

$x = \dfrac{3}{2}, -2$

Test $x = 0$: $(-3)(2) \leq 0?$ … Yes!

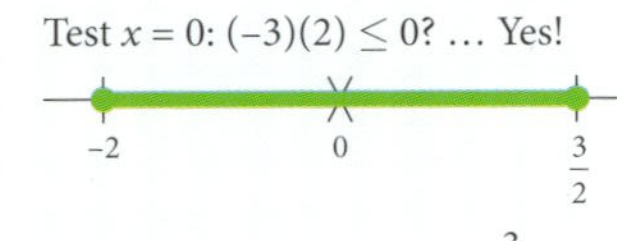

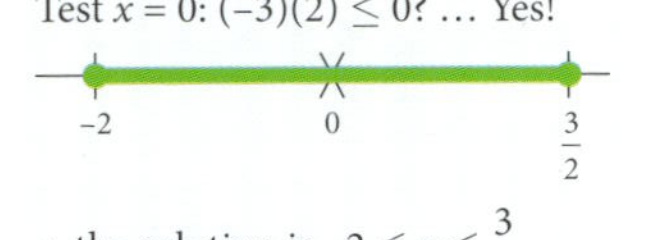

$\therefore$ the solution is $-2 \leq x \leq \dfrac{3}{2}$.

TOPIC: FUNCTIONS See ➲ *Excel* Year 11 & 12 Mathematics Extension 1 pp. 6–7

FURTHER WORK WITH FUNCTIONS

7 Inequalities involving absolute values

➔ **We consider the two cases when solving an equation** or an inequality involving absolute values.

EXAMPLE: Solve $|2x - 1| < 5$.

Solution:

$-5 < 2x - 1 < 5$

$-4 < 2x < 6$

$-2 < x < 3$

EXAMPLE:

Solve $|3x + 4| \geq 1$.

Solution:

$3x + 4 \leq -1$ or $3x + 4 \geq 1$

$3x \leq -5$ or $3x \geq -3$

$x \leq -\dfrac{5}{3}$ or $x \geq -1$

$\therefore x \leq -\dfrac{5}{3}$ or $x \geq -1$

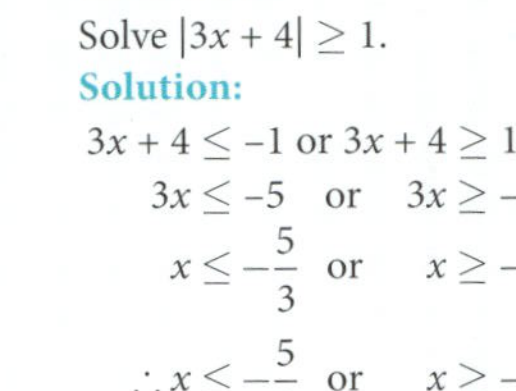

TOPIC: FUNCTIONS See ➲ *Excel* Year 11 & 12 Mathematics Extension 1 pp. 9–10

FURTHER WORK WITH FUNCTIONS

9 Domain, range and graphs of inverse functions

➔ **The domain and range of a function $y = f(x)$** are used to find the range and domain of the inverse function $y = f^{-1}(x)$

➔ **The graphs of inverse functions** reflect about the line $y = x$.

EXAMPLE: If $f(x) = (x - 2)^2$, where $x \geq 2$, find the points of intersection of $y = f(x)$ and $y = f^{-1}(x)$ and hence sketch $y = f(x)$ and $y = f^{-1}(x)$ on the same number plane.

Solution: Solve $(x - 2)^2 = x$

$x^2 - 4x + 4 - x = 0$

$x^2 - 5x + 4 = 0$

$(x - 4)(x - 1) = 0$

$x = 4, 1$

$\therefore$ the points of intersection are (1, 1) and (4, 4).
But $x \geq 2$, hence, the point of intersection is (4, 4).

TOPIC: FUNCTIONS See ➲ *Excel* Year 11 & 12 Mathematics Extension 1 pp. 10–12

FURTHER WORK WITH FUNCTIONS

11 Parametric form of a circle

➔ **The parametric form for circle $(x - h)^2 + (y - k)^2 = r^2$** is $x = r\cos t + h$ and $y = r\sin t + k$, where t is the parameter, centre (h, k) and radius r.

EXAMPLE: Find the cartesian equation of the circle represented by $x = 3\cos t - 4$ and $y = 3\sin t + 1$, and hence graph the circle.

Solution: $x = 3\cos t - 4$ and hence $\cos^2 t = \dfrac{(x+4)^2}{9}$

$y = 3\sin t + 1$ and hence $\sin^2 t = \dfrac{(y-1)^2}{9}$

Hence, $\dfrac{(x+4)^2}{9} + \dfrac{(y-1)^2}{9} = 1$, as

$\cos^2 t + \sin^2 t = 1$.

$\therefore (x + 4)^2 + (y - 1)^2 = 9$, has centre $(-4, 1)$, radius 3 units.

TOPIC: FUNCTIONS See ➲ *Excel* Year 11 & 12 Mathematics Extension 1 pp. 12–13

POLYNOMIALS

13 Division of polynomials

➔ **$P(x) = A(x).Q(x) + R(x)$** where $A(x)$ = divisor, $Q(x)$ = quotient and $R(x)$ = remainder.

EXAMPLE: Divide $P(x) = 2x^3 - 3x^2 - 4x + 2$ by $A(x) = x - 3$.

Solution:

$$\begin{array}{r} 2x^2 + 3x + 5 \\ x - 3\overline{)\,2x^3 - 3x^2 - 4x + 2} \\ \underline{2x^3 - 6x^2} \qquad\quad \\ 3x^2 - 4x \quad \\ \underline{3x^2 - 9x} \quad \\ 5x + 2 \\ \underline{5x - 15} \\ 17 \end{array}$$

How many x's in $2x^3$?
The answer is $2x^2$, which is the first term written in the quotient.
Then, $2x^2$ times $(x - 3)$ is $2x^3 - 6x^2$.
Now subtracting, gives $3x^2$.
Bring down the next term which is $-4x$. Repeat.

TOPIC: FUNCTIONS See ➲ *Excel* Year 11 & 12 Mathematics Extension 1 pp. 28–29

FURTHER WORK WITH FUNCTIONS

Graphing $y = \sqrt{f(x)}$ and $y^2 = f(x)$

2

- **Given $y = f(x)$, graph $y = \sqrt{f(x)}$.**
  - $y = f(x)$ and $y = \sqrt{f(x)}$ intersect when $y = 0$ and $y = 1$.
- **Given $y = f(x)$, graph $y^2 = f(x)$.**
  - The graph of $y = \sqrt{f(x)}$ is reflected about the x-axis.

TOPIC: FUNCTIONS See ➲ *Excel* Year 11 & 12 Mathematics Extension 1 pp. 4–6

CONTENTS (3)

76 Integration using trigonometric substitution
77 Definite integrals and substitution
78 Integration of $\sin^2 nx$
79 Integration of $\cos^2 nx$
80 Using the result $\frac{dy}{dx} = \frac{1}{\frac{dx}{dy}}$
81 Differentiation of inverse sin (arcsin) functions
82 Differentiation of inverse cos (arccos) functions
83 Differentiation of inverse tan (arctan) functions
84 Integration resulting in inverse sin (arcsin) functions
85 Integration resulting in inverse tan (arctan) functions

APPLICATIONS OF CALCULUS
86 Area between two curves
87 Volumes of solids formed by rotating about the x-axis
88 Volumes of solids formed by rotating about the y-axis
89 Volumes of solids formed by rotating regions between two curves
90 Verifying solutions of differential equations
91 Direction (slope) field of a differential equation
92 Sketching a curve on a given direction field
93 Solving differential equations of the form $\frac{dy}{dx} = f(x)$
94 Solving differential equations of the form $\frac{dy}{dx} = g(y)$
95 Solving differential equations of the form $\frac{dy}{dx} = f(x)\,g(y)$
96 Exponential growth and decay models as differential equations
97 The logistic growth model as a differential equation 1
98 The logistic growth model as a differential equation 2

STATISTICAL ANALYSIS (YEAR 12)

THE BINOMIAL DISTRIBUTION
99 Bernoulli trials
100 Applications of the Bernoulli distribution
101 Binomial distributions
102 Mean and variance of binomial distributions
103 The sample proportion $\hat{p}$
104 The normal approximation to the distribution of $\hat{p}$

CONTENTS (1)

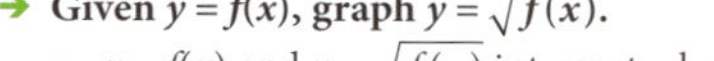
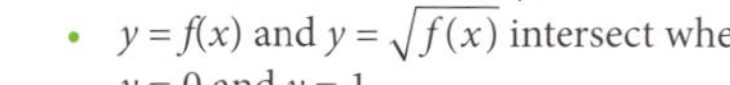
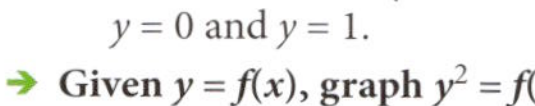
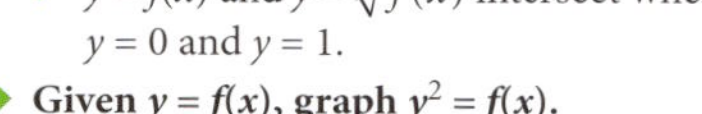
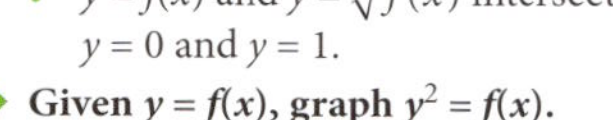
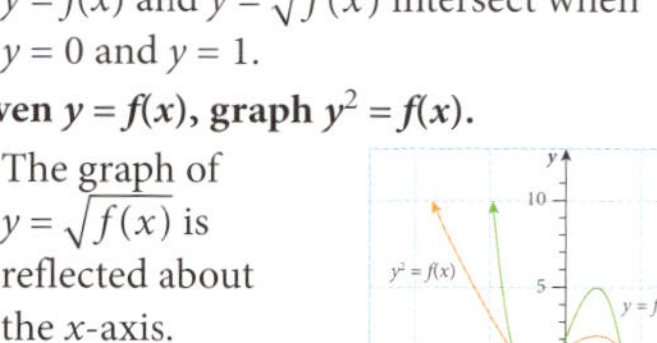

FUNCTIONS (YEAR 11)

FURTHER WORK WITH FUNCTIONS
1 Graphing $y = f(x - c)$ and $y = \frac{1}{f(x)}$
2 Graphing $y = \sqrt{f(x)}$ and $y^2 = f(x)$
3 Graphing $y = |f(x)|$ and $y = f(|x|)$
4 Graphing $y = f(x) + g(x)$ and $y = f(x).g(x)$
5 Quadratic inequalities
6 Inequalities with unknown in denominator
7 Inequalities involving absolute values
8 Finding inverse functions
9 Domain, range and graphs of inverse functions
10 Parametric form of a parabola
11 Parametric form of a circle

POLYNOMIALS
12 Polynomial terminology
13 Division of polynomials
14 Remainder theorem
15 Factor theorem
16 Sums and products of roots of polynomials
17 Multiplicity of roots
18 Graphs of polynomial functions

TRIGONOMETRIC FUNCTIONS (YEAR 11)

FURTHER TRIGONOMETRIC IDENTITIES
19 Sums and differences of angles
20 Double-angle formulae
21 The t-formulae $\left(t = \tan\frac{A}{2}\right)$
22 Trigonometric products as sums and differences 1
23 Trigonometric products as sums and differences 2

INVERSE TRIGONOMETRIC FUNCTIONS
24 Graphs of inverse trigonometric functions
25 Domain and range of inverse trigonometric functions
26 Properties of inverse trigonometric functions
27 Solving problems involving inverse trigonometric functions

CALCULUS (YEAR 11)

RATES OF CHANGE
28 Rates of change with respect to time
29 Exponential growth and decay: $N(t) = Ae^{kt}$
30 Further exponential growth and decay: $N(t) = P + Ae^{kt}$
31 Related rates of change 1
32 Related rates of change 2

COMBINATORICS (YEAR 11)

WORKING WITH COMBINATORICS
33 Fundamental counting principle (multiplication principle)
34 Factorial notation

FURTHER WORK WITH FUNCTIONS

Finding inverse functions

8

- **An inverse function** is a function obtained by expressing the dependent variable of one function as the independent variable of another; that is, if $y = f(x)$, then swap x and y to find $y = f^{-1}(x)$.
- **If the graph of $y = f(x)$** is such that any horizontal line intersects it at one point only at most, then $f(x)$ is said to be a one-to-one function and it has an inverse function.

EXAMPLE: Find the inverse function, $f^{-1}(x)$, of $f(x) = x^2 - 6x + 9$, for $x \geq 3$.
Solution: Let $y = x^2 - 6x + 9$.
Swap $y \leftrightarrow x$:

$x = y^2 - 6y + 9$
$x = (y - 3)^2$
$\sqrt{x} = y - 3$ (as domain for $f(x)$ is $x \geq 3$)
$y = \sqrt{x} + 3$
$\therefore f^{-1}(x) = \sqrt{x} + 3$

TOPIC: FUNCTIONS See ➲ *Excel* Year 11 & 12 Mathematics Extension 1 pp. 10–12

FURTHER WORK WITH FUNCTIONS

Inequalities with unknown in denominator

6

- **We cannot simply multiply by the denominator** because some values of x will make the denominator positive and other values negative.
- **First square the denominator** and then use it to multiply both sides of the inequality.

EXAMPLE: Solve $\frac{x+3}{2x} > 1$.
Solution:
Multiply both sides by $4x^2$:
$4x^2 \times \frac{x+3}{2x} > 1 \times 4x^2$
$2x^2 + 6x > 4x^2$
$2x^2 - 6x < 0$
$2x(x - 3) < 0$

Let $x(x - 3) = 0$
$x = 0, 3$
Test $x = 1$: $\frac{1+3}{2(1)} > 1$? … Yes!

0 1 3

$\therefore$ the solution is $0 < x < 3$.

TOPIC: FUNCTIONS See ➲ *Excel* Year 11 & 12 Mathematics Extension 1 pp. 7–9

FURTHER WORK WITH FUNCTIONS

Graphing $y = f(x) + g(x)$ and $y = f(x).g(x)$

4

- **Given $y = f(x)$, graph $y = f(x) + g(x)$.**
  - For each x-value, add the y-value of each function.
- **Given $y = f(x)$, graph $y = f(x).g(x)$.**
  - For each x-value, multiply the y-value of each function.
  - The x-intercepts of $y = f(x)$ or $y = g(x)$ are the x-intercepts of $y = f(x).g(x)$.

TOPIC: FUNCTIONS See ➲ *Excel* Year 11 & 12 Mathematics Extension 1 pp. 5–6

POLYNOMIALS

Remainder theorem

14

- **If a polynomial $P(x)$ is divided by $x - a$** then the remainder is $P(a)$.
  Proof: For $P(x) = A(x).Q(x) + R(x)$, let $A(x) = (x - a)$.
  $P(x) = (x - a).Q(x) + R(x)$
  As $(x - a)$ has degree 1, then $R(x)$ has degree 0 (that is, $R(x) = k$, a constant). $\therefore P(x) = (x - a).Q(x) + k$
  Subs. $x = a$: $P(a) = (a - a).Q(x) + k = k$
  $\therefore$ the remainder is $P(a)$.

EXAMPLE: Find the remainder if $2x^3 - 3x^2 + 2x + 15$ is divided by $(x + 2)$.
Solution: Let $P(x) = 2x^3 - 3x^2 + 2x + 15$
$P(-2) = 2(-2)^3 - 3(-2)^2 + 2(-2) + 15 = -16 - 12 - 4 + 15 = -17$
$\therefore$ the remainder is -17.

TOPIC: FUNCTIONS See ➲ *Excel* Year 11 & 12 Mathematics Extension 1 pp. 29–30

POLYNOMIALS

Polynomial terminology

12

- **The general form of a polynomial** is $P(x) = a_n x^n + a_{n-1}x^{n-1} + \ldots + a_1 x + a_0$, where n is a positive integer or zero and $a_n \neq 0$.
- **For the polynomial $P(x)$ above**, the degree is n, the leading coefficient is a_n and the constant term is a_0.

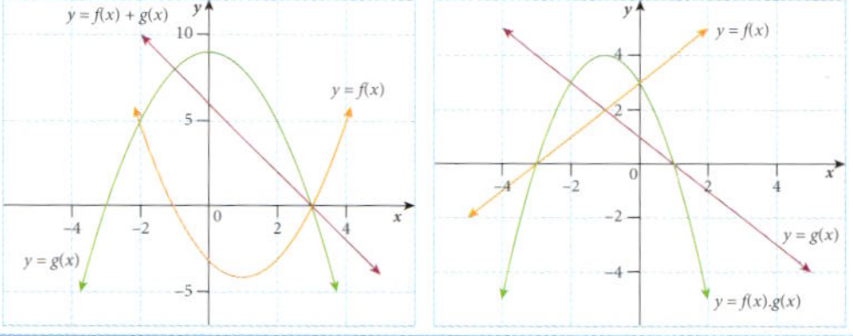

EXAMPLE:
If $P(x) = 2x^4 + 5x - 3$ and $Q(x) = 7 - 4x + 6x^2 - x^3$, for the polynomial $P(x).Q(x)$ find the:
a degree
b constant

Solution:
a The degree is $4 + 3 = 7$.
b The constant is $-3 \times 7 = -21$.

TOPIC: FUNCTIONS See ➲ *Excel* Year 11 & 12 Mathematics Extension 1 p. 28

FURTHER WORK WITH FUNCTIONS

Parametric form of a parabola

10

- **An equation involving x and y** such as $y = 3x^2 - 2x - 1$ is called a cartesian equation.
- **Parametric form** is when both x and y are written in terms of a parameter.
- **Any cartesian equation** can be written in parametric form.
- **The parametric form for parabola $x^2 = 4ay$** is $x = 2ap$, $y = ap^2$, where p is the parameter and a is a constant.

EXAMPLE: Find the cartesian equation of the parabola represented by the parametric equations $x = 12p$ and $y = 6p^2$.
$x = 12p$ … ①
$y = 6p^2$ … ②

From ①: $p = \frac{x}{12}$
Subs. in ②: $y = 6\left(\frac{x}{12}\right)^2$
$= \frac{x^2}{24}$

TOPIC: FUNCTIONS See ➲ *Excel* Year 11 & 12 Mathematics Extension 1 pp. 12–13

POLYNOMIALS
Factor theorem
15

→ **For polynomial $P(x)$, if $P(a) = 0$,** then $x - a$ is a factor of the polynomial.

Proof: For $P(x) = A(x).Q(x) + R(x)$, let $A(x) = (x - a)$.
$P(x) = (x - a).Q(x) + R(x)$

The remainder when $P(x)$ is divided by $x - a$ is $P(a)$ (from Remainder theorem). $\therefore P(x) = (x - a).Q(x) + P(a)$

But $P(a) = 0$, $\therefore P(x) = (x - a).Q(x)$

$\therefore x - a$ is a factor of the polynomial.

EXAMPLE: Show that $x - 2$ is a factor of $x^3 + x^2 - 5x - 2$.

Solution:
Let $P(x) = x^3 + x^2 - 5x - 2$

$P(2) = (2)^3 + (2)^2 - 5(2) - 2$
$= 8 + 4 - 10 - 2 = 0$
$\therefore x - 2$ is a factor.

TOPIC: FUNCTIONS

See Excel Year 11 & 12 Mathematics Extension 1 pp. 29–30

POLYNOMIALS
Multiplicity of roots
17

→ **If $P(x) = (x - a)^k.Q(x)$, and $Q(a) \neq 0$, $k > 0$,** the polynomial has k roots at $x = a$. It is said to have a root of multiplicity of k.

→ **If $P(x) = (x - a)^k.Q(x)$ has a root of multiplicity k at $x = a$,** then $P'(x)$ has a root of multiplicity $k-1$ at $x = a$.

Proof: Let $P(x) = (x - a)^k.Q(x)$, which has a root of multiplicity k.
Now, $P'(x) = k(x - a)^{k-1}.Q(x) + (x - a)^k.Q'(x)$
$P'(a) = k(a - a)^{k-1}.Q(a) + (a - a)^k.Q'(a) = 0$
$\therefore P'(x)$ has a root of multiplicity $k-1$.

EXAMPLE: A polynomial $P(x)$ has a triple root at $x = 2$. Show that $P'(2) = 0$.

Solution:
Let $P(x) = (x - 2)^3.Q(x)$

$P'(x) = 3(x - 2)^2.Q(x) + (x - 2)^3.Q'(x)$
$P'(2) = 3(2 - 2)^2.Q(2) + (2 - 2)^3.Q'(2)$
$= 0$

TOPIC: FUNCTIONS

See Excel Year 11 & 12 Mathematics Extension 1 pp. 31–32

FURTHER TRIGONOMETRIC IDENTITIES
Sums and differences of angles
19

→ **$\sin(A + B)$** $= \sin A \cos B + \cos A \sin B$
→ **$\sin(A - B)$** $= \sin A \cos B - \cos A \sin B$
→ **$\cos(A + B)$** $= \cos A \cos B - \sin A \sin B$
→ **$\cos(A - B)$** $= \cos A \cos B + \sin A \sin B$

→ **$\tan(A + B)$** $= \dfrac{\tan A + \tan B}{1 - \tan A \tan B}$ → **$\tan(A - B)$** $= \dfrac{\tan A - \tan B}{1 + \tan A \tan B}$

EXAMPLE: Find the exact value of tan 75°. Leave your answer with a rational denominator.

Solution: $\tan 75° = \tan(30° + 45°) = \dfrac{\tan 30° + \tan 45°}{1 - \tan 30° \tan 45°}$

$$= \frac{\frac{1}{\sqrt{3}} + 1}{1 - \frac{1}{\sqrt{3}} \times 1} \times \frac{\sqrt{3}}{\sqrt{3}} = \frac{1 + \sqrt{3}}{\sqrt{3} - 1} \times \frac{\sqrt{3} + 1}{\sqrt{3} + 1} = \frac{\sqrt{3} + 1 + 3 + \sqrt{3}}{3 - 1} = \frac{2\sqrt{3} + 4}{2} = \sqrt{3} + 2$$

TOPIC: TRIGONOMETRIC FUNCTIONS

See Excel Year 11 & 12 Mathematics Extension 1 pp. 42–43

FURTHER TRIGONOMETRIC IDENTITIES
The t-formulae $\left(t = \tan\frac{A}{2}\right)$
21

→ **For $t = \tan\frac{A}{2}$:**

- **$\sin A$** $= \dfrac{2t}{1 + t^2}$
- **$\cos A$** $= \dfrac{1 - t^2}{1 + t^2}$
- **$\tan A$** $= \dfrac{2t}{1 - t^2}$

EXAMPLE: Show that $\dfrac{1 - \cos\theta}{\sin\theta} = \tan\dfrac{\theta}{2}$.

Solution:

$$\text{LHS} = \frac{1 - \cos\theta}{\sin\theta} = \left(1 - \frac{1 - t^2}{1 + t^2}\right) \div \frac{2t}{1 + t^2} = \frac{1 + t^2 - 1 + t^2}{1 + t^2} \times \frac{1 + t^2}{2t} = \frac{2t^2}{2t}$$

$= t = \tan\dfrac{\theta}{2}$

$=$ RHS

TOPIC: TRIGONOMETRIC FUNCTIONS

See Excel Year 11 & 12 Mathematics Extension 1 pp. 44–45

FURTHER TRIGONOMETRIC IDENTITIES
Trigonometric products as sums and differences 2
23

→ The results on the previous card can lead to:

→ **$\cos A + \cos B$** $= 2\cos\dfrac{A + B}{2}\cos\dfrac{A - B}{2}$

→ **$\cos A - \cos B$** $= -2\sin\dfrac{A + B}{2}\sin\dfrac{A - B}{2}$

→ **$\sin A + \sin B$** $= 2\sin\dfrac{A + B}{2}\cos\dfrac{A - B}{2}$

→ **$\sin A - \sin B$** $= 2\cos\dfrac{A + B}{2}\sin\dfrac{A - B}{2}$

EXAMPLE: Find the exact value of cos 75° – cos 15°, using an appropriate difference-product identity.

Solution:
$\cos 75° - \cos 15°$
$= -2\sin\dfrac{75° + 15°}{2}\sin\dfrac{75° - 15°}{2}$
$= -2\sin 45° \sin 30°$
$= -2 \times \dfrac{1}{\sqrt{2}} \times \dfrac{1}{2}$
$= -\dfrac{1}{\sqrt{2}}$

TOPIC: TRIGONOMETRIC FUNCTIONS

See Excel Year 11 & 12 Mathematics Extension 1 pp. 45–46

INVERSE TRIGONOMETRIC FUNCTIONS
Domain and range of inverse trigonometric functions
25

→ **$y = \sin^{-1} x$ (or arcsin x)** Domain: $-1 \le x \le 1$, Range: $-\dfrac{\pi}{2} \le y \le \dfrac{\pi}{2}$.

→ **$y = \cos^{-1} x$ (or arccos x)** Domain: $-1 \le x \le 1$, Range: $0 \le y \le \pi$.

→ **$y = \tan^{-1} x$ (or arctan x)** Domain: all real x. Range: $-\dfrac{\pi}{2} < y < \dfrac{\pi}{2}$.

EXAMPLE: Find the domain of $y = 3\sin^{-1}\dfrac{2x + 1}{5}$.

Solution:
Domain: $-1 \le \dfrac{2x + 1}{5} \le 1$; $-5 \le 2x + 1 \le 5$; $-6 \le 2x \le 4$; $-3 \le x \le 2$

TOPIC: TRIGONOMETRIC FUNCTIONS

See Excel Year 11 & 12 Mathematics Extension 1 pp. 53–54

INVERSE TRIGONOMETRIC FUNCTIONS
Solving problems involving inverse trigonometric functions
27

→ **If $\alpha = \sin^{-1} x$ then $x = \sin \alpha$.**

EXAMPLE: Find the exact value of $\tan\left[\sin^{-1}\left(\dfrac{2}{\sqrt{13}}\right) + \cos^{-1}\left(\dfrac{1}{\sqrt{3}}\right)\right]$.

Solution: Let $\alpha = \sin^{-1}\left(\dfrac{2}{\sqrt{13}}\right)$ and $\beta = \cos^{-1}\left(\dfrac{1}{\sqrt{3}}\right)$.

$\therefore \sin\alpha = \dfrac{2}{\sqrt{13}}$, $\cos\beta = \dfrac{1}{\sqrt{3}}$ $\therefore \tan\alpha = \dfrac{2}{3}$, $\tan\beta = \sqrt{2}$

$\tan\left[\sin^{-1}\left(\dfrac{2}{\sqrt{13}}\right) + \cos^{-1}\left(\dfrac{1}{\sqrt{3}}\right)\right] = \tan(\alpha + \beta) = \dfrac{\tan\alpha + \tan\beta}{1 - \tan\alpha\tan\beta}$

$= \dfrac{\frac{2}{3} + \sqrt{2}}{1 - \frac{2}{3} \times \sqrt{2}} = \dfrac{2 + 3\sqrt{2}}{3 - 2\sqrt{2}}$ (by multiplying top and bottom by 3)

TOPIC: TRIGONOMETRIC FUNCTIONS

See Excel Year 11 & 12 Mathematics Extension 1 pp. 54–56

RATES OF CHANGE
Exponential growth and decay: $N(t) = Ae^{kt}$
29

→ **When the rate of change of a quantity N** is proportional to the quantity we say $\dfrac{dN}{dt} = kN$, where k is constant.

→ **If $N = Ae^{kt}$, then by differentiation,** $\dfrac{dN}{dt} = kN$. (A is the initial quantity.)

EXAMPLE: The population of a colony of rabbits over t months is given by $P = 180e^{0.347t}$. Find when the population reaches 5400.

Solution:
Let $P = 5400$: $5400 = 180e^{0.347t}$.

$e^{0.347t} = \dfrac{5400}{180}$
$e^{0.347t} = 30$
$\ln e^{0.347t} = \ln 30$
$0.347t = \ln 30$
$t = \dfrac{\ln 30}{0.347} = 9.80$ (2 dec. pl.)
$\therefore$ 9.8 months

TOPIC: CALCULUS

See Excel Year 11 & 12 Mathematics Extension 1 pp. 65–67

RATES OF CHANGE
Related rates of change 1
31

→ **The rate at which a quantity changes** is found by relating this quantity to other quantities whose rates of change are known.

→ **The chain rule** is used: $\dfrac{dy}{dt} = \dfrac{dy}{dx} \times \dfrac{dx}{dt}$.

EXAMPLE: A spherical balloon is being inflated and its radius is increasing at a constant rate of 2 cm/min. At what rate is the volume increasing when the radius of the balloon is 6 cm?

[Sphere: $V = \frac{4}{3}\pi r^3$]

Solution: $\dfrac{dr}{dt} = 2$, $\dfrac{dV}{dr} = 4\pi r^2$

$\dfrac{dV}{dr}(6) = 4\pi(6)^2$
$= 144\pi$
$\dfrac{dV}{dt} = \dfrac{dV}{dr} \times \dfrac{dr}{dt}$
$= 288\pi$ cm^3/min

TOPIC: CALCULUS

See Excel Year 11 & 12 Mathematics Extension 1 pp. 69–70

FURTHER TRIGONOMETRIC IDENTITIES

Double-angle formulae 20

- **sin 2A** = 2 sin A cos A
- **cos 2A** = $\cos^2 A - \sin^2 A$
  $= 1 - 2\sin^2 A$
  $= 2\cos^2 A - 1$
- **tan 2A** = $\dfrac{2\tan A}{1-\tan^2 A}$

EXAMPLE:
If $\tan\alpha = \frac{2}{3}$, find the value of sin 2α.
Solution:
$\therefore \sin\alpha = \frac{2}{\sqrt{13}}$ and $\cos\alpha = \frac{3}{\sqrt{13}}$
$\sin 2\alpha = 2\sin\alpha\cos\alpha$
$= 2\times\frac{2}{\sqrt{13}}\times\frac{3}{\sqrt{13}}$
$= \frac{12}{\sqrt{13}}$

TOPIC: TRIGONOMETRIC FUNCTIONS See ➲ *Excel* Year 11 & 12 Mathematics Extension 1 pp. 43–44

POLYNOMIALS

Graphs of polynomial functions 18

- $y = P(x)$ can be graphed by expressing $P(x)$ in factored form.

EXAMPLE: Sketch $y = P(x)$ if $P(x) = x^3 - 5x^2 + 8x - 4$, showing all x- and y-intercepts.
Solution: Consider
$P(2) = 2^3 - 5(2)^2 + 8(2) - 4$
$= 8 - 20 + 16 - 4$
$= 0$
$\therefore (x-2)$ is a factor.
$\therefore P(x) = (x-2)(x^2 - 3x + 2)$
$= (x-2)(x-2)(x-1)$
$= (x-2)^2(x-1)$

$$\begin{array}{r} x^2-3x+2 \\ x-2\overline{)x^3-5x^2+8x-4} \\ \underline{x^3-2x^2} \\ -3x^2+8x \\ \underline{-3x^2+6x} \\ 2x-4 \\ \underline{2x-4} \\ 0 \end{array}$$

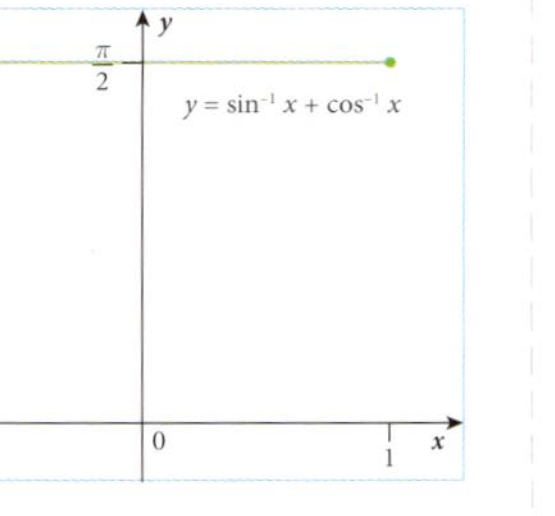

TOPIC: FUNCTIONS See ➲ *Excel* Year 11 & 12 Mathematics Extension 1 pp. 32–34

POLYNOMIALS

Sums and products of roots of polynomials 16

- **Quadratics:** If α and β are roots of $ax^2 + bx + c = 0$, $a \neq 0$ then $\alpha + \beta = -\frac{b}{a}$, $\alpha\beta = \frac{c}{a}$.
- **Cubics:** If α, β and γ are roots of $ax^3 + bx^2 + cx + d = 0$, $a \neq 0$ then $\alpha + \beta + \gamma = -\frac{b}{a}$, $\alpha\beta + \alpha\gamma + \beta\gamma = \frac{c}{a}$, $\alpha\beta\gamma = -\frac{d}{a}$.

EXAMPLE: The roots of $2x^3 + 4x^2 - 6x + 3 = 0$ are α, β and γ. Find the value of $\frac{1}{\alpha} + \frac{1}{\beta} + \frac{1}{\gamma}$.
Solution: Now, $\frac{1}{\alpha} + \frac{1}{\beta} + \frac{1}{\gamma} = \frac{\beta\gamma + \alpha\gamma + \alpha\beta}{\alpha\beta\gamma} = -3 \div -\frac{3}{2} = 2$

TOPIC: FUNCTIONS See ➲ *Excel* Year 11 & 12 Mathematics Extension 1 pp. 30–31

INVERSE TRIGONOMETRIC FUNCTIONS

Properties of inverse trigonometric functions 26

- $\sin^{-1}[\sin x] = \sin[\sin^{-1} x] = x$
- $\sin^{-1}(-x) = -\sin^{-1} x$. Hence, $y = \sin^{-1} x$ is an odd function.
- $\cos^{-1}[\cos x] = \cos[\cos^{-1} x] = x$
- $\cos^{-1}(-x) = \pi - \cos^{-1} x$. Hence, $y = \cos^{-1} x$ is not an odd function.
- $\tan^{-1}[\tan x] = \tan[\tan^{-1} x] = x$
- $\tan^{-1}(-x) = -\tan^{-1} x$. Hence, $y = \tan^{-1} x$ is an odd function.
- $\sin^{-1} x + \cos^{-1} x = \frac{\pi}{2}$

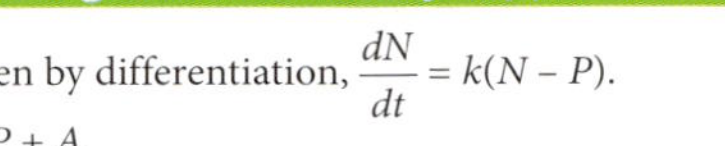

TOPIC: TRIGONOMETRIC FUNCTIONS See ➲ *Excel* Year 11 & 12 Mathematics Extension 1 pp. 54–56

INVERSE TRIGONOMETRIC FUNCTIONS

Graphs of inverse trigonometric functions 24

- $y = \sin^{-1} x$ (or arcsin x)
- $y = \cos^{-1} x$ (or arccos x)
- $y = \tan^{-1} x$ (or arctan x)

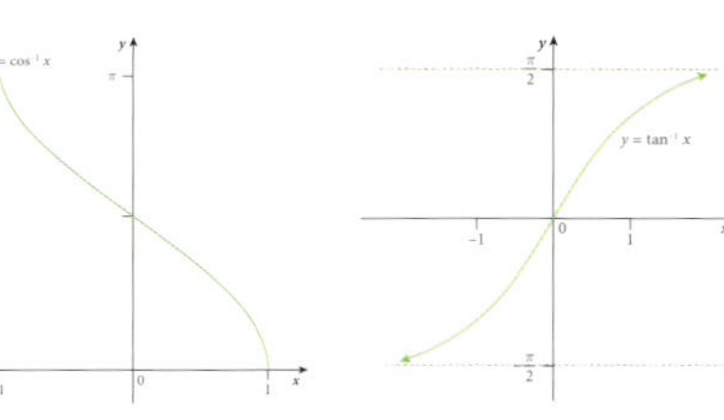

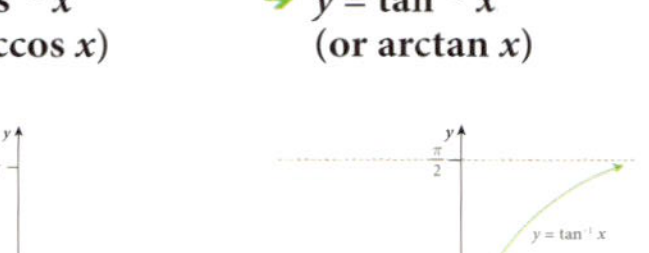

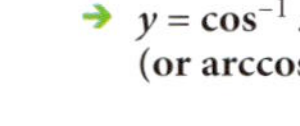

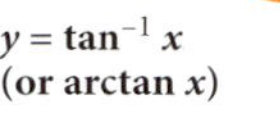

TOPIC: TRIGONOMETRIC FUNCTIONS See ➲ *Excel* Year 11 & 12 Mathematics Extension 1 pp. 53–54

FURTHER TRIGONOMETRIC IDENTITIES

Trigonometric products as sums and differences 1 22

- **cos A cos B**
  $= \frac{1}{2}[\cos(A-B) + \cos(A+B)]$
- **sin A sin B**
  $= \frac{1}{2}[\cos(A-B) - \cos(A+B)]$
- **sin A cos B**
  $= \frac{1}{2}[\sin(A+B) + \sin(A-B)]$
- **cos A sin B**
  $= \frac{1}{2}[\sin(A+B) - \sin(A-B)]$

EXAMPLE: Find the exact value of sin 105° cos 15°, using an appropriate product-sum identity.
Solution: sin 105° cos 15°
$= \frac{1}{2}[\sin(105° + 15°) + \sin(105° - 15°)]$
$= \frac{1}{2}[\sin 120° + \sin 90°]$
$= \frac{1}{2}[\frac{\sqrt{3}}{2} + 1]$
$= \frac{\sqrt{3}+2}{4}$

TOPIC: TRIGONOMETRIC FUNCTIONS See ➲ *Excel* Year 11 & 12 Mathematics Extension 1 pp. 45–46

RATES OF CHANGE

Related rates of change 2 32

- **You can use** $\frac{dV}{dt} = \frac{dV}{dr}\times\frac{dr}{dt}$, to find $\frac{dr}{dt}$ and hence $\frac{dA}{dt} = \frac{dA}{dr}\times\frac{dr}{dt}$.

EXAMPLE: Sand is being poured into a heap in the shape of a right cone at the rate of 6 cm^3/min. The height of the heap is always equal to the radius of its circular base. When the heap is 15 cm high, how fast is the area of its base increasing? [Cone: $V = \frac{1}{3}\pi r^2 h$]
Solution:
Using $\frac{dV}{dt} = \frac{dV}{dr}\times\frac{dr}{dt}$; as $\frac{dV}{dt}(15) = \pi(15)^2 = 225\pi$ $\therefore \frac{dr}{dt} = \frac{6}{225\pi}$
Now, using $\frac{dA}{dt} = \frac{dA}{dr}\times\frac{dr}{dt}$; as $\frac{dA}{dt}(15) = 30\pi$ $\therefore 30\pi\times\frac{6}{225\pi} = \frac{4}{5}$
$\therefore$ the area is increasing at $\frac{4}{5}$ cm^2/min.

TOPIC: CALCULUS See ➲ *Excel* Year 11 & 12 Mathematics Extension 1 pp. 69–70

RATES OF CHANGE

Further exponential growth and decay: $N(t) = P + Ae^{kt}$ 30

- **If $N = P + Ae^{kt}$**, then by differentiation, $\frac{dN}{dt} = k(N - P)$.
- Initial quantity is $P + A$.

EXAMPLE: A bottle of soft drink taken out of a refrigerator has a temperature of 3 °C. It is placed in a room with a constant temperature of 24 °C. If after t minutes the temperature, T °C, of the bottle is given by $T = A - Be^{-0.035t}$, find the positive values of A and B.
Solution: When $t \to \infty$, $Be^{-0.035t} \to 0$.
$\therefore 24 = A + 0$ $\therefore A = 24$
$\therefore T = 24 - Be^{-0.035t}$
When $t = 0$ and $T = 3$:
$3 = 24 - B$ $\therefore B = 21$

TOPIC: CALCULUS See ➲ *Excel* Year 11 & 12 Mathematics Extension 1 pp. 67–69

RATES OF CHANGE

Rates of change with respect to time 28

- **The rate of change** of a quantity Q with respect to time is measured by $\frac{dQ}{dt}$.
- **Displacement (x) has magnitude and direction**. Distance has magnitude only and |displacement| = distance.
- **Velocity (v, or $\dot{x}$) has magnitude and direction**. Speed has magnitude only and |velocity| = speed.
- **Acceleration** (a, or $\ddot{x}$) = $\frac{dv}{dt}$, or $\frac{d^2x}{dt^2}$.
- **Other terminology**: 'initially': $t = 0$, 'at rest': $v = 0$, 'change of direction': v changes sign, 'velocity increasing at an increasing rate': acceleration is positive.

TOPIC: CALCULUS See ➲ *Excel* Year 11 & 12 Mathematics Extension 1 pp. 63–65

WORKING WITH COMBINATORICS

33 Fundamental counting principle (multiplication principle)

- **If an event can happen in *m* different ways** and a second event can happen in *n* different ways, then those successive events can happen in *mn* different ways.

EXAMPLE: The digits 1, 2, 3, 4 and 5 are used to form 3-digit numbers where each digit is only used once.
What is the probability that a randomly chosen number is even?
As $5 \times 4 \times 3 = 60$, there are 60 possible numbers.
As the last digit is 2 or 4, and $4 \times 3 \times 2 = 24$, there are 24 even numbers.

$\text{P(even)} = \frac{24}{60} = \frac{2}{5}$.

TOPIC: COMBINATORICS See ➲ *Excel* Year 11 & 12 Mathematics Extension 1 pp. 81–82

WORKING WITH COMBINATORICS

35 Pigeonhole principle

- **If there are *n* pigeonholes** and $n + 1$ pigeons to go into them, then at least one pigeonhole must hold two or more pigeons.
- **More generally, if *n* pigeons are sitting in *k* pigeonholes**, where $n > k$, then there is at least one pigeonhole with at least $\frac{n}{k}$ pigeons in it.

EXAMPLE: What is the smallest group of people that can be assembled where it is guaranteed that at least two people have the same first and last initials in their name?
Solution: As 2 people, then $n = 2$.
Number of initial pairings $(k) = 26 \times 26 = 676$
Smallest number $= (n - 1)k + 1 = 1 \times 676 + 1 = 677$
$\therefore$ the group must have at least 677 people.

TOPIC: COMBINATORICS See ➲ *Excel* Year 11 & 12 Mathematics Extension 1 p. 83

WORKING WITH COMBINATORICS

37 Combinations

- **Combination** ${}^nC_r$, **or** $\binom{n}{r}$, **is the number of unordered selections** of r objects from a total of n different objects.
- ${}^nC_r$ **is often pronounced '*n* choose *r*'.**
- ${}^nC_r = \frac{n!}{r!(n-r)!}$

EXAMPLE: A team of four men and three women is to be chosen from a group of nine men and seven women.
Find the probability that Brittany is chosen but Daniel is not.

$\text{P(Brittany chosen, Daniel not)} = \frac{{}^6C_2 \times {}^8C_4}{{}^9C_4 \times {}^7C_3} = \frac{5}{21}$

TOPIC: COMBINATORICS See ➲ *Excel* Year 11 & 12 Mathematics Extension 1 pp. 84–85

WORKING WITH COMBINATORICS

39 Combinatorial proofs

- ${}^nC_0 = 1, {}^nC_1 = n, {}^nC_{n-1} = n, {}^nC_n = 1, {}^nC_{n-r} = {}^nC_r$

EXAMPLE: Prove that ${}^nC_k = {}^{n-1}C_{k-1} + {}^{n-1}C_k$.

Solution:
$$\begin{aligned}\text{RHS} &= \frac{(n-1)!}{(k-1)!(n-1-k+1)!} + \frac{(n-1)!}{k!(n-1-k)!} \\ &= \frac{(n-1)!}{(k-1)!(n-k)!} + \frac{(n-1)!}{k!(n-k-1)!} \\ &= \frac{k(n-1)! + (n-k)(n-1)!}{k!(n-k)!} = \frac{(n-1)![k+n-k]}{k!(n-k)!} \\ &= \frac{n(n-1)!}{k!(n-k)!} = \frac{n!}{k!(n-k)!} = {}^nC_k = \text{LHS}\end{aligned}$$

TOPIC: COMBINATORICS See ➲ *Excel* Year 11 & 12 Mathematics Extension 1 pp. 85–86

PROOF BY MATHEMATICAL INDUCTION

41 Mathematical induction relating to divisibility

- **If *M* is an integer** then $5M$ is divisible by 5, or a multiple of 5.

EXAMPLE: Use mathematical induction to prove that $3^{2n} - 1$ is divisible by 8 for all positive integers n.
Solution: Prove true for $n = 1$.
Assume true for $n = k$: $3^{2k} - 1 = 8M$, where M is an integer.
Now prove true for $n = k + 1$:
…
$\therefore$ hence the claim is true for $n = k + 1$
Since shown true for $n = 1$, is true for $n = 2, 3, 4, \ldots$ and so true for all integers $n \geq 1$.

TOPIC: PROOF See ➲ *Excel* Year 11 & 12 Mathematics Extension 1 p. 101

INTRODUCTION TO VECTORS

43 Vector terminology

- **A vector is a quantity** with magnitude (length) and direction.
- **Vectors can be written as *a*, $\underset{\sim}{a}$ or** $\vec{OA}$.
- **The negative of the vector from *A* to *B*** is the vector from B to A.
  This means $\vec{AB} = -\vec{BA}$.
- **The length of vector *AB*** is written as $|\vec{AB}|$.
- **A displacement vector** (or relative vector) is the distance and direction of a point from another e.g. $\vec{AB}$.
- $\underset{\sim}{i}$ and $\underset{\sim}{j}$ are unit vectors in the positive directions of the x-axis and y-axis respectively.
- **A scalar, λ,** is a real number and multiplies a vector. $\underset{\sim}{u}$ and $\underset{\sim}{v}$ are parallel if $\underset{\sim}{v} = \lambda \underset{\sim}{u}$, where $\lambda \neq 0$.

TOPIC: VECTORS See ➲ *Excel* Year 11 & 12 Mathematics Extension 1 p. 109

INTRODUCTION TO VECTORS

45 Operations with vectors

- **Multiplication:** Multiplying a vector by a scalar (real number) changes its length or direction. The resultant vector is parallel to the original.
- **Addition:** Two vectors $\underset{\sim}{u}$ and $\underset{\sim}{v}$ can be added geometrically by forming a parallelogram. Also, $\underset{\sim}{u} + \underset{\sim}{v} = \underset{\sim}{v} + \underset{\sim}{u}$.
- **Subtraction:** To find $\underset{\sim}{u} - \underset{\sim}{v}$, we add $\underset{\sim}{u}$ and $-\underset{\sim}{v}$.

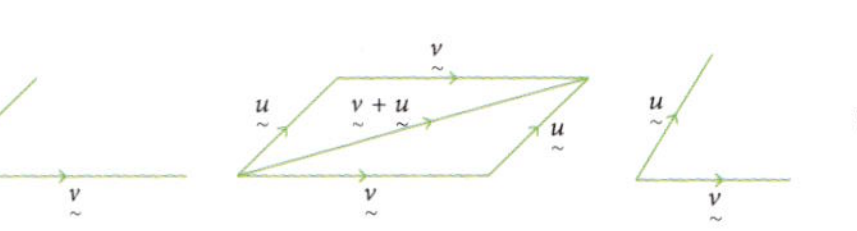

TOPIC: VECTORS See ➲ *Excel* Year 11 & 12 Mathematics Extension 1 pp. 111–113

INTRODUCTION TO VECTORS

47 Direction of vectors

- **If $\underset{\sim}{u} = x\underset{\sim}{i} + y\underset{\sim}{j}$**, then the directional angle of the vector is given as $\theta = \tan^{-1}\left(\frac{y}{x}\right)$, which is the angle formed between the vector and the horizontal in the positive direction.
- **Also, for $\underset{\sim}{u} = x\underset{\sim}{i} + y\underset{\sim}{j}$**, $x = |\underset{\sim}{u}|\cos\theta$ and $y = |\underset{\sim}{u}|\sin\theta$.

EXAMPLE: A vector $\underset{\sim}{u}$ has a magnitude of 2 and makes an angle of 60°. Express the vector in the form $\underset{\sim}{u} = x\underset{\sim}{i} + y\underset{\sim}{j}$.

Solution: Use $x = |\underset{\sim}{u}|\cos\theta = 2\cos 60° = 2 \times \frac{1}{2} = 1$

Use $y = |\underset{\sim}{u}|\sin\theta = 2\sin 60° = 2 \times \frac{\sqrt{3}}{2} = \sqrt{3}$ $\therefore \underset{\sim}{u} = x\underset{\sim}{i} + y\underset{\sim}{j} = \underset{\sim}{i} + \sqrt{3}\underset{\sim}{j}$

TOPIC: VECTORS See ➲ *Excel* Year 11 & 12 Mathematics Extension 1 pp. 120–121

NTRODUCTION TO VECTORS

49 Angle between two vectors

- $\underset{\sim}{u}.\underset{\sim}{v} = |\underset{\sim}{u}||\underset{\sim}{v}|\cos\theta$, where θ is the angle between the vectors $\underset{\sim}{u}$ and $\underset{\sim}{v}$.
- **To find the angle between two vectors θ,** use $\cos\theta = \frac{\underset{\sim}{u}.\underset{\sim}{v}}{|\underset{\sim}{u}||\underset{\sim}{v}|}$, or $\cos\theta = \frac{x_1x_2 + y_1y_2}{|\underset{\sim}{u}||\underset{\sim}{v}|}$.

EXAMPLE: P and Q are defined by the position vectors $\underset{\sim}{u}$ and $\underset{\sim}{v}$ respectively, where $\underset{\sim}{u} = 3\underset{\sim}{i} + 5\underset{\sim}{j}$ and $\underset{\sim}{v} = -4\underset{\sim}{i} + 2\underset{\sim}{j}$. If O is the origin, find the size of $\angle POQ$, to the nearest degree.

Solution: $\cos\theta = \frac{\underset{\sim}{u}.\underset{\sim}{v}}{|\underset{\sim}{u}||\underset{\sim}{v}|} = \frac{3\times(-4)+5\times 2}{\sqrt{3^2+5^2}.\sqrt{(-4)^2+2^2}} = \frac{-2}{\sqrt{34}.\sqrt{20}}$
$= -0.076\,696\,498\ldots \quad \therefore \theta = 94°$

TOPIC: VECTORS See ➲ *Excel* Year 11 & 12 Mathematics Extension 1 pp. 121–122

WORKING WITH COMBINATORICS
The binomial expansion
38

- **The coefficients of $(1 + x)^n$** can be determined using Pascal's triangle, or ${}^nC_r$.
- $(1 + x)^n = {}^nC_0 + {}^nC_1x + {}^nC_2x^2 + \ldots + {}^nC_kx^k + \ldots + {}^nC_{n-1}x^{n-1} + {}^nC_nx^n$
- $(a + x)^n = {}^nC_0a^n + {}^nC_1a^{n-1}x + {}^nC_2\,a^{n-2}x^2 + \ldots + {}^nC_ka^{n-k}\,x^k + \ldots + {}^nC_{n-1}ax^{n-1} + {}^nC_nx^n$
- **In the expansion of $(a + x)^n$**, the coefficient of x^k is ${}^nC_k$.

EXAMPLE: Find the coefficient of the fourth term in the expansion of $(1 - 2x)^5$.
Solution: The fourth term is the term involving x^3.
$(1 - 2x)^5 = \ldots + {}^5C_31^2(-2x)^3 + \ldots$
$\therefore$ the coefficient is ${}^5C_3(-2)^3 = -80$.

TOPIC: COMBINATORICS See ➲ *Excel* Year 11 & 12 Mathematics Extension 1 pp. 85–86

WORKING WITH COMBINATORICS
Permutations
36

- **Permutation ${}^nP_r$ is the number of ordered selections** of r objects from a total of n different objects.
- ${}^nP_r = \dfrac{n!}{(n-r)!}$

EXAMPLE: Express ${}^nP_2$ in factorial form and simplify.

Solution: ${}^nP_2 = \dfrac{n!}{(n-2)!} = \dfrac{n(n-1)(n-2)(n-3)\ldots}{(n-2)(n-3)\ldots} = n(n-1)$

EXAMPLE: Find the number of arrangements of four digits that can be formed using the digits 0 to 9 if each can only be used once.
Solution: ${}^{10}P_4 = 5040$

TOPIC: COMBINATORICS See ➲ *Excel* Year 11 & 12 Mathematics Extension 1 pp. 83–84

WORKING WITH COMBINATORICS
Factorial notation
34

- $\boldsymbol{n! = n(n-1)(n-2)(n-3)}$ … 3.2.1 and, by convention, $0! = 1$.
- **If n objects are placed in a row** there are $n!$ different arrangements.
- **If n objects are placed around a circle** there are $\dfrac{n!}{n} = (n - 1)!$ different arrangements.
- **The number of different ways of arranging n objects** in which x of the objects are of one kind, y of the objects are of another kind, etc., is given by $\dfrac{n!}{x!\,y!\ldots}$.

EXAMPLE: A father, a mother and their four children sit at a circular table. What is the probability that the parents do not sit together?
Solution:
Arrangements $= 5! - 4! \times 2$ $\therefore$ P(parents not together) $= \dfrac{5! - 4! \times 2}{5!} = \dfrac{3}{5}$

TOPIC: COMBINATORICS See ➲ *Excel* Year 11 & 12 Mathematics Extension 1 pp. 82–83

INTRODUCTION TO VECTORS
Expressing vectors in different forms
44

- **A component form** of a vector expresses the vector in terms of unit vectors in the form of $xi + yj$; e.g. $4i + 3j$ means 4 units in x-direction and 3 units in y-direction.
- **An ordered pair** of a vector is written with round or angle brackets; e.g. $4i + 3j$ is (4, 3) or <4, 3>.
- **A column vector** is written in the form $\begin{pmatrix}4\\3\end{pmatrix}$.

EXAMPLE: Express $\vec{OA}$ in component form.

Solution: From O to A is 7 units in x-direction and 3 units in y-direction. Hence, $\vec{OA} = 7i + 3j$.

TOPIC: VECTORS See ➲ *Excel* Year 11 & 12 Mathematics Extension 1 pp. 109–111

PROOF BY MATHEMATICAL INDUCTION
False 'proofs by induction'
42

- **The result has not been proved** if only one of the steps is proved.

EXAMPLE: Martha attempted to use mathematical induction to prove that $2 + 4 + \ldots + 2n = n(n - 1) + 2$, for $n \geq 1$. At which step (initial or inductive) did the proof fail? Support your decision with working.
Solution: …
i.e. $S_k + T_{k+1} = S_{k+1}$
i.e. $k(k - 1) + 2 + 2(k + 1) = k(k + 1) + 2 = k^2 + k + 4$
LHS $= k^2 + k + 2 + 2k + 2$ $= k^2 + 3k + 4$
RHS $= k(k + 1) + 2$ $= k^2 + k + 2$
$\therefore$ LHS $\neq$ RHS
$\therefore$ not true for $n = k + 1$.
$\therefore$ the proof fails in the inductive step.

TOPIC: PROOF See ➲ *Excel* Year 11 & 12 Mathematics Extension 1 pp. 101–102

PROOF BY MATHEMATICAL INDUCTION
Mathematical induction relating to series
40

- **Start with the initial statement**: prove true for the first value of n.
- **Then the inductive statement**: assume true for $n = k$ and then prove true for $n = k + 1$.
- **You need to establish $S_k + T_{k+1} = S_{k+1}$**, where S_k = sum to k terms, T_{k+1} = $(k+1)$th term and S_{k+1} = sum to $(k+1)$ terms.

EXAMPLE: Use mathematical induction to prove that, for $n \geq 1$, $2 + 6 + 12 + \ldots + n(n + 1) = \dfrac{n}{3}(n + 1)(n + 2)$.
Solution: Prove true for $n = 1$.
Assume true for $n = k$: $S_k = 2 + 6 + 12 + \ldots + k(k + 1)$
$= \dfrac{k}{3}(k + 1)(k + 2)$
Now prove true for $n = k + 1$: i.e. $S_k + T_{k+1} = S_{k+1}$ … and so on

TOPIC: PROOF See ➲ *Excel* Year 11 & 12 Mathematics Extension 1 pp. 100–101

INTRODUCTION TO VECTORS
Parallel and perpendicular vectors
50

- **Vectors u and v are parallel** if $u.v = \pm|u||v|$, or $u = \lambda v$.
- **Vectors u and v are perpendicular (or orthogonal)** if $u.v = 0$.

EXAMPLE: Find the value of m if the vectors:
a $u = 3i + mj$ and $v = i - 2j$ are parallel
b $u = 4i - 3j$ and $v = mi + 8j$ are perpendicular
Solution:
a If parallel, then $u = \lambda u$.
$3i + mj = k(i - 2j)$
$\therefore 3i = ki$ $\quad \therefore k = 3$
$\therefore mj = 3(-2j)$ $\quad \therefore m = -6$

b $x_1x_2 + y_1y_2 = 0$
$4 \times m + (-3) \times 8 = 0$
$4m - 24 = 0$
$m = 6$

TOPIC: VECTORS See ➲ *Excel* Year 11 & 12 Mathematics Extension 1 p. 122

INTRODUCTION TO VECTORS
Scalar (dot) product of vectors
48

- **If $u = x_1i + y_1j$, and $v = x_2i + y_2j$**, the scalar product (or dot product) of vectors is defined by $u.v = x_1x_2 + y_1y_2$.
- **Also, $u.v = v.u$; $k(u.v)$** $= (ku).v = u.(kv)$; $v.v = |v|^2$.

EXAMPLE: If $u = 3i + 2j$ and $v = 4i - 5j$, find
a $u.v$
b $u.u + v.v$
Solution:
a. $u.v = 3 \times 4 + 2 \times (-5) = 2$
b. $u.u + v.v = 3^2 + 2^2 + 4^2 + (-5)^2 = 54$

TOPIC: VECTORS See ➲ *Excel* Year 11 & 12 Mathematics Extension 1 pp. 121–122

INTRODUCTION TO VECTORS
Magnitude of vectors
46

- **The magnitude of a vector $\vec{AB}$** is the distance between points A and B.
- **If $u = xi + yj$**, then the magnitude, $|u| = |xi + yj| = \sqrt{x^2 + y^2}$, using Pythagoras.
- **Hence, $|u|^2 = x^2 + y^2$.**
- **Non-zero vector** u converts to a unit vector $\hat{u}$ by dividing by its length: $\hat{u} = \dfrac{u}{|u|}$.

EXAMPLE: If $u = i + 2j$ and $v = 3i - j$, find $|u + v|$.

Solution: $|u + v| = |i + 2j + 3i - j| = |4i + j| = \sqrt{4^2 + 1^2} = \sqrt{17}$

TOPIC: VECTORS See ➲ *Excel* Year 11 & 12 Mathematics Extension 1 pp. 120–121

INTRODUCTION TO VECTORS

Vector projection — 51

→ **The vector projection of *u*** in the direction of vector v can be expressed as $\text{proj}_v u = u_v = \frac{u.v}{|v|^2} v$

EXAMPLE: If $u = 2i + 3j$ and $v = i + 4j$, find the vector projection of u in the direction of v.

Solution: $u.v = 2 \times 1 + 3 \times 4 = 14$

$|v|^2 = 1^2 + 4^2 = 17$

$\therefore u_v = \frac{14}{17}(i + 4j)$

TOPIC: VECTORS See ➲ *Excel* Year 11 & 12 Mathematics Extension 1 pp. 122–123

INTRODUCTION TO VECTORS

Force as a vector — 53

→ **The force of an object is a vector** as it has a size, measured in Newtons, N, and a direction.

EXAMPLE: Find the magnitude and direction of the resultant force of the forces 4 N and 6 N on a particle as shown in the diagram below.

Solution: Let force vectors be u and v.

$u = 3i + 3\sqrt{3}j,\ v = 2\sqrt{3}i + 2j$

$R = (3 + 2\sqrt{3})i + (3\sqrt{3} + 2)j$

$|R| = 9.67$ (2 dec. pl.)

$\theta = 48$

$\therefore$ the resultant force is 9.67 N at an angle of 48°.

y, x, O, 6 N, 4 N, u, v, $R = u + v$, 30°, 60°

TOPIC: VECTORS See ➲ *Excel* Year 11 & 12 Mathematics Extension 1 pp. 123–125

INTRODUCTION TO VECTORS

Geometric proofs involving vectors — 55

→ In the syllabus, NESA states that students may be **required to prove** the following:

- the diagonals of a parallelogram meet at right angles if and only if it is a rhombus. (Solution involves $u \cdot v = 0$)
- the midpoints of the sides of a quadrilateral join to form a parallelogram. (Solution involves $u = \lambda u$)
- the sum of the squares of the lengths of the diagonals of a parallelogram is equal to the sum of the squares of the lengths of the sides. (Solution involves $|u + v|^2 + |v - u|^2 = 2|u|^2 + 2|v|^2$)

TOPIC: VECTORS See ➲ *Excel* Year 11 & 12 Mathematics Extension 1 p. 125

INTRODUCTION TO VECTORS

Velocity vector as a function of time — 57

→ **If the position vector is given** as $r(t) = x(t)i + y(t)j$, and $x(t)$ and $y(t)$ are differentiable, then the velocity vector is $\dot{r}(t) = v(t) = \dot{x}(t)i + \dot{y}(t)j$.

→ **Speed at time *t* is given** as $|v(t)| = \sqrt{(\dot{x}(t))^2 + (\dot{y}(t))^2}$.

EXAMPLE: A particle moves so that its position vector is given by $r(t) = 3\sin 2ti + 4\cos 2tj$. Find its speed after $\frac{3\pi}{8}$ seconds in ms^{-1}.

Solution: $v\left(\frac{3\pi}{8}\right) = 6\cos 2\left(\frac{3\pi}{8}\right)i - 8\sin 2\left(\frac{3\pi}{8}\right)j = -\frac{6}{\sqrt{2}}i - \frac{8}{\sqrt{2}}j$

$\left|v\left(\frac{3\pi}{8}\right)\right| = \sqrt{\left(-\frac{6}{\sqrt{2}}\right)^2 + \left(-\frac{8}{\sqrt{2}}\right)^2} = 5\sqrt{2}\ \therefore 5\sqrt{2}$ ms^{-1}

TOPIC: VECTORS See ➲ *Excel* Year 11 & 12 Mathematics Extension 1 pp. 135–136

INTRODUCTION TO VECTORS

Determining the position vector using the velocity vector — 59

→ **Given $v(t) = \dot{x}(t)i + \dot{y}(t)j$**, then $r(t) = x(t)i + y(t)j$, where $x(t) = \int \dot{x}(t)\,dt$ and $y(t) = \int \dot{y}(t)\,dt$.

EXAMPLE: The velocity of an object is given by $v(t) = 6t^2 i - \sin tj$ and initially $r = 2i + j$. Determine $r(t)$.

Solution: Consider $x(t) = \int 6t^2\,dt = 2t^3 + c_1$

As $x(0) = 2$,

$2(0)^3 + c_1 = 2$

$\therefore c_1 = 2$

$\therefore x(t) = 2t^3 + 2$

Consider $y(t) = \int -\sin t\,dt$
$= \cos t + c_2$

As $y(0) = 1$,

$\cos(0) + c_2 = 1$

$\therefore c_2 = 0$

$\therefore y(t) = \cos t$

$\therefore r(t) = (2t^3 + 2)i + \cos tj$

TOPIC: VECTORS See ➲ *Excel* Year 11 & 12 Mathematics Extension 1 pp. 135–136

INTRODUCTION TO VECTORS

Finding the greatest height of a projectile — 61

→ **To find maximum height** let $\dot{y}(t) = 0$ to find t, and then substitute into y.

EXAMPLE: A particle is projected from O at a speed of 60 ms^{-1} at an angle of 45° with the horizontal plane.
Given $r(t) = |v(0)|t\cos\theta\, i + (|v(0)|t\sin\theta - 5t^2)j$, find the greatest height of the ball.

Solution: Use $|v(0)| = 60$ and $\theta = 45°$.

$r(t) = 30\sqrt{2}ti + (30\sqrt{2}t - 5t^2)j$

As $y(t) = 30\sqrt{2}t - 5t^2$, then

$\dot{y}(t) = 30\sqrt{2} - 10t = 0$

$t = 3\sqrt{2}$

Substitute into

$y(t) = 30\sqrt{2}t - 5t^2$:

$y(3\sqrt{2}) = 90$

$\therefore$ greatest height of 90 m.

TOPIC: VECTORS See ➲ *Excel* Year 11 & 12 Mathematics Extension 1 pp. 136–140

INTRODUCTION TO VECTORS

The cartesian equation of the trajectory of the projectile — 63

→ **Combine the horizontal and vertical equations** into a single equation by eliminating variable t.

EXAMPLE: A golf ball is hit so that its position vector is given as $r(t) = 50t\cos\theta\, i + (50t\sin\theta - 5t^2)j$. Find the cartesian equation of the path of the ball.

Solution: $t = \frac{x}{50\cos\theta}$ in y: $y = 50\left(\frac{x}{50\cos\theta}\right)\sin\theta - 5\left(\frac{x}{50\cos\theta}\right)^2$

$= x\tan\theta - \frac{x^2\sec^2\theta}{500} = x\tan\theta - \frac{x^2(1+\tan^2\theta)}{500}$

$\therefore$ the trajectory is $y = x\tan\theta - \frac{x^2(1+\tan^2\theta)}{500}$.

TOPIC: VECTORS See ➲ *Excel* Year 11 & 12 Mathematics Extension 1 pp. 136–140

TRIGONOMETRIC EQUATIONS

Expressing in the form $r\sin(\theta \pm \alpha)$, $r\cos(\theta \pm \alpha)$ — 65

→ In each of the following cases, $r = \sqrt{a^2 + b^2}$ and $\tan\alpha = \frac{b}{a}$ (α is acute),

→ $a\sin\theta + b\cos\theta = r\sin(\theta + \alpha)$

→ $a\sin\theta - b\cos\theta = r\sin(\theta - \alpha)$

→ $a\cos\theta + b\sin\theta = r\cos(\theta - \alpha)$

→ $a\cos\theta - b\sin\theta = r\cos(\theta + \alpha)$

EXAMPLE: Find the maximum value of $f(x) = 3\sin x + 2\cos x$.

Solution: Rewrite $3\sin x + 2\cos x$ in form $r\sin(x + \alpha)$.

$r = \sqrt{3^2 + 2^2} = \sqrt{13}$

$\therefore 3\sin x + 2\cos x = \sqrt{13}\sin(x + \alpha)$.

As $-1 \le \sin(x + \alpha) \le 1$, then $-\sqrt{13} \le \sqrt{13}\sin(x + \alpha) \le \sqrt{13}$.

$\therefore$ the maximum value is $\sqrt{13}$.

TOPIC: TRIGONOMETRIC FUNCTIONS See ➲ *Excel* Year 11 & 12 Mathematics Extension 1 pp. 150–151

TRIGONOMETRIC EQUATIONS

Equations involving sums and differences — 67

→ $\sin(A \pm B) = \sin A\cos B \pm \cos A\sin B$

→ $\cos(A \pm B) = \cos A\cos B \mp \sin A\sin B$

→ $\tan(A \pm B) = \frac{\tan A \pm \tan B}{1 \mp \tan A\tan B}$

EXAMPLE: Solve $\sin 3x\cos x - \cos 3x\sin x = \frac{1}{2}$, for $0 \le x \le 2\pi$.

Solution:

$\sin(3x - x) = \frac{1}{2}$

$\sin 2x = \frac{1}{2}$

$2x = \frac{\pi}{6}, \frac{5\pi}{6}, \frac{13\pi}{6}, \frac{17\pi}{6},$

$x = \frac{\pi}{12}, \frac{5\pi}{12}, \frac{13\pi}{12}, \frac{17\pi}{12}$

TOPIC: TRIGONOMETRIC FUNCTIONS See ➲ *Excel* Year 11 & 12 Mathematics Extension 1 pp. 151–152

INTRODUCTION TO VECTORS

Position vector as a function of time — 56

- **The location of a particle** is given by its position vector $\underset{\sim}{r}$, where $\underset{\sim}{r} = x\underset{\sim}{i} + y\underset{\sim}{j}$.
- $\underset{\sim}{r}(t) = x(t)\underset{\sim}{i} + y(t)\underset{\sim}{j}$ is a vector function where r, x and y are expressed in terms of time t.

EXAMPLE: A particle moves so that its position is given by $\underset{\sim}{r}(t) = (2t-1)\underset{\sim}{i} + (t+3)\underset{\sim}{j}$. Find the distance the particle is from the origin after 2 seconds.

Solution:

$\underset{\sim}{r}(2) = (2(2)-1)\underset{\sim}{i} + (2+3)\underset{\sim}{j} = 3\underset{\sim}{i} + 5\underset{\sim}{j}$

$|\underset{\sim}{r}(2)| = \sqrt{3^2+5^2} = \sqrt{34}$

$\therefore$ the particle is $\sqrt{34}$ units from origin.

TOPIC: VECTORS — See Excel Year 11 & 12 Mathematics Extension 1 pp. 134–135

INTRODUCTION TO VECTORS

Velocity as a vector — 54

- **The magnitude of the velocity vector** gives the speed of an object and the vector direction gives its direction.

EXAMPLE: A plane is flying on a bearing of 060°T at a speed of 240 km/h. It is met by a cross wind at a bearing of 330° at a speed of 60 km/h. In the diagram the plane's path is represented by vector $\underset{\sim}{v}$ and the wind's path by vector $\underset{\sim}{w}$. Find the ground speed of the plane.

Solution:

$\underset{\sim}{v} = 120\sqrt{3}\underset{\sim}{i} + 120\underset{\sim}{j}, \quad \underset{\sim}{w} = -30\underset{\sim}{i} + 30\sqrt{3}\underset{\sim}{j}$

$\left|\underset{\sim}{v} + \underset{\sim}{w}\right| = \sqrt{(240^2 + 60^2}$

$= 247$ (nearest whole)

$\therefore$ ground speed of 247 km/h.

TOPIC: VECTORS — See Excel Year 11 & 12 Mathematics Extension 1 pp. 123–125

INTRODUCTION TO VECTORS

Displacement as a vector — 52

- **The displacement is a vector** which points from the initial position of an object to its final position.

EXAMPLE: Michael cycles 45 km in a northerly direction. He then travels 65 km in an easterly direction. Calculate the magnitude and direction of his displacement from his starting point.

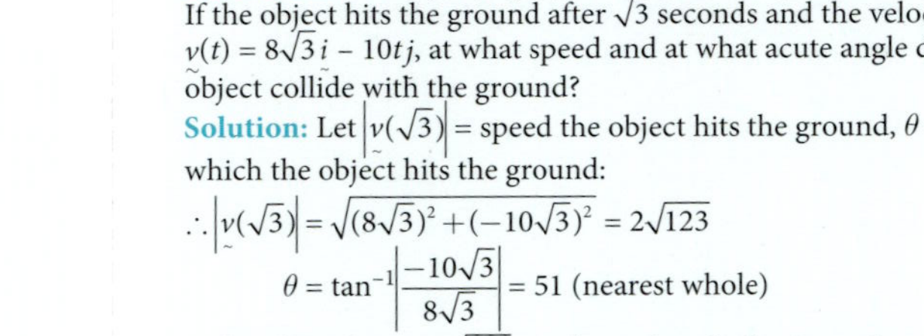

Solution:

Magnitude $= \sqrt{65^2+45^2} = 79.056\,9415\ldots = 79$ (nearest whole)

$\tan\theta = \dfrac{45}{65}$

$\theta = 34.695\,153\,53\ldots = 35$ (nearest whole)

$\therefore$ Michael's displacement is 79 km in a direction of 35° above the horizontal.

TOPIC: VECTORS — See Excel Year 11 & 12 Mathematics Extension 1 pp. 123–125

INTRODUCTION TO VECTORS

Finding the range of the projectile — 62

- **To find the range** let $y(t) = 0$ to find t, and then substitute into x.
- **Greatest range occurs** when $\theta = 45°$.

EXAMPLE: *Continue the example from the previous card.* Find the time of flight and the range of the ball when it hits the ground.

Solution:

Time of flight is $2(3\sqrt{2}) = 6\sqrt{2}$ seconds.

Substitute into $x(t) = 30\sqrt{2}t$:

$x(6\sqrt{2}) = 30\sqrt{2}(6\sqrt{2})$

$= 360$

$\therefore$ the time of flight is $6\sqrt{2}$ s and the range is 360 m.

TOPIC: VECTORS — See Excel Year 11 & 12 Mathematics Extension 1 pp. 136–140

INTRODUCTION TO VECTORS

Projectile motion concepts — 60

- **As the only force is downwards**, then $\ddot{\underset{\sim}{r}}(t) = \underset{\sim}{a}(t) = -g\underset{\sim}{j}$.
- **Using integration and** $\dot{x}(0) = \left|\underset{\sim}{v}(0)\right|\cos\theta$, $\dot{y}(0) = \left|\underset{\sim}{v}(0)\right|\sin\theta$,

  then $\underset{\sim}{v}(t) = \left|\underset{\sim}{v}(0)\right|\cos\theta\,\underset{\sim}{i} + (\left|\underset{\sim}{v}(0)\right|\sin\theta - gt)\underset{\sim}{j}$.
- **Using integration and** $\underset{\sim}{r}(0) = 0$,

  then $\underset{\sim}{r}(t) = \left|\underset{\sim}{v}(0)\right|t\cos\theta\,\underset{\sim}{i} + (\left|\underset{\sim}{v}(0)\right|t\sin\theta - \frac{1}{2}gt^2)\underset{\sim}{j}$.

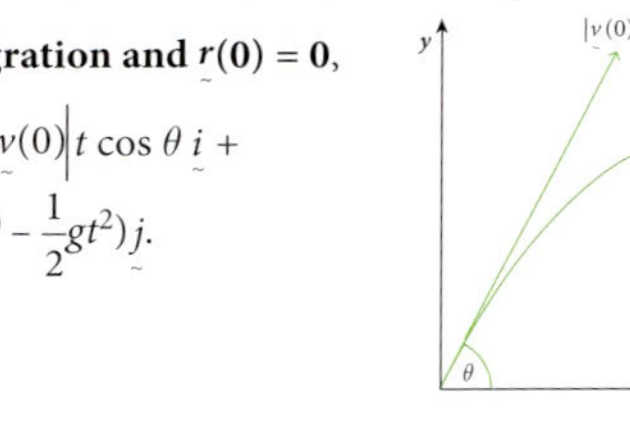
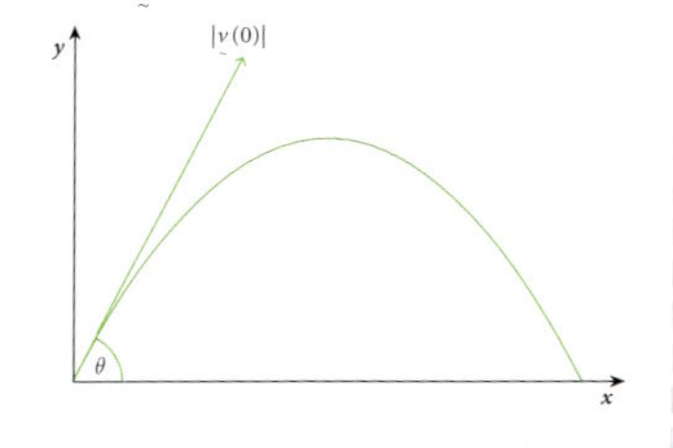

TOPIC: VECTORS — See Excel Year 11 & 12 Mathematics Extension 1 pp. 136–140

INTRODUCTION TO VECTORS

Acceleration vector as a function of time — 58

- **If the position vector is given** as $\underset{\sim}{r}(t) = x(t)\underset{\sim}{i} + y(t)\underset{\sim}{j}$, and $x(t)$ and $y(t)$ are differentiable, then the acceleration vector is $\ddot{\underset{\sim}{r}}(t) = \underset{\sim}{a}(t) = \ddot{x}(t)\underset{\sim}{i} + \ddot{y}(t)\underset{\sim}{j}$.

EXAMPLE: A particle moves so that its position vector is given by $\underset{\sim}{r}(t) = 2\sin 3t\underset{\sim}{i} + 2\cos 3t\underset{\sim}{j}$. Show that the acceleration vector is perpendicular to the velocity vector.

Solution: $\underset{\sim}{r}(t) = 2\sin 3t\underset{\sim}{i} + 2\cos 3t\underset{\sim}{j}$

$\underset{\sim}{v}(t) = 6\cos 3t\underset{\sim}{i} - 6\sin 3t\underset{\sim}{j}$

$\underset{\sim}{a}(t) = -18\sin 3t\underset{\sim}{i} - 18\cos 3t\underset{\sim}{j}$

$\underset{\sim}{a}(t).\underset{\sim}{v}(t) = (-18\sin 3t)\times 6\cos 3t + (-18\cos 3t)\times(-6\sin 3t) = 0$

$\therefore$ vectors are perpendicular.

TOPIC: VECTORS — See Excel Year 11 & 12 Mathematics Extension 1 pp. 135–136

TRIGONOMETRIC EQUATIONS

Equations involving double angles — 68

- $\sin 2A = 2\sin A\cos A$
- $\cos 2A = \cos^2 A - \sin^2 A = 1 - 2\sin^2 A = 2\cos^2 A - 1$
- $\tan 2A = \dfrac{2\tan A}{1-\tan^2 A}$

EXAMPLE: Solve $\cos 2x + \cos x = 0$, for $0 \le x \le 2\pi$.

Solution: $2\cos^2 x - 1 + \cos x = 0$

$2\cos^2 x + \cos x - 1 = 0$

$(2\cos x - 1)(\cos x + 1) = 0$

$\cos x = \dfrac{1}{2}, -1$

$x = \dfrac{\pi}{3}, \dfrac{5\pi}{3}, \pi$

TOPIC: TRIGONOMETRIC FUNCTIONS — See Excel Year 11 & 12 Mathematics Extension 1 pp. 151–152

TRIGONOMETRIC EQUATIONS

Equations of the form $a\cos x + b\sin x = c$ — 66

- **Equations in the form $a\sin x \pm b\cos x = c$** are rewritten in the form $r\sin(x \pm a) = c$.
- **Equations in the form $a\cos x \pm b\sin x = c$** are rewritten in the form $r\cos(x \mp a) = c$.

EXAMPLE: Solve, for $0° \le x \le 360°$, $\sqrt{3}\cos x + \sin x = 1$.

Solution: $\sqrt{3}\cos x + \sin x = r\cos(x - \alpha)$

$r = \sqrt{(\sqrt{3})^2 + 1^2} = 2$

$\tan\alpha = \dfrac{1}{\sqrt{3}}$

$\alpha = 30°$

$\therefore \sqrt{3}\cos x + \sin x = 2\cos(x - 30°)$

$\therefore 2\cos(x - 30°) = 1$

$\cos(x - 30°) = \dfrac{1}{2}$

$x - 30° = 60°, 300°$

$x = 90°, 330°$

TOPIC: TRIGONOMETRIC FUNCTIONS — See Excel Year 11 & 12 Mathematics Extension 1 p. 151

INTRODUCTION TO VECTORS

Angle and velocity of projectile on landing — 64

- $\theta = \tan^{-1}\left|\dfrac{\dot{y}(t)}{\dot{x}(t)}\right|$ and $\left|\underset{\sim}{v}(t)\right| = \sqrt{(\dot{x}(t))^2 + (\dot{y}(t))^2}$

EXAMPLE: An object is projected horizontally from the top of a cliff. If the object hits the ground after $\sqrt{3}$ seconds and the velocity vector is $\underset{\sim}{v}(t) = 8\sqrt{3}\underset{\sim}{i} - 10t\underset{\sim}{j}$, at what speed and at what acute angle does the object collide with the ground?

Solution: Let $\left|\underset{\sim}{v}(\sqrt{3})\right|$ = speed the object hits the ground, θ = angle at which the object hits the ground:

$\therefore \left|\underset{\sim}{v}(\sqrt{3})\right| = \sqrt{(8\sqrt{3})^2 + (-10\sqrt{3})^2} = 2\sqrt{123}$

$\theta = \tan^{-1}\left|\dfrac{-10\sqrt{3}}{8\sqrt{3}}\right| = 51$ (nearest whole)

$\therefore$ the object hits at $2\sqrt{123}$ ms^{-1} at 51° to the horizontal.

TOPIC: VECTORS — See Excel Year 11 & 12 Mathematics Extension 1 pp. 136–140

TRIGONOMETRIC EQUATIONS

69 Equations involving products as sums and differences

- $\sin A + \sin B = 2\sin\frac{A+B}{2}\cos\frac{A-B}{2}$
- $\sin A - \sin B = 2\cos\frac{A+B}{2}\sin\frac{A-B}{2}$
- $\cos A + \cos B = 2\cos\frac{A+B}{2}\cos\frac{A-B}{2}$
- $\cos A - \cos B = -2\sin\frac{A+B}{2}\sin\frac{A-B}{2}$

EXAMPLE: Solve $\sin 3x - \sin x = 0$, for $0 \le x \le \pi$.

Solution: $\sin 3x - \sin x$

$= 2\cos 2x \sin x$

$\therefore 2\cos 2x \sin x = 0$

$\cos 2x = 0 \qquad \sin x = 0$

$2x = \frac{\pi}{2}, \frac{3\pi}{2} \qquad x = 0, \pi$

$x = \frac{\pi}{4}, \frac{3\pi}{4}$

$\therefore x = 0, \frac{\pi}{4}, \frac{3\pi}{4}, \pi$

TOPIC: TRIGONOMETRIC FUNCTIONS See Excel Year 11 & 12 Mathematics Extension 1 pp. 151–152

TRIGONOMETRIC EQUATIONS

71 Expressing sin 3θ in terms of θ

- $\sin 3x = 3\sin x - 4\sin^3 x$

EXAMPLE: Show $\sin 3x = 3\sin x - 4\sin^3 x$.

Solution:

$$\begin{aligned}
\text{LHS} &= \sin 3x \\
&= \sin(2x + x) \\
&= \sin 2x\cos x + \cos 2x \sin x \\
&= (2\sin x\cos x)\cos x + (1 - 2\sin^2 x)\sin x \\
&= 2\sin x\cos^2 x + \sin x - 2\sin^3 x \\
&= 2\sin x(1 - \sin^2 x) + \sin x - 2\sin^3 x \\
&= 2\sin x - 2\sin^3 x + \sin x - 2\sin^3 x \\
&= 3\sin x - 4\sin^3 x \\
&= \text{RHS}
\end{aligned}$$

TOPIC: TRIGONOMETRIC FUNCTIONS See Excel Year 11 & 12 Mathematics Extension 1 p. 153

TRIGONOMETRIC EQUATIONS

73 Expressing tan 3θ in terms of θ

- $\tan 3\theta = \frac{3\tan\theta - \tan^3\theta}{1 - 3\tan^2\theta}$

EXAMPLE: Show $\tan 3\theta = \frac{3\tan\theta - \tan^3\theta}{1 - 3\tan^2\theta}$.

Solution: LHS = $\tan 3\theta = \tan(2\theta + \theta)$

$$= \frac{\tan 2\theta + \tan\theta}{1 - \tan 2\theta\tan\theta}$$

$$= \frac{\frac{2\tan\theta}{1-\tan^2\theta} + \tan\theta}{1 - \frac{2\tan\theta}{1-\tan^2\theta}\tan\theta}$$

$$= \frac{2\tan\theta + \tan\theta - \tan^3\theta}{1 - \tan^2\theta - 2\tan^2\theta}$$

[multiplying top and bottom by $(1 - \tan^2\theta)$]

$$= \frac{3\tan\theta - \tan^3\theta}{1 - 3\tan^2\theta}$$

= RHS

TOPIC: TRIGONOMETRIC FUNCTIONS See Excel Year 11 & 12 Mathematics Extension 1 p. 153

FURTHER CALCULUS SKILLS

75 Integration using substitution

- If $\frac{du}{dx} = g(x)$, then $dx = \frac{du}{g(x)}$.

EXAMPLE: Find $\int x^3\sqrt{x^4+3}\,dx$ using the substitution $u = x^4 + 3$.

Solution: As $u = x^4 + 3$, $\therefore \frac{du}{dx} = 4x^3 \therefore dx = \frac{du}{4x^3}$

$$\int x^3\sqrt{x^4+3}\,dx = \int x^3\sqrt{u}\,\frac{du}{4x^3} = \frac{1}{4}\int u^{\frac{1}{2}}\,du$$

$$= \frac{1}{4}\left[\frac{2u^{\frac{3}{2}}}{3}\right] + c$$

$$= \frac{\sqrt{(x^4+3)^3}}{6} + c$$

TOPIC: CALCULUS See Excel Year 11 & 12 Mathematics Extension 1 pp. 160–161

FURTHER CALCULUS SKILLS

77 Definite integrals and substitution

- **Limits of integral** are values of u.

EXAMPLE: Find $\int_3^8 \frac{x}{\sqrt{x+1}}\,dx$ using the substitution $u = x + 1$.

Solution: As $u = x + 1$, $\therefore \frac{du}{dx} = 1 \therefore dx = du$. Also, $x = u - 1$.

When $x = 8$, $u = 9$ and when $x = 3$, $u = 4$.

$$\int_3^8 \frac{x}{\sqrt{x+1}}\,dx = \int_{u=4}^{u=9} \frac{u-1}{\sqrt{u}}\,du = \int_4^9 u^{\frac{1}{2}} - u^{-\frac{1}{2}}\,du$$

$$= \left[\frac{2}{3}u^{\frac{3}{2}} - 2u^{\frac{1}{2}}\right]_4^9 = \left[\frac{2}{3}(9)^{\frac{3}{2}} - 2(9)^{\frac{1}{2}}\right] - \left[\frac{2}{3}(4)^{\frac{3}{2}} - 2(4)^{\frac{1}{2}}\right]$$

$$= 10\frac{2}{3}$$

TOPIC: CALCULUS See Excel Year 11 & 12 Mathematics Extension 1 pp. 160–161

FURTHER CALCULUS SKILLS

79 Integration of cos² nx

- $\cos 2nx = 2\cos^2 nx - 1$, and hence $\cos^2 nx = \frac{1}{2}[1 + \cos 2nx]$.

EXAMPLE: Find the exact area under the curve $y = 1 + \cos^2 x$ between $x = \frac{\pi}{3}$ and $x = \frac{2\pi}{3}$.

Solution: Area $= \int_{\frac{\pi}{3}}^{\frac{2\pi}{3}} (1 + \cos^2 x)\,dx = \int_{\frac{\pi}{3}}^{\frac{2\pi}{3}} \left(1 + \frac{1}{2}[1 + \cos x]\right)dx$

$$= \int_{\frac{\pi}{3}}^{\frac{2\pi}{3}} \left(\frac{3}{2} + \frac{1}{2}\cos 2x\right)dx = \frac{1}{2}\left[3x + \frac{1}{2}\sin 2x\right]_{\frac{\pi}{3}}^{\frac{2\pi}{3}}$$

$$= \frac{1}{4}\left[6\left(\frac{2\pi}{3}\right) + \sin 2\left(\frac{2\pi}{3}\right) - \left(6\left(\frac{\pi}{3}\right) + \sin 2\left(\frac{\pi}{3}\right)\right)\right]$$

$\therefore$ the area is $\frac{2\pi - \sqrt{3}}{4}$ units².

TOPIC: CALCULUS See Excel Year 11 & 12 Mathematics Extension 1 pp. 161–162

FURTHER CALCULUS SKILLS

81 Differentiation of inverse sin (arcsin) functions

- $\frac{d}{dx}\left(\sin^{-1}\frac{x}{a}\right) = \frac{1}{\sqrt{a^2 - x^2}}$
- $\frac{d}{dx}(\sin^{-1}[f(x)]) = \frac{1}{\sqrt{1 - [f(x)]^2}}.f'(x)$

EXAMPLE: Differentiate $\sin^{-1}(\cos x)$.

Solution: $\frac{d}{dx}\left[\sin^{-1}(\cos x)\right] = \frac{1}{\sqrt{1 - (\cos x)^2}}.-\sin x = \frac{1}{\sqrt{1 - \cos^2 x}}.-\sin x$

$$= \frac{-\sin x}{\sqrt{\sin^2 x}} = \frac{-\sin x}{|\sin x|}$$

$= -1$ when $\sin x > 0$ and 1 when $\sin x < 0$.

TOPIC: CALCULUS See Excel Year 11 & 12 Mathematics Extension 1 pp. 163–165

FURTHER CALCULUS SKILLS

83 Differentiation of inverse tan (arctan) functions

- $\frac{d}{dx}\left(\tan^{-1}\frac{x}{a}\right) = \frac{a}{a^2 + x^2}$
- $\frac{d}{dx}(\tan^{-1}[f(x)]) = \frac{1}{1 + [f(x)]^2}.f'(x)$

EXAMPLE: Find the gradient of the normal to the curve $y = 4\tan^{-1} 2x$, at the point where $x = -1$.

Solution: $\frac{dy}{dx} = 4 \times \frac{1}{1 + (2x)^2}.2 = \frac{8}{1 + 4x^2}$

$\frac{dy}{dx}(-1) = \frac{8}{1 + 4(-1)^2} = \frac{8}{5} \therefore \text{grad} = -\frac{5}{8}$

TOPIC: CALCULUS See Excel Year 11 & 12 Mathematics Extension 1 pp. 163–165

FURTHER CALCULUS SKILLS

85 Integration resulting in inverse tan (arctan) functions

- $\int \frac{a}{a^2 + x^2}\,dx = \tan^{-1}\frac{x}{a} + c$

EXAMPLE: Find $\int \frac{x - 6}{x^2 + 4}\,dx$.

Solution:

$$\int \frac{x-6}{x^2+4}\,dx = \int \frac{x}{x^2+4}\,dx - \int \frac{6}{x^2+4}\,dx$$

$$= \frac{1}{2}\int \frac{2x}{x^2+4}\,dx - 3\int \frac{2}{4+x^2}\,dx$$

$$= \frac{1}{2}\ln(x^2 + 4) - 3\tan^{-1}\frac{x}{2} + c$$

TOPIC: CALCULUS See Excel Year 11 & 12 Mathematics Extension 1 pp. 165–166

FURTHER CALCULUS SKILLS 74

Integration involving products of trigonometric functions

- $2\sin\frac{A+B}{2}\cos\frac{A-B}{2} = \sin A + \sin B$
- $2\cos\frac{A+B}{2}\sin\frac{A-B}{2} = \sin A - \sin B$
- $2\cos\frac{A+B}{2}\cos\frac{A-B}{2} = \cos A + \cos B$
- $-2\sin\frac{A+B}{2}\sin\frac{A-B}{2} = \cos A - \cos B$

EXAMPLE: Find $\int 2\sin 3x\cos x\,dx$.

Solution:

$$\int 2\sin 3x\cos x\,dx$$
$$= \int \sin 4x + \sin 2x\,dx$$
$$= -\frac{1}{4}\cos 4x - \frac{1}{2}\cos 2x + c$$

TOPIC: CALCULUS See ➲ *Excel* Year 11 & 12 Mathematics Extension 1 p. 160

TRIGONOMETRIC EQUATIONS 72

Expressing cos 3θ in terms of θ

- $\cos 3x = 4\cos^3 x - 3\cos x$

EXAMPLE: Solve $\cos 3x = \cos x$, for $-\pi \leq \theta \leq \pi$.

Solution:

$$4\cos^3 x - 3\cos x = \cos x$$
$$4\cos^3 x - 4\cos x = 0$$
$$4\cos x(\cos^2 x - 1) = 0$$
$$\cos x = 0,\ \cos x = \pm 1$$
$$x = -\frac{\pi}{2}, \frac{\pi}{2}, 0, \pi, -\pi.$$
$$\therefore x = -\pi, -\frac{\pi}{2}, 0, \frac{\pi}{2}, \pi$$

TOPIC: TRIGONOMETRIC FUNCTIONS See ➲ *Excel* Year 11 & 12 Mathematics Extension 1 p. 153

TRIGONOMETRIC EQUATIONS 70

Using *t*-formulae to solve equations

- **When we use the *t*-formulae method** to solve equations we need to check $\theta = 180°$ as a possible solution.

EXAMPLE: Using the *t*-formulae, solve $\sin\theta + \cos\theta = 1$, for $0° \leq \theta \leq 360°$. Give your answer to the nearest degree.

Solution:

$$\frac{2t}{1+t^2} + \frac{1-t^2}{1+t^2} = 1$$
$$2t + 1 - t^2 = 1 + t^2$$
$$2t^2 - 2t = 0$$
$$2t(t-1) = 0$$
$$t = 0, 1$$
$$\tan\frac{\theta}{2} = 0, 1$$
$$\frac{\theta}{2} = 0°, 45°, 180°, 225°$$
$$\theta = 0°, 90°, 360°\ (\text{as } 0° \leq \theta \leq 360°)$$

Also, check possible solution of $\theta = 180°$:
$\sin 180° + \cos 180° \neq 1$. No!
$\therefore \theta = 0°, 90°, 360°$

TOPIC: TRIGONOMETRIC FUNCTIONS See ➲ *Excel* Year 11 & 12 Mathematics Extension 1 pp. 152–153

FURTHER CALCULUS SKILLS 80

Using the result $\frac{dy}{dx} = \frac{1}{\frac{dx}{dy}}$

- $\frac{dy}{dx} = \frac{1}{\frac{dx}{dy}}$, or $\frac{dy}{dx} \times \frac{dx}{dy} = 1$

EXAMPLE: If $x = y^2$, find $\frac{dy}{dx}$.

Solution:

First, $\frac{dx}{dy} = 2y$

$$\therefore \frac{dy}{dx} = \frac{1}{2y},\ y \neq 0$$

TOPIC: CALCULUS See ➲ *Excel* Year 11 & 12 Mathematics Extension 1 pp. 162–163

FURTHER CALCULUS SKILLS 78

Integration of $\sin^2 nx$

- $\cos 2nx = 1 - 2\sin^2 nx$, and hence $\sin^2 nx = \frac{1}{2}[1 - \cos 2nx]$.

EXAMPLE: Find $\int_{\frac{\pi}{6}}^{\frac{\pi}{3}} \sin^2 x\,dx$.

Solution:

$$\int_{\frac{\pi}{6}}^{\frac{\pi}{3}} \sin^2 x\,dx = \int_{\frac{\pi}{6}}^{\frac{\pi}{3}} \frac{1}{2}[1 - \cos 2x]\,dx$$
$$= \frac{1}{2}\left[x - \frac{1}{2}\sin 2x\right]_{\frac{\pi}{6}}^{\frac{\pi}{3}}$$
$$= \frac{1}{2}\left[\frac{\pi}{3} - \frac{1}{2}\sin 2\left(\frac{\pi}{3}\right) - \left(\frac{\pi}{6} - \frac{1}{2}\sin 2\left(\frac{\pi}{6}\right)\right)\right]$$
$$= \frac{\pi}{12}$$

TOPIC: CALCULUS See ➲ *Excel* Year 11 & 12 Mathematics Extension 1 pp. 161–162

FURTHER CALCULUS SKILLS 76

Integration using trigonometric substitution

- $\sin^2 x + \cos^2 x = 1$, $\sec^2 x = 1 + \tan^2 x$, $\operatorname{cosec}^2 x = 1 + \cot^2 x$

EXAMPLE: Use the substitution $u = \tan x$, to find $\int \tan^2 x\sec^2 x\,dx$.

Solution: As $u = \tan x$, $\therefore \frac{du}{dx} = \sec^2 x$ $\therefore dx = \frac{du}{\sec^2 x}$

$$\int \tan^2 x\sec^2 x\,dx = \int u^2 \sec^2 x \frac{du}{\sec^2 x}$$
$$= \int u^2\,du$$
$$= \frac{u^3}{3} + c$$
$$= \frac{\tan^3 x}{3} + c$$

TOPIC: CALCULUS See ➲ *Excel* Year 11 & 12 Mathematics Extension 1 pp. 160–161

APPLICATIONS OF CALCULUS 86

Area between two curves

- Area $= \int_a^b f(x) - g(x)\,dx$, such that $f(x) > g(x)$.

EXAMPLE: Find the area of the region bounded by the curves $y = \frac{4}{1+x^2}$ and the line $y = 1$.

Solution:

$$\frac{4}{1+x^2} = 1$$
$$x = \pm\sqrt{3}$$
$$\text{Area} = \int_{-\sqrt{3}}^{\sqrt{3}} \left(\frac{4}{1+x^2} - 1\right)dx$$
$$= 2\left[4\tan^{-1} x - x\right]_0^{\sqrt{3}}$$
$$= 2\left[4\tan^{-1}\sqrt{3} - \sqrt{3} - (4\tan^{-1} 0 - 0)\right]$$
$$= 2\left(\frac{4\pi}{3} - \sqrt{3}\right)$$

$\therefore$ the area is $2\left(\frac{4\pi}{3} - \sqrt{3}\right)$ units^2.

TOPIC: CALCULUS See ➲ *Excel* Year 11 & 12 Mathematics Extension 1 pp. 176–177

FURTHER CALCULUS SKILLS 84

Integration resulting in inverse sine (arcsin) functions

- $\int \frac{1}{\sqrt{a^2 - x^2}}\,dx = \sin^{-1}\frac{x}{a} + c$

EXAMPLE: Evaluate $\int_{-\frac{1}{3}}^{\frac{1}{3}} \frac{1}{\sqrt{4 - 9x^2}}\,dx$.

Solution:

$$\int_{-\frac{1}{3}}^{\frac{1}{3}} \frac{1}{\sqrt{4-9x^2}}\,dx = \int_{-\frac{1}{3}}^{\frac{1}{3}} \frac{1}{\sqrt{9\left(\frac{4}{9} - x^2\right)}}\,dx = \frac{1}{3}\int_{-\frac{1}{3}}^{\frac{1}{3}} \frac{1}{\sqrt{\left(\frac{4}{9} - x^2\right)}}\,dx$$
$$= \frac{1}{3}\left[\sin^{-1}\frac{3x}{2}\right]_{-\frac{1}{3}}^{\frac{1}{3}} = \frac{1}{3}\left[\sin^{-1}\frac{3\left(\frac{1}{3}\right)}{2} - \sin^{-1}\frac{3\left(-\frac{1}{3}\right)}{2}\right] = \frac{\pi}{9}$$

TOPIC: CALCULUS See ➲ *Excel* Year 11 & 12 Mathematics Extension 1 pp. 165–166

FURTHER CALCULUS SKILLS 82

Differentiation of inverse cos (arccos) functions

- $\frac{d}{dx}\left(\cos^{-1}\frac{x}{a}\right) = \frac{-1}{\sqrt{a^2 - x^2}}$
- $\frac{d}{dx}(\cos^{-1}[f(x)]) = \frac{-1}{\sqrt{1 - [f(x)]^2}} f'(x)$

EXAMPLE: If $y = \sin^{-1}\frac{x}{2} + \cos^{-1}\frac{x}{2}$, find $\frac{dy}{dx}$.

Solution:

$$\frac{dy}{dx} = \frac{1}{\sqrt{4-x^2}} + \frac{-1}{\sqrt{4-x^2}}$$
$$= 0$$

TOPIC: CALCULUS See ➲ *Excel* Year 11 & 12 Mathematics Extension 1 pp. 163–165

87 Volumes of solids formed by rotating about the x-axis

➔ **If $y = f(x)$ is rotated about the x-axis between $x = a$ and $x = b$,** the volume of the solid formed is given as $V = \pi\int_a^b y^2\,dx$.

EXAMPLE: The diagram shows the region bounded by $y = \frac{1}{(x+1)^2}$, the x-axis, the y-axis, and the line $x = 2$. The region is rotated about the x-axis to form a solid. Find the volume of the solid.

Solution: $y^2 = \frac{1}{(x+1)^4} = (x+1)^{-4}$

$$V = \pi\int_0^2 (x+1)^{-4}\,dx = \pi\left[\frac{(x+1)^{-3}}{-3}\right]_0^2 = \frac{26\pi}{81} \quad \therefore \frac{26\pi}{81}\text{ units}^3.$$

TOPIC: CALCULUS See ➲ *Excel* Year 11 & 12 Mathematics Extension 1 pp. 177–179

89 Volumes of solids formed by rotating regions between two curves

➔ **If the area between $y = f(x)$ and $y = g(x)$ is rotated about the x-axis** between $x = a$ and $x = b$, the volume of the solid formed is given as $V = \pi\int_a^b ([f(x)]^2 - [g(x)]^2)\,dx$.

➔ **If the area between $y = f(x)$ and $y = g(x)$ is rotated about the y-axis** between $y = a$ and $y = b$, the volume of the solid formed is given as $V = \pi\int_a^b ([f(y)]^2 - [g(y)]^2)\,dy$.

EXAMPLE: The area between the curves $y = x^2$ and $y = \sqrt{x}$ is rotated about the x-axis. Find the exact volume of the solid formed.

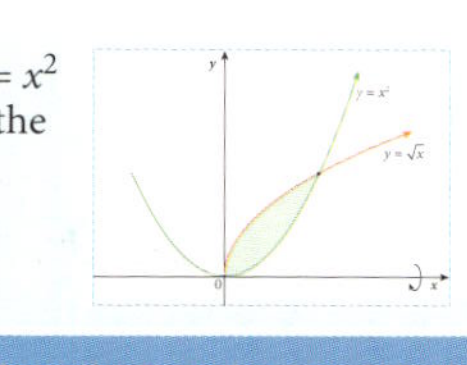

Solution:

$$V = \pi\int_0^1 (\sqrt{x})^2 - (x^2)^2\,dx = \frac{3\pi}{10} \quad \therefore \frac{3\pi}{10}\text{ units}^3.$$

TOPIC: CALCULUS See ➲ *Excel* Year 11 & 12 Mathematics Extension 1 pp. 182–185

91 Direction (slope) field of a differential equation

➔ **A first-order differential equation** is an equation involving a first derivative, e.g. $\frac{dy}{dx} = 5x + 2$, $x^2 + 2y\frac{dy}{dx} = 6$, etc.

➔ **A direction field (or slope field)** is a graphical representation of the tangent lines to the solutions of a first-order differential equation.

EXAMPLE: A direction field representing a differential equation is shown. Which of these could be the equation?

A $\frac{dy}{dx} = \frac{x}{1+y^2}$ B $\frac{dy}{dx} = \frac{x}{2y}$

C $\frac{dy}{dx} = \frac{-y}{2x}$ D $\frac{dy}{dx} = \frac{x}{1-y^2}$

Solution: A

TOPIC: CALCULUS See ➲ *Excel* Year 11 & 12 Mathematics Extension 1 pp. 196–200

93 Solving differential equations of the form $\frac{dy}{dx} = f(x)$

➔ **If $\frac{dy}{dx} = f(x)$, then $\int dy = \int f(x)\,dx$.** Hence, $y = \int f(x)\,dx$.

EXAMPLE: Find the general solution of the differential equation $2x\frac{dy}{dx} - 3x^2 = 5$.

Solution:

$$2x\frac{dy}{dx} = 3x^2 + 5$$
$$\frac{dy}{dx} = \frac{3x^2}{2x} + \frac{5}{2x}$$
$$y = \int \frac{3x}{2} + \frac{5}{2x}\,dx$$
$$y = \frac{3x^2}{2} + \frac{5}{2}\log_e x + c$$

TOPIC: CALCULUS See ➲ *Excel* Year 11 & 12 Mathematics Extension 1 pp. 200–201

95 Solving differential equations of the form $\frac{dy}{dx} = f(x)\,g(y)$

➔ **If $\frac{dy}{dx} = f(x)\,g(y)$, then $\int \frac{1}{g(y)}\,dy = \int f(x)dx$.**

EXAMPLE: Solve the differential equation $\frac{dy}{dx} = \frac{-x}{1+y^2}$, with the condition $y(-1) = 1$.

Solution:

$$(1 + y^2)\,dy = -x\,dx$$
$$\int 1 + y^2 dy = \int -x\,dx$$
$$y + \frac{y^3}{3} = -\frac{x^2}{2} + c$$

Substitute $y(-1) = 1$:

$$1 + \frac{1^3}{3} = -\frac{(-1)^2}{2} + c$$
$$c = \frac{11}{6}$$
$$\therefore y + \frac{y^3}{3} = -\frac{x^2}{2} + \frac{11}{6}$$
$$\therefore 3x^2 + 2y^3 + 6y - 11 = 0$$

TOPIC: CALCULUS See ➲ *Excel* Year 11 & 12 Mathematics Extension 1 p. 202

97 The logistic growth model as a differential equation 1

➔ $\frac{dP}{dt} = kP(M - P)$, where the rate of change of the population P is proportional to both the size of P and the difference between the maximum value of P and P at any given time t.

EXAMPLE: A remote location with no feral animals was fenced and used to set up a breeding population of Tasmanian devils. Initially 24 devils were released in the area which is expected to support a population of 120 devils. After 3 years, the population has grown to 60.

Use the logistic model $\frac{dP}{dt} = kP(M - P)$ to show $P = \frac{30e^{120kt}}{1+0.25e^{120kt}}$ where $k = \frac{\log_e 4}{360}$.

[Hint: Use the result $\frac{1}{P(120-P)} = \frac{1}{120P} + \frac{1}{120(120-P)}$]

TOPIC: CALCULUS See ➲ *Excel* Year 11 & 12 Mathematics Extension 1 p. 205

99 Bernoulli trials

➔ **Characteristics of Bernoulli trials** are:

- **each trial has only two possible outcomes**: 'success' and 'failure'
- **P(success) = p and P(failure) = $q = 1 - p$** and this is consistent for every trial
- **trials are independent**.

➔ **X = number of successes in independent Bernoulli trials.** If there is only 1 trial, then the probability distribution table is detailed below:

x	0	1
$P(X = x)$	$1 - p$	p

➔ **Mean** = $E(X) = \mu = 0(1 - p) + 1(p) = p$

➔ **Variance** = $Var(X) = \sigma^2 = E[X^2] - [E(X)]^2 = \{0^2(1 - p) + 1^2(p)\} - p^2 = p(1 - p)$

TOPIC: STATISTICAL ANALYSIS See ➲ *Excel* Year 11 & 12 Mathematics Extension 1 pp. 217–218

101 Binomial distributions

➔ **$X \sim$ Bin(n, p) means random variable X** is distributed binomially in n trials and P(success) = p.

➔ **$P(X = r)$ is used to denote the probability** that there are r successes in n trials.

➔ $P(X = r) = {}^nC_r p^r (1-p)^{n-r}$

EXAMPLE: A dice is rolled 30 times. Let X be the discrete random variable that describes the number of times a number greater than 3 is thrown. What is the probability of rolling a number greater than 3 on exactly 20 occasions? Give your answer correct to four decimal places.

Solution: P(greater than 3) = 0.5

Using X~Bin(30, 0.5): $P(X = 20) = {}^{30}C_{20}(0.5)^{20}(1-0.5)^{10} = 0.0280$ (4 dec. pl.)

TOPIC: STATISTICAL ANALYSIS See ➲ *Excel* Year 11 & 12 Mathematics Extension 1 pp. 218–220

103 The sample proportion $\hat{p}$

➔ **In a sample**, the 'success' proportion is recorded as $\hat{p}$, pronounced 'p hat'.

➔ As long as $np \geq 10$ and $n(1 - p) \geq 10$ then $\hat{p}$ is said to be **normally distributed**.

➔ Mean $(\mu_{\hat{p}}) = \frac{np}{n} = p$

➔ Standard deviation $(\sigma_{\hat{p}}) = \frac{\sqrt{np(1-p)}}{n}$ or $\sqrt{\frac{p(1-p)}{n}}$.

EXAMPLE: Given $n = 30$ and $p = 0.2$, find the standard deviation of $\hat{p}$, correct to three decimal places.

Solution: $\sigma_{\hat{p}} = \frac{\sqrt{np(1-p)}}{n} = 0.073$ (3 dec. pl.)

TOPIC: STATISTICAL ANALYSIS See ➲ *Excel* Year 11 & 12 Mathematics Extension 1 pp. 221–223

APPLICATIONS OF CALCULUS

Sketching a curve on a given direction field

92

- A solution curve uses the gradients as a path.

EXAMPLE: A slope field representing the differential equation $\frac{dy}{dx} = \frac{x}{1+y}$ is shown below.

Sketch the solution curve of the differential equation corresponding to the condition $y(0) = 1$.

Solution:

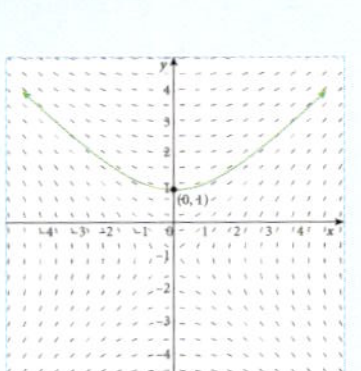
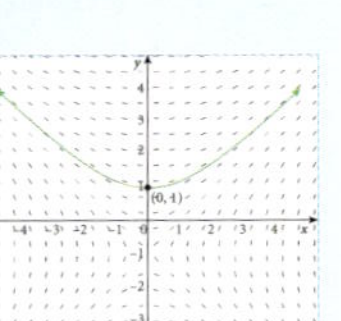
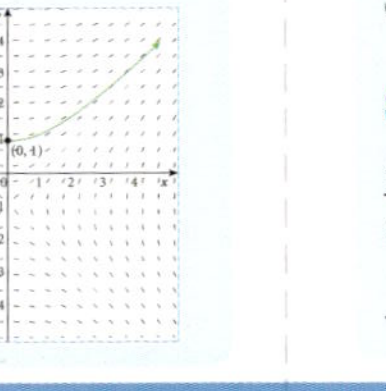

TOPIC: CALCULUS See ➲ *Excel* Year 11 & 12 Mathematics Extension 1 pp. 196–200

APPLICATIONS OF CALCULUS

Verifying solutions of differential equations

90

- **A differential equation** is an equation involving derivatives, e.g. $\frac{dy}{dx} = 2x + 3$, $\ddot{y} + 2\dot{y} = 6$, etc.
- **Use differentiation to verify a solution** of a differential equation.

EXAMPLE: Verify that $y = 2 + e^{3x}$ is a solution to the differential equation $\frac{d^2y}{dx^2} - 2\frac{dy}{dx} - 3y + 6 = 0$.

Solution: $\frac{dy}{dx} = 3e^{3x}$ $\quad \frac{d^2y}{dx^2} = 9e^{3x}$

$$\frac{d^2y}{dx^2} - 2\frac{dy}{dx} - 3y + 6 = 9e^{3x} - 2(3e^{3x}) - 3(2 + e^{3x}) + 6$$
$$= 9e^{3x} - 6e^{3x} - 6 - 3e^{3x} + 6 = 0$$

$\therefore y = 2 + e^{3x}$ is a solution.

TOPIC: CALCULUS See ➲ *Excel* Year 11 & 12 Mathematics Extension 1 p. 196

APPLICATIONS OF CALCULUS

Volumes of solids formed by rotating about the y-axis

88

- **If $y = f(x)$ is rotated about the y-axis between $y = a$ and $y = b$,** the volume of the solid formed is given as $V = \pi\int_a^b x^2\,dy$.

EXAMPLE: The region bounded by the y-axis, the curve $y = -1 + \sqrt{x}$ and the line $y = 1$ is rotated about the y-axis to form a solid. Find the volume of the solid.

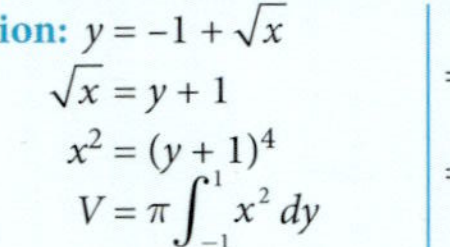

Solution: $y = -1 + \sqrt{x}$

$$\sqrt{x} = y + 1$$
$$x^2 = (y+1)^4$$
$$V = \pi\int_{-1}^{1} x^2\,dy$$
$$= \pi\int_{-1}^{1}(y+1)^4\,dy$$
$$= \pi\left[\frac{(y+1)^5}{5}\right]_{-1}^{1}$$
$$= \frac{\pi}{5}\left[(1+1)^3 - (-1+1)^3\right] = \frac{8\pi}{5}$$

$\therefore$ the volume is $\frac{8\pi}{5}$ units^3.

TOPIC: CALCULUS See ➲ *Excel* Year 11 & 12 Mathematics Extension 1 pp. 180–182

APPLICATIONS OF CALCULUS

The logistic growth model as a differential equation 2

98

$$\frac{dP}{dt} = kP(120 - P)$$
$$\int\frac{1}{P(120-P)}\,dP = \int k\,dt$$
$$\int\frac{1}{120P} + \frac{1}{120(120-P)}\,dP = kt + c$$
$$\frac{P}{120-P} = Ae^{120kt}, \text{ for constant } A, \text{ where } A = e^{120c}$$
$$P = \frac{120Ae^{120kt}}{1+Ae^{120kt}}$$

Substitute $P = 24$, $t = 0$:

$$24 = \frac{120Ae^{120k(0)}}{1+Ae^{120k(0)}} \quad \therefore A = 0.25$$
$$\therefore P = \frac{120 - 0.25e^{120kt}}{1+0.25e^{120kt}} = \frac{30e^{120kt}}{1+0.25e^{120kt}}$$

Substitute $P = 60$, $t = 3$:

$$60 = \frac{30e^{120kt}}{1+0.25e^{120kt}} \quad \therefore k = \frac{\log_e 4}{360}$$

**for complete solution see p. 205 of the study guide

TOPIC: CALCULUS See ➲ *Excel* Year 11 & 12 Mathematics Extension 1 p. 205

APPLICATIONS OF CALCULUS

Exponential growth and decay models as differential equations

96

- $\frac{dP}{dt} = kP$
- $\frac{dP}{dt} = k(P - A)$

EXAMPLE: The population P of a city at time t years after a certain date increases at a rate proportional to the population at that time. Construct and solve a differential equation for P in terms of t.

Solution: $\frac{dP}{dt} \sim P$

$$\frac{dP}{dt} = kP$$
$$\frac{1}{P}\,dP = k\,dt$$
$$\int\frac{1}{P}\,dP = \int k\,dt$$
$$\ln|P| = kt + c$$
$$P = e^{kt+c}$$

$P = Ae^{kt}$, for constant A, where $A = e^c$.

TOPIC: CALCULUS See ➲ *Excel* Year 11 & 12 Mathematics Extension 1 pp. 202–204

APPLICATIONS OF CALCULUS

Solving differential equations of the form $\frac{dy}{dx} = g(y)$

94

- **If $\frac{dy}{dx} = g(y)$, then $\int\frac{1}{g(y)}\,dy = \int dx$.**

EXAMPLE: Solve the differential equation $\frac{dy}{dx} = 1 - \frac{2y}{3}$, with the condition $x = 3$, $y = 1$.

Solution: $\frac{dy}{dx} = \frac{3-2y}{3}$

$$\int\frac{3}{3-2y}\,dy = \int dx$$
$$-\frac{3}{2}\ln|3 - 2y| = x + c$$

Substitute $y(3) = 1$:

$$c = -3$$
$$-\frac{3}{2}\ln|3-2y| = x - 3$$
$$\ln|3-2y| = -\frac{2}{3}(x-3)$$
$$3 - 2y = e^{\frac{6-2x}{3}}$$
$$y = \frac{1}{2}\left(3 - e^{\frac{6-2x}{3}}\right)$$

TOPIC: CALCULUS See ➲ *Excel* Year 11 & 12 Mathematics Extension 1 pp. 201–202

THE BINOMIAL DISTRIBUTION

The normal approximation to the distribution of $\hat{p}$

104

- A table for a normal distribution can be used to determine probabilities in the sample.

EXAMPLE: In Australia, 17% of the adult population smoke every day. A survey of 150 shoppers at a shopping centre was conducted. Estimate the probability that more than 20% of those surveyed smoke every day. **Solution:**

As $n = 150$ and $p = 0.17$

$$\mu_{\hat{p}} = \frac{np}{n} = 0.17$$
$$\sigma_{\hat{p}} = \frac{\sqrt{np(1-p)}}{n}$$
$$= 0.0307 \text{ (4 dec. pl.)}$$
$$z = \frac{x - \mu}{\sigma}$$
$$= 0.98 \text{ (2 dec. pl.)}$$

Now, from a probability table, $P(z < 0.98) = 0.8365$.

$\therefore P(z > 0.98) = 1 - 0.8365 = 0.1635$. The probability is 0.1635.

TOPIC: STATISTICAL ANALYSIS See ➲ *Excel* Year 11 & 12 Mathematics Extension 1 pp. 221–223

THE BINOMIAL DISTRIBUTION

Mean and variance of binomial distributions

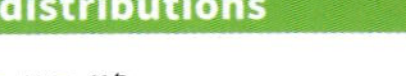
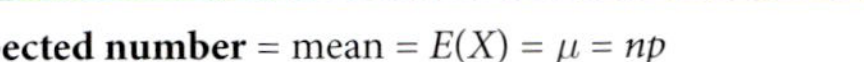
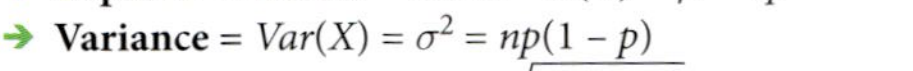

102

- **Expected number** = mean = $E(X) = \mu = np$
- **Variance** = $Var(X) = \sigma^2 = np(1 - p)$
- **Standard deviation** = $\sigma = \sqrt{np(1-p)}$

EXAMPLE: A binomial random variable Y has a mean of 8.28 and a variance of 4.4712.

a Find the probability of success, p.

b Find the number of trials, n.

Solution:

a $E(X) = np = 8.28$

$Var(X) = np(1 - p) = 4.4712$

$8.28(1 - p) = 4.4712$

$p = 0.46$

b Substitute into $np = 8.28$:

$n(0.46) = 8.28$

$n = 18$

TOPIC: STATISTICAL ANALYSIS See ➲ *Excel* Year 11 & 12 Mathematics Extension 1 pp. 220–221

THE BINOMIAL DISTRIBUTION

Applications of the Bernoulli distribution

100

- **The geometric probability distribution** represents the number of failures before a success in a series of Bernoulli trials.
- **If there are x failures before a success**, then $P(X = x) = q^x p$.

EXAMPLE: A bag contains red, blue and green balls. Balls are randomly chosen from the bag, the colour noted and then the balls returned to the bag. The probability of choosing a red ball is 0.15. Let X be the number of times a ball is drawn from the bag until a red ball is selected. Find $P(3 \le X \le 6)$, correct to four decimal places.

Solution: Let $p = 0.15$, $q = 0.85$.

$P(X = 3) + P(X = 4) + P(X = 5) + P(X = 6)$

$= (0.85)^3(0.15) + (0.85)^4(0.15) + (0.85)^5(0.15) + (0.85)^6(0.15)$

$= 0.2935$ (4 dec. pl.)

TOPIC: STATISTICAL ANALYSIS See ➲ *Excel* Year 11 & 12 Mathematics Extension 1 pp. 218–220